Engineering for Agricultural Drainage

Engineering for Agricultural Drainage

by

HARRY BURGESS ROE, C.E.

Professor Emeritus of Agricultural Engineering, Formerly Head, Section of Soil Moisture Relations, Department of Agriculture, and Professor of Irrigation and Drainage in the College of Engineering, University of Minnesota; Life Fellow, American Society of Agricultural Engineers

&

QUINCY CLAUDE AYRES, C.E.

Professor of Agricultural Engineering, Iowa State College and Assistant to the President; Member, American Society of Civil Engineers, Fellow, American Society of Agricultural Engineers; Past President, Iowa Engineering Society; Captain, Civil Engineer Corps, United States Naval Reserve

BIOTECH

2011

BIOTECH BOOKS®

Delhi - 110 035

Roe, Harry Burgess (b. 1876–)
Ayres, Quincy Claude (b. 1891–)

ISBN 81-7622-218-6
ISBN 978-81-7622-218-1

Published by : **BIOTECH BOOKS®**
1123/74, Deva Ram Park, Tri Nagar,
DELHI – 110 035
Phone: 27383999, (M) 09868812760
E-mail: biotechbooks@yahoo.co.in

Showroom : 4762-63/23, Ansari Road, Darya Ganj,
NEW DELHI - 110 002
Phone: 23245578

Printed at : **Chawla Offset Printers**
New Delhi – 110 052

PRINTED IN INDIA

To that large group of men
in the various phases of agriculture
earnestly seeking the best and improving practice
in the control of soil moisture in crop production,
this book is respectfully dedicated.

PREFACE

Since about 1935 interest in drainage for agricultural benefit in the United States has greatly increased, largely because of the work of the U.S. Soil Conservation Service in the improvement of farm practices. One of the most commendable results of the work of this agency has been the great increase in research in farm drainage theory and practice, so that advance in essential scientific knowledge has forged far ahead of the printed texts and reference books on the subject. This situation has caused an insistent demand among the workers for a new and up-to-date book on farm drainage. It is the intended mission of this book to meet that demand for students, for farmers, and for all agricultural technicians, whether engineers, agriculturists, or farm managers.

Drainage on a large scale began in the United States in the early days of the twentieth century, but for several decades, in the minds of most people, it was looked upon merely as an aid to disposition of excess rainfall flowing away under natural law, without any special connection with science other than the most elementary hydraulics. Recent research, however, has clearly demonstrated that good practice in farm drainage is closely allied to soil science, agronomy, horticulture, farm management, and economics, as well as being an absolutely essential accompaniment of continuous and successful irrigation. Therefore, more than any earlier book on drainage, this book lays special stress on soil and soil-moisture characteristics and behavior, crop-plant characteristics and growth, and on the successful and continuous practice of irrigation as related to drainage. It presents in complete form the latest developments in scientific drainage design with special chapters on soil character and behavior, drainage of irrigated lands, moisture control on agricultural peat lands, and on quality of drain tile and methods required in manufacture to secure permanence and protection from natural enemies.

This book is not merely the product of the limited erudition and experience of its authors, although they are both lifetime specialists in the theory and practice of drainage. Necessarily, in such a book, numerous sources of information are drawn upon, and this is the product of the efforts, contributions, and counsel of a large number of men and groups in Federal and state agencies of government, in industry, among other authors and publishers, and in the private professional fields.

This helpful group includes men in all the following agencies and fields: the Soil Conservation Service from coast to coast, the Bureau of Reclamation and the Corps of Engineers in the Federal service, the executive departments and the agricultural and engineering experiment stations in the various state services across the country, the manufacturers of drainage machinery and other farm equipment in industry, engineers in private practice in the professional field, and other authors and publishers.

The authors wish to express thanks to John Wiley and Sons, Inc., to the McGraw-Hill Book Company, Inc., and to other publishers and their authors for permission to use copyrighted material under their various controls.

Space does not permit a complete list of the men who have helped build this book, but the authors are deeply grateful for their unstinted cooperation and numerous contributions, and special effort has been made to accord full individual recognition in the body of the text in footnotes and the list of references. In spite of the most scrupulous attention to detail, however, it is probable that occasional credit omissions have occurred. For these and other errors characteristic of first editions, the authors ask indulgence.

A special debt of gratitude is due the following men for their generous aid: to Don Kirkham and Orville Hosmer for their valuable technical counsel and practical, up-to-date contributions; to Walter T. Madsen for his valuable counsel and data in the construction and contracting fields; to Dalton G. Miller for his excellent contribution of Chap. 11 on drain-tile manufacture, tests, and quality; to Harry F. Blaney and William W. Donnan for their thorough review of Chap. 13 on Drainage of Irrigated Lands and their constructive suggestions for its improvement; and to Fred H. Larson who contributed throughout of his time and thought as counselor and adviser on organization, essential content, clarity and correctness of presentation, and as critical reader of the entire manuscript by chapters as the work progressed.

It is the intention and hope of the authors to present a book which will be found genuinely helpful everywhere in the agricultural field, among students as a text in classwork, among technical men of all kinds in the agricultural field, and among farmers, as a reliable reference book and guide. They especially invite constructive criticism and suggestions for improvement in later editions and will greatly appreciate notice of any errors that may be discovered.

HARRY BURGESS ROE
QUINCY CLAUDE AYRES

CONTENTS

LIST OF TABLES

CHAPTER 1

INTRODUCTION

Only about one-fourth of the earth's surface, or approximately 49,000,000 square miles, is land. At least one-half of this area is too cold or rocky to support any kind of plant growth, and half of the remainder is too dry or infertile. Thus only about 12,000,000 square miles (one-sixteenth of the earth's crust) is available for growing vegetation. A large portion of this must be dedicated permanently to forests, shrubs, and grasses, because of great steepness, inaccessibility, and other reasons. Much of the land that is left over must receive artificial conditioning before it is ready to produce cultivated crops. This supplementary preparation may include one or more of the following: (1) drainage, (2) prevention of overflow on low land, (3) land clearing (removal of stumps and stones), (4) irrigation of dry land, and (5) retarding runoff from undulating land to conserve soil and water. All these operations have an important place in the field of soil and water conservation. Wherever such undertakings have been judiciously conceived and faithfully carried out, they have contributed enormously to the well-being of mankind. This book deals primarily with the first and second of these functions.

THE NATURE, PLACE, AND FUNCTION OF DRAINAGE IN AGRICULTURE

1.1. Definition of Drainage. In its broadest sense, land drainage, in the humid regions, is the removal of free water, both from the land surface and from the soil of the root zone of plants. In its more scientific sense in modern farming, drainage is understood as that control of the elevation of the ground-water table within the root zone which gives best results in the production of crops. In the arid regions, drainage is only one of the agencies for such control.

1.2. Purpose of Drainage. The purpose of drainage is to remove all free water from the surface of cropped fields and from the soil of the root zone as quickly as practicable after it accumulates. This is necessary because persistent surface floodwater soon drowns upland plant life, while in the soil itself the roots of most useful plants will not penetrate free water. If free water rises around them and stands there very long, it seriously interferes with the root functions and quickly injures and often kills the plant. In irrigated regions, while the ultimate purpose is the same as in humid regions, subsurface drainage is frequently needed for relief of artesian pressure, interception of seepage, and removal of alkali from the root zone.

1.3. History of Land Drainage. The history of land drainage reaches into remote antiquity. The Greek historian Herodotus in 400 B.C. refers to its application to the Nile Valley in ancient Egypt. Many Roman writers (Cato, Virgil, and others) mention drainage channels and the practice of drainage in some of its phases in most of the early eastern nations. Irrigation and drainage were well understood and applied; hence, today, the modern Persians have efficient systems although they themselves have long since lost the art. Interesting quotations from these authors are given in a book on land drainage by John H. Klippart [69].

Spectacular accomplishments of the Old World, frequently mentioned by drainage authorities, are the Fens of England and Haarlem Lake of Holland. The unqualified success of these projects has doubtless had some influence in the initiation of similar enterprises in America. Encouraged by the success with Haarlem Lake, the Dutch government, in 1920, launched and is now developing the mighty Zuider Zee project, which when completed, will open for settlement 538,000 acres and accommodate 300,000 people—more than ten times the area and more than twenty times as many people as the Haarlem Lake project. The project will extend the coast line into the sea by dikes built under water, and the enclosed area will be pumped dry. Five such enclosures, or *polders*, are contemplated east and north of Amsterdam, varying in size from 50,000 to 140,000 acres each. The land to be reclaimed lies 1 to 15 feet below the level of the North Sea. Internal drainage will be accomplished by means of subsidiary dikes, ditches, pumping plants, and tide gates. The original estimated cost of $210,000,000 (nearly $400 an acre) includes construction of 20 miles of main dikes as well as the pumping plants, auxiliary structures, and special methods of caring for the soil and preparing it for crops. In the economy of Holland this is not considered an exorbitant charge. The undertaking is being financed by the government and was originally expected to be completed gradually over a period of 20 years. Owing to unforeseen incidents it is not now expected to be completed until in the early 1960's. When complete it will add 7 per cent to the total area of the Netherlands and 10 per cent to its agricultural area.

The history of nearly every European country contains records of finished drainage projects which, though smaller in single units and, therefore, less spectacular than those just mentioned, still are projects of considerable national importance. The Pontine Marshes of Italy and the projects of La Geronde and Forez in France are notable examples of this kind. During the years 1936 to 1938, the Germans spent $320 million to reclaim new land and announced in 1938 that another $400 million was to be expended for land reclamation and increased use of fertilizers.

1.4. Drainage in the United States. More than a century ago (1835), the first drain tile were laid in the United States. These were handmade tile and were installed on the farm of "the father of tile drainage," John Johnston, of Geneva, N.Y.; 13 years later, 856 acres in Central Park, New York City, was successfully drained.

From these early beginnings the spread of land drainage was rapid, first in a westerly direction through the East Central and Middle Western states but later covering the entire humid region of the North, East, Middle West, and South, and finally extending to the great irrigated regions of the arid West.

The following developments during the past century have had a distinctly stimulating effect on the progress of land drainage:

1. The invention of a drain-tile machine and its introduction into the United States, in 1848.
2. The enactment, early in the twentieth century, of state laws providing the legal machinery for the operation of community outlet projects.
3. The perfection of excavating machinery during the latter part of the nineteenth century.
4. The establishment of numerous drain-tile factories throughout the country.
5. The formation of state, sectional, and national associations of interested landowners, engineers, and others for the interchange of experiences and ideas.
6. The work of state and national research organizations, which by painstaking investigations have made important contributions to our store of scientific knowledge.
7. The labor of extension engineers and other workers of state colleges and of engineers of the U.S. Department of Agriculture who have, by actual demonstration in the field, convinced many communities of the value of drainage.
8. The publication of numerous circulars and bulletins on the subject which, together with personal correspondence and the use of audiovisual aids, have done much to spread information and stimulate activity.

Successful drainage improvements perform their service in an inconspicuous manner. Being largely buried out of sight and functioning silently, they attract little attention; hence it is difficult for the present generation of Americans to visualize the original condition of large areas of what is now our best productive land. It is interesting to contrast present conditions with those described in early accounts of pioneer adventures such as in the book by Keating [66] published in 1825 under the title "Long's Expedition to the Source of St. Peters River in 1823." The following passage is typical of the experiences reported:

> Near to this house we passed the state line which divides Ohio from Indiana. The distance from this to Fort Wayne is 24 miles without a settlement; the country is so wet that we scarcely saw an acre of land upon which a settlement could be made. We traveled for a couple of miles with our horses wading through water, sometimes to the girth. Having found a small patch of esculent grass (which from its color is known here by the name of bluegrass), we attempted to stop and pasture our horses, but this we found impossible on account of the immense swarms of mosquitoes and horse flies, which tormented both horses and riders in a manner that excluded all possibility of rest.

After reading this description, and many others like it, it is easy to see why complete drainage was early recognized as fundamental in providing an environment suitable for human habitation as well as conditions necessary for the growth of crops. Without drainage it is literally true that much of the most prosperous agricultural area in the country could not successfully have been settled, nor would highways, railways, rural schools, sewerage systems, or the many other individual and community activities which constitute civilization have been possible.

1.5. United States Drainage Census. The first census of drainage in the United States was taken by the Federal government in 1920 and since then has been repeated at 10-year intervals. Complete results are available in printed form and at nominal cost on application to the Superintendent of Documents, Washington, D.C.

In attempting an analysis of these statistics, one should keep in mind the differentiation between drainage on farms (private drainage) and community enterprises (public drainage), which was clearly recognized and observed.

Drainage on farms is direct drainage which the individual farmer carries out privately as part of the development of his farm. The outlet of such a system may be either into a natural channel or into a public or private artifically constructed waterway. Information concerning farm drainage is collected by regular census enumerators at the same time and in the same manner—by personal canvass—that other agricultural data are gathered.

On the other hand, statistics regarding community enterprises are concerned only with outlet projects where groups of landowners combined to secure potential benefits impossible of attainment by individual effort. Most of these districts were organized and conducted under state laws providing specific regulations governing their procedure. Since the records on file at the various county seats are required to contain all the details of each such project, the compilation of essential data is a relatively simple matter for the special enumerators appointed for the purpose. Drainage districts are empowered to perform only that kind of service for the individual which he is unable to perform for himself, *i.e.*, to provide easily accessible outlets. The obligation of the district to the individual ends here. Obviously, little or no benefit accrues to the owner until, by installing lateral drains emptying into the public outlet, he makes it serve him.

The Federal Census of Drainage [23,24] gives, for each county in each state, the number of farms reporting drainage and the total acreage drained in each; the number of drainage enterprises; the area drained in each, and the capital invested therein; the type of drainage; the total length of open and of tile drains; and many minor details pertaining to each of these main divisions. A summary of each of these leading items, by states, is also given. The published report of the Drainage Census for 1940 [23] also gives a synopsis of the drainage laws of each of the several states. A copy of the final report of each decennial census is available to interested parties for inspection in most public libraries and in the libraries of state agricultural colleges or, as already stated, by purchase from the Superintendent of Documents, Washington, D.C.

1.6. The 1950 Federal Census. The 1950 United States census of drainage [24] is the largest and most detailed of the four as yet taken. It reported more than one and one-half times as many acres of drained land as did the 1920 census. It listed a total of 103 million acres of drained land as against 87 million reported in the 1940 census. This is an increase during the decade of 18.4 per cent.

None of the Federal censuses of drainage included individual drained areas of less than 500 acres. These are very numerous and would increase the grand total of drained lands far beyond the total official figures of the 1950 census just mentioned.

The specific purposes of land drainage are, in the main, threefold:

1. Reclamation and improvement of land.
2. Protection against overflow.
3. Seepage relief from irrigation ditches or for control or prevention of alkali formation.

In the East and Middle West the farmers need drainage mainly for the first two purposes; in the Western states they need it more largely for the third purpose.

Up to the present the miles of open ditches far exceed the miles of tile drains, the larger part of which are in the states of the north-central Mississippi Basin.

1.7. Nonagricultural Drainage. With the beginning of the era of industrial expansion in the last half of the nineteenth century many new applications were found for drainage. One of these was the safeguarding of foundations. It was known that the bearing power of soils of all kinds was greatly reduced by the presence of excessive moisture. Diversion of surface water and drainage of ground water from the vicinity of bridge abutments and building foundations soon came to be one of the most important services rendered by drainage.

Drainage to protect structures finds perhaps its largest application under railroads and highways. Because of their physical character, modern rail lines, pavements, and smooth roads would be unknown if means of preserving their stability and protecting them against washout had not been developed through drainage.

In the municipal field, drainage is important from several standpoints. The control and quick removal of surface and ground water is essential to personal comfort and public health; to the uninterrupted flow of commerce, industry, and business in general; and to the protection of all kinds of public improvements.

Other needs for drainage are found on athletic fields, golf courses, race tracks, aviation fields or airports, and similar developments where special use results in high value and large justifiable expenditures for drainage.

Drainage for the prevention of malaria has a well-established technique. The transmission of the disease has been traced to a night-biting mosquito whose breeding habits have been learned, and methods to prevent its propagation, largely by drainage, are known. Together with house-screening and medical elimination of the acute disease from infected persons who serve as malaria carriers, such drainage is of high economic value in settled areas.

Drainage for abatement of pest varieties of mosquitoes, not involving public health consideration, also has a special technique. Generally only very shallow lowering of the ground water horizon is necessary or economically justifiable.

Well-constructed projects have proven of high economic value through increased values of residential property, increased field-worker efficiency, and increased milk yield of dairy herds.[1]

1.8. Extent and Importance of the Agricultural Drainage Problem in the United States. Analysis of all available official records shows that the original total of all agricultural land in the United States requiring reclamation by drainage or that would be appreciably benefited thereby was about 216,000,000 acres, or about 24 per cent of all potentially agricultural land in the country. This 216 million acres can be broken down approximately into the following groups:

[1] Report of the Water Planning Committee, National Resources Board, Dec. 1, 1934, p. 341.

Further discussion of nonagricultural drainage is beyond the scope of this book.

Types of land	Approx. area, millions of acres	Approx. % all agricultural land in the United States
Humid-region lands lying mostly east of the 103d meridian, potentially suitable for some type of agriculture, but originally too wet for such use until reclaimed by drainage		
Swamp or periodically swamp	50	6
Periodically overflowed	30	3
Tidal marsh	8	1
Group total	88	10
Already reclaimed from all 3 of above types	53	6
Areas in above region in production from beginning, without drainage, but that would be greatly benefited thereby	60	7
At least partially drained irrigated lands in the 17 arid Western states	14	1
Grand total	215	24

This analysis shows the importance of land drainage in agriculture, especially when it is remembered how necessary it is to increase food production to meet the needs of a rapidly growing population.

BENEFITS OF DRAINAGE

Good practice generally involves both surface and subsurface drainage in proper proportions (see Fig. 1.1) as the best performance of subsurface drains depends on adequate overland disposal of surface water. In the beginning the cost of such drainage, where needed, was considered justified in general terms by making possible the utilization of otherwise unavailable fertility.

As time went on the many benefits of drainage came to be recognized and enumerated in specific form. These may logically be grouped under three general headings: (1) benefits on the individual farm, (2) community benefits from drainage, and (3) benefits of drainage from a regional viewpoint.

1.9. Benefits on the Individual Farm. The benefits of drainage on the individual farm are as follows:

1. Flooding and consequent loss of seeds and fertilizers are largely eliminated.

2. Without extending the farm boundaries, good drainage adds to the best acres on the farm by bringing under cultivation permanently ponded areas, swamps, and running sloughs the beds of which are of great potential fertility but too wet in their original state to cultivate and to produce crops (see Figs. 1.1 to 1.5).

3. This elimination of ponds and swamps renders fields more easily and cheaply worked as these originally wet areas are obstructions around or across which farm equipment must be moved with great difficulty.

4. Drainage dries the fields earlier in spring and sooner after rains, thus making possible earlier spring planting, more prompt cultivation following

FIG. 1.1. A boggy pasture in northern Iowa, too wet for cultivated crops. The "hummocks" are typical of such a condition.

rains, and earlier maturing of crops in the fall. In this connection the following experiences are of interest.

At University Farm, St. Paul, Minn., after drainage was installed the farm foreman was able to begin the spring work 10 days to 2 weeks earlier than before the fields were drained. Scientists at the North Dakota Agricultural Experiment Station have given similar testimony and added that the men are able to start work on the tile-drained fields 2 to 4 hours earlier after a heavy rain than on the undrained areas. Various farmers throughout the entire watershed of the Mississippi River, and in other states, who have tile drained their farms have frequently made similar statements.

According to unpublished data by Arny, formerly of the Minnesota Agricultural Experiment Station, results over a period of 3 years show losses as follows from seeding being delayed 30 days later than normal:

Kind of grain	Days late planting	Percentage of loss of normal crop
Barley	30	33.9–47.8
Oats	30	44.5–56.3
Wheat	30	67.7
Wheat	20	45.0–56.8

Delays less than 30 days show results consistent with the above. Therefore, it is apparent that earlier thawing out and drying of the ground in the spring due to drainage, where this amounts to 1 to 2 weeks, tends to prevent loss of final crop much as follows:

Kind of grain	Limits of percentage of saving from loss	Approximate average saving from loss for all grains, per cent
Barley	10.2–20.5	
Oats	12.6–25.2	27
Wheat	17.8–35.7	

Assuming about equal value of these three crops, the average saving from loss would be about 27 per cent.

5. It is a matter of common observation wherever tile drainage is practiced that the better physical condition of the soil promotes vigorous root growth and more hardy plants. The result is a full maturing of crops—particularly corn—1 to 2 weeks earlier than on undrained land, with a consequent noticeable rise in yield and quality of crop.

The benefits just enumerated are confirmed and magnified by the following supplementary benefits.

6. By removing free water from the soil, tile drainage prevents surface puddling, thus preserving the good granular structure so conducive to maintaining good tilth and resisting splash erosion. It opens the soil pores to greater depths and provides protection against drought damage

in the two following ways: by increasing the storage capacity of the soil for the form of soil moisture used by plants (the capillary moisture), and by greatly extending the possible feeding ground for plant roots thus inducing deeper rooting. Given this opportunity different types of crop

FIG. 1.2. Two cultivated fields in widely separated parts of the United States in need of both surface and subsurface drainage. *Above:* An Iowa field. (*Courtesy of Mason City Brick and Tile Company.*) *Below:* A field in the eastern United States. (*USDA Farmers' Bul. Courtesy of John G. Sutton.*)

plants will root 2 to 6 feet deep (see Table 10.7, page 264). By the same action drainage opens the soil to a freer passage of air (better aeration) which results in warming it up earlier in the spring and keeping it warm later in the fall thus tending to ward off late spring and early fall frosts in the cold agricultural regions.

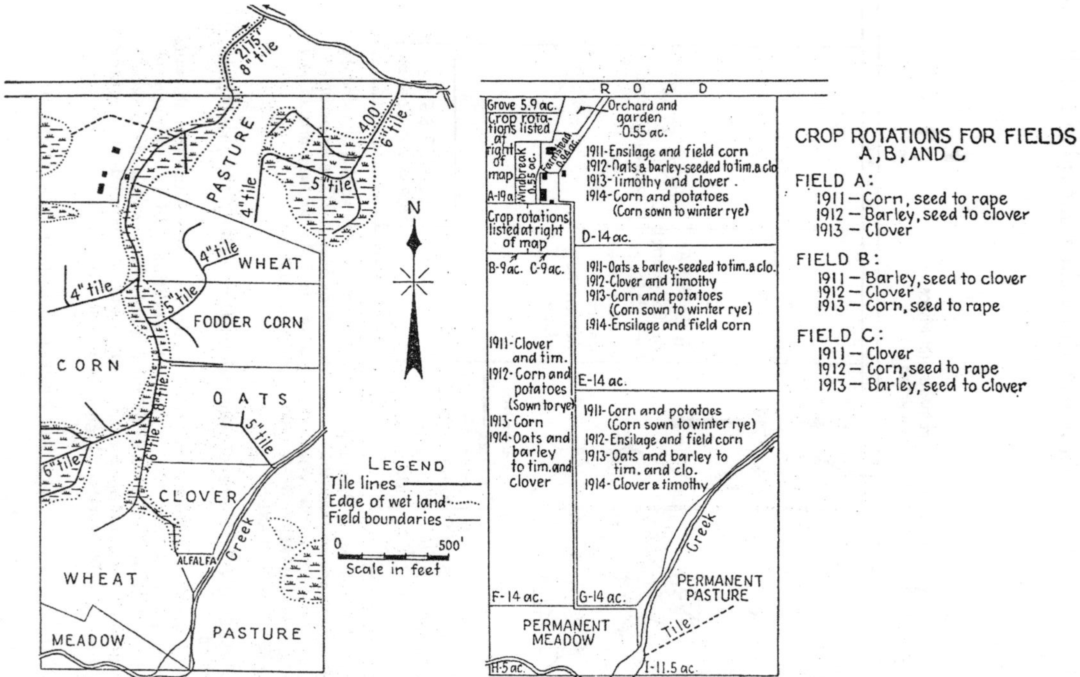

An 80 acre farm in Southern Minnesota showing the swampy irregular condition of the fields before drainage and the tile system installed in 1909

The same farm as the foregoing showing the uniformity of the fields as made possible by the drainage, in the final cropping plan perfected in 1914

A TYPICAL TILE DRAINAGE SYSTEM FOR ROLLING LAND

NOTE: The income from this farm increased primarily as a result of drainage, from \$994.00 in 1908 to \$2455.00 in 1912

FIG. 1.3. Comparative maps of an 80-acre farm in Scott County, Minnesota, before and after drainage.

7. On properly drained fields evaporation is slowed and baking and cracking of the soil are greatly reduced.

8. Through tile drainage, also, the chemical condition of the soil is made more conducive to sturdy plant growth by the removal of alkali and

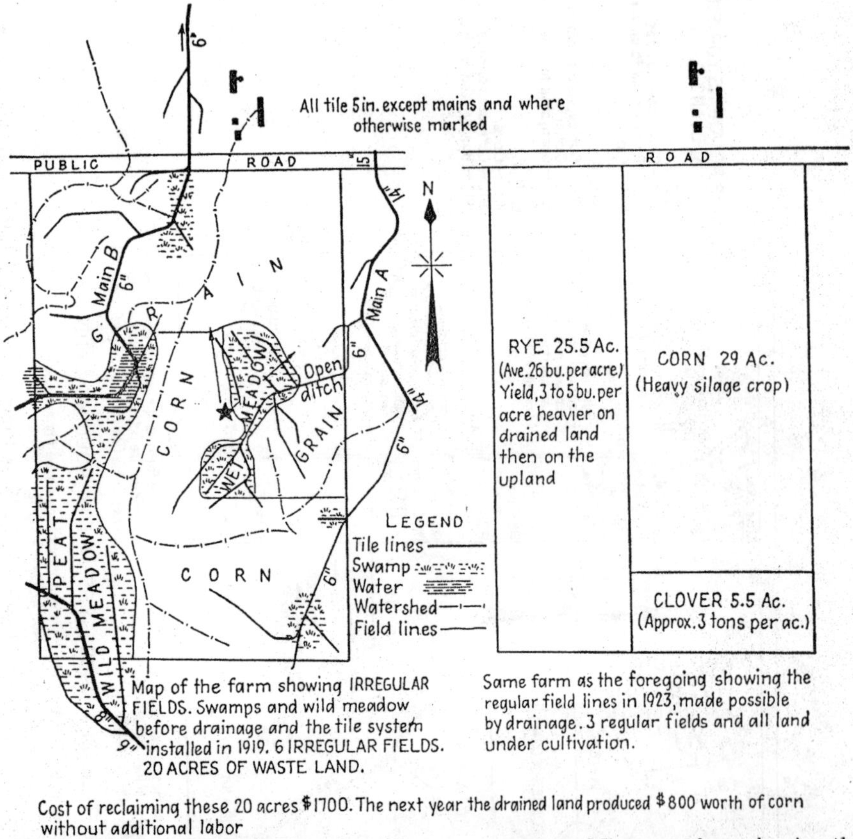

FIG. 1.4. Comparative maps of a 60-acre unit on a southern Minnesota farm showing the improving influence of tile drainage. *Left:* The wet conditions before drainage and the tiling system that was installed; 20 wet acres producing only $200 worth of low-grade hay. *Right:* The same area completely cropped with regular fields 3 years later; production on the originally wet 20 acres, $800 worth of high-grade corn (see Fig. 1.5).

acids and by inducing more vigorous bacterial action resulting in stimulated decay of organic matter and resultant nitrification.

9. By keeping the soil pores open throughout the entire year, tile drainage increases both infiltration and permeability rates of the soil, enabling the absorption of a greater amount of rainfall and melted snow and thus reducing surface erosion. Furthermore, especially in latitudes subject to heavy freezing, this continuously open condition of the soil helps to check heaving by frost and consequent winterkilling.

10. All these benefits result in marked increases in returns from the farm investment through production of larger and better quality crops, lessening of the cost of field operations, and elimination of losses resulting from poor drainage (see Figs. 1.5 to 1.7). Testimony from all parts of the United States indicates that the earnings on a farm-drainage investment normally range between 15 and 65 per cent per year, with a normal aver-

FIG. 1.5. Two views taken 4 years after tile was installed, from starred point in Fig. 1.4 with directions faced in taking pictures indicated by arrows. *Above:* Across the upland. *Below:* Across the drained slough. Note much better crop on the lowland.

age of about 25 per cent. In occasional rare cases it has been known to run as high as several hundred per cent. There is always a widely varying increase in the farm income resulting from tile drainage without appreciable increase in the cost of production.

11. The foregoing discussion fully justifies the statement frequently made by both farm managers and agricultural economists that the initial cost of drainage may logically be added at once to the predrainage value of the farm.

12. Finally, drainage improves sanitary conditions on the farm and

Fig. 1.6. Lush growth of soybeans on a well-drained field in Marshall County, Iowa.

Fig. 1.7. Cattle grazing on former marsh land in Union County, Mississippi. Ditch is well maintained by grazing as a part of a larger pasture. (*USDA Farmers' Bul.* 2047. *Courtesy of John G. Sutton.*)

greatly improves its appearance, making it a more wholesome and pleasanter place to live.

1.10. Community Benefits from Drainage. From the community standpoint there are also definite benefits from drainage, some direct, others indirect. The most notable of these follow. More than any other requirement, drainage makes possible the building and maintaining of good rural highways which, in turn, stimulate the following facets of civic progress:

1. The rural postal and express service, the value of which in modern rural life can scarcely be overestimated.
2. Stabilization of the farm business by making readily available better and more distant markets, which in turn induce a better-balanced selection of crops to be produced.
3. Consolidated rural schools, which provide better facilities for education both in physical equipment and competent instruction.
4. Improved facilities for rural churches.
5. More widely extended social intercourse and relaxation in and between rural communities.
6. Reduction of breeding places for mosquitoes and other disease-bearing pests by removal of stagnant water from shallow pools and crevices.

TABLE 1.1. SUMMARY OF ESTIMATED ACREAGES AND YIELDS ON 35 TYPICAL DRAINAGE JOBS PLANNED BY THE SOIL CONSERVATION SERVICE AND EXECUTED UNDER ITS DIRECTION DURING 1943

Crops	Acres		Average yields per acre		Net total increase in production due to drainage
	Before jobs started	After drainage completed	Before jobs started	After drainage completed	
Corn	7,632	8,484	28 bu	46 bu	175,100 bu
Oats	1,387	915	33 bu	48 bu	1,500 bu
Barley	465	950	32 bu	40 bu	22,740 bu
Wheat	2,010	2,892	16 bu	21 bu	29,370 bu
Sorghum (grain)	200	150	30 bu	50 bu	1,500 bu
Soybeans (for beans)	4,021	4,570	18 bu	27 bu	52,200 bu
Hay	2,705	3,737	1.3 ton	2.9 ton	7,530 ton
Cotton	5,380	5,622	0.64 bales	0.79 bales	999 bales*
Rice	25	25	43 bu	61 bu	450 bu
Potatoes	2,498	2,748	123 bu	162 bu	137,900 bu
Vegetables	1,427	1,587	$55	$71	$34,200
Pasture	2,752	2,586	743†	1,165†	968,000†
	30,502	34,266			

* Cotton also produced increase of 440 tons seed valued at $45 per ton.

† Pounds of milk equivalent.

1.11. Benefits of Drainage from a Regional Viewpoint. From a regional viewpoint the benefits accruing from drainage are clearly indicated by Sutton [145] reporting on drainage rehabilitation work of the U.S. Department of Agriculture Soil Conservation Service for a typical year of activity. In this report particular attention is called to 35 selected jobs scattered over 12 central, southern, and eastern states in which 33,992 acres in 450 farms were benefited. The entire work of the Service in this field covered a much greater total of benefited acres and many more than 450 farms. The 35 jobs covering the 450 farms mentioned were selected because they show in a clear-cut manner the definite and appreciable benefits accruing from well-considered and properly executed farm-drainage projects. The striking results on these 33,992 acres are shown in Table 1.1, adapted from Sutton's report. It is pointed out that much of the increased production on these lands was due to improved farming practices which could not, however, have been effectively invoked without the primary aid of bettered drainage conditions.

POPULAR FALLACIES CONCERNING DRAINAGE

Several organizations as well as a good many private individuals in this country are traditionally opposed to drainage development for various but largely erroneous reasons. It is true there have been occasional ill-advised and ill-fated projects, but these are not nearly numerous enough to justify the sweeping denunciations often heard. For each publicized failure at least a dozen outstanding successes can be cited.

1. It is claimed that drainage is an extensive destroyer of the essential breeding places and habitat of wild life and is carried out simply to serve the selfish interests of unscrupulous promoters who wish to reclaim and reap the profits from the sale or mass cropping of the drained lands, whereas there is already too much land in agricultural use in this country resulting in huge agricultural surpluses. Furthermore, it is said, most of the lands thus drained, especially sloughs and ponded areas, are nearly or wholly worthless for farming after surplus water is removed.

The facts are far otherwise. Most of the drainage activity in the United States is by individual farmers or groups of farmers and is carried out for the purpose of improving their farms and farming communities to take advantage of the legitimate benefits discussed in the preceding section. With millions of human beings dying yearly from starvation in some regions of the world, it is clear that it is not agricultural surpluses but failure to provide and practice an effective world distribution that is the problem. This clearly is not a result of too much drainage.

Moreover our own population is increasing by leaps and bounds and much land ill-suited to farming needs to be replaced by permanently fertile areas. It is a fact well supported by experience that, barring high-acid peat swamps, almost without exception the beds of swamps and ponded areas, when drained and properly farmed, are the most fertile spots in agricultural regions.

In 1950 the President of the United States appointed a special commission to consider water-resources policy. That part of the Commission's report [156] relating to land reclamation is based on the premise of an expanding economy. Present crop surpluses are seen as temporary. Assuming a population of 190 million by 1975, the Commission estimates that approximately 100 million new cropped acres must be put into cultivation. This estimate is based on the further assumption that neither the nation's dietary habits nor crop production *per acre* will undergo material change.

Definitely in line with the foregoing statement, in 1951, the Engineers' Joint Council[1] [43] published a contribution entitled "Principles of a Sound National Water Policy" from page 182 of which the following significant pronouncement is quoted: "It is reasonably possible to secure increased agricultural production through land drainage at less cost than in any other way, and the public benefit is equally large."

A survey conducted in 1952 by a privately financed fact-gathering agency, The Twentieth Century Fund,[2] supported the general conclusion that "agricultural production in the United States could be increased 50 per cent by soil conservation and land development programs."

Wild life properly fills a large place in the interests of mankind. There are ample large regions in our own country and also in most other countries of the world well suited as breeding areas and as a habitat of wild life, which are at the same time too rough or otherwise unfit for agricultural use, to provide all needed habitat for desirable wild life. The growing policy of our state and national governments is sufficiently broad to safeguard these areas as wild-life refuges and thus provide for the continuity of all desirable forms of such life without official interference with normal development of farms and farming areas.

2. It is further claimed that drainage lowers regional ground-water tables permanently beyond the reach of plant life, thereby not only destroying wild-life habitat but rendering arid vast areas of potentially fertile lands in humid regions and greatly reducing or entirely destroying their productive capacity, increasing liability to wind erosion, and wiping out the natural beauty of the landscape.

[1] The Engineers' Joint Council is made up of engineers from all the old-line national engineering societies including the A.S.C.E., the A.S.M.E., the A.S.E.E., and all the others—at least five in all—an organization of conservative but constructive temper and influence in the American economy.

[2] As reported in the *Central Constructor*, official publication of Associated General Contractors of Iowa, July, 1952.

Here, too, the facts are otherwise. The depth of regional water tables in humid areas is a function of the influence of nature's drainageways, the rivers. Nature in the form of weather and climate governs in this case, a work for the most part far beyond the ability of man to control. These regional natural channels are in general many times deeper than man-made outlet drainage ditches. The vast fluctuations that occur in the depth of ground-water tables in humid regions from time to time are the result of cumulative seasons of low rainfall and of continuous evaporation, not of man's drainage operations. The truth of this statement was well illustrated in northern Minnesota during the drought period from 1929 to 1937 when the ground-water level dropped to 10 or 12 feet below the bottoms of the deepest drainage outlet ditches. This situation was soon reversed by the heavy rains of the spring and summer of 1938 which refilled the soil-storage reservoir and heavily flooded the land surface, destroying crops over hundreds of square miles of the same area.

Artificial drainage over most of the farm area in humid regions seldom exceeds 4 feet, and much of it is not over 3 feet deep. This is the minimum essential for the growth of crops, the root systems of most of which normally reach a depth ranging from 36 to 48 inches when drainage conditions permit.

Scientific studies over extensive areas before and after installation of drainage projects have shown that the regimen of major streams is but little affected. The probable explanation is that the speed-up of surplus surface runoff in drainage ditches is counterbalanced by the creation of temporary storage in the soil itself.

3. The opponents of drainage further claim that it is a menace to the sturdy growth and life of trees, hence is inimical to forest development, at the same time greatly increasing the fire hazard in forest areas.

Scientific research fails to support these claims. Extended studies in both Europe and America have clearly established the fact that forest drainage greatly increases the amount and improves the quality of merchantable timber in those kinds of trees commonly used for lumber. An excellent example of this is found in the studies reported by Averill and McGrew [8] fully discussed in a subsequent chapter of this book to which the reader is referred for detailed facts. This study further shows that open ditches in swamp forests do not, as a rule, appreciably lower the ground-water level at distances greater than 35 to 50 feet from the ditch on either side and that the distance from the ditch on either side at which the drainage effect *disappears* varies from 50 to 940 feet, with an average of only 330 feet (see Fig. 6.13).

4. Another fallacy sometimes heard relates to the effect on climate of man's obliteration of shallow lakes, ponds, and sloughs. It is said that

evaporation from such local water surfaces formerly produced abundant local rainfall which, after drainage, ceases to occur. Studies of rainfall patterns over long periods of 50 years or more do not bear out this contention. It is an established meteorological fact that only about 10 per cent of the normal rainfall in humid regions is of local origin and that this small proportion is only partly affected by drainage.

CHAPTER 2

SOILS IN RELATION TO DRAINAGE

ORIGIN AND NATURAL GROUPS OF SOILS

2.1. Introduction. All soils are composed of three forms of matter: solid, liquid, and gaseous. According to their origin soils fall into two great groups, namely, *organic or cumulose soils,* usually known as *peat* and/or *muck,* and *inorganic soils* usually called *mineral soils* which constitute far the larger of the two groups. Peat and muck soils are of such importance and are so scattered over the world that an entire chapter is later devoted to their general character, variations in composition, and handling, including drainage, in agriculture.

Except for a small amount of decayed vegetable matter (*humus*) the solid matter in mineral soils is a mass of fine rock particles (the *soil grains*) and/or clusters called *granules* which give *structure* to the soil. The spaces between the soil grains and between the granules or both are called the *soil pores.* The pores contain the gaseous element (*air*) and the liquid element (*water*). It is the portion of this soil water free to move under the pull of gravity that gives rise to the subdrainage problem in agriculture.

MINERAL SOIL GROUPS

The effect of the wide differences in soil characteristics on subdrainage is so marked as to make proper presentation of the subject difficult without a review of the soil origins and more important soil properties.

Mineral soils are classified according to their origin into the following four general groups: (1) residual, (2) glacial, (3) alluvial, and (4) marine, each with its own peculiar drainage characteristics.

2.2. Residual. Residual soils are those which have usually been formed by the weathering and disintegration of native rocks in place. The destructive action of many natural forces, exerted through countless centuries of time, has resulted in both a physical and chemical decomposition of material which now covers a large area of the United States. As the rocks from which such soils are formed differ widely in mineral composition, a complex variety has been created which is difficult to classify, but includes all classes from sands to clays. Hence, the natural drainage conditions of residual soils vary from good to very poor.

In sections of humid climate, the original soil constituency has been materially modified by the leaching out of the more soluble minerals and by accretion of organic matter through successive periods of growth and decay of vegetation.

2.3. Glacial. Thousands of years ago, the entire northern portion of the temperate zone of North America was covered with an immense mantle of ice varying in thickness from a few feet near the edges to a mile or more at its center.

By reason of tremendous internal pressure and the action of gravity, this glacial mass was capable of moving several feet per day, accumulating fragments of rock and detritus on its underside, overriding hills, and gouging out valleys. Great quantities of pulverized material became imprisoned under the ice and were carried southward until temperatures were encountered high enough to melt the ice more rapidly than it could advance. A large portion of the imprisoned material was then carried still farther south and there deposited by gushing streams from melting ice. Still larger portions of this detritus were deposited where it lay as the ice cake thawed and the glacier retreated. Authorities on glaciation now agree that the ice age was not one unbroken invasion and retreat of the ice cap but consisted of five distinct periods, the last of which occurred more than 10,000 years ago. This accounts, in part, for the present existence of subsurface stratification or successive layers of pervious and impervious matter which affects so vitally many drainage problems.

Most of the glacial deposit is too coarse and heavy to be susceptible to extensive movement by either wind or water. This heavier material, variously known as *glacial till*, *glacial drift*, or *boulder clay*, still remains where it was originally deposited. These glacial till soils mostly fall into two subgroups commonly thought of as the *impervious* group, composed mostly of heavy clay, gravel, and boulders, with badly impeded natural drainage and not responding readily to underdrainage; and the *pervious* group composed largely of sand, silt, gravel, and boulders and responding readily to underdrainage.

Lake Soils. Frequently glacial lakes were formed against the face of the glacier both in its advance and in its retreat. These lakes finally disappeared with the complete withdrawal of the glacier, but during their existence their beds accumulated a considerable blanket of fine fertile soil deposited by streams flowing into them or dropped from the face of the glacier itself. These soil blankets are designated as glacial lake soils. They generally lack sufficient outlet for natural drainage, but as a rule they respond readily to subdrainage when outlets are provided.

Loess Soils. It is thought that the ice age was followed by an epoch of aridity, accompanied by strong westerly winds causing large quantities of dry detritus to be picked up and transported great distances. In this manner, the formation of the loess soils of the central United States and the adobe soils of the arid Southwest is explained. This weathering of glacial deposits of fine wind-blown material is a secondary modification resulting in the loess soils which are widely distributed in certain areas of the country. On account of their great extent and their susceptibility to washing, the loess soils are often listed as a major division. The loess, adobe, and lake soils are all very fertile, of exceptionally good tilth, and respond readily to underdrainage.

2.4. Alluvial. As the name implies, *alluvial* soils owe their origin to the eroding and transporting capacity of water. Without going into a detailed discussion, it may be remarked that the power of streams to carry fine silt in suspension and to roll coarser particles along their bottoms depends wholly on the velocity with which the water travels. A slight increase in velocity results in a greatly accelerated transporting ability, and, conversely, if the velocity is checked from any cause whatever, suspended silt is likely to be deposited. Since the gradient or slope of streams is always much steeper in their upper reaches than it is near their mouths, the rate of travel of water is rapid in the former case and sluggish in the latter. Consequently, a large quantity of soil is annually cut away from higher lands, carried down by the streams to lower lands, and there deposited in a thin sheet as the water spreads out over its natural flood plain. The sediment thus deposited is invariably found assorted, with heavier particles near the stream bank and the finer silt farther out, causing a formation typical of all overflow valleys with the surface slope away from the watercourse. That part of the suspended silt not thus deposited is carried out to sea and contributes to the gradual raising of the ocean bed. Narrow blankets of alluvia, laid down in the manner just described, are found along practically every stream not closely confined between steep hills or mountains. Alluvial deposits, from a geological standpoint, are the youngest of all soils. They are still in the formative stage except in those localities where floods are confined between levees.

Such soils are surpassed by none in fertility, but the topography where they occur is generally flat, low, and in need of drainage.

2.5. Marine. Along the entire Atlantic and Gulf seaboard in the eastern and southeastern portions of the United States, there is a wide stretch of territory which nature has reclaimed from the ocean. This soil is derived from the older residual types to the west and north and has originated partly by a vast upheaval of the ocean's bed in some prehistoric age. Owing to long-sustained agitation by waves, marine soils occur in relatively large, well-rounded particles, such as sand and sandy loam, the coarse *texture* of which permits easy underdrainage where adequate outlets can be secured. Though saline in character near the coast line and of flat topography, these soils, when drained, are capable of supporting a wide range of diversified crops.

PHYSICAL COMPOSITION OF SOILS

There is a close relationship between soil character, water requirements of plants, and good irrigation, drainage, and soil-erosion-control practice. The amount of surface exposed by the soil particles and structural granules within the root zone is a paramount consideration in agriculture, for on it largely depends not only the ability of a soil to retain moisture in the form essential to plants, but also the possibility of removal through drainage of the excess water harmful to plants and the adhesive and cohesive qualities of the soil particles so essential in control of soil erosion. The finer the *texture* of the soil (that is, the smaller the soil particles), the greater the total area exposed in any given volume of soil; hence the greater the frictional resistance to removal of excess water through underdrainage. Conversely, the coarser the soil *texture* (that is, the larger the soil particles), the less the total area exposed in any given volume of soil; hence the less the capillary attraction to moisture and the less the frictional resistance to flow of water through the soil into tile drains under the influence of gravity. This partly explains why clay soils drain out excess water so much more slowly than do sands or sandy soils.

Also the finer the soil the greater the feeding area for the plant because of the larger surface area, and the consequent extension of the capillary moisture film gives a greater opportunity for plant-food solution. Generally, however, coarse-textured soils are less readily eroded than are those of finer texture owing to the greater interlocking of the soil particles and to the greater ease with which water passes through the coarse soil.

The following geometrical illustration will aid in explaining the rapid increase in surface area with reduction of the diameter of the soil grains in a given volume of soil.

Consider a hollow cubical box, 1 foot on an edge, inside. This will just hold, packed in uniform order and tangent to its sides, numbers of spheres, as shown in the accompanying table.

Number of spheres	Diameter of spheres, in.	Total volume of spheres, cu ft	Total surface area of spheres, sq ft
1	12	$\frac{1}{6}\pi$*	π
8	6	$\frac{1}{6}\pi$	2π
64	3	$\frac{1}{6}\pi$	4π
512	$1\frac{1}{2}$	$\frac{1}{6}\pi$	8π
4,096	$\frac{3}{4}$	$\frac{1}{6}\pi$	16π
32,768	$\frac{3}{8}$	$\frac{1}{6}\pi$	32π
etc.			

* $\pi = 3.1416$.

From this table it clearly appears that, with the diameter reduced to that of a clay particle of 0.005 millimeter, or about 2/10,000 inch, the total surface area would be relatively prodigious and the resulting friction against it correspondingly great.

2.6. The Soil Separates. On a purely theoretical physical basis soils are classified according to the size or limited range of sizes of the rock particles of which they are composed. The range of size of particles generally used in the United States to delimit each of these theoretical classes is that early adopted by the former Bureau of Chemistry and Soils in the

TABLE 2.1. SOIL SEPARATES

Separate	Diameter of grains, mm*	Separate	Diameter of grains, mm
Fine gravel	2.0 –1.0	Very fine sand	0.10 –0.05
Coarse sand	1.0 –0.5	Silt	0.05 –0.005
Medium sand	0.5 –0.25	Clay	0.005 –0.0001
Fine sand	0.25–0.10	Colloids	0.0001–0.000001

* 1 mm = 0.03937 in.

U.S. Department of Agriculture. The various classes thus established, which serve as a definite basis for scientific considerations but never occur in nature except by the merest accident, are called the soil separates. They are listed in Table 2.1 along with the limiting range of size of particles constituting each class.

2.7. Natural Field Soils. Natural field soils, however, will be found to be a mixture of widely varying percentages of several separates, usually some organic material, and various salts or other chemical compounds. Hence there results a vast number of possible classes of natural field soils such as are commonly spoken of as very sandy loam, medium sandy loam, silt loam, silty clay loam, clay loam, etc. Table 2.2 gives the approximate limits in the proportions of the physical soil separates as they may occur in 10 of the commonly recognized field soils. This table indicates the great number of possible combinations and the extreme variability possible even within any single class. The ranges here shown are in reasonably close agreement with those given by Lyon and Fippin and other authors.

2.8. Classification of Field Soils. Soils may be classified in a number of ways, but the following taken from the *U.S. Bureau of Soils Bulletin* 96 describes the method used by that bureau in classifying soils in the United States:

The soils are brought together into groups on the basis of certain characteristics of the soils themselves. each group constituting a series.

A *soil series* may be defined, therefore, as a group of soils having the same range in color; the same character of subsoil, particularly as regards color and structure; broadly, the same type of relief and drainage; and a common or similar origin (*e.g.*, Wooster series and Olympic series). A subdivision within a series, *e.g.*, Olympic silty clay loam, is a type. Soils may be more practically grouped on the basis of texture also, and a group of soils all having the same texture belong to the same class. A *soil class*, therefore, includes all soils having the same texture, such as sands, clays, and loams.

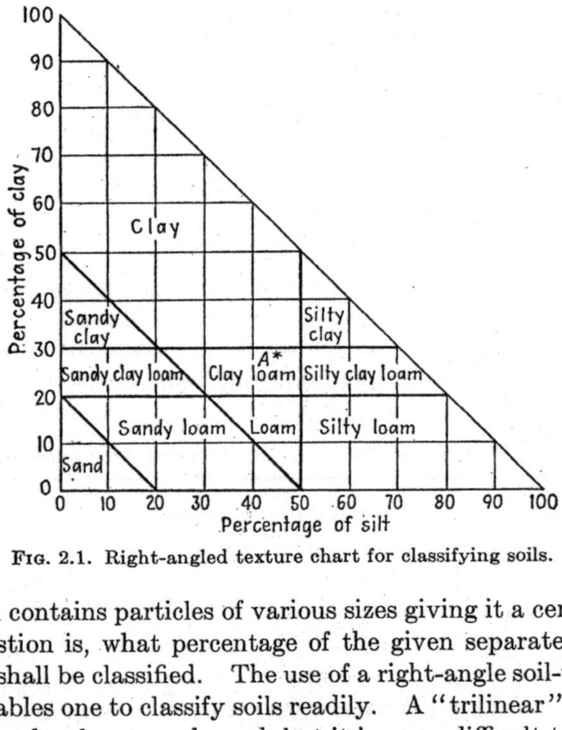

FIG. 2.1. Right-angled texture chart for classifying soils.

A field soil contains particles of various sizes giving it a certain texture, and the question is, what percentage of the given separates determines how the soil shall be classified. The use of a right-angle soil-texture chart (Fig. 2.1) enables one to classify soils readily. A "trilinear" (equilateral-triangle) chart has been much used, but it is more difficult to read. The same values are obtained with both charts.

Example. A soil contains 28 per cent clay, 45 per cent silt, and 27 per cent sand (by difference).

This may be located as point *A* in Fig. 2.1, and its classification found to be clay loam.

Translating the chart into tabular form, the values are those given in Table 2.2.

The average percentage of separates present in soils as they commonly occur is shown in Table 2.3, which is taken from *Bulletin* 50 of the Engineering Experiment Station, Iowa State College.

One of the most significant facts shown by Table 2.3 is that the surface exposed by the soil particles in a true clay soil is between two and three to four times as great as that in a loam or a sandy-loam soil. This shows still more clearly the reason for the very slow movement of water into and through clay soil as compared with a sandy soil, an important consideration relative to underdrainage.

TABLE 2.2.* MECHANICAL COMPOSITION OF THE PRINCIPAL SOIL CLASSES

Soil classes	Limits in the proportions of the soil separates		
	Sand, %	Silt, %	Clay, %
Sand	80–100	0– 20	0– 20
Sandy loam	50– 80	0– 50	0– 20
Silt loam	0– 50	50–100	0– 20
Loam	30– 50	30– 50	0– 20
Silty clay loam	0– 30	50– 80	20– 30
Sandy clay loam	50– 80	0– 30	20– 30
Clay loam	20– 50	20– 50	20– 30
Silty clay	0– 20	50– 70	30– 50
Sandy clay	50– 70	0– 20	30– 50
Clay	0– 50	0– 50	30–100

* Davis, R. O. E., and H. H. Bennett, *USDA Bur. Soils Cir.* 319, Table 1.

TABLE 2.3.* AVERAGE PHYSICAL COMPOSITION OF COMMON SOILS

Soils	Percentage of separate present							Per cent pore space	Surface area in 1 cu ft, sq ft	Acres
	Fine gravel	Coarse sand	Medium sand	Fine sand	Very fine sand	Silt	Clay			
Coarse sand	5.0	13.0	27.0	30.0	11.0	8.5	5.5	40.0	40,500	0.93
Medium sand	5.0	13.0	20.0	32.5	14.0	9.0	6.5	41.8	44,500	1.02
Sandy loan	5.0	10.0	11.0	26.0	11.0	22.0	15.0	51.0	66,600	1.53
Fine sandy loam	2.0	2.5	5.0	20.0	27.5	32.0	11.0	50.0	62,000	1.42
Silt loam	1.0	1.5	2.5	6.0	11.0	56.0	22.0	53.0	104,000	2.39
Clay loam	3.0	5.0	5.0	12.0	10.0	28.0	37.0	54.0	136,000	3.12
Clay	0.0	2.0	2.5	5.5	7.0	37.0	46.0	56.0	142,000	3.29

* Compiled from data given in "The Principles of Soil Management," by Lyon and Fippin, The Macmillan Company, New York, 1911.

2.9. Only Major Limits in Fundamental Soil Classification Needed. The long-established system of classifying soils as being in the Marshall, Carrington, Yolo Series, and the like, as, for example, Marshall silt loam, Carrington loam, Yolo clay, and so on, means little or nothing to the practical planner or operator of the land. It gives little or no clue, except through some lengthy index or guide, to the physical character, stratification, and needed treatment of the soil, or more broadly, of the land in farming practice.

In most of the practical considerations of soil and soil-moisture relations in agriculture only the four major classifications of agricultural soils as the sands, the sandy loams, the silt loams, the clays or clay loams, and the limited but definitely named variations of each of the four (*e.g.*, fine

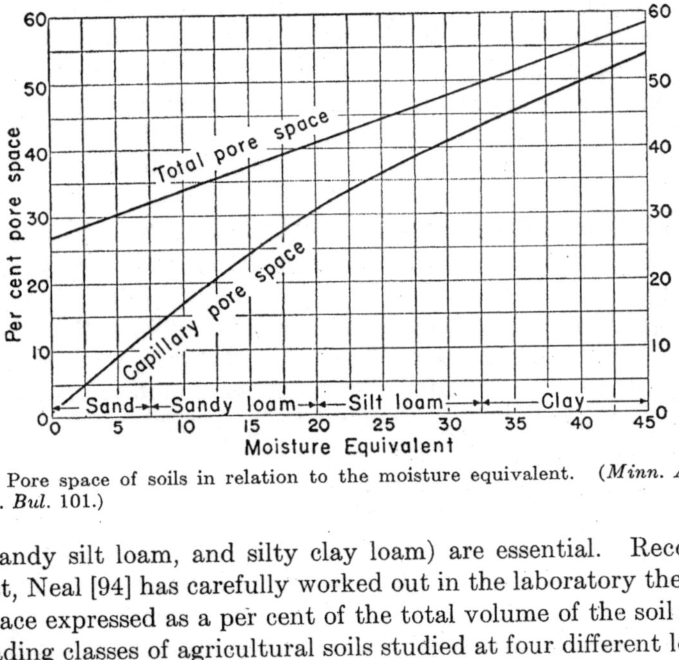

FIG. 2.2. Pore space of soils in relation to the moisture equivalent. (*Minn. Agr. Expt. Sta. Tech. Bul.* 101.)

sand, sandy silt loam, and silty clay loam) are essential. Recognizing this fact, Neal [94] has carefully worked out in the laboratory the normal pore space expressed as a per cent of the total volume of the soil in these four leading classes of agricultural soils studied at four different locations in Minnesota. The results of these studies are epitomized in Fig. 2.2, which also shows the proportion of that pore space occupied by the capillary water and the moisture equivalent.[1] The scheme of relationships illustrated in Fig. 2.2 will be the basis of discussion, henceforth, in this book, concerned with topics or problems involving the physical composition of soils in relation to soil moisture.

2.10. A Practical, Descriptive Classification of Soils. The following descriptions of soil classes, adapted from the "Engineering Handbook for Farm Planners: Region III, Upper Mississippi Valley" of the Soil Conservation Service [141], is presented here for its simplicity, clarity, and practical usefulness in field examinations and discussions. All the commonly accepted soil classes are included except peat and muck, to which

[1] For a definition of moisture equivalent, see Sec. 3.12.

Chap. 14 is devoted. The order of presentation is largely the relative position in which the different classes usually occur from the surface down.

1. *Silt Loam.* Contains a moderate amount of the finer grades of sand and a little clay, over half the total being silt. When dry it may appear cloddy, but lumps are readily broken and, when pulverized, feel soft and floury. When wet, all runs together and puddles. Either dry or moist it will form casts not broken when freely handled. Squeezed between thumb and finger it will not ribbon but gives a broken appearance.

2. *Silty Clay Loam.* Contains a moderate amount of clay and the fine grades of sand, over half the total being silt. When dry it appears cloddy, the lumps breaking with difficulty. Pulverized it feels soft and floury, not readily distinguished from silt loam. When wet it is slightly plastic and can be made into casts freely handled without breaking. Squeezed between thumb and finger, it forms a ribbon which breaks readily. The surface of the ribbon will be flaky rather than smooth as in clay.

3. *Sandy Loam.* Contains much sand but enough silt and clay to make it somewhat coherent. Individual sand grains readily seen and felt. Squeezed, when dry, it forms a cast which readily falls apart. Squeezed when wet, it forms a cast that will bear careful handling without breaking.

4. *Clay Loam.* A fine-textured soil which breaks into clods or lumps, hard when dry. When moist, pinched between thumb and finger it forms a thin ribbon which breaks readily, barely sustaining its own weight. The moist soil is plastic and will form a cast that bears much handling. When kneaded in hand it does not crumble readily but tends to become a heavy, compact mass.

5. *Sandy Clay.* Contains a little silt and a moderate amount of sand, but at least half of the total is clay. Dry, it forms hard clods which break into small granules under force. This texture is best seen in moist condition. Pinched between thumb and finger it forms a thin ribbon with surface rough and gritty due to sand particles.

6. *Clay.* A fine-textured soil that forms very hard clods when dry but is quite plastic and usually sticky when wet. Pinched between thumb and finger when moist it will form a long flexible ribbon.

7. *Sand.* Loose, composed of single grains, readily seen or felt. Squeezed in the hand when dry it falls apart on release of pressure. Squeezed when moist it forms a cast that crumbles when touched.

8. *Fine Gravel.* Resembles sand but individual grains range in size up to that of an ordinary garden pea.

9. *Coarse Gravel.* Resembles fine gravel but particles are larger, usually an inch or less. May range up to size of a hen's egg. Larger material would interfere with a trenching machine in drainage work.

10. *Slate and Shale.* Finely stratified or laminated beds of fine-

textured soil, as silt or clay. Slate is much more dense and compact than shale.

11. *Glacial Drift* (*Impervious*). Includes glacial deposits, both stratified and unstratified; largely clay, gravel, and boulders; may include some silt and sand.

12. *Glacial Drift* (*Pervious*). Includes glacial deposits, both stratified and unstratified; largely sand, silt, gravel, and boulders; contains small amounts of clay.

2.11. Colloids. Most soils contain a portion of finely divided, jelly-like (colloidal) material. It is most abundant in the clays, in which it is an important element, limiting and often wholly destroying their permeability. On this account it is a serious handicap in the agricultural handling of clay soils and especially so in their drainage. Common experience concerning the nature of clay colloids and their behavior in drainage and in agriculture is as follows:

The colloidal particles are wholly ultramicroscopic, varying in size from one ten-thousandth of a millimeter down nearly to zero. In any appreciable volume of soil they aggregate an enormous surface area, have great absorptive capacity which serves as a measure of soil colloids, swell enormously when wet, and shrink and crack when dry. In contact with alkaline bases they will form certain loose electrochemical combinations that enable them to remain in suspension in water indefinitely, making leaching out practically impossible. Although in a calcium clay these combinations may cause flocculation, resulting in a friable, easily worked condition of the soil, in a sodium clay (black alkali being the sodium salt present) the effect will be the opposite, causing deflocculation which, stimulated by concentration of the soil solution, tends to "plug" the soil completely against the passage of water. Hence, subdrainage of clay soils becomes largely impossible except under continued frequent repetition of wetting and drying or chemical treatment and cultural practices

TABLE 2.4. PORE SPACE AND WEIGHT PER CUBIC FOOT OF COMMON SOILS

Kind of soil	Per cent pore space (from Fig. 2.2)	Approximate theoretical weight, lb per cu ft	Approximate weight of natural soils, lb per cu ft
Coarse sand..........	28	121	These will be 80 to 92 per cent, or an average of 85 per cent, of theoretical, owing to presence of organic matter
Medium sand........	30.5	117	
Average sandy loam...	36.5	105	
Average silt loam.....	45	91	
Average clay loam....	51	84	
Average clay.........	55	77	

which include abundant use of barnyard manure or other humus-producing material. These last two practices tend to break up the amorphous, pasty condition of the soil and render it more friable and permeable.

2.12. Weight and Apparent Specific Gravity of Dry Soils. The weight per cubic foot of most soil-forming solid rock approximates 166 pounds. Water weighs 62.5 pounds per cubic foot. Hence the specific gravity of most solid rock approximates 166/62.5, or 2.65. Unless otherwise distinctly specified or qualified, the expression *dry soil* in this discussion means *oven-dry soil*, that is, soil which has been heated in an oven held at or slightly above 212 degrees Fahrenheit until the weight of the soil becomes constant—a point at which all the soil moisture has been driven out of the soil. Theoretically at least, this dry weight is easily determined in the laboratory for any given field soil by securing an undisturbed field sample thereof of known volume (preferably 1 cubic foot) and obtaining its weight after it has been fully dried in an oven at or above 212 degrees. The dry weight of 1 cubic foot of the soil divided by the weight of 1 cubic foot of water gives the specific gravity of the dry soil. This specific gravity of the dry field soil is called the *apparent specific gravity* A_s of that soil. That is, it is the ratio of the combined weight of the rock-soil particles, the organic matter present in the soil, and the gases in the pores of the soil to the weight of an equal volume of water. The unit dry weight (see Table 2.4) and the apparent specific gravities of different types of field soils will vary widely as opposed to the specific gravity of the solid soil-forming rocks. The apparent specific gravity is a convenient and an almost necessary factor in dealing with the moisture problems of any given soil.

2.13. Pore Space in Soils. The influence of the pore spaces on water movement, water retention, and water storage in soils is fully as important as that of the superficial area exposed to friction of the moving water. The pores are much larger, individually, in the coarser soils than they are in the finer soils. Therefore they will hold larger columns of water whose weight tends to a much freer and more rapid water movement, under the influence of gravity, than through the finer soils. The pore space is of vital importance to growing plants because it must accommodate both the water and the gases required by the plants, as well as the plant roots, without any of the three entirely crowding out the others.

2.14. Total Pore Space of Soils. The volume of the pore space in any given field soil, whether expressed in volumetric units or as a per cent of the total volume of the soil, is an important factor in all irrigation problems and in some drainage problems because, with certain qualifications later discussed, it is a practical measure of the maximum water-holding capacity of the given soil. If the apparent specific gravity A_s of the given soil is known and R_s is the specific gravity of the parent rock, the pore space on a percentage basis S is given by the equation

$$S = 100\left(1 - \frac{A_s}{R_s}\right) \tag{1}$$

If R_s is considered constant at 2.65, this equation becomes

$$S = 100\left(1 - \frac{A_s}{2.65}\right) \tag{2}$$

At least in theory the volume of pore space in a given soil is readily obtained in the laboratory in the following manner: Obtain an undisturbed sample of the field soil of known volume in a close-fitting container. Soak the sample (held in the container to prevent loss of any of the soil itself into the water) for 24 to 48 hours. Weigh the saturated sample, container and all. Then dry the sample in the container to oven-dry condition, and weigh again in the container. The difference in weights will be the weight of water that the saturated soil contained. This value, divided by 62.5, will give the volume of this water in cubic feet.

The maximum water-holding capacity of a given soil, expressed as a per cent of the dry weight of the soil, is the weight of the water in the saturated sample of field soil divided by the dry weight of the sample.

PROBLEMS

1. Name seven potentially agricultural soils according to their origin.

(*a*) Which of these are most widely distributed in the United States?

(*b*) Which are potentially the most easily drained?

2. What is meant by the "soil separates"?

(*a*) Which of these are most conducive to good drainage?

(*b*) Which are the *greatest* impediments thereto?

3. Where would you classify soil that analyzed as follows: 1% coarse sand, 2% medium sand, 4% fine sand, 60% silt, and 33% clay?

What would be its probable drainability?

4. What would be the average classification, according to the moisture equivalent, to a depth of 4 ft when the moisture equivalent tested as follows: top 6 in. 28, next 18 in. 22, next 6 in. 15, and last 18 in. 33?

(*a*) Would this soil drain easily? Explain your answer.

5. If the solid parent rock of a given soil weighs 166 lb per cu ft and the A_s of the soil is 80%, what is the dry weight of the soil per cubic foot neglecting the organic matter?

6. Name five leading benefits from drainage to the individual farm.

(*a*) Explain how each directly affects the farm income.

CHAPTER 3

WATER PROPERTIES OF SOILS

SOIL PERMEABILITY

O'Neal [99] and Uhland and O'Neal [151] define permeability as "the capacity of a soil to transmit water and air." High permeability makes good drainability. Resistance to flow of water varies nearly as the wetted area of the walls of the conducting channels. In a soil the channels are the soil pores. It has earlier been shown that the finer the soil grains, the greater their total superficial area and the smaller the pores, while, conversely, the larger the soil grains, the larger the pores and the less their total superficial area resisting the flow of water and air. The size and interrelation of the soil pores, therefore, has far greater influence on the freedom of soil-water movement than does their *total* volume. Hence, generally speaking, the more porous a soil, the less permeable it is. It thus appears that *texture* is one of the leading factors determining soil permeability and susceptibility to drainage. However, it is not the only such factor, for the soil grains, forming into clusters of varying types, give to the soil *structure*, which is another agency determining the size of soil pores and, therefore, influencing the drainability of the soil.

3.1. The Soil-structure Relation. *Texture* in a given soil is practically constant, but its *structure* may vary widely within comparatively narrow limits of land area, and it may be changed by either natural or artificial agencies. Such changes may improve or may injure both the tilth and the drainability of the soil. Good tilth conditions and a high permeability accompany a coarse structure with large pore spaces, but a fine and compact structure tends toward a puddled condition and impeded drainage. This, in turn, often results in a waterlogged condition in which soil structure may collapse completely and leave the soil wholly unworkable.

Confronted in planning its agronomic, erosion-control, and drainage programs with need of more complete knowledge of the influence of soil on tilth and especially on permeability, the Soil Conservation Service carried out intensive studies in eight North Central and seven Southeastern states. Results of these studies as presented by O'Neal, may be summarized substantially as follows:

Permeability can be expressed quantitatively as water-transmission rate through a column of soil under specific conditions, usually in inches per hour. (It is also commonly expressed as centimeters per square centimeter per hour or as gallons per square foot per day.)[1] Tests were made on saturated core samples of field soils of specific dimensions, with the least possible disturbance of their volume and structure. Each horizon below the surface layer should be thus tested to the desired depth.

Seven permeability classes and the corresponding percolation rates have been established, for use in the field, as follows:

1. Very slow: less than 0.05 inch per hour.
2. Slow: 0.05 to 0.2 inch per hour.
3. Moderately slow: 0.2 to 0.8 inch per hour.
4. Moderate: 0.8 to 2.5 inches per hour.
5. Moderately rapid: 2.5 to 5.0 inches per hour.
6. Rapid: 5.0 to 10.0 inches per hour.
7. Very rapid: 10.0 or more inches per hour.

There were 96 soil horizons studied on which permeability measurements are available. Predominant types of structure (see Fig. 3.1*a* and *b*) and other characteristics, all shown in Table 3.1, were viewed. Space limitations preclude extended discussion of details, but a study of the table, which is generally looked upon as a reliable permeability and subdrainage guide, will show interesting variations and trends. Pertinent and helpful clues, given by O'Neal, to evaluation of *P* are presented as follows:

> Structure is the most significant but not the sole important factor;
>
> Relation between length of horizontal and vertical axes and direction and amount of overlap of aggregates are important factors;
>
> Texture alone is not a reliable clue although heavy texture is sometimes associated with class 2 and light texture with class 6;
>
> Mottling is not a reliable clue;
>
> Durability of structural aggregates in some soils, and size and number of visible pores or direction of easiest fracture seem correlated with permeability.

Of 38 estimates through these clues, 36 were correct. One incorrect case was off only one class and the other only two classes.

However, it is warned that *all* factors must be considered singly and together; no one alone will serve; and strict adherence to clues observed is essential to success. Even then this indirect method of determining *P* has its drawbacks and limitations: (1) It does not clearly indicate whether the permeability found is vertical, or horizontal, or the direct resultant of

[1] This value is the coefficient of permeability *P* which is defined as "the rate of flow of water through a unit cross-sectional area under unit head." Some scientists prefer the term "hydraulic conductivity" instead of permeability. It is frequently expressed as k, K_s, or K. Until such time as there is more general agreement the authors prefer the symbol *P*.

the two, for the given position in the drawdown curve. This is left to the judgment of the practitioner. (2) Therefore, wide experience in its use and trained judgment are essential.

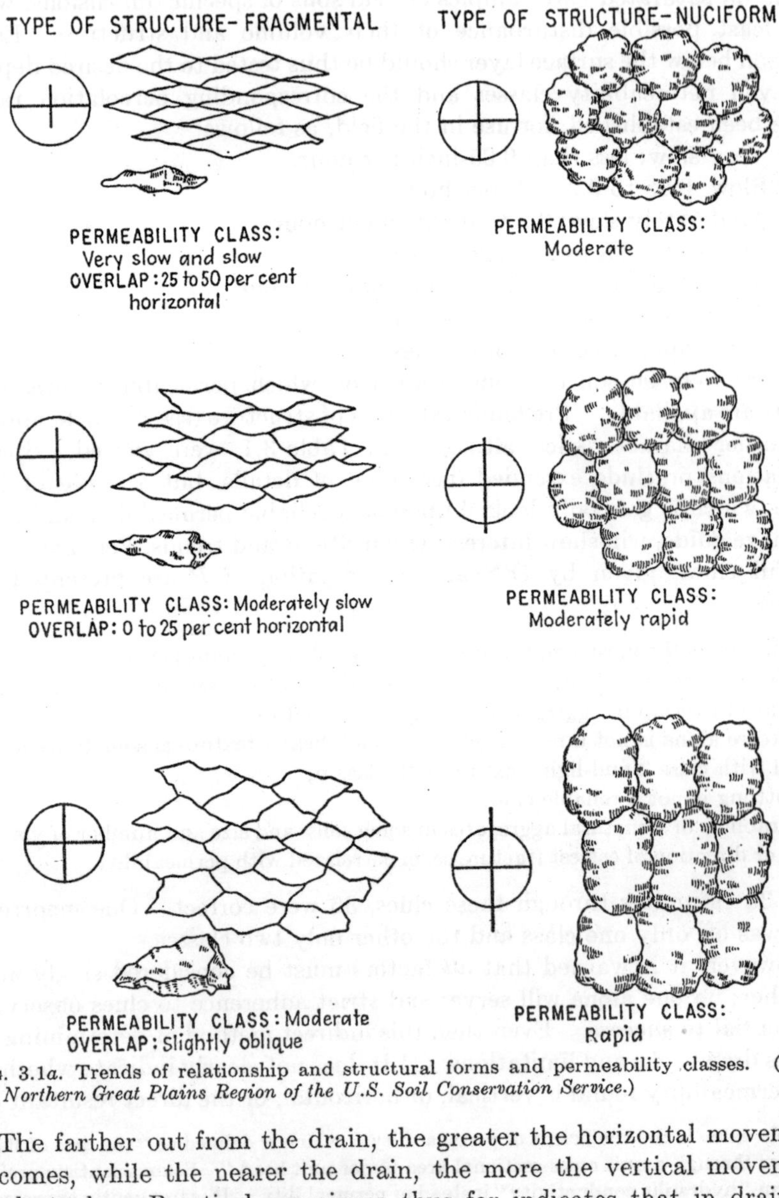

FIG. 3.1*a*. Trends of relationship and structural forms and permeability classes. (*From the Northern Great Plains Region of the U.S. Soil Conservation Service.*)

The farther out from the drain, the greater the horizontal movement becomes, while the nearer the drain, the more the vertical movement tends to be. Practical experience thus far indicates that in drainage considerations the horizontal component of P should be used. This is reasonable because, in reaching the drain, the water moves many times farther, horizontally than vertically. But the ratio of the horizontal permeability to the vertical never becomes constant but varies widely between different areas because of the wide variation in soil texture, structure, and salt and humus content. Hence, however reliable this indirect method of determining P may *seem* to be, without occasional check *quantitative* determinations, such estimates may prove misleading.

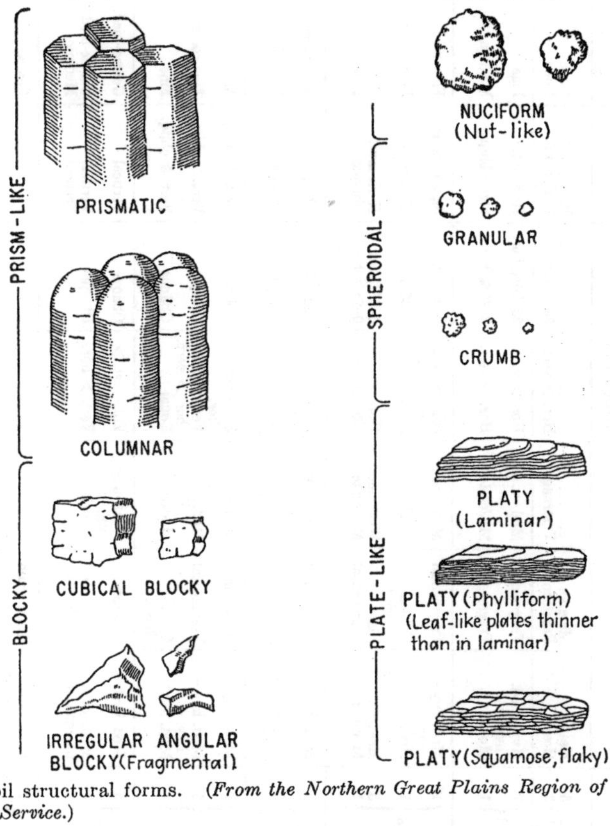

FIG. 3.1*b*. Soil structural forms. (*From the Northern Great Plains Region of the U.S. Soil Conservation Service.*)

QUANTITATIVE DETERMINATIONS

There are several methods of quantitative determination of P, both for field and for laboratory use, of varying reliability. Space limitations preclude a complete discussion. The Report on Drainage Investigations in Irrigated Areas of the Imperial Valley, California, by Donnan, Aronovici, and Blaney [34], hereinafter called the Imperial Valley Report, indicates that the following three methods are the most reliable and readily used.

FIELD MEASUREMENT

The first is a direct measurement *in the field*, activated by pumping from a small well through the aquifer and reaching to the barrier level,

TABLE 3.1.* RELATIONSHIP BETWEEN MEASURED PERCOLATION RATES AND SOIL CHARACTERISTICS

Permeability class	Range of measured percolation rates, in./hr	Number of horizons	Texture	Structure: Type	Structure: Grade	Structure: Relation of horizontal and vertical axes	Structure: Overlap	Direction of easiest natural fracture	Size of visible pores	Mottling	Remarks
1. Very slow	0.01–0.04	7 { 5	Heavy	Massive	....	....	....	Horizontal	None	Strong	
		2	Heavy	Fragmental	V. strong	Hor. > vert.	50% hor.	Horizontal	None	Strong	
2. Slow	0.06–0.20	8 { 7	Hvy.–mod. hvy.	Fragmental	Mod.–v. stg.	Hor. > vert.	25–50% hor.	Horizontal	Fine	Mod. stg.	
		1	Mod. hvy.	Fragmental	Moderate	Hor. > vert.	0–25% hor.	Slgt. oblique	Medium	Moderate	Slightly cemented
3. Moderately slow	0.20–0.79	19 { 12	Mod. hvy.–hvy.	Fragmental	Mod.–mod. stg.	Hor. > vert.-eql.	0–25% hor.	Slgt. oblique	Fine–med.	Strong	
		1	Heavy	Cubical blocky	Weak	Equal	0–25% hor.	Slgt. oblique	Fine	Slight	
		5	Heavy	Platy	Weak–mod.	Hor. > vert.	0–25% hor.	Horizontal	Fine	Moderate	
		1	Medium	Fragmental	Moderate	Hor. > vert.	0–25% hor.	Oblique	Fine	Moderate	Moderately cemented
4. Moderate	0.90–2.18	12 { 7	Mod. hvy.–hvy.	Fragmental	Strong.–v. stg.	Eql.–hor. > vert.	Slgt. oblique	Oblique	Med.–fine	Strong	
		5	Med.–mod.–hvy.	Nuciform	Weak	Equal	Oblique	Oblique	Med.–fine	Slight	
5. Moderately rapid	2.58–4.73	11 { 7	Med.–hvy.	Nuciform	Mod.–stg.	Eql.–vert. > hor.	Oblique	Vert. & obliq.	Med.–coarse	None–slgt.	
		3	Mod. hvy.–hvy.	Fragmental	Moderate	Vert. > hor.–hor. > vert.	Oblique	Vert. & obliq.	Fine-coarse	Strong	
		1	Mod. hvy.	Cubical blocky	Moderate	Equal	Oblique	Vert. & obliq.	Fine	Strong	
6. Rapid	5.12–9.77	12 { 5	Mod. lt.–hvy.	Nuciform	Weak–mod. stg.	Eql.–vert. > hor.	Oblique	Vertical	Coarse–med.	None–mod.	
		3	Mod.–hvy.	Fragmental	Moderate	Vert. > hor.	Oblique	Vertical	Coarse-med.	None–slgt.	
		4	Lt.-mod. lt.	Single grain	....	....	....	Vertical	Medium	....	Smaller sand grains than 7
7. Very rapid	10.15–100+	27 { 21	V. lt.–mod. lt.	Single grain	....	....	....	....	Coarse	None–slgt.	Larger sand grains than 6
		3	Mod. lt.–mod. hvy.	Nuciform	Weak.–mod.	Vert. > hor.	Oblique	Vertical	Coarse–med.	None–slgt.	
		3	Mod. hvy.–hvy.	Fragmental	Weak–mod.	Vert. > hor.	Oblique	Vertical	Coarse–med.		

NOTE: Hvy. = heavy, mod. = moderate, med. = medium, lt. = light, v. = very, stg. = strong, eql. = equal, slgt. = slight, hor. = horizontal, vert. = vertical, obliq. = oblique.

* Table 1 of "Soil Characteristics Significant in Evaluating Permeability" by Alfred M. O'Neal. Published in *Soil Science*, Vol. 67, No. 5, May, 1949. Republished by permission.

measuring the discharge and the position of the drawdown curve. From Fig. 3.2 the formula for P in gallons per square foot per day is readily

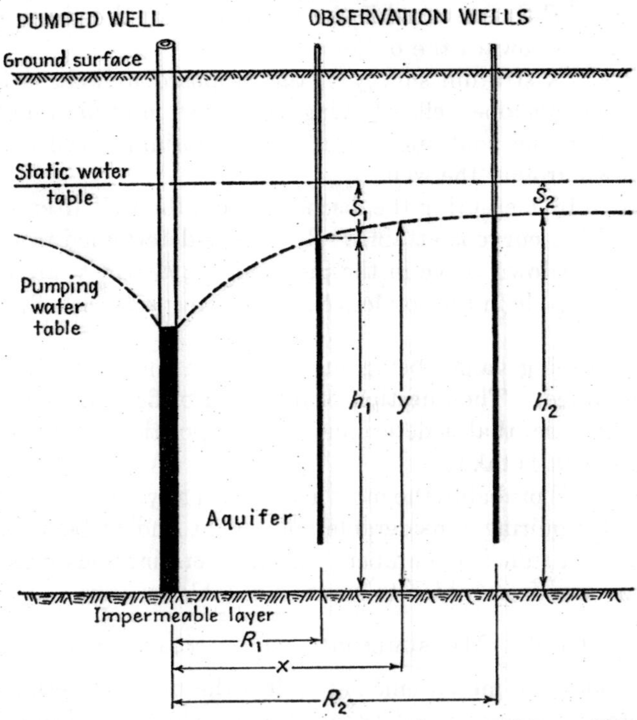

FIG. 3.2. Diagram of factors involved in calculating soil permeability coefficient from field experiment. (Figure 5 of the Imperial Valley Report.)

found in this case to be

$$P = \frac{2.303Q \log_{10} (R_2/R_1)}{\pi(h_2 + h_1)(S_1 - S_2)} \tag{3}$$

where P is the required coefficient of permeability
- 2.303 is the reduction factor from Naperian logarithms to common logarithms
- Q is the discharge from the well, cfs
- R_1 and R_2 are the horizontal distances of the observation wells from the axis of the pumped well, ft
- h_1 and h_2 are the corresponding heights from the surface of the impermeable layer to the pumping water table at the corresponding observation wells, ft
- S_1 and S_2 are the vertical distances between the static and the pumping water table at the given observation wells, ft

x and y are the coordinates of any point on the pumping water table referred to the base of the pumped well on the surface of the impermeable layer as axis

The value of P thus secured is the true value for the given location as the water moves toward the outlet (the well).

Procedure. A stratum survey of the proposed well site. Sinking of a small gravel envelope well. Laying out a system of piezometers about the well. Pumping continued until an equilibrium curve is secured in the drawdown around the well.

Q is secured by measuring the discharge from the well after equilibrium in the drawdown curve is attained; $S_1 - S_2$ is determined from the position of the drawdown curve in the piezometers; h_1 and h_2 are equal level of static water table in the soil less S_1 and S_2, respectively; R_1 and R_2 are measured.

All the foregoing values being substituted in the above formula, P is readily computed. The question of direction of flow in each individual case does not arise: that is determined by the position of the curve where the measurement is taken.

This method is probably the most accurate of any, but it has the serious drawback of requiring considerable equipment and of being very time-consuming, as a sufficient number of such determinations must be made over a given area from which to secure a reliable average value of P.

Direct Measurement in the Laboratory

Direct laboratory measurement requires the use of a permeameter, of which there are a number of types, among which the Imperial Valley Report indicates that the discharging permeameter and the falling-head permeameter give the most satisfactory results. With either a direct measurement of P may be obtained, either vertical or horizontal according to the sample used.

The horizontal P must be used in design; hence the local ratio between the vertical and the horizontal P must be obtained in order to reduce all values obtained, to the horizontal. Where the variation between the vertical and the horizontal values of P is great, horizontal cores are sometimes secured by digging a pit that enables direct securing of such cores. A considerable percentage of cores thus taken horizontally furnishes for the given region a safe basis for computing a reliable ratio between the horizontal and the vertical values of P.

3.2. Laboratory Manuals the Guide. A permeameter should be set up and operated only by or under the close supervision of a highly trained and experienced laboratory technician. Laboratory guides or manuals, apart from general reference books, are available at testing laboratories, furnishing comprehensive. detailed instructions for use of the permeameter. Such technical details are outside the scope of this book, but there are available in the literature a number of scientific reports and articles giving well-illustrated information on such equipment. Among these are the Imperial Valley Report [34], the informative contribution by Wenzel [160], that by Meinzer and Fishel [82], and Jenkins's manual [64].

However, attention is called to the requirements for securing the undisturbed field samples to be used in the permeameter. They must be of

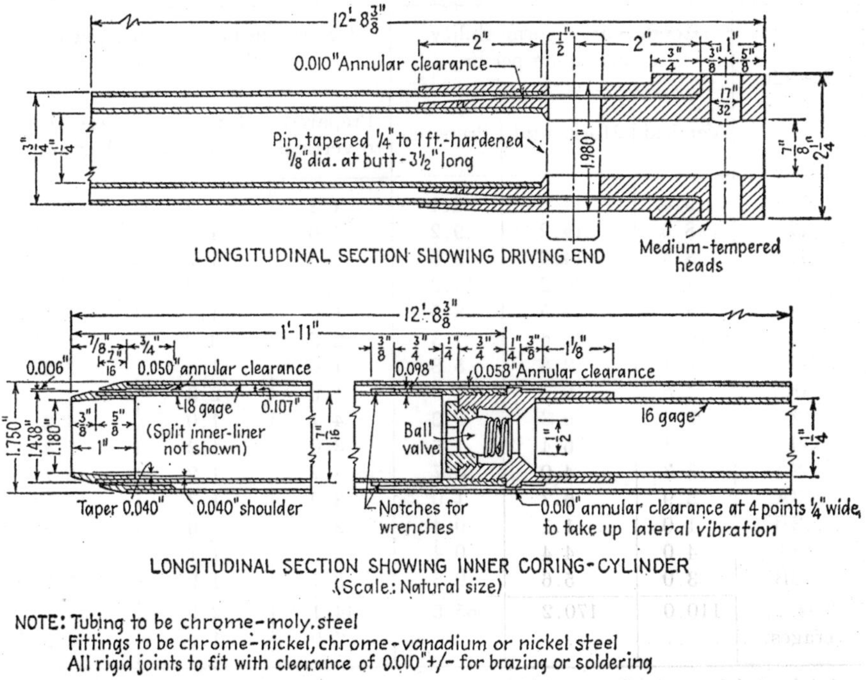

FIG. 3.3. Imperial Valley soil-sampling tube. *(Courtesy of Soil Conservation Service.)*

a size that exactly fits the permeameter soil tube and must be secured with a strictly minimum disturbance of the samples. This requires a specially designed field sampling tube. Figure 3.3 illustrates such a device driven by an air hammer which was designed by Blaney and Taylor [15] and discussed fully by Donnan, Aronovici, and Fox [35], who state:

> The tube must bring out a sample that represents accurately the texture and exhibits the structure of the soil at the exact location from which it is taken; hence there must be little distortion of the core and no mixture of the strata. . . . It must be possible to operate the equipment with relative ease and to obtain samples from a depth of 12 to 18 feet as rapidly as can be done by any other method.

It is further stated that the most satisfactory samples are taken when the sediment is moist but not wet.

3.3. Permeameter Types for Specific Tests.[1] The discharging permeameter was used most generally in the Imperial Valley investigation because it was found the most suitable for sediments of the aquifer group

TABLE 3.2.* RELATIONSHIP BETWEEN P MEASUREMENTS BY PERMEAMETER OF VERTICAL, HORIZONTAL, AND PREPARED SAMPLES ON A MISCELLANEOUS GROUP OF IMPERIAL VALLEY SOILS

Sample number	Coefficient of permeability, cc per sq cm per hr			Correction factor to nearest first decimal place		
	Vertical	Horizontal	Prepared	Prepared to vertical	Horizontal to vertical	Prepared to horizontal
123	9.4	11.5	8.1	1.2	1.2	1.4
125	8.9	14.2	9.2	1.0	1.6	1.5
122	8.4	14.0	3.5	2.4	1.7	4.0
329	21.3	26.2	14.0	1.5	1.2	1.9
149	8.1	11.5	8.5	1.0	1.4	1.4
311	10.7	31.6	4.3	2.5	3.0	7.3
313	9.1	12.8	5.8	1.6	1.4	2.2
46E	4.8	10.9	4.9	1.0	2.3	2.2
312	9.0	9.3	2.2	4.1	1.0	4.2
339	5.1	6.4	1.6	3.0	1.2	3.8
338	3.3	4.0	0.7	4.7	1.2	5.7
213A	3.9	6.7	1.0	3.9	1.7	6.7
213B	1.0	1.1	0.4	2.5	1.0	2.8
212A	4.0	4.4	0.4	10.0	1.1	11.0
212B	3.0	5.6	0.8	3.7	1.9	7.0
Totals.....	110.0	170.2	65.5	44.1	22.9	63.1
Averages...	.....	.....	....	2.9	1.6	4.2

* Adapted from the Imperial Valley Report, Table 5.

of that area. However, occasions may arise where it is desirable to know the approximate permeability of the aquiclude or barrier groups. For these the falling-head type is the only practical apparatus. Undisturbed specimens, only, should be used for this latter type of measurement.

3.4. Water Supply. Regardless of the type of permeameter the selection of the water for use is very important. The original degree of flocculation must be maintained. Distilled water would quickly wash out the salts and greatly reduce P due to deflocculation. For example, Colorado River water fits Imperial Valley conditions.

3.5. Evaluation of Permeameter Measurements. Results with the permeameter are accurate, but the method has the serious drawbacks of

[1] See Appendix B for description of the two types of permeameter mentioned in this paragraph.

consuming much time and requiring complicated apparatus. Further, the results present no definite relationship between vertical and horizontal values of P. No generally applicable solution to this problem has yet appeared except comparison of actual results in a few scattered cases. To be generally reliable that method would require an extremely large number of comparisons owing to great variation in soil conditions everywhere. This method of securing correction factors, as between either vertical values of P or those from prepared disturbed samples and the necessary horizontal values, is illustrated for a small number of cases in Table 3.2. The table shows 15 samples of Imperial Valley soils chosen at random from a specific area where all three P values, vertical, horizontal, and prepared, were determined by permeameter, both vertical and horizontal cores having been secured for the purpose. This table not only shows the ratios for individual items of the horizontal to each of the other two values but also the corresponding average ratio for all 15 items. These ratios, both individual and average, should be applied with great discretion because of the limited number of cases on which they are based.

Indirect Measurement in the Laboratory

The excessive cost of direct measurement of P in time and equipment has greatly stimulated search for a shorter, reliable indirect method. Various studies along this general line have not, as a rule, worked out satisfactorily. However, the Imperial Valley investigations produced a method, based on a possible correlation of the silt-plus-clay percentage and P, which is recommended in the Report as relatively simple, speedy, and giving a fair degree of accuracy. It involves mechanical analysis of the soils by both the Bouyoucos and the sieve-analysis method. The Report discusses the method substantially as follows:

A large number of mechanical analyses and measurements of P values on prepared samples showed a definite correlation between the silt-plus-clay per cent (grain particles smaller than 0.05 millimeter) and the P values. Correlation was most consistent in the P range between 1.0 and 30.0 cubic centimeters per square centimeter per hour; good from 0.3 to 70.0 cubic centimeters per square centimeter per hour; less definite much above or below these limits. Sediments with fine sand (0.25 to 1.0 millimeter) predominant were most consistent in the pattern, with those sediments where sands smaller than 0.10 millimeter predominated falling in the same correlation when the silt-plus-clay content was less than 50 per cent. With higher clay content permeability fell below normal. Sand of 0.25-millimeter size predominating materially *raised* the permeability. Although a general correlation still existed, the change in permeability with increase in silt-plus-clay was much greater.

The Report stated, "Figure 3.4 (A and B)[1] illustrates these relationships. The curve drawn in part A and reproduced in part B was statistically established by the analysis of the points illustrated. A highly significant correlation was established ($r = 0.85$ when $n = 83$)"

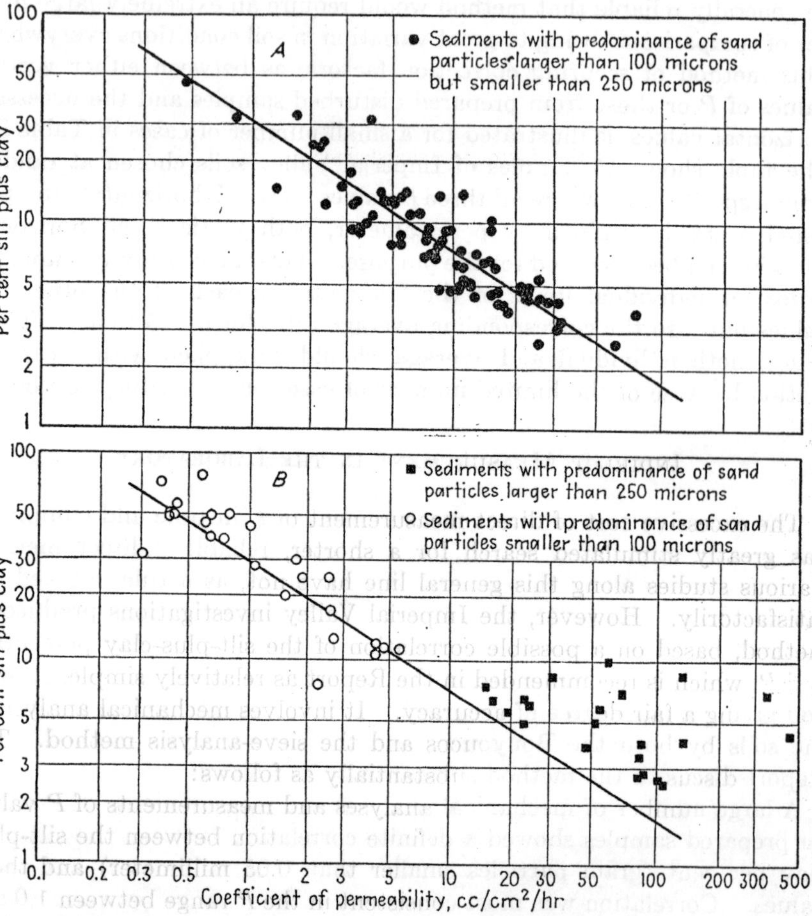

FIG. 3.4. Ratio between silt plus clay content and coefficient of permeability. (Figure 18 of the Imperial Valley Report. *Courtesy of Soil Conservation Service.*)

In the double analysis, that is, by both the Bouyoucos hydrometer and the sieve-analysis methods, for convenience in final computation of the percentages of the different groups of separates, 100-gram samples are recommended as the best size to use. Both methods of mechanical analysis being virtually standardized, any regular soils laboratory is equipped to handle them. Therefore detailed discussion of the methods is outside the scope of this book. The Imperial Valley Report gives the

[1] Figure 18 (A and B) of the Imperial Valley Report.

remaining steps of this indirect method of determining P substantially as follows:

3.6. Estimating the Coefficient of Permeability. Upon the completion of the mechanical analysis, the results are classified on the basis of the

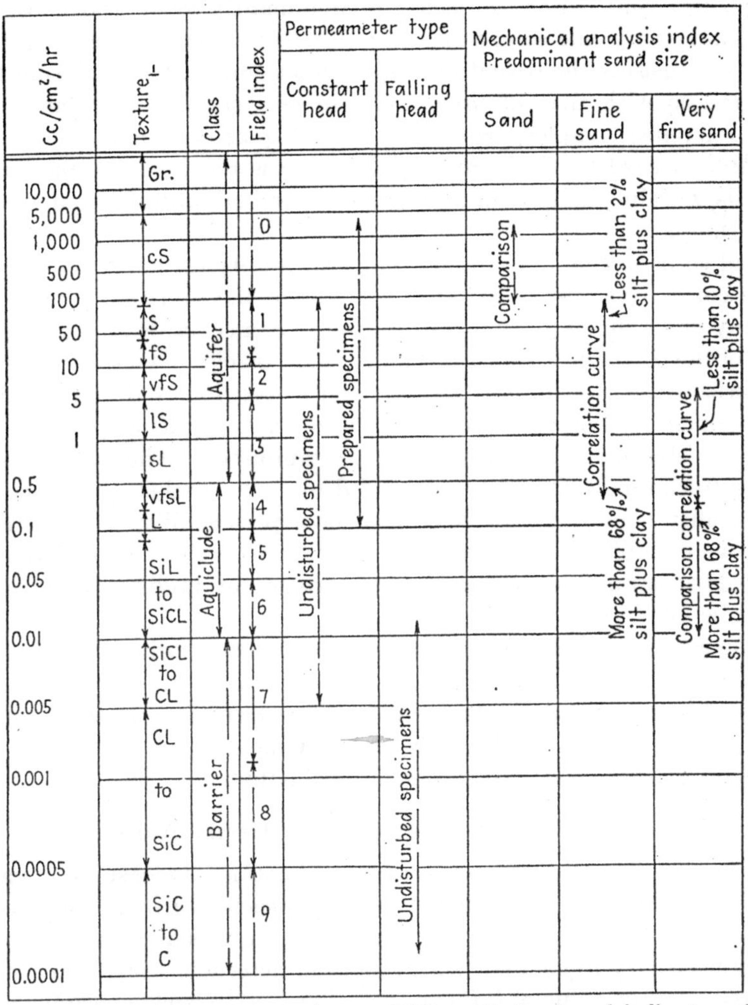

FIG. 3.5. Comparison of coefficient of permeability with direct and indirect method of measurement. (Figure 20 of the Imperial Valley Report. *Courtesy of Soil Conservation Service.*)

predominance of the sand grains. Inspection of the results of the mechanical analysis will show what proportion is greatest, that of the sand, of the fine sand, or of the very fine sand.

One of two methods is then used to estimate the permeability as indicated in Fig. 3.5. In the first method the upper silt-plus-clay limit is 68 per cent, the exact figure in a given case being determined from Fig. 3.4 which shows the ratio between the silt-plus-clay content and the P values. When fine sand is dominant, the silt-plus-clay limits lie between 2 and 68 per cent. When the very find sand is dominant, the limits lie between 10 and 68 per cent. For sediments predominantly sand and those predominantly very fine sand with *more than* 68 per cent silt-plus-clay content, one must use the *second method* which is a comparison of the mechanical analysis with that of *previously measured* samples. A card file of these reference samples, based on their silt-plus-clay content, must be available, a card for each sample, on which is recorded the mechanical analysis, P value, method of measurement, and laboratory number of the given sample.

In all cases in the first method the permeability estimate made by use of the correlative curve (Fig. 3.4) is the P of the prepared sample, and in many cases in the second method the reference cards also give the P value of the prepared samples. Therefore, a correction factor to establish the horizontal P value must be provided as previously described in the measurement by direct methods. Accuracy is dependent upon the quality of the stratum survey and adequate sampling. Whenever possible, indirect techniques should be supported by a few undisturbed specimen permeability runs. All direct coefficient of permeability measurements should be carefully run and recorded in the master file. Complete notes of the structural and textural characteristics of undisturbed samples should be kept for reference purposes.

Its sponsors feel that the indirect method of measuring P values just described is a fairly reliable timesaver over the more direct methods, but at best, it is an indirect method and would, therefore, seem to be subject to the same limitations and safeguards recommended for the method presented by O'Neal, discussed earlier in this chapter.

Average Values of P for a Given Area

While it is, as a rule, essential in the application of measured P values to drainage problems to secure many such values for variable sections in a given area or for variable strata in the same area, it is also often desirable or even necessary to combine such groups of determinations into a single weighted average value of P for the entire area. In such a case the average P value should be weighted with respect to per cent of total depth represented by any given stratum, the individual P_v value in cubic centimeters per square centimeter per hour for the given stratum, the conversion factor from vertical to horizontal, and the ratio to reduce P_v value to gallons per square feet per day, as shown in Table 3.3 from the Imperial Valley Report.

Table 3.4 shows values of P as determined by means of a variable-head

TABLE 3.3.* ILLUSTRATION OF COMPUTATION OF AVERAGE P FOR AN 80-ACRE FIELD SHOWING, BY BORINGS, SIX DISTINCT STRATA
(Borings 8 to 10 feet deep)

(1) Stratum	(2) Per cent	(3) Permeability index, cc/cm²/hr	(4) Conversion factor, vertical to horizontal	(5) Conversion factor, cc/cm²/hr to gal/ft²/day	(6) Product of columns (2) to (5)
1	32.0	4.01	2	5.27	1,352
2	18.0	1.22	2	5.27	231
3	18.0	0.43	3	5.27	122
4	12.0	0.19	4	5.27	48
5	11.5	0.095	4	5.27	23
6	8.5	0.017	4	5.27	3
Totals.....	100.0	.....	..	....	1,779

$$\text{Weighted average field } P = \frac{1{,}779}{100} = 17.79, \text{ or practically } 17.8$$

* From unnumbered tabulation, p. 88 of the Imperial Valley Report.

TABLE 3.4. VALUES OF P ON PREPARED SATURATED SAMPLES OF 52 SURFACE SOILS AND 51 CORRESPONDING SUBSOILS FROM VARIOUS REGIONS IN MINNESOTA

Soil classes	Number of samples tested	P range, lin in. per hr	Weighted* average P, lin in. per hr	
Very fine sandy loam........	2	0.048 – 0.027	0.088	Surface soils
Fine sandy loam............	4	0.053 – 0.573	0.212	
Sandy loam...............	7	0.013 – 0.900	0.120	
Loamy fine sand...........	1	....	0.336	
Loamy sand...............	1	....	0.078	
Loam.....................	9	0.015 – 0.266	0.093	
Silt loam.................	16	0.014 – 0.081	0.043	
Silty clay loam............	4	0.009 – 0.027	0.019	
Clay loam................	1	....	0.004	
Silty clay................	5	0.008 – 0.039	0.016	
Clay....................	2	0.002 – 0.017	0.006	
Sand....................	2	18.567 –18.970	18.701	Subsoils
Loamy very fine sand.......	1	....	0.776	
Loamy fine sand...........	1	....	4.890	
Fine sand................	1	....	1.517	
Loamy sand...............	4	0.714 – 2.635	1.110	
Sandy loam...............	1	....	0.051	
Loam....................	17	0.002 – 2.799	0.215	
Silt loam.................	15	0.005 – 0.065	0.030	
Silty clay loam............	9	0.0001– 0.0330	0.0079	

* Averages were weighted on the basis of the number of runs of each sample, whether triplicate or more.

discharging permeameter, in triplicate, of 52 prepared samples of saturated surface soils and corresponding 51 prepared samples of saturated subsoils in Minnesota. A few of the more variable and sensitive samples were put through five to nine tests instead of merely in triplicate. All values given in the table are averages of the replications in each case, whether triplicate or more. These few items show the wide variations that occur even in the same soil. Nevertheless these values may be found useful in making preliminary designs and estimates of tile drainage in that and similar regions where glaciated soils are the rule.

FIELD MEASUREMENTS OF PERMEABILITY BELOW SHALLOW WATER TABLES

Some authorities insist that permeability cannot be made to correlate successfully with any criterion other than direct measurement in the soil stratum itself. Starting with the work of Kirkham in 1945 [68] practical

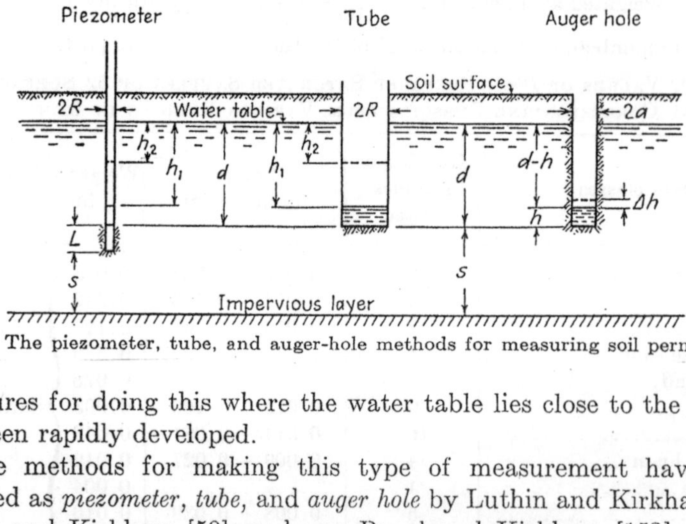

FIG. 3.6. The piezometer, tube, and auger-hole methods for measuring soil permeability.

procedures for doing this where the water table lies close to the surface have been rapidly developed.

Three methods for making this type of measurement have been described as *piezometer, tube,* and *auger hole* by Luthin and Kirkham [73], Frevert and Kirkham [50], and van Bavel and Kirkham [153], respectively. Salient features of the three methods are shown in Fig. 3.6 with the symbols used in the formulas of Reeve and Kirkham [118].

In both the piezometer and tube methods unperforated metal pipes are tightly inserted in apertures bored in the soil to the depth desired. The pipes are open at both ends. When in operating position the piezometer pipe stops short of the bottom of the hole, leaving an open cavity 4 inches high and about 2 inches in diameter. In the tube method the pipe casing is driven flush with the bottom of the hole so that the measured permeability is accented vertically. The piezometer method accents horizontal permeability primarily and is best adapted to stratified soils where permeability is apt to vary in layers at different depths below the surface. The piezometer is also practical for use at greater depths than either of the other two methods.

The auger-hole method is the utmost in simplicity, it being nothing more than an open uncased hole that permits water to seep in freely all around the sides and bottom. The rate of inflow is a composite measure of permeability at the site of the hole, taking into account all the vagaries to which natural soils are subject, such as worm holes, cracks, decomposed roots, and embedded rocks. A small hand pump is an essential part of the equipment for all three methods. It is used to flush out the holes preparatory to taking measurements and partially to empty the holes at the start of a series of readings. Permeability is determined by the rate at which water rises in the holes.

Practical apparatus and a simplified procedure for applying the piezometer and auger-hole methods are described in detail by Johnson, Frevert, and Evans in *Agricultural Engineering* [65] for May, 1952. This article constitutes in large part the basis for the following discussion, together with the accompanying formulas, drawings, and graphs. As the same basic formula applies to both the piezometer and tube methods, only the details of the piezometer and auger-hole methods are included.

3.7. Piezometer Method. Remove surface litter and sod. Using a 1$^{15}/_{16}$-inch auger, bore a hole 6 inches deep. Insert a thin-walled 2-inch pipe. With the auger inside the pipe, bore another 6 inches, remove the auger, drive the pipe, and insert the auger—repeat the cycle in 6-inch increments until the desired depth is reached, being careful to leave an uncased cavity 4 inches high at the bottom. The purpose of this procedure is to ensure a tight fit and minimum disturbance of the natural condition of the soil.

Set up the apparatus as indicated in Fig. 3.7. Insert the hose attached to the hand pump and flush out the cavity several times to cleanse soil voids and ensure reproducible and consistent readings. Allow the water to rise in the tube to the level of the water table in the surrounding soil. After the water has stopped rising, place the platform in position as shown in Fig. 3.7 and connect the apparatus.[1] Lower the probe and record the depth of the water table L_r below the top of the pipe. Remove the apparatus, pump most of the water from the hole, and replace. Record position h_1 and note time of reading t_1. Raise the probe a measured distance, at least 0.1 foot. When rising water makes contact, note

[1] The apparatus consists of a platform and pipe housing to allow insertion of a flexible insulated electrical conductor or probe with exposed end. On the platform is a fishing reel for easy manipulation of the conductor, a battery, resistors, and a milliameter. The circuit is arranged to deflect the milliameter when the exposed end of the conductor touches the water.

time t_2 and record second probe position h_2. At least three increments of rise and time should be taken at each tube position.

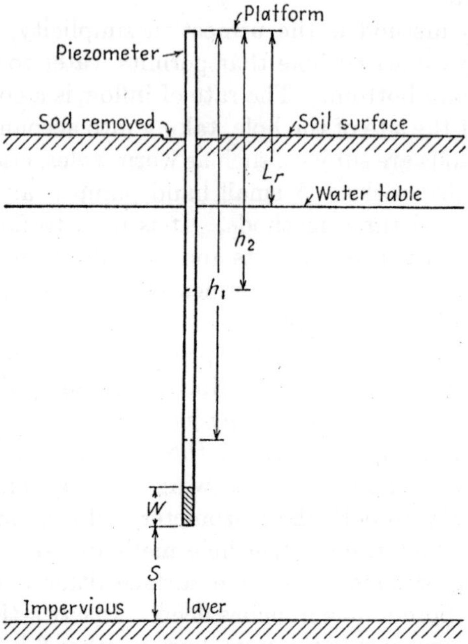

FIG. 3.7. Diagram of method followed and symbols used in applying piezometer tubes for permeability measurement.

The permeability formula applicable to this procedure with the symbols of Fig. 3.7 is

$$P = \frac{\pi R^2 \log_e \dfrac{(h_1 - L_r)}{(h_2 - L_r)}}{A(t_2 - t_1)} \tag{4}$$

where R is the inside radius of the tube, in.

$\log_e$ is the natural logarithm of the ratio shown ($\log_e N = 2.303 \log_{10} N$)

A is a linear dimensional coefficient which is determined by the geometry of the flow system into the cavity below the tube, in.

Other symbols in the formula are obvious from Fig. 3.7 and the foregoing text.

For any given size of tube and cavity π, R^2, and A may be combined as a constant. The A factor for a cavity 4 inches long and $1^{15}/_{16}$ inches in diameter is 17 inches if S, the distance from the bottom of the cavity to a possible impervious layer or barrier, and the length of pipe below the water table both exceed 8 inches. An error of less than 8 per cent, plus or minus, is introduced if the length of pipe below the water table remains equal to or greater than W, and S is equal to or greater than $\frac{1}{2}W$ (see Fig. 3.7). For computation of the A factor corresponding to other dimensions of cavity and tube the reader is referred to Ref. 118.

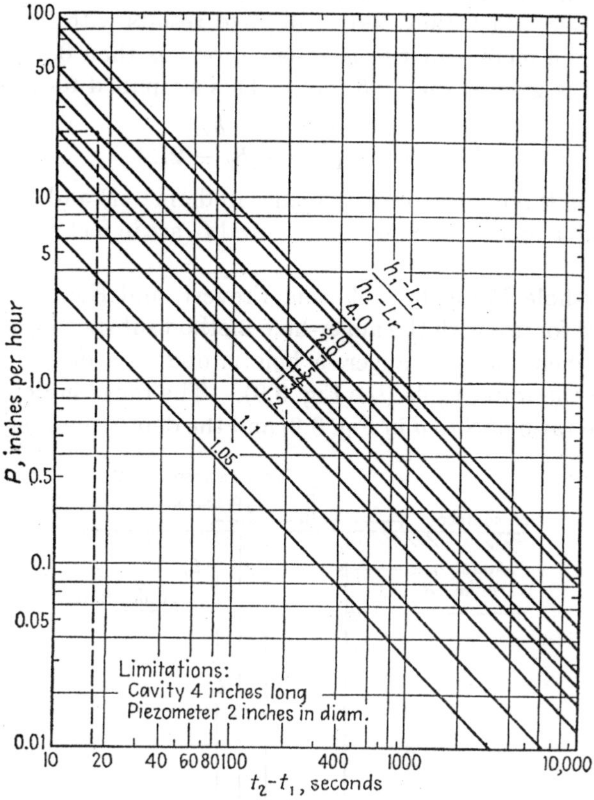

FIG. 3.8. Graph for solution for coefficient of permeability by the piezometer method.

Combining the above values of π, R^2, and A and converting seconds to hours, Eq. (4) reduces to

$$P = \frac{625 \log_e \dfrac{(h_1 - L_r)}{(h_2 - L_r)}}{t_2 - t_1} \tag{4a}$$

which yields a P value in inches per hour.

Johnson, Frevert, and Evans [65] present a graphical solution of Eq. (4a) which is reproduced as Fig. 3.8. An example of the use of Fig. 3.8 follows.

Let $L_r = 1.25$ ft, $h_1 = 3.22$ ft, $h_2 = 2.39$ ft, $t_1 = 0$ sec, $t_2 = 17$ sec Then $(h_1 - L_r)/(h_2 - L_r) = 1.73$. Enter the horizontal axis at 17 seconds, follow up vertically to a point proportionately between diagonal lines marked 1.7 and 2.0, and opposite this point read $P = 22$ inches per hour on the vertical axis. From page 33 this is seen to be an extremely rapid rate and denotes a very open soil.

The same authors propose an approximate formula for quick calculations in the field which they state will yield results with less than 1 per cent error if the increment rise is equal to or less than 0.2 foot and the average head is greater than 0.5 foot. This simplified formula is

$$P = \frac{1{,}330(h_1 - h_2)}{(t_2 - t_1)(h_1 + h_2 - 2L_r)} \tag{4b}$$

Assuming that R remains 1 inch (2 inches in diameter) and W is 4 inches, P is expressed in inches per hour if h_1, h_2, and L_r are in feet and t_1 and t_2 are in seconds.

3.8. Auger-hole Method. Johnson, Frevert, and Evans [65] describe the procedure for use with the auger-hole method substantially as follows:

Using a 4-inch-diameter auger, a hole is dug to a depth of 12 to 30 inches below the water table (see Fig. 3.9). The range in depth is determined by the ratio of S to d in Fig. 3.9, and the depth must remain within

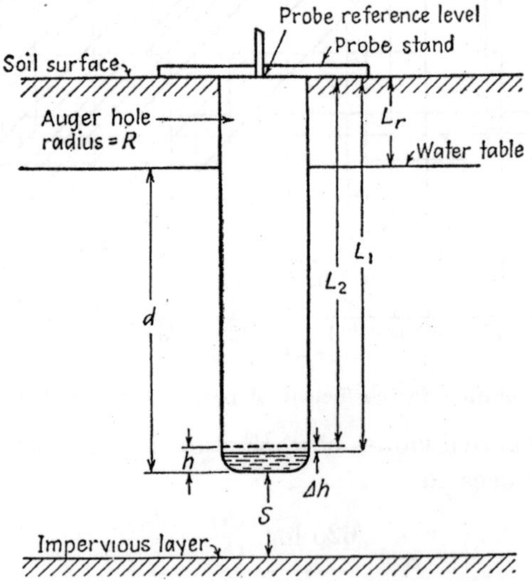

FIG. 3.9. Diagram of method followed in permeability determination by the auger-hole procedure.

the limits indicated in Fig. 3.10. If no impermeable layer exists, there probably would be little need for systematic underdrainage and hence no need for the method.

The depth of the hole would depend to some extent on the depth at which a permeability reading is desired, even though this results in a composite permeability. Seepage water is pumped from the hole two or three times to minimize the effect of puddled soil on the walls. The water table is then allowed to come to equilibrium, which may require as long as three days in very impermeable soil.

The same measuring probe may be used with this method as with the piezometer tube method if a stand is provided. The measuring probe and stand are placed solidly over the center of the hole. The probe is lowered the distance L_r

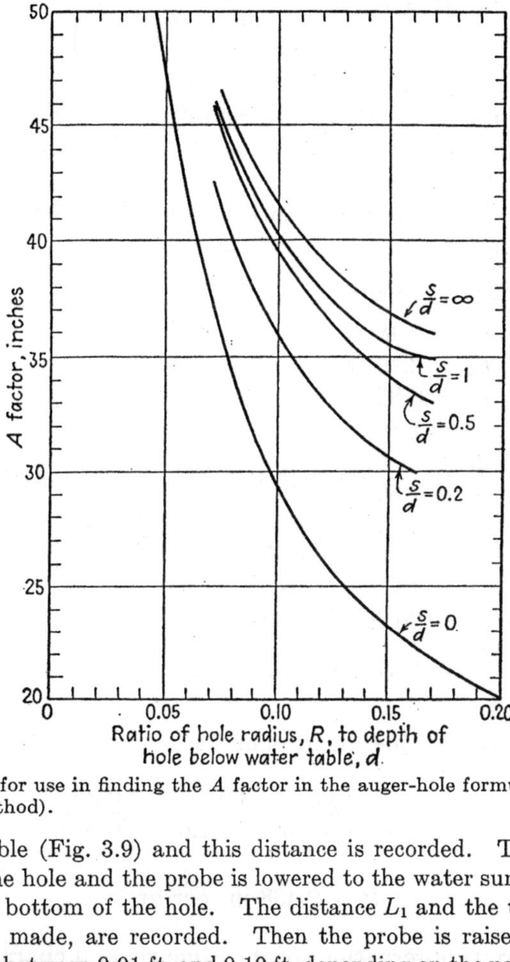

FIG. 3.10. Graph for use in finding the A factor in the auger-hole formula (does not apply to piezometer method).

to the water table (Fig. 3.9) and this distance is recorded. The water is then pumped from the hole and the probe is lowered to the water surface about 0.1 to 0.2 ft. from the bottom of the hole. The distance L_1 and the time t_1, when the measurement is made, are recorded. Then the probe is raised a distance Δh, which should be between 0.01 ft. and 0.10 ft. depending on the permeability of the soil. The water is then permitted to rise the distance Δh to L_2. When the water strikes the probe the time t_2 is recorded. The probe is raised an equal distance, Δh, again and the same procedure repeated. Three or four readings are taken over successive equal increments of rise of the water in the hole, and the time at the end of each increment is recorded. The time required for each increment of rise should be nearly the same. All readings should be taken before the water rises to the level where $h/d = 0.2$, where h is the average height of water above the bottom of the hole during any increment of rise and d is the depth of the hole below the water table. This facilitates the use of the formula which follows. At least three readings of the rate of rise should be made in each hole.[1]

The formula applying to the foregoing conditions is

$$P = \frac{\pi R^2 (L_1 - L_2)}{A(t_2 - t_1)\left(\dfrac{L_2 + L_1}{2} - L_r\right)} \tag{4c}$$

all the symbols of which are obvious from Figs. 3.9 and 3.10. To express P in inches per hour it is important to note that the units must be those indicated in the following example.

Let it be assumed that the following values have been recorded from field observations:

$$\begin{aligned}
\pi R^2 &= 12.57 \text{ sq in.} \\
L_1 &= 2.48 \text{ ft} \\
L_2 &= 2.46 \text{ ft} \\
L_r &= 0.83 \text{ ft} \\
d &= 1.80 \text{ ft} \\
\frac{R}{d} &= \frac{2}{1.80 \times 12} = 0.09 \text{ (for use in Fig. 3.10)} \\
\frac{S}{d} &= 1.0 \text{ (for assumed conditions)} \\
A &= 42 \text{ in. (from Fig. 3.10)} \\
t_2 - t_1 &= 30 \text{ sec} = 0.0083 \text{ hr} \\
\frac{L_2 + L_1}{2} &= 2.47 \text{ ft}
\end{aligned}$$

Substituting these values in formula (4c)

$$P = \frac{12.57 \times 0.02}{42 \times 0.0083 \times 1.64} = \frac{0.251}{0.572}$$

or

$$P = 0.438 \text{ in. per hr}$$

Where numerous observations of the same general character need to be reduced, nomographs may be easily constructed, within certain limiting conditions, to relieve the drudgery of repetitious computations. Johnson, Frevert, and Evans [65] present one such nomograph which gives a value of $P = 0.46$ inch per hour for the above stated conditions.

[1] Quoted from Ref. 65.

SOIL-MOISTURE RELATIONS

The capacity of soils to *hold* moisture bears a close relation to their textural and structural characteristics, though it is difficult to test this capacity without destroying their original structure. Whenever a soil is saturated with water for a considerable period of time, it becomes "puddled," and any structure it may have possessed is broken down.

FORMS OF SOIL WATER

There are three forms of soil water.

3.9. The Hygroscopic Moisture. The hygroscopic moisture, or moisture of condensation, is that portion of the moisture in the soil so closely associated with the soil grains that it cannot be used by plants. It is therefore often called the *useless water* of the soil. It stands in such a close physical relation to the soil that it can be driven out only by the application of heat to a soil which, upon cooling, quickly reabsorbs its normal quota from the air. In other words, soil, such as dust, which is apparently quite dry, still contains hygroscopic moisture, in an amount dependent upon the area of soil-grain surface exposed to the air and upon the temperature and relative humidity of the surrounding atmosphere. Hence it bears no relation to drainage or soil erosion and probably not to irrigation. Like the other forms of soil moisture the amount of the hygroscopic water in any soil is usually expressed as a per cent of the dry weight of the soil. It bears a fairly constant relation to the total capillary capacity of the given soil, being about one-third of the latter, especially among the finer-textured soils, as the silt loams, clay loams, and clays.

3.10. The Capillary Moisture. The capillary moisture is the only form of soil moisture used by plants and essential to plant life in absorbing the plant food from the soil and in assimilating it and moving it to the various parts of the plant. It is therefore often called the *useful water* of the soil. It takes the form of a thin film around each soil grain or granule and moves by capillarity both vertically upward against gravity and laterally. Capillary water cannot be removed from the soil by drainage, but only by evaporation, by transpiration by plants, or by heat. In most soils, when capillary water is present to capacity, about two-thirds of it is available for plant growth.

Maximum Capillary Capacity of Soils. The maximum capillary capacity of a soil, which is the maximum amount of moisture that the soil will hold against gravity, is the same as the maximum field capacity of a well-drained soil. Owing to the fact that the maximum capillary capacity is practically the optimum moisture content both for plant growth and for good tilth, it is an important factor in both irrigation and drainage. It can be determined for any given soil only by the following experimental method unavoidably subject to errors in the carrying out of the process: An undisturbed field sample of known volume of the given soil is saturated, then allowed to drain freely until such time as it ceases to drip free water. It is then weighed in the container and, after being oven-dried, is again weighed in the container. The difference in the two weights is the maximum capillary capacity of the soil by weight. This difference divided by the dry weight of the soil (the total final weight less the weight of the container) is the maximum capillary capacity expressed as a per cent of the dry weight of the soil. The average of determinations in triplicate which do not vary widely may be considered reliable. However, according to Adams [1], it is usually more satisfactory in practice, with the fine-textured soils but not always with the sands, to assume that the maximum capillary capacity is the same as the moisture equivalent, which is defined in a later paragraph and which can more readily be determined with less chance of serious error.

3.11. The Gravitational Water. As the name implies, *gravitational water* is that which is free to obey the laws of gravity. If there is more than enough water to saturate a soil to its maximum capillary capacity, the excess will lie in the pores and gradually work its way downward to an outlet, or else it will accumulate until all the pores are entirely filled and the soil becomes supersaturated or puddled. In such a condition, the structure tends to break down, air is driven out of the pore spaces, and most upland vegetation is soon suffocated and drowned. On this account free water is known as the *harmful water* of the soil. Obviously, the most favorable conditions occur when the soil grains are crumbly and friable and are holding as much capillary moisture as possible, while the spaces between are filled with warm moist air and other gases. These conditions are brought about only when the surplus gravitational water has a chance to escape into a permeable subsoil or, failing that, into drains placed in

TABLE 3.5. GRAVITATIONAL WATER THAT MAY BE DRAINED FROM SATURATED SOILS*

Kind of soil	Total pore space, per cent	Capillary pore space, per cent	Pore space occupied by gravitational water, per cent	Depth of gravitational water per ft of depth, in.
Coarse sand	28	5	23	2.75
Medium sand	30.5	10.5	20	2.40
Average sandy loam	36.5	22.5	14	1.70
Average silt loam	45	37.5	7.5	0.90
Average clay loam	51	45	6	0.75
Average clay	55	50.5	4.5	0.55

* Computed from Fig. 2.2 and Table 3.2.

the subsoil for this purpose. In fact, the first and most direct purpose of agricultural drainage is the removal of free water from the root zone of plants as quickly as possible after it appears there, before it causes appreciable harm to plants.

Table 3.5 shows approximate total amounts of gravitational water that may be drained from completely saturated soils.

SOIL-MOISTURE CHARACTERISTICS

A number of soil-moisture designations are generally accepted and extensively used by investigators in the field of soil-moisture relations in agriculture. In presenting these attention is called to the facts that, while most of them are fairly closely interrelated, the relations herein shown are seldom mathematically exact and care and good judgment must be used in passing from one to the other, the coefficient of relation being, in most cases, a variable one within a more or less definite range. Only those characteristics of value in the drainage field will be discussed here.

3.12. The Moisture Equivalent (M_e). The moisture equivalent is the moisture, expressed as a per cent of the dry weight of the soil, that a specified small portion of the given soil will retain against a centrifugal force of 1,000 times the force of gravity exerted on it for a period of 40 minutes. Its value varies with different soils, being low in sand or sandy soils but very high in clays. Researches of a number of soil scientists and engineers, among whom may be mentioned Alway, Israelsen, McLaughlin, Neal, and Veihmeyer, show that the moisture equivalent closely approximates the maximum capillary capacity.

Veihmeyer, Israelsen, and Conrad [154] found considerable variation in the value of the moisture equivalent of any given soil according to the size and initial moisture condition of the sample used in the determination: that the smaller the sample, the larger the percentage of water retained; that for precision in its determination it is necessary to measure the sample by weight rather than by volume; that the value obtained will vary with the compactness and depth of the sample in the centrifuge cup and with the time the sample has been kept moist. However, it has been the experience of the Minnesota station that uniform weight of sample, compacting it well by jarring the container vigorously on the table top and leaving it in the moist closet until ready to centrifuge, gives satisfactory results if the speed of the centrifuge is carefully regulated. Triplicate values determined successively under these conditions for any given soil have generally been found so nearly equal as to serve as a reliable practical measure for a number of other conditions not so readily determined as is the moisture equivalent, but of great importance in soil-moisture considerations. In addition to its close relationship to maximum field and capillary capacity it is also a practical measure of the optimum moisture content for various types of crops, and the wilting coefficient, later discussed, is frequently expressed as approximately a given percentage of the moisture equivalent of the soil concerned. In tile-drainage design the moisture equivalent is the basis of the first practical scientific index worked out for spacing and depth of the drains according to soil type and now used in various parts of the country by the Soil Conservation Service (see Chap. 10).

However, important as is the moisture equivalent in dealing with soil-moisture problems, attention is called to the fact that its determination requires a laboratory setup not usually readily available to the rural practitioner; hence some unit bearing a close and definite relationship to it and more readily determined at any location is desirable. Recognizing this need, Neal [94] offers the following short cuts not requiring rare and intricate equipment:

A definite relationship exists between the moisture equivalent and the upper and lower plastic limits of the soil (see Sec. 3.15). The plastic limits are determined according to the Atterberg method.

3.13. The Lower Plastic Limit. The *lower plastic limit* is the wetness at which the soil can barely be rolled into a wire under the fingers. At this lower limit when the size of the wire is about the same as that of a lead in a common lead pencil, it will break into segments about ⅛ inch long. To obtain the lower plastic limit, take about 100 grams (4 ounces) of air-dry soil and add water until the soil can be rolled into a wire as before described. If too much water is added, add more dry soil. When the proper moisture condition is reached, weigh the moist soil, thoroughly dry it in an oven at 100 degrees centigrade (212 degrees Fahrenheit), and then weigh the dry soil. The difference in weight thus obtained divided by the weight of dry soil is the *lower plastic limit.*

3.14. The Upper Plastic Limit. *The upper plastic limit* is the wetness at which the soil, in layers 1.5 centimeters (⅝ inch) thick, just ceases to show a tendency to flow. A V-shaped furrow in a cake of soil in a dish will just exhibit flowing by closing at the bottom when assisted by violent jarring. If the soil is too wet, which will be indicated by the complete closing of the furrow when assisted by violent jarring, add more dry soil and thoroughly mix. When the proper condition is reached, weigh the wet soil, thoroughly dry in an oven at 100 degrees centigrade (212 degrees Fahrenheit), and then weigh the dry soil. The difference in weight thus obtained divided by the weight of the dry soil is the *upper plastic limit.*

Baver [11] states that in the operation of the Atterberg method different investigators tend to vary considerably in results obtained with the same soil, owing to the difference in the jarring force applied by each. To eliminate this undesirable personal influence, he offers a new method as follows:[1]

Roberts has suggested the use of small glass cups for making these measurements. These cups are made from ordinary 100 cc graduate cylinders which have been cut off at the 5 cc mark. A V-shaped groove, 4 mm deep and 4 mm

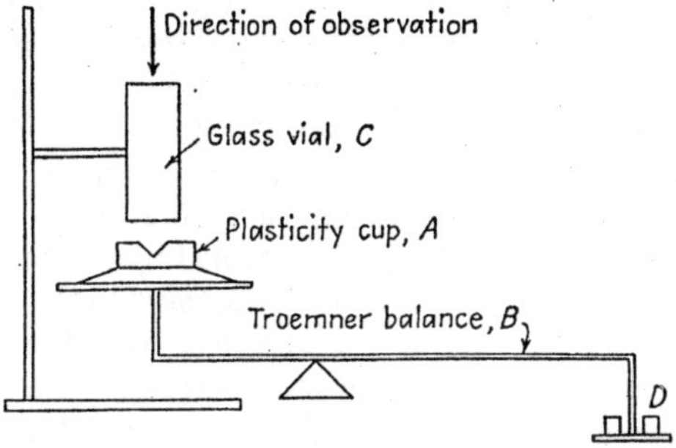

FIG. 3.11. Baver's apparatus for determining the upper plastic limit of soils. (*Jour. Am. Soc. Agron., Vol.* 24, *No.* 9.)

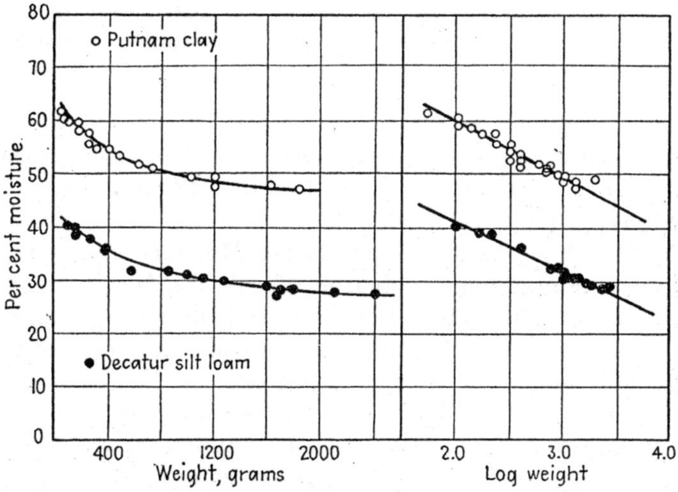

FIG. 3.12. The relation of the moisture content of the soil to the force required to produce flow. (*Jour. Am. Soc. Agron., Vol.* 24, *No.* 9.)

wide at the top, is cut into the top of the cylinder. This standardizes the size of the groove which is cut into the soil.

The experimental arrangement for determining the upper plastic limit is shown in Fig. 3.11. The soil is thoroughly kneaded and mixed into a thick paste . . . and transferred to the glass cup, *A*. The soil is firmly squeezed into the cup and

[1] The quotation and Figs. 3.11 and 3.12 are reprinted by permission from "The Use of the Troemner Balance for Measuring the Upper Plastic Limit of Soils" by L. D. Baver, *Jour. Agron.*, Vol. 24, No. 9, September, 1932. Figures 3.11 and 3.12 are Figs. 1 and 2 of Baver's paper.

the surface is made level with the top of the cup. The groove is cut into the soil paste by using a porcelain or wooden spatula which is so beveled that it fits the groove in the cup. After the groove has been clearly cut, the cup is placed on the Troemner balance, *B*. A glass vial, *C*, the diameter of which is slightly smaller than the inside diameter of the cup, is lowered until it just touches the surface of the soil. The weight, *D*, is then placed on the balance which exerts a given force on the soil in the cup and causes the groove to flow together. Observations of the amount of flow are made from above through the bottom of the glass vial. The moisture content at which the soil will barely flow together is called the upper plastic limit. It was found that a weight of 1000 grams on the balance exerted sufficient force to give results comparable with those obtained by the hand method. . . . Since a weight of 1000 grams on the balance apparently gives values corresponding to those obtained by the hand method, one may choose either of two ways in making the determination. First, the moisture content can be so adjusted that flow is obtained with the use of 1000 grams. Second, one can mix up the soil into a paste, determine what weight is necessary to make the soil flow, and then plot the moisture content against the logarithm of the weight. By using three or four different mixtures a straight line can be drawn through the points and the moisture content corresponding to the logarithm 3 (logarithm of 1000) would be that of the *upper plastic limit*. . . . The data show that there is a very good agreement between the two methods. Of course, using 1000 grams as a constant weight and adjusting the moisture content would be more accurate.

That the relationship between the force applied and the per cent moisture in the soil is logarithmic is shown by Fig. 3.12 plotted from the data on which Baver's paper is based.

3.15. Ratios of the Plastic Limits to the Moisture Equivalent. As a result of the determinations made from 49 subsoils, Neal found the average lower plastic limit to be 70 per cent and the average upper plastic limit to be 120 per cent of the moisture equivalent. These results agree with those of a number of other investigators.

3.16. The Per Cent of Clay. It is impractical to think of sands and very sandy soils as being plastic. It is equally impractical to determine either their plastic limits or their moisture equivalents with satisfying reliability. Therefore, in dealing with the moisture problems of such soils, determination of the clay content which may readily be found, as in the case of the moisture equivalent, only with a laboratory setup usually not easily available to the rural practitioner, is outside the scope of this book. Those wishing to become acquainted with the method are referred to the discussion thereof by Neal [94]. It is worth noting that a fair average ratio of the clay content of soils to the moisture equivalent has been determined by Neal to be 1.25, but this ratio must not be confused with the figure that represents the ratio of the upper plastic limit to the moisture equivalent. The first is a mere numerical ratio between two abstract per cents of different kinds of quantity, namely, water and clay. The latter is a definite ratio between values of two like quantities—both moisture.

It is wise not to try to use per cent of clay as the drainage index for any soil higher in the fine-texture scale than a very sandy loam with a moisture equivalent appreciably in excess of 10. However, it must often be used as the index in the case of very sandy soils, but these do not usually require such detail of exact design, relative to moisture, as do the finer soils. In such a case the rural practitioner for whom well-equipped laboratories are difficult of access is advised to decide, as definitely as he can with the means at his command, just where in the sandy-soil scale the given soil lies.

Summary of Relationships between the Soil-moisture Characteristics

The general relationships between the soil-moisture characteristics are shown schematically in Fig. 3.13. However, the various percentage

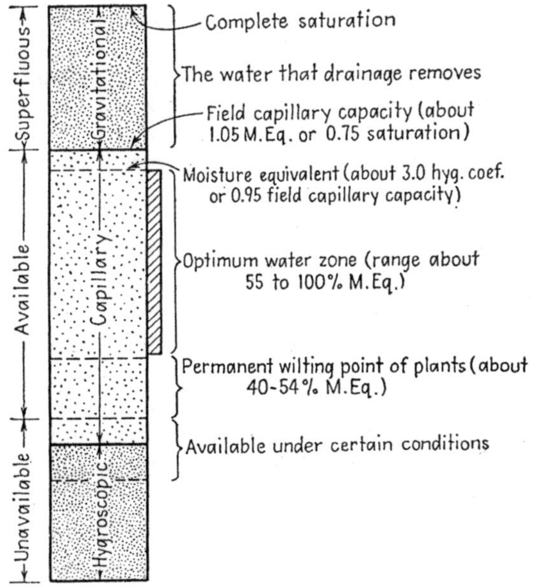

Fig. 3.13. Diagram illustrating the approximate relative magnitude of soil-moisture constants.

relationships there shown should be used with caution because they are at best only approximate comparisons of a mass of values all of which vary within rather wide limits in different soils, even of the same general class.

Table 3.6, based on the relationship of each other unit to the moisture equivalent, will be found convenient in problems involving the moisture content of soils, muck and peat soils not included.

TABLE 3.6. APPROXIMATE RANGE OF SOIL-MOISTURE CONSTANTS FOR EACH OF THE FOUR MAJOR SOIL CLASSES

Soil type	Per cent dry weight of the soil				
	Apparent specific gravity	Moisture equivalent	Lower plastic limit	Upper plastic limit	Per cent of clay
Sands	1.90–1.67	2.5– 7.5	1.8– 5.3	3.0– 9.0	3.1– 9.4
Sandy loams	1.67–1.51	7.5–20.5	5.3–14.4	9.0–24.6	9.4–25.6
Silt loams	1.51–1.30	20.5–33.0	14.4–23.1	24.6–39.6	25.6–51.2
Clays	1.30–1.08	33.0–50.0	23.1–35.0	39.6–60.0	51.2–62.5

Soil type	Per cent of total volume of the soil				
	Per cent pore space	Moisture equivalent	Lower plastic limit	Upper plastic limit	Per cent of clay
Sands	28–33	4.8–12.5	3.4– 8.8	5.8–15.0	6.0–15.6
Sandy loams	33–42	12.5–31.0	8.8–21.7	15.0–37.2	15.6–38.3
Silt loams	42–51	31.0–43.0	21.7–30.1	37.2–51.6	38.3–53.8
Clays	51–59	43.0–54.0	30.1–37.8	51.6–69.8	53.8–67.5

INFILTRATION

Infiltration is the movement of water through the soil surface into the soil body. It is not generally considered to be an active influence below the first foot of depth, which is about the plane where permeability considerations begin. Therefore, it is not considered a factor of appreciable influence in underdrainage, but in runoff and, therefore, in surface drainage and in soil erosion it is an important and active factor.

The entire topic is too extensive to permit of a complete discussion in this book, but the following outline of a few of the highlights may be helpful in further consideration.

3.17. Relation to Runoff. The ratio of runoff to rainfall is governed by those agencies which cause detention of a part of the rain water where it falls. These agencies are mainly two, vegetative cover and infiltration, the first of which stimulates the latter. On bare soil or that which has only a sparse vegetative cover, infiltration is practically the difference between the rainfall and the runoff from any given storm or storm period. Therefore, the runoff is strongly influenced by the local rate of infiltration.

3.18. Factors Influencing Rate of Infiltration. The rate and total amount of infiltration from any given storm or any other source of water, as, for example, irrigation water, are the important considerations, and both are difficult to measure. In fact, about the only practical method, assuming that the rate and total amount of rainfall or other water applied to the surface are or can be measured, is also to measure the rate and amount of runoff from the same storm or application of irrigation water. This can be done by well-established methods. The difference between rainfall and runoff is usually considered to be the infiltration except for the water removed by evaporation and transpiration, both of which are practically negligible for the short period of a single rain, but which may have to be reckoned with in the case of a protracted storm period or irrigation operation.

The rate of infiltration is influenced by many local factors, the most noticeable of which are as follows:

1. *Soil Texture and Structure.* Water moves into and through coarse-textured soils such as the sands and sandy loams and through coarse-structured soils much more quickly than through silts, clays, and other soils with compact or laminated structure.

2. *Soil Admixtures.* Large humus content stimulates infiltration; alkali such as sodium salts, causing deflocculation and puddling, checks and may even entirely stop infiltration. Certain experimental chemicals improve it.

3. *Natural Slope.* On level fields or those with a low rate of fall, where the runoff is sluggish, the rate of infiltration tends to be greater than on steeper slopes where more rapid runoff allows less time for passing water to soak in.

4. *Vegetative Cover.* A complete cover crop such as alfalfa or bluegrass sod tends markedly to increase the infiltration rate both through detention of the water by the top growth and by its protection against the retarding effect of impact of intense rainfall noted in item 7.

5. *Ponding.* Ponding also tends, markedly, to increase the infiltration rate by holding water on the surface, thus giving more time and opportunity for infiltration wherever ponding occurs.

6. *Antecedent Moisture Condition of the Soil.* A fairly dry antecedent condition of the surface permits much more ready infiltration than a wet surface soil in which the soil pores are already full and prevent or at least retard further infiltration.

7. *Rainfall Intensity.* The impact of intense rainfall tends to spread out the plastic clay and colloidal elements in the surface thus clogging the pores and retarding infiltration, whereas a quiet steady rain or a slow-moving blanket or stream of irrigation water has no such effect. This retarding effect of rain impact can be minimized by vegetal cover (see item 4).

8. *Quality of Underdrainage.* Good underdrainage, by keeping the pores open to the depth of the root zone, tends to induce a high infiltration rate, which poor drainage retards.

Rate of infiltration is usually expressed in inches per hour or, over a considerable period of time, is shown as an infiltration-rate curve, the general standard form of which is one lobe of an equilateral hyperbola.

3.19. Infiltration Rates. In any given case, under the influence of some or all of the factors just discussed, the rate of infiltration from any source of water decreases with the elapsed time from the beginning of the storm, usually very rapidly during about the first half hour, and in the course of several hours it tends to become constant and the curve to become asymptotic (parallel) to the X (horizontal) axis as in the equilateral hyperbola.

Owing to the wide variation in attendant conditions in different regions throughout the country, it is impossible to show tables of infiltration rates generally applicable over any extended area, and none such will be offered; but the following individual cases may be of interest and serve as a rough guide in dealing with problems that may arise in the experience of the reader.

Experiments by Neal [95] on a laboratory prepared bed of Putnam silt loam showed average results as follows:

Carefully controlled direct measurements of infiltration rates on a normally dry soil, with measured amounts of water on a fixed area and rates of infiltration exactly measured, with time taken by stop watch gave the following: 1.64 inches first 10 minutes; 0.20 inch third 10 minutes; 0.45 inch second 30 minutes (average per 10-minute period 0.15 inch); 0.57 inch second hour (average per 10-minute period 0.095 inch); 0.59 inch third hour (average per 10-minute period 0.098 inch).

The same soil, saturated, gave corresponding results as follows: 0.90 inch first 10 minutes; 0.23 inch third 10 minutes; 0.38 inch second 30 minutes (average per 10-minute period 0.13 inch); 0.17 inch second hour (average per 10-minute period 0.028 inch); 0.19 inch third hour (average per 10-minute period 0.032 inch).

Measurements were also taken on the same soil bed by difference between artificial rainfall and runoff, rainfall intensities varying from 0.9 to 4.0 inches per hour, surface slopes from 0 to 16 per cent; average moisture in surface inch, 15.6 per cent; and at 1- to 4-inch depth, 25.6 per cent. Average infiltration rates per minute for successive 10-minute periods were as follows: for first 10-minute period, 0.028 inch; for second 10-minute period, 0.011 inch; for third 10-minute period, 0.007 inch; for second 30-minute period by 10-minute sections, 0.004 inch; and the same for the second hour.

The foregoing figures all follow closely the standard form of infiltration curve, the rate tending to become constant after the first 30 minutes.

Ellison [42] experimented with four different soils spread in a uniform layer 1 inch thick on prepared laboratory beds over an 18-inch base of well-drained sand. The purpose was to test the effect of impact of intense rainfall with drops of constant diameter falling at velocities ranging from zero to 20.4 feet per second. Results showed average rates of infiltration in inches per hour from 15 tests on the four soils as follows: at the end of 5 minutes, 3.3 inches; at the end of 15 minutes, 2.2 inches; at the end of 30 minutes, 2.1 inches; at the end of 1 hour, 2.2 inches.

The A.S.C.E. Hydrology Handbook [59], after R. E. Horton, gives the range of values of the infiltration capacity (f_1) after 1 hour of continuous rainfall for three broad classes of bare soils of high, medium, and low infiltration capacity, as shown in Table 3.7.

TABLE 3.7.* VALUES OF f_1 (INFILTRATION CAPACITY SHOWN BY STANDARD CURVE AT TIME 1 HOUR) FOR BARE SOILS

Soil Group	*Range in f_1, In. per Hr*
High	0.50–1.00
Intermediate	0.10–0.50
Low	0.01–0.10

* Table 9 of Chap. 2 on Infiltration.

The three broad soil groups are described in their main features substantially as follows:

The High Group. The high group is mostly sandy soils but includes other open-structured soils such as the friable silt loams.

TABLE 3.8.* COVER FACTORS

Cover†		Range in value of cover factor‡
Type	Condition	
Permanent (forest and grass)	Good	3.0–7.5
	Medium	2.0–3.0
	Poor	1.2–1.4
Close-growing crops	Good	2.5–3.0
	Medium	1.6–2.0
	Poor	1.1–1.3
Row crop	Good	1.3–1.5
	Medium	1.1–1.3
	Poor	1.0–1.1

* Covers referred to are described in the text.

† Ratio of f_1 for given soil with cover to f_1 for same soil without cover.

‡ Table 10 of Chap. 2 on Infiltration.

The Medium Group. The medium group includes soils lying between the lower side of the high group and the higher side of the low group, being largely the typical loams of the better agricultural regions.

The Low Group. The low group includes the clays, clay loams, and other soils of dense structure.

In Table 3.8 the same publication gives values corresponding to the cover conditions of the problem in hand as a factor by which to multiply the selected value from Table 3.7 for bare soils, to obtain the *infiltration capacity* for the given type of cover. "The adjectives 'good,' 'medium' and 'poor' are descriptive solely of the effect of the *cover* on infiltration capacity." The range in these factor values provides both for a real variation in cover and an allowance for error.

Description of Covers in Table 3.8

Good: humus 1 in. or more thick; soil not excessively eroded previously; area not overgrazed.

Medium: humus 1 in. to ¼ in. thick; soil not excessively eroded previously; area not overgrazed.

Poor: humus less than ¼ in. thick; soil previously eroded; area grazed too intensively.

Absorption (which is really infiltration in the larger view) may be increased by improving the infiltration rate of the soil and by impounding the water where it falls, thus increasing the time of contact or the absorption opportunity. Under exceptionally favorable circumstances it may be possible to induce absorption of the entire rainfall, in which case there will be no runoff.

That such a possibility is within reason in humid regions has been demonstrated by the experiments of G. W. Musgrave [93] who by the use of lysimeters and experimental plots measured the infiltration rates of two soils in their natural undisturbed state. Marshall silt loam and Shelby silt loam were the two types studied, and all the natural vagaries to which soils are subject, such as nonuniform texture, varying structure and the presence of root cavities, and worm holes, were taken into account.

Musgrave's results are shown graphically in Fig. 3.14. An analysis of Fig. 3.14 shows that in the first 15 minutes the infiltration rate upon the Marshall exceeded that upon the Selby silt loam by 0.37 inch. In the first 30 minutes the difference was 0.62 inch; 45 minutes, 0.88 inch; 1¼ hours, 1.30 inches; and within 5½ hours the difference was 3.82 inches. Assuming that 1.5 inches of rain can be held on the surface where it falls by contour listing or basins and that the critical rainfall rate is as indicated, the chart shows that excess runoff will occur on Shelby silt loam after about 20 minutes (where the rainfall curve intersects the Shelby curve) and will continue at an increasing rate for the duration of the rain;

whereas, on the Marshall there will be no runoff because, after an initial "boost" of 1.5 inches, the infiltration rate exceeds the rainfall rate throughout. The total amount of runoff on the Shelby, in this hypothetical study, is represented by the area included between the rainfall curve and the Shelby curve, and the total reserve capacity of the Marshall is similarly represented by the area between the Marshall and the rainfall curves; likewise the total difference in absorptive capacity for the 6-hour period is the area between the Marshall and the Shelby curves.

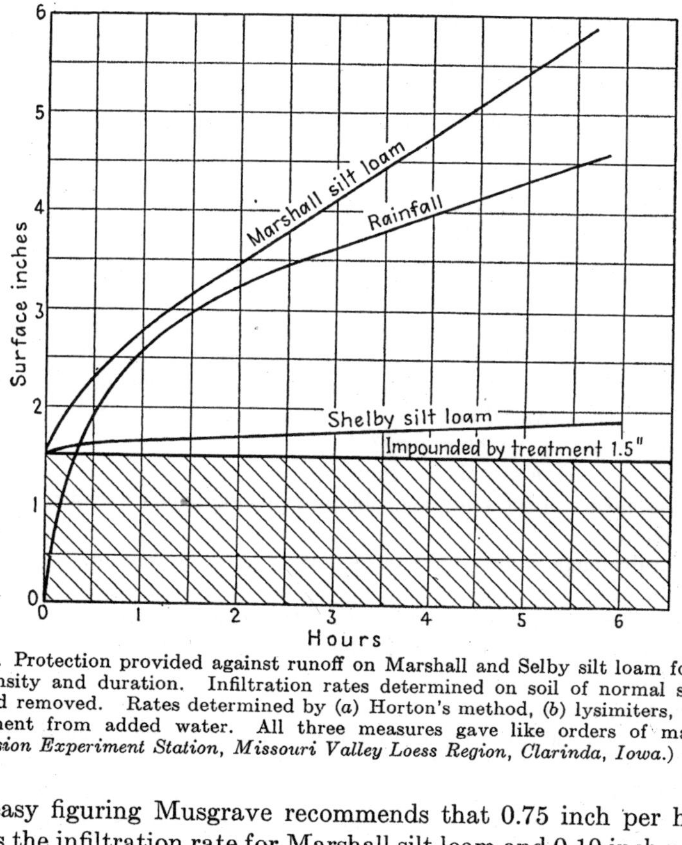

FIG. 3.14. Protection provided against runoff on Marshall and Selby silt loam for rain of rare intensity and duration. Infiltration rates determined on soil of normal structure, moist, sod removed. Rates determined by (*a*) Horton's method, (*b*) lysimiters, (*c*) direct measurement from added water. All three measures gave like orders of magnitude. (*Soil Erosion Experiment Station, Missouri Valley Loess Region, Clarinda, Iowa.*)

For easy figuring Musgrave recommends that 0.75 inch per hour be taken as the infiltration rate for Marshall silt loam and 0.10 inch per hour for Shelby silt loam.

The infiltration values herein discussed cover a small number of the possible highlights from among the innumerable variations that may present themselves in practice. They should, therefore, be used with extreme caution in cases that are not exactly parallel. To the practitioner making use of them it is strongly recommended that, before so doing, he consult the full references from which they are taken, namely:

1. "Hydrology Handbook of the American Society of Civil Engineers," Chap. 2, published by the ASCE, adopted in 1949.

2. "An Approach toward a Physical Interpretation of Infiltration Capacity," by R. E. Horton, *Proceedings*, Soil Science Society of America, Vol. 5, 1940, pp. 399–417.

3. "Soil Erosion Studies—Part III: Some Effects of Soil Erosion on Infiltration and Surface Runoff," by W. D. Ellison, *Agricultural Engineering*, Vol. 28, No. 6, June, 1947, pp. 245–248.

4. "The Effect of the Degree of Slope and Rainfall Characteristics on Runoff and Soil Erosion," by Jesse H. Neal, *University of Missouri Agricultural Experiment Station Research Bulletin* 280, April, 1938.

Particular emphasis should be put on the first two which give a comprehensive and enlightening discussion of the entire topic of infiltration.

PROBLEMS

1. In what terms may permeability of soil be expressed?

(*a*) Define *permeability coefficient.*

2. What general types of structure have a permeability that is conducive to good drainage?

(*a*) What types markedly impede good drainage?

3. Describe a reliable method for direct measurement of permeability in the field that does not require a considerable amount of scientific apparatus.

4. What are the physical essentials of a soil sample the permeability of which is to be determined in the laboratory?

(*a*) How may such a sample be secured?

(*b*) What feature of the sampling tube prevents change of density as the result of compression?

5. In tile drainage which is the most essential, the horizontal or the vertical permeability rate?

(*a*) How may the one be reduced to the other close enough for practical use?

6. What is the essential value of the moisture equivalent in soil-moisture considerations?

(*a*) What are its limitations?

(*b*) What is its relation to the lower plastic limit; to the upper plastic limit; to the per cent of clay in a soil?

7. In a given silt loam whose moisture equivalent, on a dry weight basis, is 25, on a volume basis is 36 and equal to 80% of the total pore space, what does it amount to in the first 4 ft of depth of soil in inches of depth of water?

8. The total pore space of a given soil was found by analysis to be 48% of the total volume. What was the probable moisture equivalent and the probable physical class of the soil? What were the probable values of its following features: lower plastic limit, upper plastic limit, per cent of clay, probable tile drainability, and probable infiltration rate of the bare soil?

CHAPTER 4

RAINFALL AND RUNOFF

RAINFALL

4.1. Importance of Precipitation in Drainage. Precipitation is of vital importance in drainage both as regards its distribution during the growing season and its character during any given storm period. These considerations largely determine the kind and extent of drainage works necessary for the quick removal of excess moisture at those times when it tends to be inimical to crop development. In the humid regions practically all the water carried off as runoff is that from precipitation. In the irrigated regions the rainfall, usually very light, is the source of only a limited part of the runoff, much of which is waste water from irrigation. On the Pacific slope, where intense rainstorms occur at high altitudes when snow is on the ground, it is customary when estimating runoff to add to the rainfall a small percentage for snow melt, but in any case, precipitation is a leading factor in determining the size of channels needed to carry the runoff. It is generally recognized that, in all but very special cases, works designed to carry runoff from design storms[1] will have sufficient capacity to handle the lesser volume from thaws. It is, therefore, common practice to direct precipitation studies largely toward a determination of rainfall characteristics.

4.2. Precipitation Characteristics. The volume, distribution, and character of precipitation are all important features in any phase of the soil-moisture problem in agriculture.

The quantity of rainfall available for the use of any given large region is markedly influenced by the following geographical conditions:

1. Proximity to the ocean or large inland waters.
2. Mountain ranges in the path of moisture-laden winds.
3. Location on or near the path of cyclonic storms.

These geographical conditions each produce rainstorms of different types in duration, volume, and intensity according to the season.

4.3. Precipitation Records. The distribution of precipitation varies greatly over the earth's surface. In modern times, the importance of precipitation history has been generally recognized in civilized countries and in none more than in the United States where a country-wide con-

[1] A *design storm*, often improperly called a critical rain, is one which, during its intense period, will result in a rate of runoff of maximum volume per unit time which it is economically feasible to provide for in control channels.

tinuous record is constantly kept by the Weather Bureau of the U.S. Department of Commerce, which maintains about 5,000 observation stations distributed all over the country. The various state agricultural colleges as well as numerous other public and private agencies maintain weather-observation stations where daily records are made and kept in permanent files. Until recent years, it has been customary to publish rainfall records in this country only in terms of total monthly and annual precipitation expressed in inches of depth. Average monthly rainfall, covering a period of about 20 years, is given in Table 4.3 for a considerable number of locations throughout the United States.

4.4. Annual Precipitation. A special map of the United States, prepared by the U.S. Geological Survey and showing the amount and distribution of mean annual precipitation as determined from all reliable records to date, is represented in Fig. 4.1. The irregular lines superimposed on the map join points of equal annual precipitation in amounts as shown by the numerals in the lines, which represent yearly totals in inches. These numerals are average or mean values for the entire period of record, and the total for any single year may vary widely therefrom; in a period of 100 years a single year's record may vary as much as 50 per cent above or below the mean. There is nothing in these data to indicate how the total annual precipitation is distributed throughout the year by months or by seasons. Such data are available from publications of the U.S. Weather Bureau when needed for special studies. As will be noted from Fig. 4.1, the annual precipitation in the United States varies from 4 inches or less in certain parts of the West to 60 or 70 inches on the Atlantic slope and from 20 to 40 inches in the Mississippi Basin to 120 inches or more in two or three limited spots over the northwest coast slope.

4.5. Precipitation Cycles. The notion that annual precipitation is cyclic in character, with periods of several excessively wet years followed by periods of excessively dry years with consistent regularity, has no foundation in the factual record which makes it quite clear that the vagaries of precipitation history are wholly within the realm of capricious nature and vast meteorological changes which are as yet imperfectly understood.

In general, the records indicate that the following aspects of rainfall are approximately true:

1. Intense storms usually cover only small areas and are of short duration.
2. Storms lasting several days cover large areas and fall at low intensity.
3. The magnitude and intensity of any storm bears a direct relation to its average frequency of recurrence.

There does not appear to be any close relationship between the total annual rainfall at a given locality and the number or magnitude of intense storms likely to occur in a given time period.

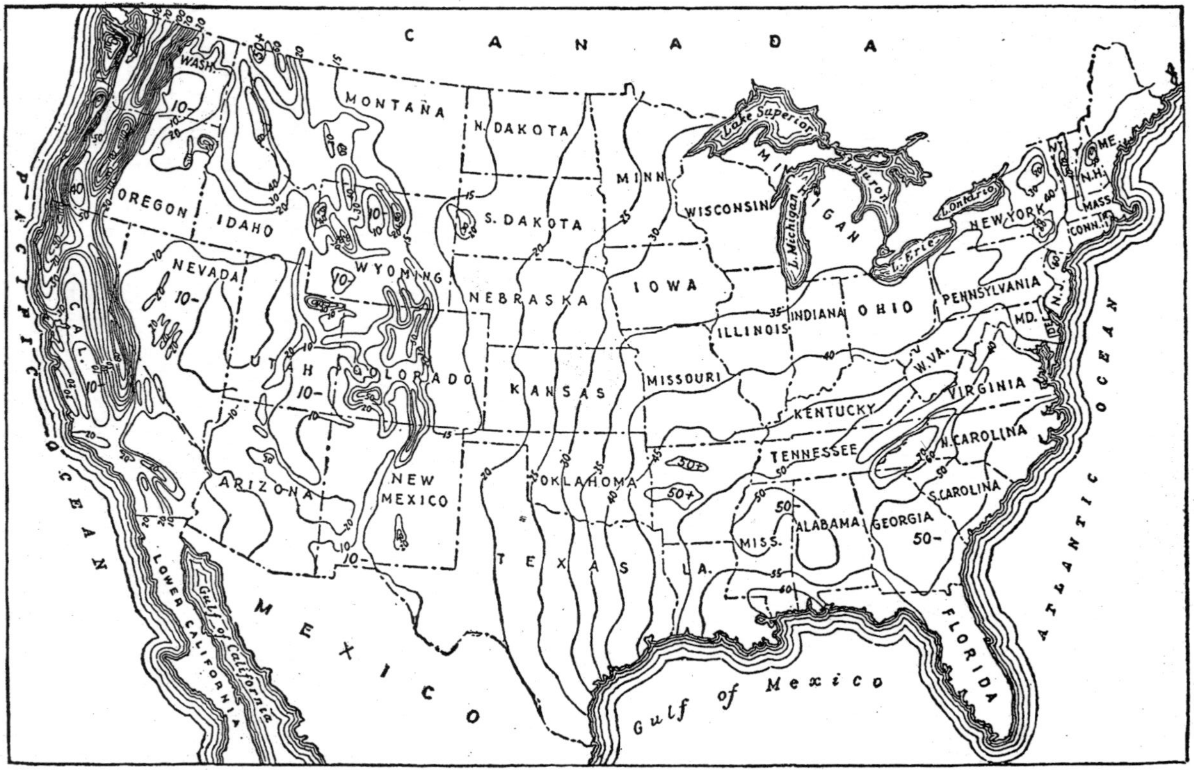

FIG. 4.1. Map of the United States showing mean annual precipitation. (*Courtesy of U.S. Geological Survey.*)

4.6. Short, Intense Single Rainstorms. Short, intense single rainstorms are a marked characteristic of the Mississippi River Basin especially during the summer months. They occur less frequently east of the Appalachians, seldom on the Great Plains and in the intermountain region, and almost not at all on the Pacific slope. In the regions where they are common, such storms are one of the most serious factors in farm-drainage considerations because they are apt to occur at any time during the growing season, when, over a relatively limited area, an immense volume of water falls in a very short time—often, in fact, in a few minutes up to an hour or two. Especially in such cases the character and amount of resultant control activity necessary, following the storm, depend upon several factors such as the character of the surface soil, its relative dryness or wetness (moisture content), the slope of the surface, the roughness of the topography, and the character of the vegetal covering; for all these things are strong influences in determining how much of the rainfall will run off quickly over the surface into streams and lakes, how much of it will soak into the soil and be taken up by plant growth, and how much of it will continue to percolate through the soil into tile drains or into other subdrainage or deep underground storage provided by nature.

A study, the results of which are shown in Table 4.1, of rainfall data for various drainage basins of the United States as compiled by Yarnell [165] shows that, on the average, in any given rainstorm in which the length of the intense period did not exceed 1 hour, about 65 to 95 per cent of the rain falls during the intense period alone and the length of the intense period averages between 20 and 40 per cent of the total duration of the entire storm. It will further be noted that in most cases east of the Rocky Mountains the variations from the weighted average values are not great.

At such times the water from rainfall accumulates on the land faster than it can run off through natural channels and ditches or soak into the soil. The tendency, then, is for the water to impound in puddles of various sizes and depths on the surface, with a resulting further tendency to drown out patches of a crop before the excess water can be removed. The problems of flood removal and of soil drainage are to design means of removing temporary excesses quickly enough and in such a manner that damage to crops shall be reduced to a minimum.

4.7. Steady, Long-continued Storms of Low Intensity. Steady rainstorms lasting several days and covering large areas at low intensities are as characteristic of the Pacific slope and, to a less degree, the Atlantic slope as are the short, intense storms of the Mississippi Basin. This type of storm may last 1 day to a week, a length of 4 or 5 days being common experience, the total rain falling in a single one or even two of those days frequently amounting to 3 to 6 inches. Figure 4.2 shows clearly, by

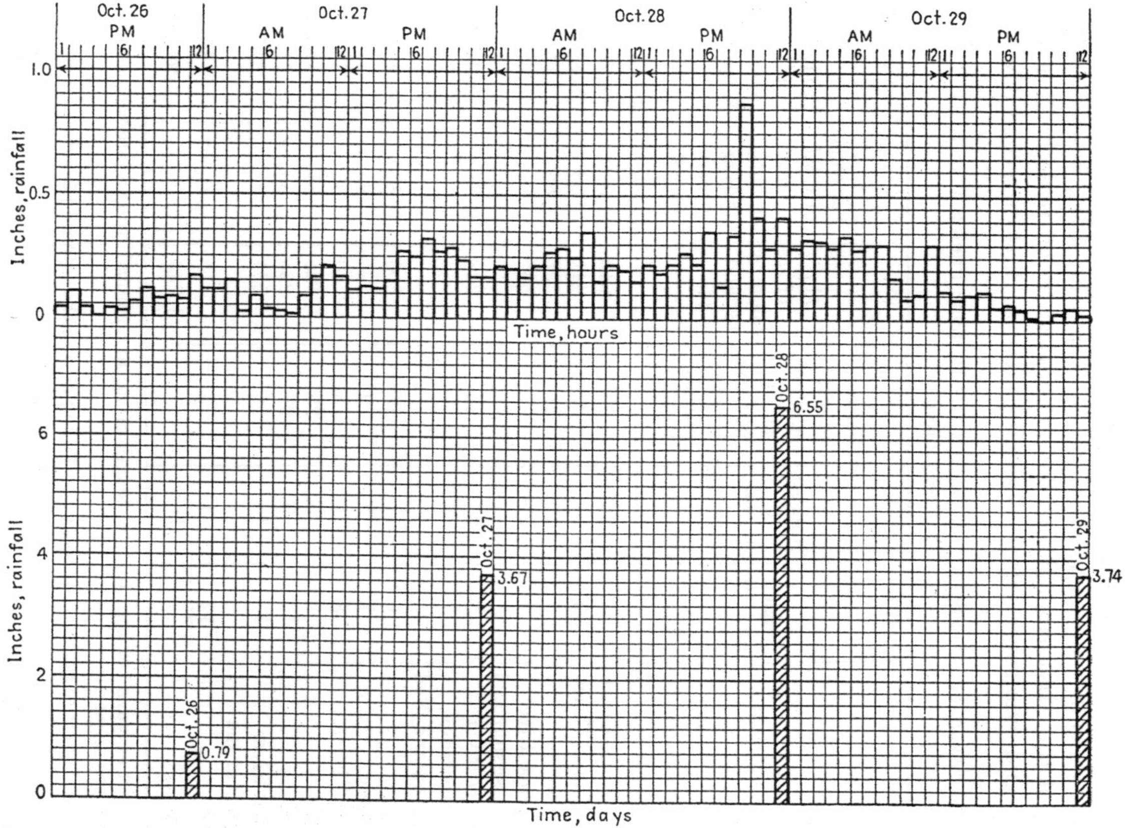

FIG. 4.2. Complete hourly and daily record of a 3½-day typical rainstorm of low intensity, representative of the Pacific slope, Kerby, S.W. Oregon, Oct. 26–29, 1948. (*Drawn from the U.S. Weather Bureau Record. Courtesy of A. L. Sharp, Soil Conservation Service, Pacific Region.*)

hours, a true record of one such rainstorm typical of its class. The drainage problems raised by a storm of this type are comparable in importance and in details of solution to those presented by the short, intense rainstorms of the Mississippi Basin; but the critical runoff period will, as a rule, extend over several days, beginning after the first or second day of

TABLE 4.1. AVERAGES FOR STORMS IN WHICH THE INTENSE PERIOD WAS LESS THAN 1 HOUR LONG

(1) Name of basin	(2) Region number	(3) Number of observations	(4) Per cent of total length of storm occupied by the intense period	(5) Product of columns (3) and (4)	(6) Per cent of total rain falling during intense period	(7) Product of columns (3) and (6)
North Atlantic Coast	1	20	32	640	87	1,740
South Atlantic Coast	2	21	21	441	88	1,848
Ohio River	3	11	27	297	88	968
St. Lawrence River	4	21	27	567	87	1,827
Upper Mississippi and Missouri River	5	29	22	638	81	2,349
Lower Mississippi, Arkansas, and Red River	6	17	18	306	83	1,411
Rio Grande	7	5	39	195	95	475
Colorado River	8	4	15	60	85	340
Salt Lake	9	5	25	125	83	415
Pacific slope	10	9	25	225	68	612
Columbia River	11	13	18	234	65	845
Red River of the North	12	1	15	15	74	74
Totals	..	156	..	3,743	..	12,904

Weighted average per cent of total length of storm occupied by the intense period = 3,743/156 = 24.0%.

Weighted average per cent of total rain falling during the intense period = 12,904/156 = 82.7%.

the storm and continuing about an equal time after the storm has ended instead of only a fraction of an hour or an hour or two as is usually the case with the intense storms of short duration.

4.8. Frequency of Occurrence of Storms of Given Intensity. Records of the rainstorms of known great intensity and/or duration constitute one of the safest guides in designing requisite control works for removal of excess water from the land before it does appreciable damage to crops.

Modern weather science has developed the practice of classifying such heavy storms according to their probable frequency of occurrence (Potter [105a]). For example, how often, in a given region, would a rainfall of ½ inch in 30 minutes be likely to occur? Or the question might be stated this way: What is the maximum intense rainfall of a duration of 15 minutes that is apt to occur in the given region every 10 years? Or it might be asked: How often is an average rainfall rate of ¼ inch per hour apt to occur in the given region for a given duration of rainfall?

4.9. Frequency Records of Short, Intense Storms. Figures 4.3 to 4.22 taken from reference 165 show 20 expectation frequencies for short,

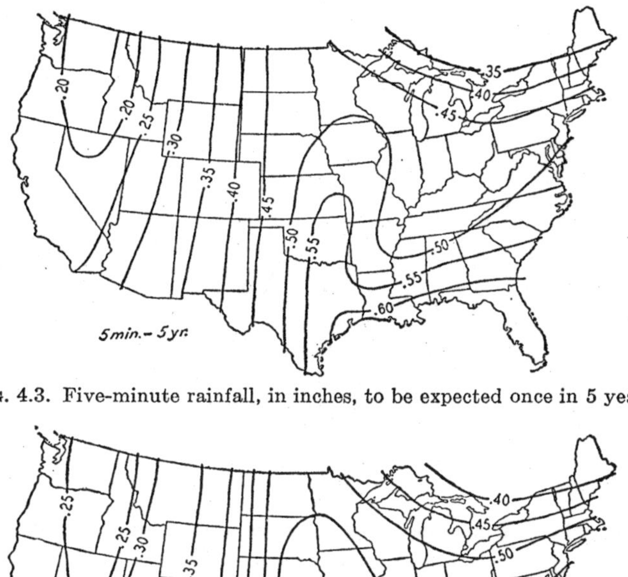

FIG. 4.3. Five-minute rainfall, in inches, to be expected once in 5 years.

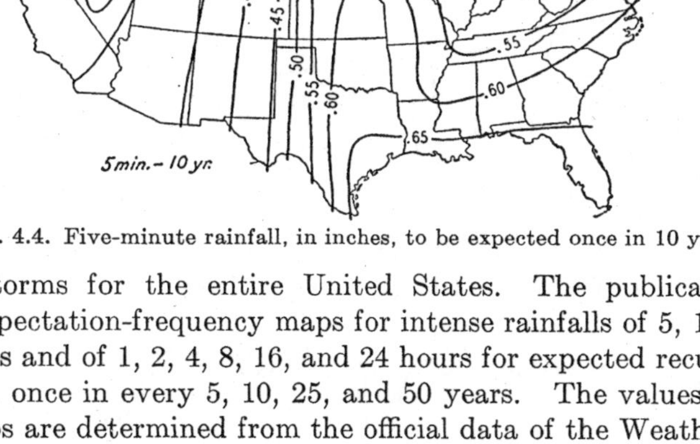

FIG. 4.4. Five-minute rainfall, in inches, to be expected once in 10 years.

intense storms for the entire United States. The publication gives similar expectation-frequency maps for intense rainfalls of 5, 10, 15, and 30 minutes and of 1, 2, 4, 8, 16, and 24 hours for expected recurrences in each case, once in every 5, 10, 25, and 50 years. The values shown on these maps are determined from the official data of the Weather Bureau on individual storms shown in the first part of the same publication.

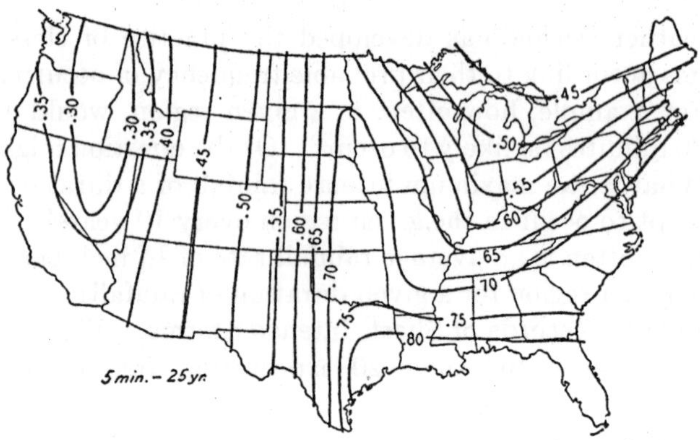

FIG. 4.5. Five-minute rainfall, in inches, to be expected once in 25 years.

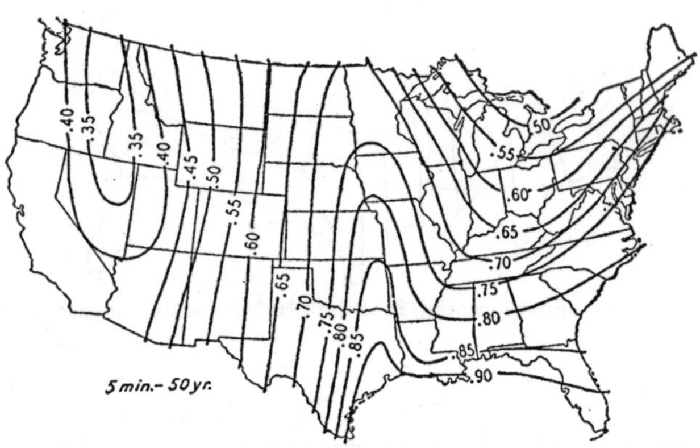

FIG. 4.6. Five-minute rainfall, in inches, to be expected once in 50 years.

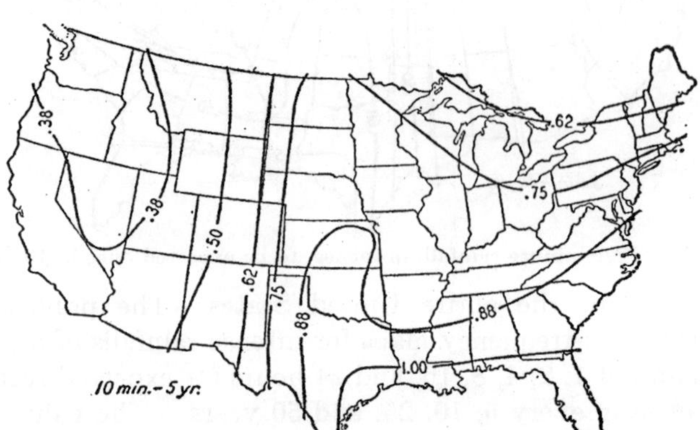

FIG. 4.7. Ten-minute rainfall, in inches, to be expected once in 5 years.

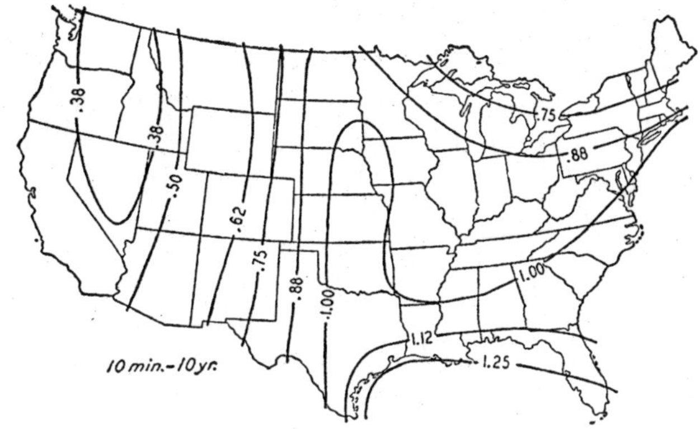

FIG. 4.8. Ten-minute rainfall, in inches, to be expected once in 10 years.

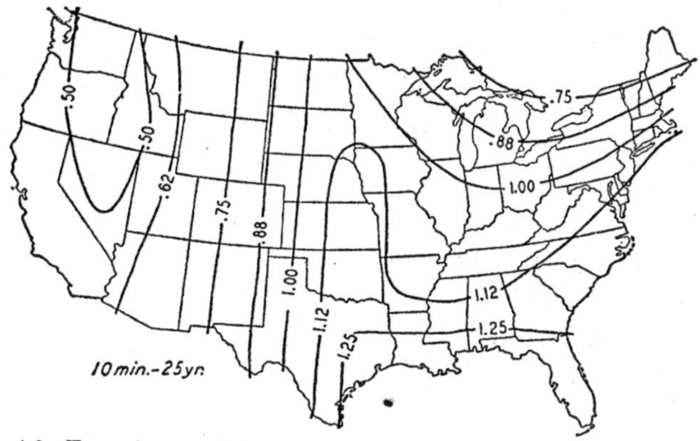

FIG. 4.9. Ten-minute rainfall, in inches, to be expected once in 25 years.

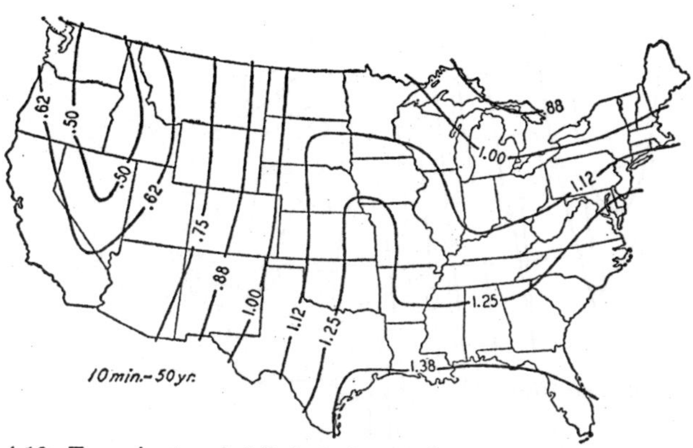

FIG. 4.10. Ten-minute rainfall, in inches, to be expected once in 50 years.

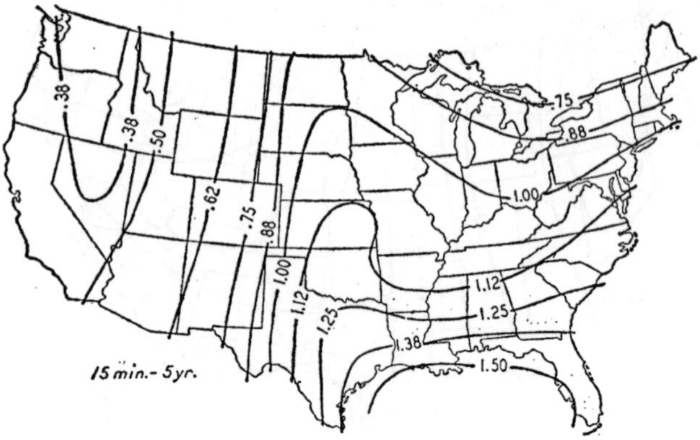

FIG. 4.11. Fifteen-minute rainfall, in inches, to be expected once in 5 years.

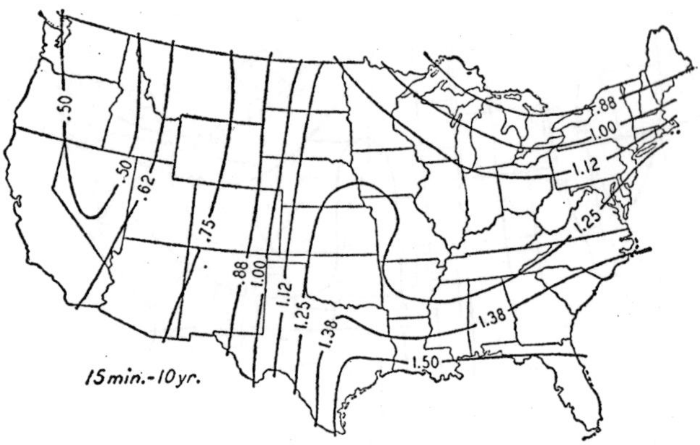

FIG. 4.12. Fifteen-minute rainfall, in inches, to be expected once in 10 years.

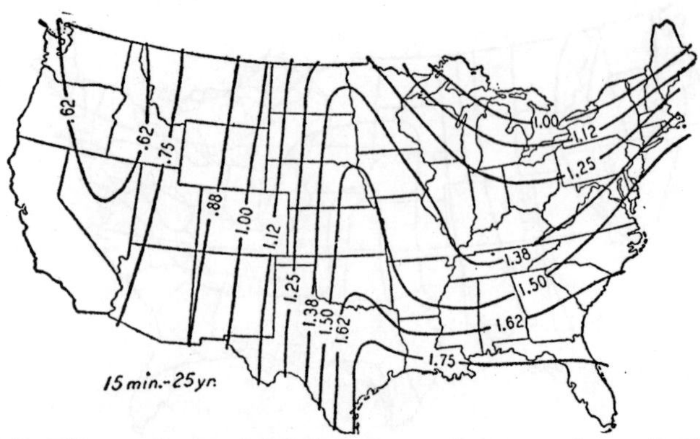

FIG. 4.13. Fifteen-minute rainfall, in inches, to be expected once in 25 years.

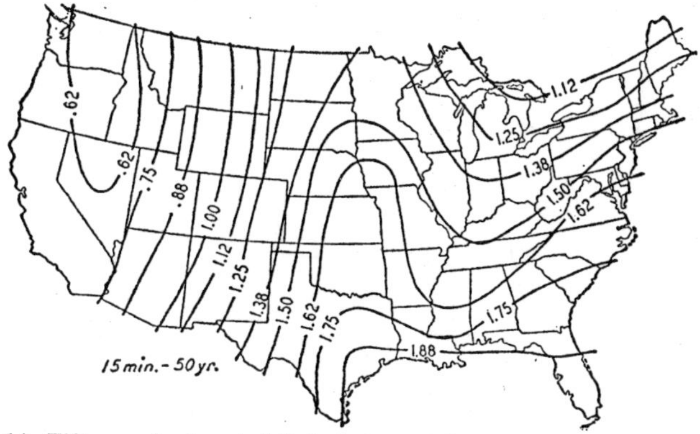

FIG. 4.14. Fifteen-minute rainfall, in inches, to be expected once in 50 years.

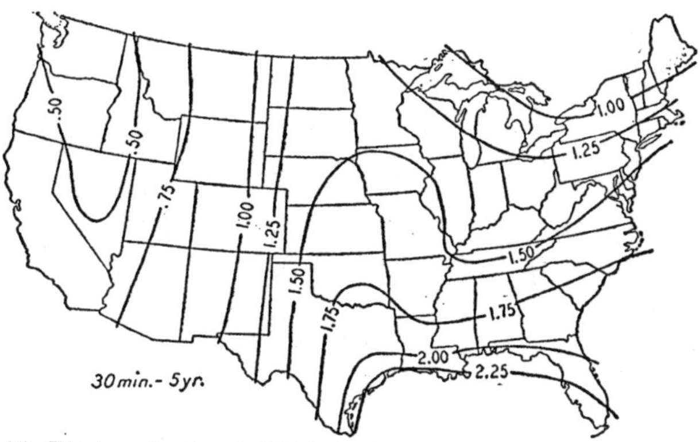

FIG. 4.15. Thirty-minute rainfall, in inches, to be expected once in 5 years.

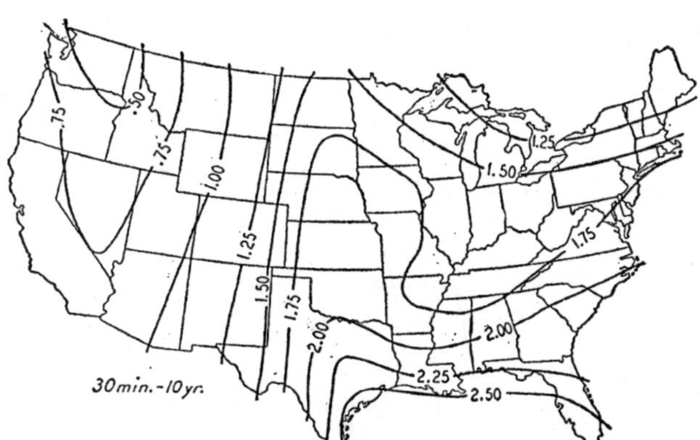

FIG. 4.16. Thirty-minute rainfall, in inches, to be expected once in 10 years.

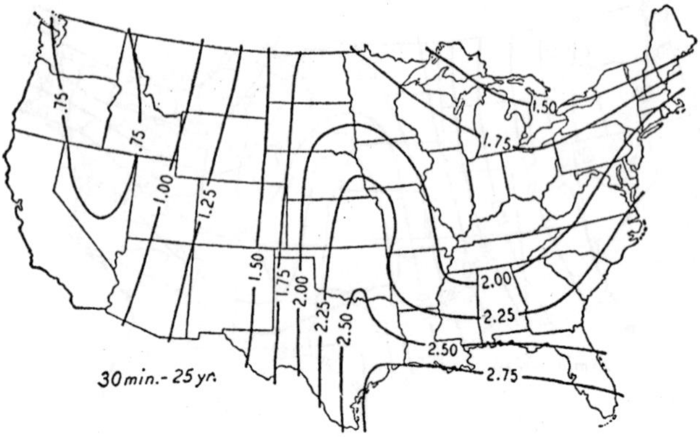

FIG. 4.17. Thirty-minute rainfall, in inches, to be expected once in 25 years.

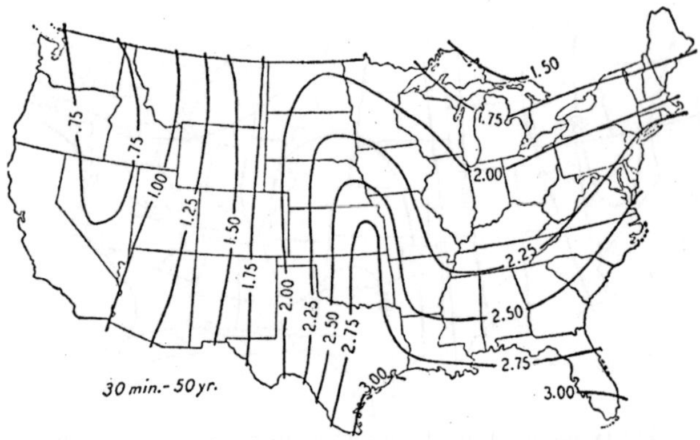

FIG. 4.18. Thirty-minute rainfall, in inches, to be expected once in 50 years.

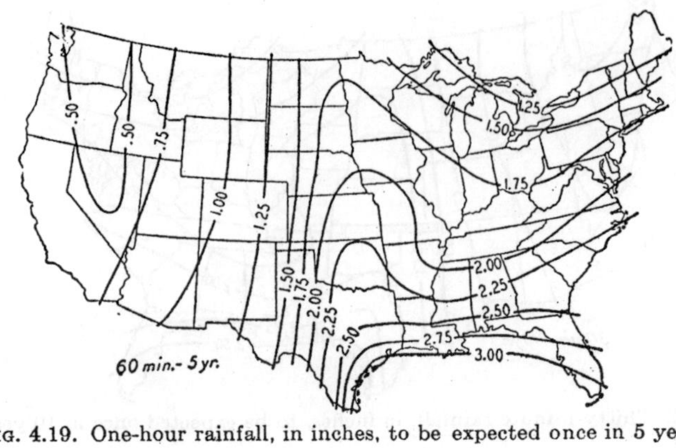

FIG. 4.19. One-hour rainfall, in inches, to be expected once in 5 years.

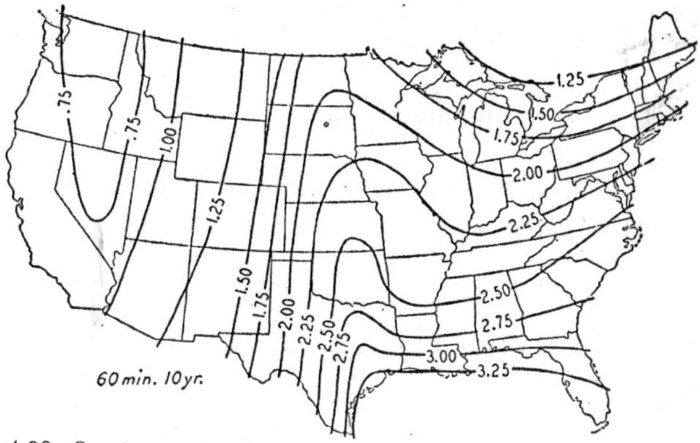

FIG. 4.20. One-hour rainfall, in inches, to be expected once in 10 years.

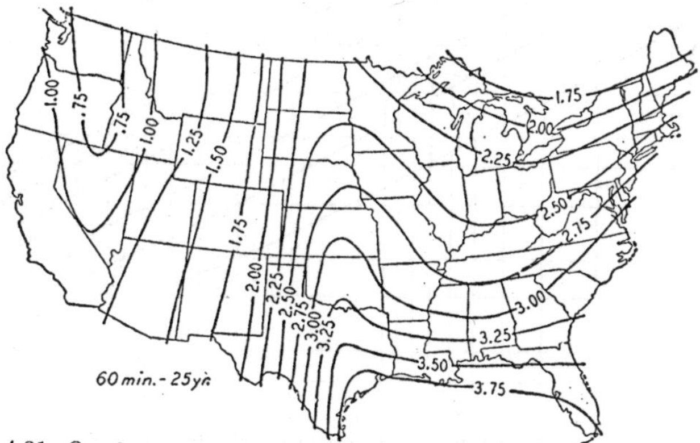

FIG. 4.21. One-hour rainfall, in inches, to be expected once in 25 years.

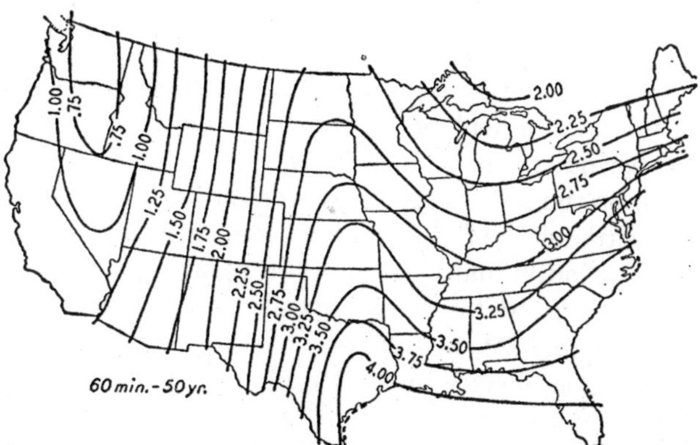

FIG. 4.22. One-hour rainfall, in inches, to be expected once in 50 years.

The facts being thus known or reliably predicted regarding any given type of rainstorm or rainfall period, the effective means of removal of excess, as they apply to any given region, may be correlated with the rainfall data as a guide to the proper design of necessary artificial controls of drainage and runoff.

4.10. Frequency Records of Long-continued Storms of Low Intensity. Original records of frequency of long-continued rainstorms of low intensity

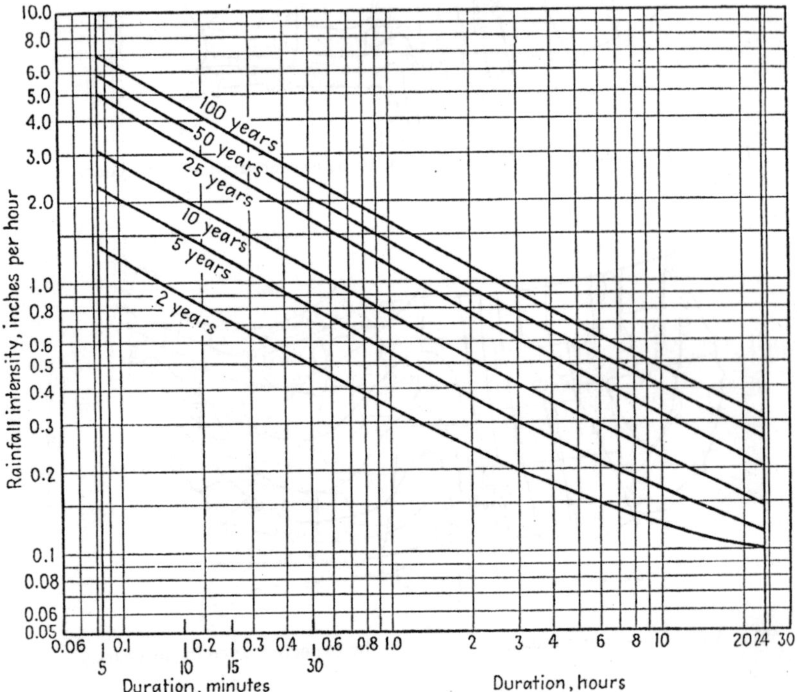

FIG. 4.23. Rainfall intensity-duration curves; years of record 1897–1939, Portland, Oreg. (*Courtesy of A. L. Sharp, Soil Conservation Service, Pacific Region.*)

are available in the public archives, but not in the most convenient form. Still they are as essential in dealing with such storms as are the frequency records for short-period storms of high intensity. Hence, it is necessary to depend either upon the few compiled records for limited areas, available in the regional records of the Soil Conservation Service and other interested agencies, or on individual local effort in securing or uncovering, in Weather Bureau files, the detailed records of such storms. Three local charts are shown in Figs. 4.23, to 4.25, and by the courtesy of A. L. Sharp of the Soil Conservation Service in the Pacific region.

Of great interest in this connection is the long-time study of about 1,000 major flood-producing storms, a study initiated by the U. S. Army

Corps of Engineers in 1937. Up to December, 1949, printed sheets had been issued containing detailed data on the maximum average depth of rainfall, area covered, and duration of 398 such storms scattered over the entire United States. These sheets were assembled in a loose-leaf binder under the title "Storm Rainfall in the United States," and they are available from the Chief of Engineers, Department of the Army, Washington, D.C.

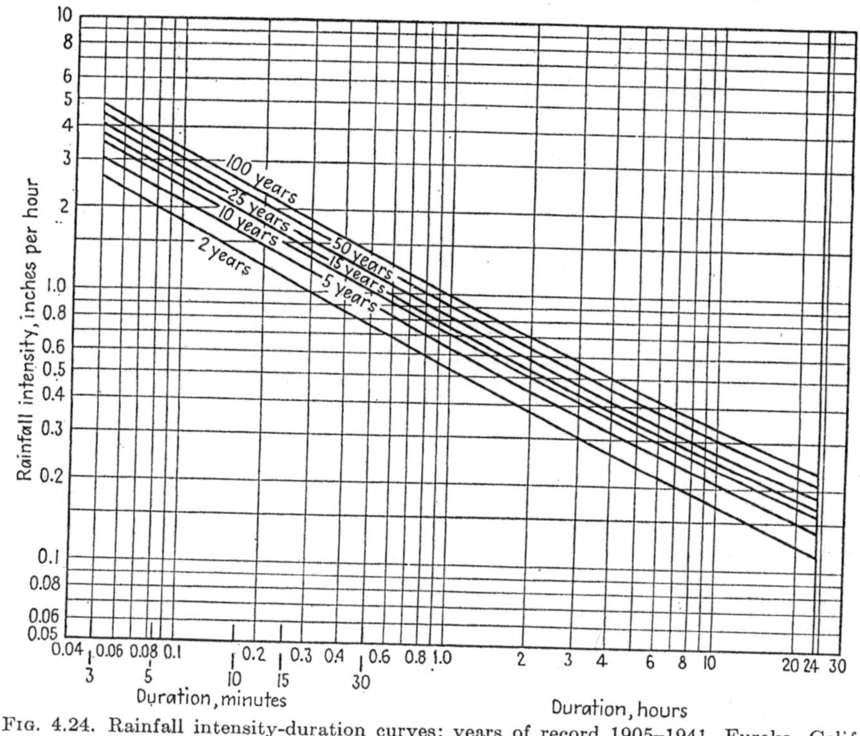

FIG. 4.24. Rainfall intensity-duration curves; years of record 1905–1941, Eureka, Calif. Developed from monthly maxima depths for selected time intervals. (*Courtesy of A. L. Sharp, Soil Conservation Service, Pacific Region.*)

4.11. Long-time Records Essential. The records of total daily rainfall of some of the older Weather Bureau stations reach back over 100 years, but these are very few. A short-term record is of little value in planning works or developments seriously affected by rainfall, because short periods are deceptive and unreliable as bases for prediction of probable future occurrences. A 1-year record is of practically no value, a 5-year record of little value, a 10-year record is somewhat better, a 50-year record is usually good and about the best that can be expected. A number of the largest and best stations are now equipped to secure continuous weather records by means of delicate automatic recorders which give the exact history of each individual rain, including beginning and ending time and the character of the storm, that is, its rate during the relatively short period of high intensity, and where, within the storm limits, this intense period occurs. Long period records of this type are not yet available as the continuous automatic recorders began to come into general use only about 1930.

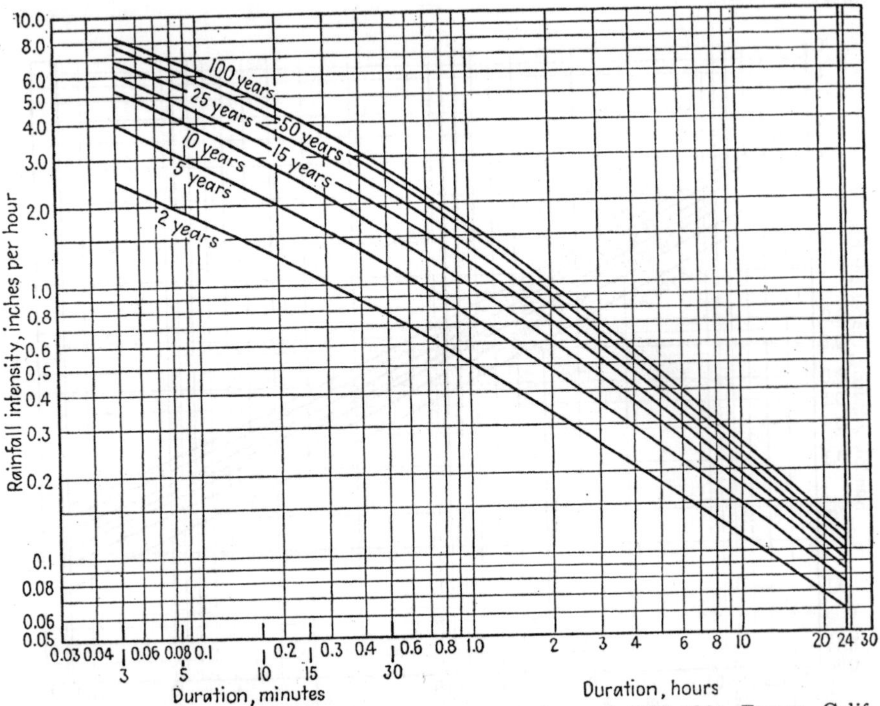

FIG. 4.25. Rainfall intensity-duration curves; years of record 1903–1941, Fresno, Calif. Developed from monthly maxima depths for selected time intervals. Short time durations corrected for longer period of record from available data on daily precipitation, 1888–1941. (*Courtesy of A. L. Sharp, Soil Conservation Service, Pacific Region.*)

4.12. Measurement of Rainfall. The measurement of rainfall is a relatively simple matter. The unit used is inches of depth of water falling in a given period of time. The commonly used unit by the interested general public is inches per year. However, the agricultural scientist and the engineer must consider the rainfall in much shorter periods of time such as 24 hours or even a fraction of an hour, which latter rate is of especial and necessary interest in dealing with runoff problems. The standard rain gage is the device most commonly used to measure simply the total depth of rainfall per day or per hour. This gage consists of a galvanized-iron cylinder 8 inches in diameter and 24 inches high, with a concave funnel top equipped with a brass knife-edge periphery finished exactly to the 8-inch diameter. The spout of the funnel leads into a circular inner tube of brass whose cross-sectional area is just one-tenth of the area of the cross section of the larger cylinder at the knife-edge. This device concentrates the water falling within the knife-edge and magnifies its depth in the brass cylinder just ten times over the true depth. Hence great accuracy in measurement is possible. The gage is shown in Fig. 4.26. The measuring scale is a dry red-cedar rule graduated in inches and tenths of an inch, which represent in actual depth of rainfall 1/10 and 1/100 inch, respectively. The rule is sufficiently slender to be slipped readily

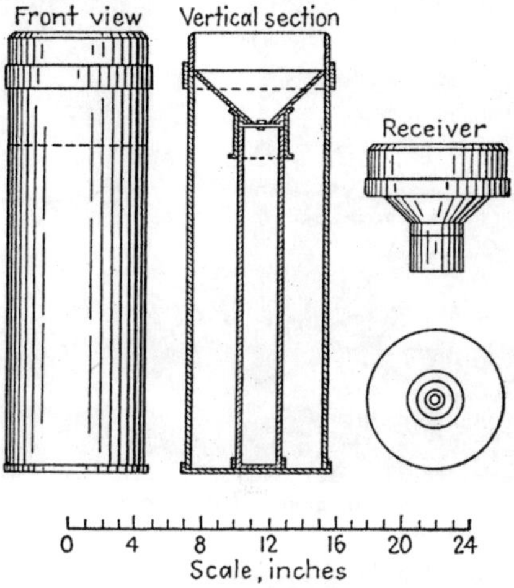

FIG. 4.26. Standard rain gage.

through the opening in the funnel and down into the inner brass tube. If this is done quickly and the rule is as quickly withdrawn, the mark between the wet and the dry portion is sharp and distinct and is easily read.

The measurement of intensities and of duration of intense period, however, requires much more complicated equipment of a delicate character and costing considerable money. Such an instrument is equipped with a continuous-recording device that shows, on a continuous chart on a revolving drum, the varying rainfall of the storm over a given period of time, as, for example, 1 hour or 24 hours. From this chart there may be determined, with sufficient accuracy for practical purposes, the rate of precipitation for any given small time interval during the storm. One such instrument is shown in Fig. 4.27, and a 24-hour chart, on which is traced the record for May 29, 1943, at University Farm, St. Paul, Minn., is shown in Fig. 4.28.

The type of record that may be obtained for a long-period storm of low intensity is shown in Fig. 4.2, plotted from an hourly statistical record.

The time unit in such a record need not usually be smaller than 1 hour. This storm-history kind of record of short, intense and long-continued low-intensity rainstorms is essential in determination of the runoff from such storms, as it is intimately related to the "time of concentration" of

FIG. 4.27. Continuous-recording rain gage. *Left:* General view of complete installation. *Right:* Detail of interior mechanism.

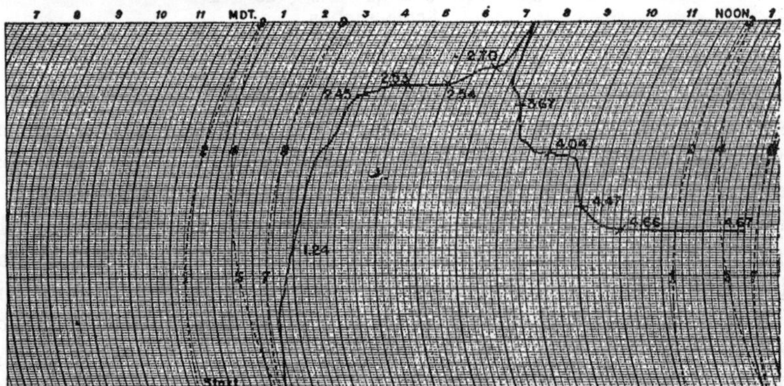

FIG. 4.28. Record of the rain of May 29, 1943, at University Farm, St. Paul, Minn., recorded by the rain gage shown in Fig. 4.27. (A typical automatic record of an intense storm.)

runoff, later discussed, which, in turn, is a vital factor in the design of the outlet works.

4.13. Measurement of Snowfall. The measurement of snowfall is a different and more complicated matter. Where very heavy snowfall, such as is common occurrence in mountain regions, makes the simple methods of its measurement, discussed below, slow and impractical, the problem can readily be met by employing the device used of late years in the mountain snow surveys of the arid West. The description of this equipment and its use is so long as to be beyond the general scope of this book, but it is fully covered by Marr [79] in a government publication available to anyone interested.

In humid and semihumid regions, where the snowfall is of relatively minor importance, its measurement for single storms, simply as a portion of the total annual precipitation, is the principal consideration. The U.S. Weather Bureau [152] methods here quoted are probably the most satisfactory.

> During the winter season . . . the overflow attachment, only, of the rain gage should be exposed . . . as a snow gage. Remove the receiver and measuring tube, as these parts cannot be used for measuring snow, and even if rain should occur, it is very apt to be frozen while in the measuring tube, generally bursting it and rendering it worthless or highly inaccurate.
>
> *First Method.* The snowfall collected in the overflow attachment is measured after placing the vessel in a warm room until the snow is melted. The water is then carefully poured into the measuring tube and measured just as though it were rainfall.
>
> *Second Method.* The first method is objectionable, because it often requires considerable time, and is liable to be inaccurate, owing to the loss of snow or water by evaporation. The following plan is much better . . . : Take the overflow into the room and pour into it carefully one measuring tube full to the brim with water, preferably warm water. This, in general, will mostly melt, or at least reduce to a very fluid slush, a considerable snowfall. The measuring tube should be again carefully filled to the brim from the melted contents of the overflow and emptied; whereupon the remaining water in the overflow should be carefully measured in the measuring tube, thus giving quickly and easily the depth of melted snow.
>
> The amount of snow collected in the overflow of the rain gage is likely to be greatly deficient when the wind blows during the snowfall. In such cases it will be much better if the observer will discard the snowfall in the overflow, empty it out and cut out a section of the snow in an open place where the depth truly represents the precipitation. This section is to be taken by plunging the empty overflow, mouth downward, in the snow so as to cut out a cylindrical portion the size of the overflow and the depth of the snow. By the use of a thin board . . . it will not be difficult to gather up the complete section of the snow inside the overflow, after which it should be reduced to slush and measured as already explained.

RUNOFF

Mead defines runoff as "that portion of the rainfall that is not absorbed by the deep strata, utilized by vegetation or lost by evaporation and which finds its way into the streams as surface or subsurface flow."

Hence runoff is closely allied to *evaporation*, *transpiration*, and *deep seepage*, as all four of these agencies are resultants of precipitation. The three last named all limit the residue of the rainfall to be removed by runoff. Hence it is important to evaluate all these agencies.

4.14. Evaporation. Evaporation is the natural process by which water is changed from the liquid to the gaseous state. It goes on continuously from both water and land surfaces and varies greatly with climate, temperature, wind velocity, humidity of the atmosphere, rainfall, and character of the exposed surface. In regions of high temperature and heavy rainfall, the amount of evaporation, expressed in inches of depth of water, is an appreciable factor in the disposal of rainfall. In temperate regions such as the northern Mississippi Valley, however, even during the hottest summer days, evaporation is negligible as its average *daily* total from land surfaces, in Minnesota, for example, is less than 0.07 inch, although its average *annual* total from the land surface there is about one-third of the total annual rainfall. Evaporation from water surfaces is usually several times greater (about 1.5 to 4.5, average about 2.5) than that from land surfaces in the same region. Crops are considerably affected by evaporation from the land surface but not appreciably by that from water surfaces. Methods of its measurement are exacting and will not be discussed here. Some evaporation records for scattered locations will be found in almost any standard work on hydrology. When exact data are not available regarding evaporation from land surfaces for any given region, fairly reliable figures may be obtained from Adolph F. Meyer's evaporation chart (see Fig. 4.29). In the use of the chart, however, Meyer [83] warns:

> To the values of evaporation in inches of depth per month, as taken off the curve, a coefficient must be applied to reduce these quantities to actual evaporation from the given watershed. This coefficient ranges from about .95 to 1.25 for most watersheds of the Northwest and for similar ones elsewhere. Very sandy watersheds may require a coefficient as low as .60 and very impervious flat watersheds may require a coefficient in excess of 1.25. The coefficient to be used depends on topography, vegetal cover, soil, subsoil, humidity and wind. An extremely high coefficient of evaporation (in excess of 1.25) would result from flat topography devoid of vegetation, moderately pervious, shallow soil underlain with impervious subsoil or rock, low humidity and high wind velocity. An extremely low coefficient (less than .95) would result from rugged topography, bare sandy soil underlain with rock, high humidity and low wind velocity or extremely sandy soil. Between these limits the working values will be found. With a little experience one can select coefficients for different watersheds with considerable accuracy.[1]

[1] Reprinted by permission from "Elements of Hydrology" by Adolph F. Meyer, published by John Wiley & Sons, Inc.

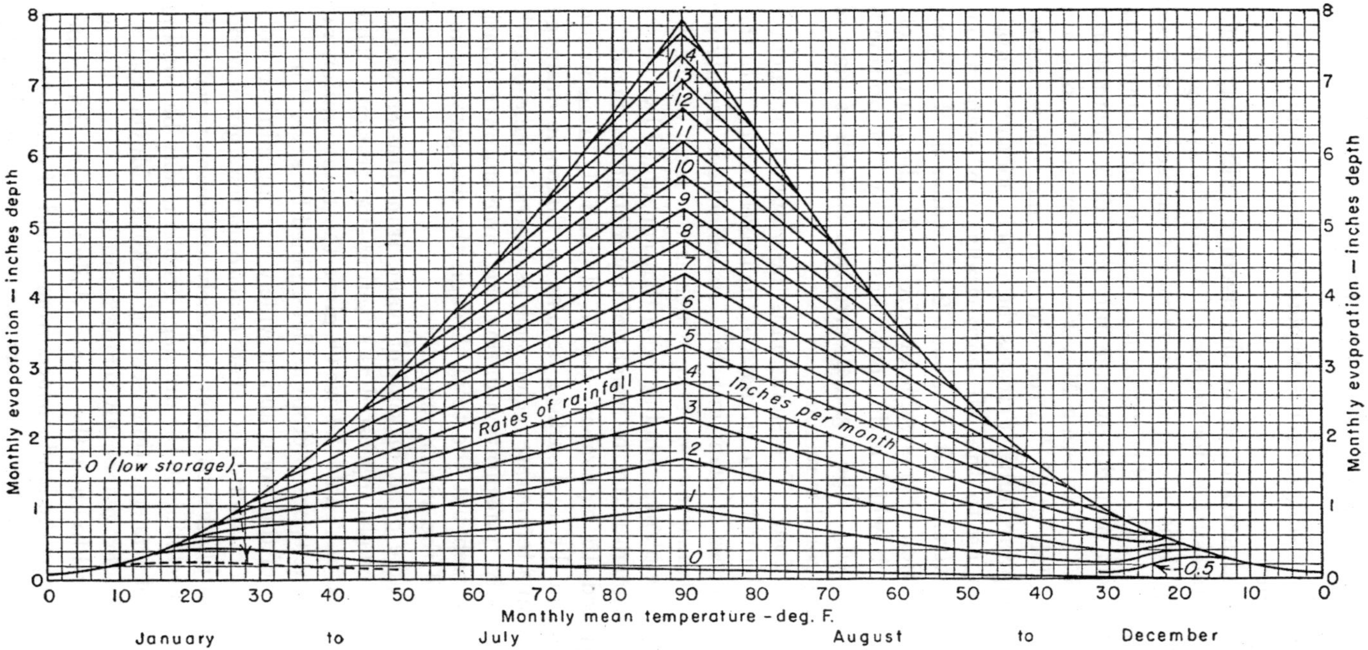

FIG. 4.29. Evaporation from land areas for various temperatures and rates of rainfall. *(Reproduced by permission from "Elements of Hydrology" by Adolph F. Meyer, published by John Wiley & Sons, Inc.)*

The method of using Meyer's evaporation chart is illustrated in the following example:

Suppose it is desired to know the average daily evaporation from ordinary farm-land surface for the month of July for average temperature and rainfall conditions in central Minnesota, expressed in inches of depth over that surface. For this region the average July temperature is 68 degrees Fahrenheit, and the average July rainfall is about 4 inches (see Tables 4.2 and 4.3). Entering Meyer's chart on the temperature base for the month of July at 68 degrees Fahrenheit and following the vertical from this point upward until it crosses the 4-inch monthly-rainfall sloping line, thence running to the left horizontally, the evaporation axis is crossed on the extreme left at about 2.15, which is the approximate monthly evaporation. This value divided by 31, the number of days in July, gives 0.07 inch, which is the approximate daily evaporation from the land surface desired. The regional-correction coefficient may safely be assumed as unity for this region.

4.15. Transpiration. Transpiration is the natural movement of water through the root hairs to the stomata of the leaves and on into the atmosphere, in the process of plant growth. Its amount is usually expressed in terms of the transpiration ratio, which is the ratio of the weight of water used by the plant, including both that consumed by the plant and that passing through it, to the weight of dry matter produced.

When reduced to equivalent depth of rainfall, transpiration, like evaporation from the land surface, becomes negligible in drainage-design considerations in a temperate region such as central Minnesota where, for example, the daily transpiration during the three summer months averages about 0.07 inch but the total through the five growing months is about one-third of the annual rainfall.

Even in the same plant or kind of plant, transpiration varies widely with the conditions already listed as affecting evaporation. This fact is clearly illustrated in Table 4.4 which gives the transpiration ratio of a number of different types of crops as determined by several investigators in different parts of the world and in widely varying climatic conditions.

As in the case of evaporation, where exact data for a given region and a particular type of plant are lacking, a closely approximate value, from the standpoint of its relation to the drainage problem, may be obtained from the base curve of transpiration (see Fig. 4.30) by Meyer [83] when the time of year and average monthly temperature are known. For example, suppose it is desired to know the average daily transpiration, expressed in inches of depth over the land surface, from a field of corn or sweet clover for the month of August for average temperature conditions in central Minnesota. For this region the average temperature for August is 68 degrees Fahrenheit. Entering Meyer's transpiration chart on the temperature base for August to November at 68 degrees Fahrenheit and following vertically upward through this point until the transpiration curve is crossed and then from this intersection horizontally to the right, the transpiration axis is crossed at about 2.10 inches, which is the approximate monthly transpiration. This value divided by 31 gives about 0.07 inch as the average daily transpiration for that time of year at an average temperature of 68 degrees Fahrenheit.

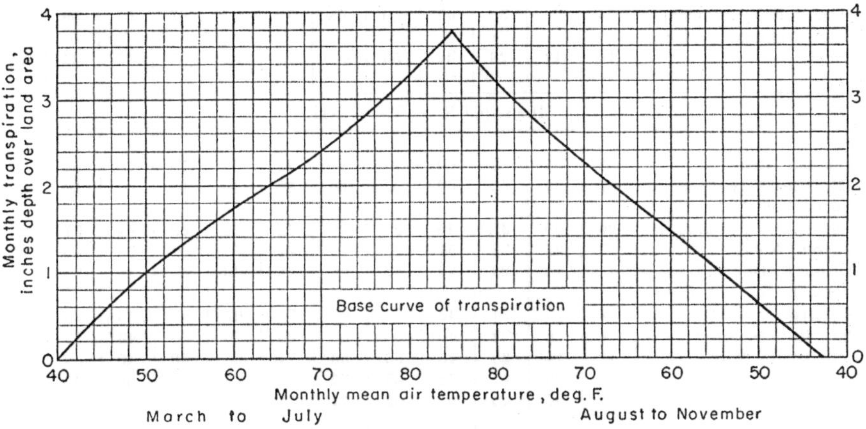

FIG. 4.30. Base curve of transpiration. (*Reproduced by permission from "Elements of Hydrology" by Adolph F. Meyer, published by John Wiley & Sons, Inc.*)

4.16. The Transpiration Ratio Reduced to Depth of Rainfall. In practice, especially in irrigation, it is necessary to reduce the seasonal water requirement in terms of the transpiration ratio to depth of water that must be applied at the surface either by rainfall or by artificial watering. This transition also involves a knowledge of the accepted weights of the commercial units of the cured crops, the moisture retained by the cured crop, and the dry-weight ratio of the tops, or marketable portion, of the crop to the roots, or nonmarketable portion. Table 4.5 gives approximate values of these three factors for a limited number of the common field crops. In order to make correct use of these factors in accordance with established methods in agricultural science, it should be remembered that moisture content of soils and evaporation and transpiration ratios are all computed on the basis of the dry weight of the soil and of the solid matter in the crops, respectively, whereas the moisture content in a cured crop is taken as the given per cent of the *total weight of the cured crop* for the given area.

To illustrate:

1. If the moisture content of a given soil is 20 per cent and the oven-dry weight of 1 cubic foot of that soil is 90 pounds, the total moisture in the soil in pounds per cubic foot is 0.20 × 90, or 18 pounds.

TABLE 4.2. AVERAGE MONTHLY TEMPERATURES, DEGREES FAHRENHEIT
(From the "Agricultural Atlas," Government Printing Office, Washington, D.C., 1936)

Station	Jan.	Feb.	Mar.	Apr.	May	June	July	Aug.	Sept.	Oct.	Nov.	Dec.
Atlantic Coast region												
Ft. Myers, Fla.	63	64	69	72	77	79	81	81	80	75	69	64
Norfolk, Va.	41	42	48	57	67	74	79	77	72	62	51	43
Trenton, N.J.	32	31	41	51	62	70	76	74	67	57	45	35
Albany, N.Y. (eastern)	24	24	33	47	59	68	72	70	63	51	39	28
Maine (approx. south central)	16	17	28	42	53	61	67	66	57	48	34	22
Mississippi River Valley region												
Houston, Tex.	53	55	63	69	75	81	83	83	79	70	61	54
Oklahoma City, Okla.	37	39	50	60	67	76	80	80	73	62	50	39
Scotts Bluff, Nebr. (western)	26	26	35	46	55	66	72	71	62	49	37	28
Lincoln, Nebr.	23	25	38	51	62	72	77	75	67	55	40	29
Huron, S. Dak.	11	14	29	46	56	66	72	70	61	48	31	19
Bismark, N. Dak. (south central)	7	10	24	43	54	64	70	67	58	45	28	15
Park Rapids, Minn.	4	6	22	40	52	63	68	65	56	43	27	11
Collegeville, Minn.	12	14	28	45	56	66	71	68	60	48	32	18
Springfield, Ill.	27	29	41	53	63	72	77	75	67	56	42	32
Columbus, Ohio	29	30	40	51	62	71	75	73	67	55	42	32
Rocky Mountain region												
Phoenix, Ariz.	52	56	60	67	74	85	89	88	82	70	60	51
Roswell, N. Mex.	40	43	51	58	67	75	78	76	70	58	47	39
Fort Collins, Colo.	26	27	36	45	54	63	68	68	59	48	36	27
Pocatello, Idaho	27	29	38	46	54	63	71	69	60	48	38	27
Helena, Mont.	20	23	32	44	51	60	67	66	56	45	33	25
Pacific Coast region												
San Diego, Calif.	54	55	57	59	61	65	68	70	68	64	60	56
San Francisco, Calif.	50	52	54	55	57	59	58	59	61	60	56	51
Eugene, Oreg.	40	43	46	50	55	60	66	65	58	53	43	41
Waterville, Wash. (central)	21	26	35	45	53	60	67	67	57	46	33	24
Centralia, Wash.	38	41	45	49	54	59	64	64	58	52	45	40
Gulf Coast region												
Montgomery, Ala.	41	51	58	65	73	80	82	81	76	66	56	49
Macon, Ga.	47	49	57	64	72	79	81	80	75	65	54	47
Louisville, Miss.	46	47	56	63	70	78	79	80	75	65	54	47
Alexandria, La.	40	51	60	67	74	80	82	82	77	66	57	50

TABLE 4.3. AVERAGE MONTHLY RAINFALL, INCHES
(From the "Agricultural Atlas," Government Printing Office, Washington, D.C., 1936, except as shown in footnote)

Station	Jan.	Feb.	Mar.	Apr.	May	June	July	Aug.	Sept.	Oct.	Nov.	Dec.
Atlantic Coast region												
Ft. Myers, Fla.	2.0	2.3	1.8	2.1	4.5	9.0	8.5	8.5	6.8	3.8	1.5	1.7
Norfolk, Va.	2.4	4.2	3.6	3.1	3.5	4.0	5.3	5.0	3.8	2.8	2.4	3.1
Trenton, N.J.*	3.3	3.0	3.5	3.4	3.3	3.7	4.2	5.3	3.8	3.0	2.6	3.2
Albany, N.Y. (eastern)	2.0	2.3	2.6	2.6	2.7	3.2	3.3	3.2	2.7	2.6	2.2	1.8
Orono, Me. (south central)	3.6	3.1	3.7	3.1	3.2	3.2	3.2	3.1	3.7	3.7	2.2	2.2
Mississippi River Valley region												
Houston, Tex.	2.8	3.3	3.4	3.7	4.5	3.9	4.0	3.9	4.0	3.9	3.7	4.5
Oklahoma City, Okla.	1.0	1.2	1.9	3.3	5.2	4.0	3.0	2.7	2.5	2.2	2.1	1.7
Lincoln, Nebr.*	0.6	0.9	1.2	2.2	3.6	4.0	3.6	3.7	3.4	1.9	1.4	0.8
Scotts Bluff, Nebr. (western)	0.5	0.7	0.8	1.6	2.8	2.7	2.1	1.4	1.2	0.8	0.6	0.6
Huron, S. Dak.	0.4	0.5	1.0	2.0	3.1	3.6	2.8	2.4	1.8	1.4	0.7	0.6
Bismarck, N. Dak. (south central)	0.4	0.6	0.8	1.2	2.5	3.3	2.3	1.8	1.5	1.1	0.7	0.6
Park Rapids, Minn.	0.7	0.6	2.0	2.3	3.2	4.2	4.0	4.0	3.1	2.0	1.2	0.7
Collegeville, Minn.†	0.7	1.1	1.6	2.2	2.2	3.1	2.4	3.5	1.4	1.0	0.6	0.7
Springfield, Ill.	2.0	1.8	3.1	3.1	3.8	3.6	4.0	3.3	3.8	3.5	2.3	1.9
Columbis, Ohio‡	2.9	2.1	3.4	3.6	3.0	4.0	4.3	3.0	3.0	2.4	2.3	2.6
Rocky Mountain region												
Phoenix, Ariz.‡	0.8	0.9	0.7	0.5	1.0	0.0	1.5	1.0	0.8	0.4	0.7	0.9
Roswell, N. Mex.	0.4	0.4	0.3	0.8	1.2	2.0	2.3	1.8	1.8	1.4	1.1	0.6
Ft. Collins, Colo.	0.4	0.6	1.0	2.4	2.6	1.8	2.0	1.4	1.4	1.0	0.4	0.3
American Falls, Idaho	1.5	1.3	1.4	1.4	1.7	1.2	0.7	0.6	0.7	1.2	1.3	1.2
Helena, Mont.*	0.7	0.6	0.8	1.0	2.1	2.1	1.1	0.9	1.2	0.9	0.6	0.7
Pacific Coast region												
Riverside, Calif.	2.4	1.9	2.0	0.8	0.3	0.1	0.1	0.1	0.2	0.6	0.8	1.4
San Francisco, Calif.	4.8	3.5	3.3	1.3	0.7	0.3	0.2	0.3	0.8	1.1	2.2	1.7
Eugene, Oreg.*	5.6	4.8	4.0	2.7	2.3	1.5	0.4	0.5	1.9	2.8	5.8	5.8
Waterville, Wash.‡	1.5	1.1	0.8	0.7	0.5	1.0	0.5	0.5	0.7	0.8	1.4	1.6
Centralia, Wash.*	6.6	5.0	4.8	2.9	2.4	1.9	0.6	1.0	2.2	3.7	7.2	7.5
Gulf Coast region												
Montgomery, Ala.	4.4	5.6	5.0	4.8	3.6	3.6	4.3	3.2	2.7	2.2	3.0	4.3
Macon, Ga.‡	3.8	4.5	3.7	3.5	3.0	4.0	5.0	4.8	3.0	2.5	2.4	3.7
Louisville, Miss.‡	5.0	5.2	6.1	5.0	4.1	4.0	4.5	4.2	3.0	2.6	3.3	5.5
Alexandria, La.‡	5.1	5.1	5.5	5.0	4.5	4.5	5.5	4.0	4.0	3.4	4.3	6.4

* From U.S. Department of Agriculture yearbook for 1941, "Climate and Man."
† Average of actual records of the Weather Bureau, 1931 to 1940.
‡ Approximations only, from monthly isohyetal maps of the "Agricultural Atlas."

TABLE 4.4. WATER REQUIREMENTS OF CROPS*
(Transpiration ratio as determined by various early investigators)

Kind of crop	Lawes and Gilbert, England	Leather, India	Briggs and Shantz, Colorado	King, Wisconsin	Widtsoe, Utah	Thom and Holtz,† Washington state	Hellriegel, Germany	Von Sellhorst, Germany	Wollny, Germany
Hay and forage crops									
Alfalfa	...	...	1,068	...	...	446			
Clover (red)	251	...	.....	577	...	...	330		
Clover (sweet)	...	...	709	...	...	484			
Millet	...	...	275	...	...	...	...	...	447
Rape	...	...	441	...	...	...	337	...	912
Grain crops									
Barley	258	468	539	464	...	320	297	365	774
Buckwheat	...	...	578	...	...	...	...	371	646
Corn	...	337	369	271	386	249	...	...	233
Flax	...	807							
Oats	...	469	614	504	...	352	401	...	665
Rice	...	811							
Rye	...	...	724	...	...	...	377	386	
Wheat	235	554	507	...	546	432	359	333	
Root crops									
Onions	...	...	.....	...	...	235			
Potatoes	...	...	448	385	...	167	...	281	
Sugar beets	...	...	377	...	497	262			
Miscellaneous crops									
Beans	214								
Chick-peas	...	818							
Horse beans	...	...	.....	...	...	...	263		
Lupine	...	...	.....	...	...	...	373		
Mustard	...	496	.....	...	...	...	...	...	843
Peas	235	563	800	477	843	420	292	...	416
Sorghum	...	437	306	...	...	240			
Sugar cane	...	212							
Sunflowers	...	...	.....	...	...	...	...	...	490

* German results from "The Water Requirements of Plants," *USDA Bur. Plant Ind. Bul.* 285, November, 1913, by Briggs and Shantz.

† Thom and Holt from "Irrigation Principles and Practices," 1932, by Orson W. Israelsen, reprinted by John Wiley & Sons, Inc., published by permission. All other data after investigator quoted.

2. If the transpiration ratio of alfalfa is 700, for every oven-dry pound of solid matter produced 700 pounds of water is transpired by the crop in the process of its growth.

But

3. If the moisture content of the cured alfalfa is 12 per cent, this is interpreted to mean that 12 per cent of the total cured weight is water; hence, for 1 ton (2,000 pounds) of cured alfalfa, the water content in pounds is 0.12 × 2,000, or 240.

TABLE 4.5. ESSENTIAL ITEMS IN RELATION TO WATER REQUIREMENTS OF CROPS

Type of crop	Standard or accepted weight, lb		Approximate per cent of water in cured crops	Approximate dry-matter ratio of tops to roots
	Per bushel	Per ton		
Alfalfa	..	2,000	10	15
Barley	47	.....	9	14
Buckwheat	50	.....	12	
Clover (red)	..	2,000	13	
Clover (sweet)	..	2,000	9	
Corn	54	.....	12	15
Oats	32	.....	9	11
Potatoes	60	.....	79	
Rye	56	.....	9	
Sorghum	..	2,000	12	15
Sugar beets	..	2,000	84	
Wheat	60	.....	14	11

4.17. Deep Seepage. In most soils of humid regions moisture disposal by deep seepage is not particularly active except when sandy strata or laminated or seamed rock formations closely underlie the surface soil. On this account deep seepage is usually negligible in soil-moisture considerations in the humid regions. Percolation to tile drains is not classified as deep seepage. In the irrigated regions, where the soils are apt to be much coarser and deeper than in the humid regions, deep seepage is sufficiently active so that it frequently becomes a serious factor in soil-moisture considerations. Measurement of deep seepage is difficult and can usually be accomplished with acceptable accuracy only by taking from the total of available water the sum of evaporation, transpiration, runoff, and the residue of impounded water, this difference being that removed by deep seepage.

Runoff must be reckoned with under two main types, surface runoff and subsurface runoff. Either type is a leading element in drainage considerations, especially in design. Proper design of open channels is determined largely by surface-runoff considerations.

Factors Influencing Runoff

The leading factors influencing rate and volume of surface runoff are size, geology, shape, and topographic character (including slope) of the watershed; character of the rainfall; extent and character of the vegetal cover; character and condition of the surface soil; general drainage conditions over the watershed; and seasonal and climatic factors. These factors also affect the time of concentration.

The factors affecting subsurface runoff are the same as those just listed for surface runoff and, in addition, include the character of the soil and subsoil and the extent and effectiveness of the subdrainage.

4.18. Time of Concentration. The time of concentration is the time required for a drop of water from a given rainstorm, falling upon the portion of the watershed farthest from the outlet and moving along the course which it would naturally follow, to reach the outlet. It is all-important because it determines the duration of the rainstorm that will produce maximum runoff. It is especially applicable on small watersheds.

4.19. Size of Watershed. Other conditions remaining the same, the larger the watershed, the longer will be the time required for the total runoff from any given storm or storm period to disappear and, especially on long narrow watersheds, the longer the time of concentration will tend to be. In fact, so influential is the matter of the size of watershed, both from an engineering and an economic standpoint, that it has been found necessary, in dealing with runoff and flood control, to consider watersheds in three separate groups according to size, namely: (1) *large watersheds*, such as the entire drainage basin of a great river like the Mississippi, the Columbia, or the Colorado, embracing hundreds or even thousands of square miles and intimately affecting not only agriculture but also practically all other types of human endeavor; (2) *medium watersheds* such as the drainage basins of relatively small river branches or independent small streams, embracing 1 or 2 up to 50 or 100 square miles, and usually not so closely and widely associated with the varying types of human activity, but quite closely associated with agriculture and the land-drainage problem; and (3) *small watersheds* among which are prominently included small agricultural drainage areas embracing a few up to 1,000 acres.

Large areas will usually have a much longer total runoff period than will small ones because the water must travel much farther to get to the outlet channels, there is much more of it, and the time of concentration will tend to be much longer. On this account, as will appear in later sections, time of concentration, which is an important consideration in dealing with runoff from small areas, is practically eliminated, as a distinct factor, from formulas for runoff from large areas, it being negligible in comparison with the factors of size and character of watershed.

4.20. Geology of the Watershed. In rough, craggy areas or where the basic rock is loosely laminated, largely permeable as in lava formations, or broken by excessive fissures and cracks, the flood runoff is apt to be much slowed down and reduced by deep seepage over that from areas of opposite geologic character.

4.21. Topographic Character of the Watershed. The shape of watershed also influences the runoff. A long, narrow tract with the ditch on the narrow side will have a smaller rate of runoff than would the same tract with the ditch on the long side.

Topography has a great deal to do with the rate of runoff, because it determines, to a large extent, the length of time required for water to concentrate or pile up at a given point. In very hilly land having mixed sharp slopes or where the general slope is appreciable, the runoff is much greater and more rapid than in flat lands.

4.22. Character of Rainfall. Both the rate and total amount of runoff tend to increase with the increase in the intensity of rainfall and with the extent of the area actually covered by the storm. Maximum discharge occurs at the instant when water from all parts of the watershed first reaches the outlet, and this means that the "critical rain" (from an economic design standpoint) is one which continues for a length of time equal to the time of concentration. The more intense the rate of rainfall, the shorter the time it lasts; hence, for a watershed of a given size and shape, the more the time of concentration can be prolonged, the lower the rate of runoff which must be dealt with from a given storm.

4.23. Extent and Character of the Vegetal Cover. The amount and kind of vegetal cover on the area under consideration have a direct bearing on the runoff. Lands covered with rank vegetation detain and absorb more water than those having little or none.

It is a general belief that forested lands with a heavy litter have a greater influence in retarding runoff than do any other types of vegetation, but, as pointed out by Pickels [104], there is a wide difference of opinion on this point among foresters, engineers, and soil scientists. Pickels cites quite as eminent authority against the effectiveness of forests in retarding of runoff as for it, both in this country and abroad. This seems to indicate that there is no hard and fast rule and that the extent of the influence of forests and forest litter is widely variable, depending on local conditions of topography, geology, precipitation, and snow blanket.

4.24. Character and Condition of the Surface Soil. The kind of soil and the condition that it is in have a direct bearing on the amount of runoff. A sandy soil will have a higher rate of infiltration before there is any runoff than will clay soil; while in the case of rock or a very rocky soil the surface will be more impervious, and the runoff will generally be greater. The condition of the soil surface also has much influence on the amount of runoff. If the surface of the soil is rough, there will not be so much runoff as if it were smooth. Farmers often take advantage of this fact by fall plowing to leave their fields more or less rough in order that the soil may check the runoff and absorb as much as possible of the winter rains. Fertile surface soils containing considerable humus have high infiltration capacity; hence they usually do not allow the water to run off so rapidly or to the same extent as do the less fertile soils. Furthermore, if the ground is saturated at the time a given storm begins, infiltration will be checked and the runoff from a storm will be much greater than it would have been if the soil had been dry.

In connection with this and the preceding section the reader should review the discussion of *permeability* and *infiltration* in the preceding chapter.

4.25. Existing Drainage Conditions. Runoff from heavy precipitation will be more continuous through deep channels of ample capacity than through shallow ones that slow runoff in times of heavy flood by reason of widely overflowing flat lands adjacent thereto. The more ample the provision for artificial drainage, the more evenly distributed and the more uniform will be the runoff. This is especially true regarding tile drains, whose tendency is to distribute the runoff more uniformly over a wide area, thus reducing flood peaks. This fact is well established by the investigations of Woodward and Nagler [162] conducted over the extensive tile-drained area of northern Iowa.

The type of outlet into which the runoff goes from the various areas also influences the rate of that runoff. If the outlets are unimpeded and of ample capacity, the water gets away more quickly than if they are poor.

Storage basins so located that they receive a considerable amount of water which would otherwise be discharged into open channels or onto low-lying agricultural lands tend to reduce the amount of runoff through drainage channels. The chief use of storage basins, however, is in flood prevention and in irrigation.

4.26. Seasonal and Climatic Factors. Especially in cold climates the seasons affect the amount of evaporation, transpiration, vegetal cover, snow, and ice, and soil influences vary to some extent with climate and season.

Southern watersheds and those lying in the more frequent storm paths, especially where the direction of flow is practically the same as that of the storm movement rather than across the storm path, are generally subject to more severe and protracted rainstorms and consequently greater flood runoff than in geographical locations where the opposite conditions prevail.

Frequency of Recurrence of Intense Rainstorms and Resultant Floods

Rainstorms producing floods of given magnitudes and consequently their resulting peak floods tend to occur, roughly according to intensity, with certain fairly consistent frequencies. Theoretically, a rainstorm of such intensity as to produce the greatest peak flood ever known in a given watershed may occur on any size of watershed, but such a storm must cover the entire watershed during its period of intensity. It is a well-established fact in hydrologic science that intense rainstorms of limited duration and covering limited areas, with their resultant peak floods, recur with much greater frequency than do the larger storms causing larger and more widespread floods. Consequently in flood-control activities it has been developed as effective practice to group peak-flood recurrences together with the three size groups of watersheds earlier discussed: for example, peak floods of a given degree of magnitude, occurring once in 25, 50, 100, 500 years, etc., are classed with the *large* watersheds; those occurring with frequencies of once in 10, 15, 25, 50, and possibly up to 100 years are more closely associated with the *medium*-sized watersheds; those with frequencies of once in a fraction of a year, 1, 2, 5, 10, 25 or even 50 years tend to occur in greater numbers in the *small* watersheds.

4.27. Selection of Frequency for a Given Case. When human life is not imperiled, it is generally considered economically impractical to provide for runoff from the most intense rainfall on record in any given watershed, for it may easily be more profitable to take a reasonable chance of loss from damage occurring only once over a long period of years rather than to make the heavy investment in control works required to provide for infrequent storms of extreme intensity. Hence the first step in planning runoff control is to decide on a calculated risk or the flood frequency to be provided against. For example, in agriculture, where the affected area in most cases falls in the *small* watershed, it is common practice to use a frequency of 10 years, in the more important cases 25 years, and in rare cases 50 years. The relation of flood frequency to storm intensity and time of concentration is discussed in a later section under Estimating Runoff.

4.28. Regional Selection of Frequency. The active factors which should normally influence selection of frequencies are related to the type of project and tend to vary somewhat according to the region. Hence it is wise for regional design agencies to set up, for design purposes, a group of such factors considered to fit the regional conditions. Illustrative of this procedure is the following table of frequency determining factors from the Soil Conservation Service "Engineering Handbook for the Pacific Region" [142].

Frequency to be used in design is controlled by such considerations as loss of life and damage to valuable property in event of failure; initial cost; utility; and permanency of construction. Suggested frequencies are as follows:

1.	Temporary gully structures, brush, burlap, pipe and wire	5 years
2.	Semi-permanent structures, rock-wire, timber	10 years
3.	Small permanent structures, masonry or concrete	15–25 years
4.	Large permanent structures, masonry or concrete	50–100 years
5.	Terraces or diversions on cultivated fields	10 years
6.	Terrace outlet structures	10 years
7.	Earth dams	50–100 years
8.	Shallow stock ponds	10–25 years
9.	Revetments	25–50 years
10.	Channels, depending on potential damage as balanced against cost.	

Measurement of Runoff

The problem of runoff is so to control floods in any given watershed as to prevent them, as far as possible, from doing damage. The first essential step in such control in each case is to measure the volume of runoff that must be taken care of. Where the waters have been or can be led into a definite channel whose physical characteristics are determinable, direct measurement of the runoff through the stream is relatively easy by standard mathematical or mechanical means. These means include weirs, the Parshall flume, the current meter, and the standard formulas (such as the Chézy-Kutter and Manning formulas) for flow of water in open channels, all of which are discussed in Chap. 5. It is not always possible, however, to lead the runoff from rainfall on a given watershed into convenient nearby streams for measurement by any of the mathematical or mechanical methods just listed. Hence, in many cases, some wholly empirical method of estimating the runoff must be resorted to as discussed in the following section.

Estimating Runoff

Through many years study of the physical characteristics of watersheds from which it has been possible accurately to measure the runoff under various rainfall conditions, engineers have been able to evolve empirical formulas for estimating runoff from any watershed with practical accuracy. It has been found necessary, however, for reliable results to deal with watersheds in the three separate size groups earlier described.

4.29. Large Watersheds. In modern times the solution of the runoff and flood problem on *large* watersheds is usually accomplished through the development of hydrographs and comparison with other known cases of fairly similar character. This brings them properly into the field of the science of hydrology which is a field in its own right, the discussion of which is mainly inappropriate in a book of this character, although a brief discussion of the application of the unit hydrograph to runoff problems is given in Appendix A.

4.30. Medium Watersheds. For medium watersheds it has been found possible to develop reliable empirical formulas of an engineering character. On such watersheds peak floods usually result either from long-continued rainstorms of moderate intensity lasting several days, or from shorter successive storm periods separated by time intervals so small as to preclude any marked change in surface or attendant conditions within any such single interval. In such a case, intensity of precipitation has little effect because such a rainfall usually covers only a small portion of the watershed at any one time; whereas the runoff may continue for several hours, days, or even longer after the given rainstorm. Therefore, the principal factors affecting runoff from the medium watershed are topography, character and extent of vegetal cover, and normal infiltration rate of the soil. In most of the accepted runoff formulas, therefore, rate of runoff Q is expressed as some function of the area in square miles and an arbitrary factor representing watershed characteristics.

With few exceptions, each of the formulas for runoff from medium watersheds has been developed to fit the needs of a particular situation and type of watershed and is, therefore, of limited application. As the problem of floods on medium watersheds frequently concerns industry, commerce, transportation, and civic situations quite as much as it does agriculture, an extended discussion of these many formulas is inappropriate in this book which is concerned, chiefly, with agricultural areas. Therefore, further discussion of runoff from medium watersheds will be based on the formula by Meyer [83] which, although designed especially for Minnesota conditions, seems generally adaptable elsewhere without need for extensive and detailed field determinations. This formula follows:

$$Q = CC_1A^{0.6} \tag{5}$$

where Q is the maximum flood flow from the entire watershed, cfs
A is the area of the watershed, sq. miles
C is a coefficient embracing watershed characteristics
C_1 is a frequency coefficient

The values suggested by Meyer for the coefficients C and C_1 are given in Tables 4.6 and 4.7.

In view of the stated limited purpose of this formula, discretion should be exercised in its use.

TABLE 4.6. RUNOFF COEFFICIENTS C (FOR MEYER'S FORMULA)*

Character of drainage basin	Coefficients		
	Sandy soil	Loam	Clayey soil
1. Very flat agricultural or timber land with some marshes and swamps	.35	.40	.50
2. Relatively flat agricultural or timber land with some marshes and ponds	.45	.50	.60
3. Gently rolling agricultural or timber land full of lakes, ponds, and marshes connected by poorly defined water courses	.50	.60	.75
4. Relatively flat agricultural or timber land of fairly uniform slope, without lakes and ponds	.60	.70	.85
5. Slightly undulating agricultural or timber land without lakes or ponds; or distinctly rolling to hilly agricultural or timber land, with lakes and ponds	.70	.80	1.00
6. Gently rolling agricultural or timber land without lakes and ponds	.85	1.00	1.25
7. Distinctly rolling to hilly agricultural or timber land without lakes and ponds; or hilly agricultural or timber lands with steep slopes and lakes, ponds, and marshes in valleys	1.10	1.50	2.00
8. Hilly agricultural or timber land with steep slopes barely admitting of cultivation; without lakes, ponds, or marshes	2.25	3.00	4.00
9. Very hilly timber or brush-covered land, slopes too steep for cultivation; ravines and gullies with occasional small ponds or marshes	3.50	4.50	6.00
10. Very hilly timber or brush-covered land with some rock outcropping; ravines and gullies, and occasional small ponds or marshes	5.00	6.00	8.00
11. Very hilly to rugged country with much rock outcropping; scattered timber; occasional small ponds and marshes	9.00	10.00	12.00
12. Rugged to precipitous rocky country with practically no soil cover; small timber and brush; ravines and gullies; no lakes, ponds or marshes to retard runoff	....	15.00	

NOTE: The available information on the subject does not indicate that forests have any material effect upon the extreme flood flow. They have a tendency to reduce the ordinary flood flow somewhat. In northern Minnesota forests have aggravated spring floods by retarding the melting of snow until warm April rains set in.

In determining the proper coefficient to use for a given drainage basin of diverse characteristics, subdivide the basin into approximately similar areas, select the proper coefficient for each subdivision, and compute the average coefficient applicable to the entire area.

In general, it may be stated that lakes, ponds and marshes have a most pronounced retarding effect upon the runoff, resulting in low coefficients. A region may be hilly—even rocky—but between the hills and rock outcrops there may be lakes, ponds, marshes, bogs or muskegs, which greatly retard the flow of water to the main stream, even though the rainfall rapidly runs off from the hills into the valleys where the lakes, ponds and marshes lie. Drainage basins having a relatively uniform slope in one direction, particularly if gullied, would require the use of large coefficients.

* Reprinted by permission from "Elements of Hydrology" by Adolph F. Meyer, published by John Wiley & Sons, Inc.

TABLE 4.7. FREQUENCY COEFFICIENTS C_1 (FOR MEYER'S FORMULA)

For a Flood of Magnitude to Be Expected	*Coefficient*
Once in 10 years	0.85
Once in 25 years	1.00
Once in 100 years	1.40

4.31. Small Agricultural Watersheds. On small watersheds, single intense rainstorms or rainstorms of several days duration at low intensities often cover the entire watershed; hence the time element enters any suitable formula for maximum rate of runoff through the intensity of rainfall I. On this account, formulas for runoff from large or moderately large watersheds have been found by engineers unsatisfactory for small watersheds. Since the difference in character between the small watershed and the larger types has been recognized, few formulas for runoff from the former have been developed.

The Rational Formula. Following a report, in 1929, of the Committee on Runoff from Agricultural Lands, of the American Society of Agricultural Engineers, by Ramser, Norton, and Ellison [113], engineers generally adopted for estimating storm-water runoff from agricultural areas what is known as the *rational formula:*

$$Q = CIA \tag{6}$$

where Q = maximum rate of runoff, cfs

C = runoff coefficient (partially a coefficient of imperviousness) representing the ratio of the rate of runoff to the rate of rainfall, dependent on the following conditions and factors, which also affect the time of concentration of runoff from the watershed, mentioned below, under I: (1) the characteristics of the watershed, embracing topography, shape and vegetative cover; (2) length, contour, condition, and fall of the watercourse concerned; (3) the active opportunities for surface storage of flood water; (4) character of the surface soil, its degree of saturation and consequent normal infiltration rate

I = rainfall intensity in cubic feet per second per acre, which unit happens to be approximately identical with the rate of rainfall in inches per hour during the period of intensity. This period is often about equal to the time of concentration of runoff for the watershed, which has already been defined as the time required for a drop of water from a given rainstorm, falling upon the portion of the watershed farthest from the outlet, moving along the course it would normally follow, to reach the outlet

A = watershed area. acres

This equation was proposed by C. E. Ramser [112] as a result of measurements of rates of runoff and rainfall from small agricultural areas. The degree of accuracy with which design runoff can be computed from the rational formula is, of course, no greater than that of the least accurate element, but even so it is far superior to empirical formulas which are inflexible and make no adequate provision for the many variables involved.

In the present state of knowledge the elements most difficult to evaluate are proper values for the coefficient C and the time of concentration which determines the rainfall rate to be selected for a particular locality. The watershed area can be taken from a map or measured in the field.

Ramser suggests control values for C as given in Table 4.8.

TABLE 4.8. VALUES OF C FOR USE IN FORMULA $Q = CIA$

Kind of watershed	Slope, per cent	Values of C
Cultivated, rolling	5–10	0.60
Cultivated, hilly	10–30	0.72
Pasture, rolling	5–10	0.36
Pasture, hilly	10–30	0.42
Timber, rolling	5–10	0.18
Timber, hilly	10–30	0.21

These values are a composite measure of all the variable factors affecting runoff. They should be considered as somewhere near the upper average limit and as applying to relatively impervious soils. To be on the safe side they are recommended for use unless a particular soil is definitely known to be exceptionally absorptive.

Accuracy of the time of concentration is dependent upon the accuracy with which the route of travel and velocity of flow can be judged.

This is not such a difficult task as it may at first appear, since the line of travel is usually in a more or less well-defined channel or channels wherein distance of travel is the chief variable as between watersheds subject to erosion. This being true, and if the shape is not abnormal, times of concentration become almost direct functions of watershed sizes. At least the values given in Table 4.9 may be used as a general guide.

As will be noted, the figures given represent minimum times of concentration for channel grades approximating 5 feet fall per 100 feet. Since soil and cover variables, and to some extent prevailing watershed slopes, are taken care of in the values assigned to the runoff coefficient C, the interval representing the time of concentration is useful chiefly in selecting the proper rainfall rate for a particular set of conditions. Any increase that can be induced in the time of concentration will, of course, not only allow greater opportunity for absorption but will automatically reduce the rate of rainfall which results in maximum design runoff and hence will also reduce the rate of runoff.

TABLE 4.9. TIME OF CONCENTRATION AS AFFECTED BY SIZE OF WATERSHED

Size of watershed, acres	Time of minimum concentration, min	Size of watershed, acres	Time of minimum concentration, min
1	1.4	300	29
3	3.0	400	35
5	3.5	500	41
10	4.0	600	47
20	4.8	700	53
30	8.0	800	60
50	12.0	900	67
100	17.0	1,000	75
200	23.0		

4.32. Illustration of the Use of the Rational Formula. To illustrate the use of the rational formula on unterraced land [9] let it be assumed that the maximum rate of runoff, not likely to be exceeded more than once in 10 years, is wanted for an area of 20.7 acres of hilly agricultural land near Jackson, Tenn. A fairly well defined channel leads from the upper reaches of the watershed to the point of exit, a distance by channel of 1,200 feet, at an average rate of fall of 4.89 feet per 100 feet. The velocity of the water flowing in the channel is estimated, by trial with Manning's formula (see next chapter), at 245 feet per minute, which makes the time of concentration 5 minutes (practically the same as the value given in Table 4.9 for a watershed of 20 acres).

Reference to Fig. 4.4 shows that 0.55 inch of rainfall is to be expected in central Tennessee during a period of 5 minutes on an average of once in 10 years. This is at a rate of 6.6 inches per hour.

The soil on the tract is described as Lexington silt loam and is in a badly eroded state because of continuous cultivation in cotton for many years without special protection. The subsoil is quite impervious. From a reconnaissance survey the tract has been classified as 10 acres or 48 per cent of hilly, cultivated land, 7.7 acres or 38 per cent of hilly pasture, and 3 acres or 14 per cent of hilly timber. Multiplying these percentages by the value of C for their respective conditions, as shown in Ramser's tabulation, Table 4.8 preceding, and adding gives 0.54 as the value of the runoff coefficient.

Substituting in $Q = CIA$ gives

$$Q = (0.54)(6.6)(20.7)$$

or

$$Q = 73.6 \text{ cfs}$$

which is the design rate of runoff for the conditions assumed.

For more rapid and convenient computation than that offered by reference to the rainfall intensity-frequency maps, Ramser [112] presents Fig. 4.31 from which, after the time of concentration has been determined, the value of I may be read directly as follows: Follow up the 10-year frequency ordinate to its intersection with the 5-minute duration curve, thence to the left, horizontally to the intersection with the rainfall-intensity axis which is at 6.1, the required value of I. This makes Q equal 68.2 cubic feet per second, a variation of 7.3 per cent from the result

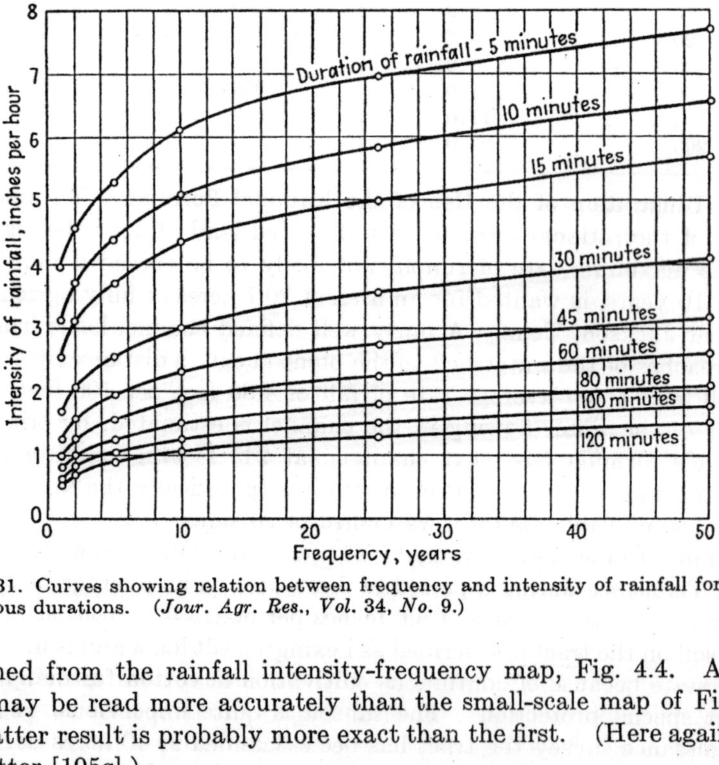

Fig. 4.31. Curves showing relation between frequency and intensity of rainfall for storms of various durations. (*Jour. Agr. Res., Vol.* 34, *No.* 9.)

obtained from the rainfall intensity-frequency map, Fig. 4.4. As Fig. 4.31 may be read more accurately than the small-scale map of Fig. 4.4, this latter result is probably more exact than the first. (Here again refer to Potter [105*a*].)

4.33. Variation in Values of *C* in Different Regions. The values of C thus far given in this discussion of the rational formula were worked out particularly for the Mississippi River Basin. They will not necessarily be the same for other regions where rainfall, topographic, soil, and climatic conditions are different; hence, for such other regions, such as, for example, the Pacific slope, another set of values of C to fit the established area classifications must eventually be worked out in each case. This had not been fully accomplished at the time this book was written, but it is in process in various regions throughout the country. The existing lack of regional coefficients, however, in no way affects the general validity of the formula.

4.34. The Soil Conservation Service Method of Region III. The value of C in the rational formula in any given case rests largely on the judgment of the engineer. Two men seldom agree on it. Hence, in their

TABLE 4.10.* WATERSHED CHARACTERISTICS FOR DETERMINING RUNOFF COEFFICIENT

Designation of watershed characteristics	Runoff-producing characteristics			
	100 extreme	75 high	50 normal	25 low
Relief	(40) Steep, rugged terrain, with average slopes generally above 30 %	(30) Hilly, with average slopes of 10 to 30 %	(20) Rolling, with average slopes of 5 to 10 %	(10) Relatively flat land, with average slopes of 0 to 5 %
Soil infiltration	(20) No effective soil cover, either rock or thin soil mantle of negligible infiltration capacity	(15) Slow to take up water; clay or other soil of low infiltration capacity, such as heavy gumbo	(10) Normal; deep loam with infiltration about equal to that of typical prairie soil	(5) High; deep sand or other soil that takes up water readily and rapidly
Vegetal cover	(20) No effective plant cover; bare or very sparse cover	(15) Poor to fair; clean-cultivated crops or poor natural cover; less than 10 % of drainage area under good cover	(10) Fair to good; about 50 % of drainage area in good grassland, woodland, or equivalent cover; not more than 50 % of area in clean-cultivated crops	(5) Good to excellent; about 90 % of drainage area in good grassland, woodland, or equivalent cover
Surface storage	(20) Negligible; surface depressions few and shallow; drainage-ways steep and small; no ponds or marshes	(15) Low; well-defined system of small drainage-ways; no ponds or marshes	(10) Normal; considerable surface-depression storage; drainage system similar to that of typical prairie lands; lakes, ponds and marshes less than 2 % of drainage area	(5) High; surface-depression storage high; drainage system not sharply defined; large flood-plain storage or a large number of lakes, ponds or marshes

* Courtesy of the Soil Conservation Service, Region III.

"Engineering Handbook for Farm Planners," the Soil Conservation Service, Region III [141], offers a method which bases the runoff coefficient on classification of four watershed characteristics readily determinable by field inspection. The method has grown out of unpublished studies by

TABLE 4.11.* RUNOFF CHART FOR DESIGN OF GRASS WATERWAYS BASED ON 10 YEARS' FREQUENCY AND RAINFALL FACTOR OF 1.0

Drainage area, acres	Watershed characteristics										
	30	35	40	45	50	55	60	65	70	75	85
	Cubic feet per second										
4	5	7	8	9	10	11	13	15	17	20	24
6	6	8	10	11	13	16	18	21	24	28	34
8	7	9	12	13	17	20	24	28	32	37	44
10	8	10	14	16	21	25	30	35	40	46	53
12	9	11	16	19	24	29	35	41	48	54	62
14	10	13	18	22	28	33	40	47	55	62	71
16	11	15	20	25	31	38	45	53	62	70	80
18	12	17	22	28	34	42	50	59	79	78	89
20	14	19	24	31	38	46	55	65	76	86	98
25	16	22	28	36	46	55	66	78	90	103	118
30	18	25	32	42	53	64	77	91	105	120	139
35	20	28	36	47	60	73	87	104	120	137	159
40	22	32	41	53	68	82	98	117	135	154	179
45	24	36	46	59	76	91	109	130	150	171	200
50	26	40	51	65	83	100	119	142	165	188	220
60	30	45	59	76	97	117	139	165	193	221	259
70	34	50	66	87	110	133	159	188	221	253	298
80	38	55	74	97	123	149	179	211	248	285	338
90	42	60	81	107	136	166	199	234	274	317	378
100	46	65	88	117	150	182	218	256	300	348	418
120	53	75	102	135	175	213	256	300	351	410	487
140	60	85	116	154	200	244	293	344	401	471	556
160	67	95	129	172	224	275	329	388	451	529	625
180	73	105	142	190	248	305	364	432	500	585	694
200	79	115	155	207	271	334	398	476	550	640	764
220	85	124	168	224	294	363	432	516	598	693	832
240	91	132	180	241	317	392	464	556	645	746	899
260	97	141	192	257	340	421	496	595	691	799	964
280	102	149	204	273	362	450	528	634	736	852	1,028
300	108	158	216	289	384	480	560	672	781	905	1,090
320	114	167	228	305	407	505	591	709	824	956	1,150
340	119	175	240	321	429	530	623	746	867	1,007	1,210
360	124	184	253	337	451	555	654	783	909	1,058	1,270
380	129	192	265	355	473	580	684	820	952	1,109	1,330
400	135	200	277	371	494	605	713	856	994	1,160	1,390
420	141	208	288	386	515	629	741	891	1,035	1,209	
440	146	216	299	401	535	653	769	926	1,076	1,257	
460	151	224	310	416	554	677	797	961	1,117	1,303	
480	156	232	321	432	572	700	825	996	1,158	1,348	
500	161	239	332	447	590	723	853	1,030	1,199	1,392	
520	166	246	342	462	607	746	881	1,064	1,238	1,435	
540	171	253	352	476	624	769	909	1,098	1,277	1,478	
560	176	260	362	490	641	791	937	1,131	1,315	1,521	
580	181	267	372	504	658	813	965	1,164	1,350	1,563	
600	185	274	382	518	675	835	993	1,197	1,386	1,605	

* Courtesy of the Soil Conservation Service, Region III.

TABLE 4.12.* RUNOFF CHART FOR DESIGN OF PERMANENT STRUCTURES BASED ON 50 YEARS' FREQUENCY AND RAINFALL FACTOR OF 1.0

Drainage area, acres	Watershed characteristics												
	25	30	35	40	45	50	55	60	65	70	75	85	100
	Cubic feet per second												
4	5	7	10	11	13	14	15	18	21	24	28	34	36
6	6	8	11	14	15	18	23	25	30	34	40	48	51
8	7	10	13	17	18	24	28	34	40	45	52	62	66
10	8	11	15	19	24	29	35	43	50	58	66	75	81
12	9	12	17	22	28	34	41	50	59	68	78	88	97
14	10	14	19	25	32	39	47	57	68	78	90	101	112
16	11	15	21	28	36	44	53	64	76	87	100	114	127
18	12	17	24	31	40	49	59	71	84	96	110	126	142
20	13	19	26	33	43	53	64	77	92	105	120	138	155
25	16	22	31	40	52	64	78	93	110	128	146	168	190
30	18	25	36	47	60	75	91	109	128	149	172	198	225
35	20	28	41	53	68	86	104	124	146	170	196	228	260
40	22	31	45	59	76	97	117	139	164	190	221	257	295
45	24	34	49	65	84	107	130	154	182	210	245	286	330
50	26	37	53	71	92	117	142	168	200	230	267	315	365
60	29	43	62	82	108	136	165	197	234	270	314	366	435
70	32	49	70	93	123	155	188	226	268	310	360	425	505
80	36	54	78	104	137	173	211	255	301	350	406	478	570
90	40	59	85	114	151	193	233	281	334	389	448	530	635
100	43	64	92	124	165	210	255	307	365	426	490	590	700
120	49	74	107	144	192	246	300	359	427	500	574	690	825
140	55	84	121	164	218	281	345	410	487	570	658	790	950
160	61	94	134	183	243	315	390	460	547	640	740	890	1,070
180	67	103	147	202	268	349	430	510	607	710	820	990	1,190
200	73	111	160	220	292	383	470	560	667	780	900	1,080	1,310
220	78	119	173	237	316	416	510	610	724	845	980	1,170	1,420
240	83	127	186	254	340	449	550	660	781	910	1,060	1,260	1,530
260	88	135	199	271	363	481	590	704	838	970	1,134	1,350	1,640
280	93	143	211	288	386	513	630	748	894	1,030	1,207	1,440	1,750
300	98	151	223	305	409	545	670	792	950	1,090	1,280	1,530	1,860
320	103	159	235	322	432	576	705	836	1,000	1,150	1,350	1,620	1,970
340	108	167	247	339	455	604	740	880	1,050	1,210	1,420	1,710	2,080
360	113	174	259	356	478	633	775	924	1,100	1,270	1,490	1,800	2,190
380	118	182	270	373	501	662	810	967	1,150	1,330	1,560	1,890	2,290
400	123	190	281	390	523	691	845	1,010	1,200	1,390	1,630	1,980	2,390
420	128	197	292	405	544	720	880	1,048	1,248	1,448	1,694	2,064	2,490
440	133	204	303	420	565	748	915	1,086	1,296	1,506	1,758	2,148	2,590
460	137	211	314	435	586	776	950	1,124	1,344	1,564	1,822	2,232	2,690
480	141	218	325	450	607	803	985	1,162	1,392	1,622	1,886	2,316	2,790
500	145	225	335	465	628	830	1,020	1,200	1,440	1,680	1,950	2,400	2,885
520	149	232	345	480	648	856	1,050	1,238	1,486	1,734	2,010	2,476	2,980
540	153	239	355	495	668	882	1,080	1,276	1,532	1,788	2,070	2,552	3,075
560	157	245	365	510	688	908	1,120	1,314	1,578	1,842	2,130	2,628	3,170
580	161	250	375	525	708	934	1,140	1,352	1,624	1,896	2,190	2,704	3,265
600	165	255	385	540	728	960	1,170	1,390	1,670	1,950	2,250	2,780	3,360

* Courtesy of the Soil Conservation Service, Region III.

H. L. Cook[1] and is adapted to small agricultural areas in *USDA Farmers' Bulletin* 1859. The characteristics are outlined in Table 4.10. Determination of runoff to fit two sets of conditions, indicated in the headings, are provided for in Tables 4.11 and 4.12. The following example and its solution are taken from the handbook mentioned and illustrate the method.

ILLUSTRATIVE EXAMPLE

Existing Conditions. (1) Area, 93 acres comprising 60 acres of strip-cropped land and 33 acres of pasture; (2) average slope, 6 per cent; (3) soil, average loam; (4) waterways, well defined.

Design Solution. From Table 4.10, the sum of the watershed characteristics Σw is evaluated as follows:

Relief, for average slope of 6 per cent. 12
Soil infiltration for average loam. 10
Vegetal cover—10 for strip-cropped area and 5 for pasture, interpolating $= \dfrac{(10 \times 60) + (5 \times 33)}{93} =$ 8
Surface storage, drainage well defined. 15
Σw 45

Then from Table 4.12, for 50-year frequency, $A = 90$, $w = 45$, and rainfall factor $= 1.0$; normal runoff $Q_1 = 151$.

This value, representing the joint effect of A and Σw, must be multiplied by the regional rainfall factor. This is defined by Cook as follows: "the ratio of the runoff-producing capacity of the average storm at any point in the United States to that at some selected point." It is based on studies of Yarnell's contributions [165]. The assurance is given that, "over most of the United States, the rainfall factors taken from the curves shown on the rainfall factor map (Fig. 4.32) do not deviate greatly from the ratio

$$\frac{\text{1 hour, 100 year, intensity at the selected point}}{\text{1 hour, 100 year, intensity in the central U.S.}}$$

Determined from Yarnell's map, the curve marked 1.0 is the basis of comparison with all other regions."

In the illustrative example, if the structure to be designed is in Ashland County, Wisconsin, Fig. 4.32 shows R to be 0.7: whence

$$Q = 151 \times 0.7 = 106.$$

Figure 4.33 enables a graphical determination of Q. In the illustrative example just given, proceed as follows:

[1] Irrigation Programs Officer, Office of the Secretary, U.S. Department of Agriculture, Washington, D.C.

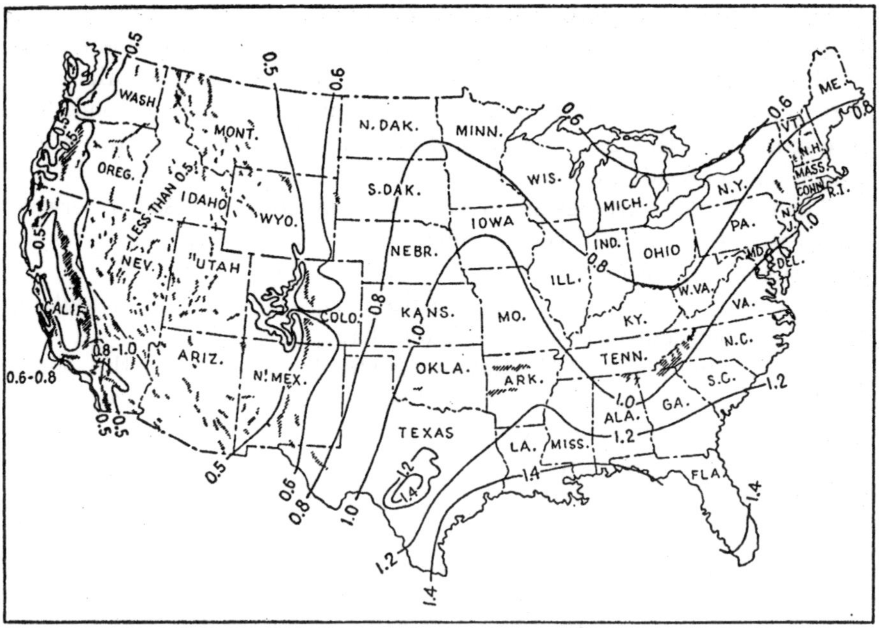

FIG. 4.32. Map of United States showing rainfall factors for use in estimating flood flows. (*Courtesy of H. L. Cook of the U.S. Department of Agriculture.*)

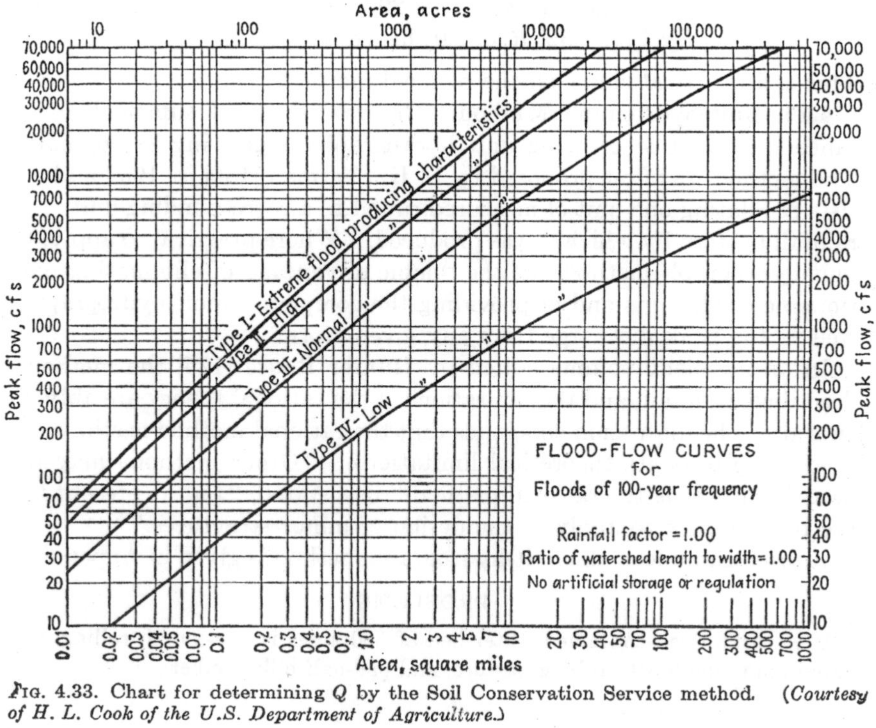

FIG. 4.33. Chart for determining Q by the Soil Conservation Service method. (*Courtesy of H. L. Cook of the U.S. Department of Agriculture.*)

On the area scale at the top find $A(= 93$ acres). Through this point follow the ordinate *down* until it intersects the type curve about one-fourth of the way from type curve III toward type curve IV; thence horizontally, left to intersection with the peak flow axis in cubic feet per second. Q is the reading at this intersection and is somewhat less than 110—a good check on the result by Table 4.12.

The following warnings are given: (1) for design of grassed waterways, use Table 4.11 for a 10-year frequency; (2) for capacities of permanent structures, use Table 4.12 for a 50-year frequency.

It is convenient to know that results from Table 4.11 are about uniformly 72 per cent of those from Table 4.12.

4.35. Comparison with the Rational Method. Following is the solution by the Soil Conservation Service method of the illustrative example given of the rational method.

Relief, hilly..........................	30
Soil infiltration......................	15
Vegetal cover $\dfrac{(10 \times 15) + (7.7 \times 10) + (3 \times 6)}{20.7} =$	12
Surface storage......................	11
Σw..................................	68

By Table 4.11 and area 20.7, $Q_1 = 73.2$

By Fig. 4.32, $R = 1.0$; hence $Q = 73.2$

This is a practical check on the rational method.

4.36. Unit-hydrograph Method. On watersheds of sufficient importance the practice of using unit hydrographs is growing in popularity. The method was first proposed by L. K. Sherman [136] in 1932 and consists of relating one unit of measured runoff to a corresponding unit of the rainfall pattern by which it was produced. Where practical to apply, the unit-hydrograph method has the double advantage of eliminating questionable coefficients and of providing the complete runoff hydrograph as well as the peak discharge from any given storm.

In an excellent volume edited by Hunter Rouse [125] the method is described in detail, and the limitations of the rational theory are ably discussed. The discussion concludes that the rational method is still widely used in spite of its theoretical limitations. "Other methods may ultimately supersede the rational method," it is stated, "but at present they are too complex to apply in the design of minor structures."

A brief example of the unit-hydrograph method is given in Appendix A.

PROBLEMS

1. A certain self-contained watershed of 500 acres is one-fourth rolling cultivated land, one-fourth rolling pasture, and one-half hilly timber.

(*a*) If the total runoff from a given storm is 700 cfs, what is the value of I in the rational formula?

(*b*) What does the result obtained in Prob. 1(*a*) amount to in inches of depth of rainfall in 24 hr, over the entire watershed?

2. The area of a given tract is a self-contained watershed of 400 acres, medium sandy loam with some gravelly material widely distributed throughout, the area being divided as follows: 10 acres steep rock outcrop, 15 acres hilly open pasture, 75 acres rolling woodland, 100 acres rolling to hilly permanent brush pasture, 100 acres rolling permanent meadow, and 100 acres gently rolling, well-tilled plow land. A 3½-in. rain fell on this area in just 1 hr.

(*a*) Determine the proper value of C for it in the rational formula.

(*b*) Assuming the footnote values in Table 4.1, page 72, as correct and applicable here, compute the probable length of the period of intense rain.

(*c*) If ¼ in. of the rain fell in the first 15 min up to the beginning of the intense period, compute the apparent values of I and Q by ordinary arithmetical methods without the aid of a runoff formula and compare the result with that obtained under 2(*b*).

3. (*a*) Compute the value of Q in Prob. 2 by Meyer's formula, on a 10-year frequency basis.

(*b*) Explain the marked discrepancy between this result and that obtained in Prob. 2 itself.

4. (*a*) For a 20-min rainfall occurring once in 10 years in eastern Colorado, what would be the intensity in inches per hour?

(*b*) If such a storm happened on a field where the distance from the head of the field run to the outlet was 1,500 ft and the total drop in elevation 45 ft, what would be the time of concentration through a natural V-shaped channel with side slopes of 1 on 20 through a heavily rolling pasture?

5. Solve Prob. 4 for Q by both the rational method and the Soil Conservation Service method. Do the results compare closely enough to be considered to be a good check on each other?

6. What is the average daily evaporation from ordinary farm land surface for the month of June, with average rainfall and temperature conditions, expressed in inches of depth of rainfall for eastern Nebraska?

7. Using Meyer's formula for runoff from medium watersheds, his evaporation chart, and a United States map drawn to a reliable scale, together with such data as are needed from the 1940 U.S. census of drainage, determine the probable flood discharge of the Mississippi River just above its junction with the discharge of the Ohio River Basin, for a total rainfall of 6 in. over a period of a week, occurring once in 25 yrs.

8. Name the watershed characteristics which are used by the Soil Conservation Service to determine design runoff. (These estimates are used to determine the value of C in the rational formula for runoff.)

9. The design rate of runoff for an area of 20.7 acres in Tennessee has been calculated as 73.6 cfs. If the design rainfall rate is taken as 6.6 in. per hr, what is the value of the runoff coefficient? What is a reasonable time of concentration for the given conditions?

CHAPTER 5

FLOW AND MEASUREMENT OF WATER

This book is not a treatise on hydraulics. However, the solution of drainage problems involves dealing with moving water and consequently with some elementary principles of hydraulics. Therefore, for convenience, this chapter will be devoted to a brief discussion of those devices and formulas which have been adapted to every-day practice and have a direct relationship to problems in the drainage field.

FLOW OF WATER

5.1. Units of Measure. The flow of water involves both linear and volumetric quantity as well as rates of flow or distance in unit time. The following units of measure are in general use. Certain common equivalents are also listed.

Linear Units:

Inch
Foot or decimals of same
Mile or its quarter or half (1 mile = 5,280 feet)

Units of Area:

Square foot
Acre (43,560 square feet)

Units of Volume:

Cubic inch
Cubic foot (1 cubic foot of water weighs 62.4 pounds)[1]
U.S. standard gallon (231 cubic inches; there are 7.48 gallons in 1 cubic foot, and 1 gallon of water weighs 8.34 pounds)
Acre-inch—a depth of 1 inch of water over 1 entire acre (an acre-inch = 3,630 cubic feet = 27,154 gallons)
Acre-foot—a depth of 1 foot of water over 1 entire acre (an acre-foot = 43,560 cubic feet of water = 12 acre-inches)

Units of Velocity:

Feet per second (v) or per minute, per hour, or per day

[1] Maximum density of water occurs at a temperature of 39.2 degrees Fahrenheit at which 1 cubic centimeter weighs 1 gram or 1 cubic foot weighs 62.36 pounds. Most handbooks give the normal weight of water as 62.4 pounds as shown above in Sec. 5.1.

Quantity-rate Units:

Cubic feet per second, or second-feet, abbreviated cfs (1 cubic foot per second is nearly equivalent to 1 acre-inch per hour or 7½ gallons per second or 450 gallons per minute)

Acre-inches per hour or per day

Acre-feet per day (1 acre-foot per 24-hour day = 226 gallons per minute)

Gallons per minute, abbreviated gpm

1 inch of rainfall in 24 hours = 0.042 cubic foot per second per acre

The water which flows in open channels and pipes is usually measured in cubic feet per second. This unit means that 1 cubic foot of water passes a certain point in the conduit every second.

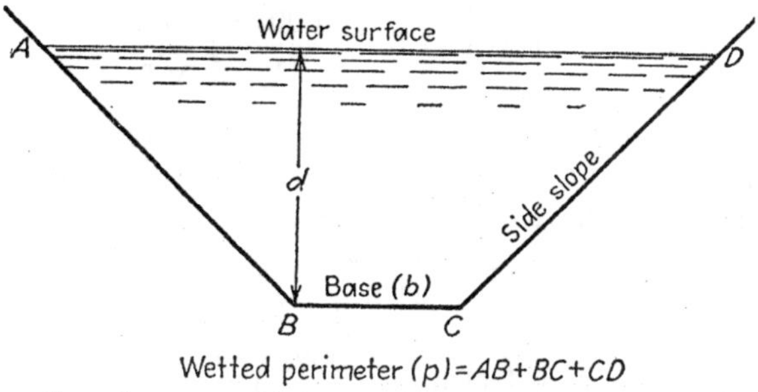

FIG. 5.1. Cross section of a trapezoidal channel, showing the relation between the base b, the depth d, and the wetted perimeter p.

5.2. Elements and Symbols of Formulas for Water Flow. The reader is reminded that any formula for flowing water, regardless of the type of conduit, involves the following elements:

a—the area of the cross section of the stream at right angles to the line of flow.

v—the velocity of flow in feet per second.

s—the hydraulic slope or simply the slope of the stream, which equals h/l, in which h is the drop of the hydraulic grade line in length l; that is, it is the sine of the angle of the slope (of the water surface in open channels).

r—the hydraulic radius a/p, a constant for any given channel and depth of flow, being the ratio of the cross-sectional area a of the stream to the wetted perimeter p (see Fig. 5.1).

q—the discharge of the stream in cubic feet per second, which equals av.

n—the factor of friction of the flowing water against the sides and bottom of the conduit and of eddying due to irregularities in successive cross sections or to obstructions in the channel. (Usually called the roughness coefficient.)

5.3. Formulas for Rates of Flow. One of the oldest and best-known flow formulas called the *Chézy*[1] *formula*, after the engineer who developed it,

$$v = C\sqrt{rs} \tag{7}$$

in which C proved to be an arbitrary constant dependent upon size and depth of stream, character of slope, and nature of the surface of the conduit.

When it was found that the value of C in the Chézy formula was not constant for all types of surfaces, Ganguillet and Kutter,[2] by a long series of experiments, evolved the following general value for C:

$$C = \frac{\dfrac{1.811}{n} + 41.6 + \dfrac{0.0028}{s}}{1 + \left(41.6 + \dfrac{0.0028}{s}\right)\dfrac{n}{\sqrt{r}}} \tag{8}$$

in which n is the coefficient of roughness of the walls of the channel or conduit. The value of C given in Eq. (8) when substituted for C in the Chézy formula gives the well-known *Chézy-Kutter formula* for velocity:

$$v = \frac{\dfrac{1.811}{n} + 41.6 + \dfrac{0.0028}{s}}{1 + \left(41.6 + \dfrac{0.0028}{s}\right)\dfrac{n}{\sqrt{r}}}\sqrt{rs} \tag{9}$$

This formula has stood the test of 100 years of general use throughout the civilized world and is still the standard, supported by a long list of values of n determined experimentally for surfaces of all kinds (see Table 5.1). Complete values of C have also been worked out for many conditions of slope, velocity, shape of cross section of stream, and character of surface of the channel. A few of these are given in Table 5.2.

The Chézy-Kutter formula is especially suited for use in channels where the side friction and irregularities of cross section are pronounced, as in the beds of natural streams.

A more workable and satisfactory formula, also carefully determined, experimentally, for artificial channels of uniform cross section and smooth sides, is *Manning's formula*

$$q = av = a\,\frac{1.486}{n}\,r^{2/3}s^{1/2} \tag{10}$$

in which the various letters represent, in general, the same factors as in the Chézy-Kutter formula. However the value of n is not the same in the two formulas throughout its entire range; hence, for convenience, the

[1] A French engineer of the late eighteenth century.

[2] Two Swiss engineers of the nineteenth century.

TABLE 5.1.* HORTON'S VALUES OF n
(To be used with Kutter's and Manning's formulas)

Surface	Best	Good	Fair	Bad
Riveted and spiral steel pipe	0.013	0.015†	0.017†	
Vitrified sewer pipe	{0.010 0.011	0.013†	0.015	0.017
Common clay drainage tile	0.011	0.012†	0.014†	0.017
Concrete pipe	0.012	0.013	0.015†	0.016
Wood-stave pipe	0.010	0.011	0.012	0.013
Plank flumes:				
Planed	0.010	0.012†	0.013	0.014
Unplaned	0.011	0.013†	0.014	0.015
With battens	0.012	0.015†	0.016	
Concrete-lined channels	0.012	0.014	0.016	0.018
Semicircular metal flumes, smooth	0.011	0.012	0.013	0.015
Canals and ditches:				
Earth, straight and uniform	0.017	0.020	0.0225†	0.025
Winding, sluggish canals	0.0225	0.025†	0.0275	0.030
Dredged earth channels	0.025	0.0275†	0.030	0.033
Canals with rough, stony beds, weeds on earth banks	0.025	0.030	0.035†	0.040
Earth bottom, rubble sides	0.028	0.030†	0.033†	0.035
Natural stream channels:				
1. Clean, straight bank, full stage, no rifts or deep pools	0.025	0.0275	0.030	0.033
2. Same as (1), but some weeds and stones.	0.030	0.033	0.035	0.040
3. Winding, some pools and shoals, clean.	0.033	0.035	0.040	0.045
4. Same as (3), lower stages, more ineffective slope and sections	0.040	0.045	0.050	0.055
5. Same as (3), some weeds and stones	0.035	0.040	0.045	0.050
6. Same as (4), stony sections	0.045	0.050	0.055	0.060
7. Sluggish river reaches, rather weedy or with very deep pools	0.050	0.060	0.070	0.080
8. Very weedy reaches	0.075	0.100	0.0125	0.150

* Abridged by permission from "Handbook of Hydraulics" by Horace W. King, published by McGraw-Hill Book Company, Inc.

† Values commonly used in designing.

symbol n' is commonly used in Manning's formula in place of n[1] (see Table 5.5).

[1] Scobey [127] points out that, for small channels or those with smooth interior surfaces and for values of n from 0.012 to 0.020, results by the two formulas are practically identical; but for values of n above 0.020 and for large channels or those of excessive hydraulic roughness there is a marked divergence in results from the two formulas. King [67] in Table 92 gives comparative values of n and n' applicable to a wide range of conditions.

Tables 5.3 and 5.4 are included as aids in solution of Manning's formula.

TABLE 5.2.* VALUES OF C FOR USE IN THE CHÉZY-KUTTER FORMULA

Slope s	Unit rate of fall	Fall rate per mile, ft	n	Hydraulic radius r, ft												
				0.10	0.20	0.40	0.80	1.00	2.00	3.00	3.28	4.00	6.00	10.0	20.0	50.0
				C values												
0.00005	1 in 20,000	0.264	0.010	67	87	109	133	140	164	178	181	187	199	212	228	245
			0.012	52	68	88	107	114	135	148	151	156	168	181	196	213
			0.020	26	35	46	59	64	79	88	91	95	105	116	131	148
			0.030	16	21	28	37	40	51	59	60	64	72	82	96	112
			0.035	13	18	24	31	34	44	50	52	56	63	72	85	101
			0.040	11	15	20	27	29	38	44	46	49	56	64	77	93
0.0001	1 in 10,000	0.528	0.010	78	98	119	140	147	168	178	C has the same value for all slopes when r = 3.28 ft	186	195	205	216	227
			0.012	60	76	95	114	120	138	149		155	164	174	185	197
			0.020	30	39	50	63	67	81	89		94	102	111	122	134
			0.030	17	23	31	39	42	53	59		64	71	78	89	100
			0.035	14	19	25	33	35	45	51		55	61	69	79	91
			0.040	12	16	22	28	31	39	45		49	54	62	71	83
0.0002	1 in 5,000	1.056	0.010	85	105	125	145	151	170	179		185	193	201	210	220
			0.012	65	83	100	118	123	140	149		155	162	170	180	189
			0.020	32	42	53	65	69	82	89		94	100	108	117	126
			0.030	18	25	32	41	44	54	59		63	69	76	85	94
			0.035	15	21	27	34	37	45	51		55	60	67	76	85
			0.040	12	17	23	29	32	40	45		48	53	60	68	78
0.0004	1 in 2,500	2.112	0.010	89	110	129	148	154	170	179		184	191	199	207	215
			0.012	69	87	104	121	125	141	149		154	161	168	176	184
			0.020	34	44	55	67	70	83	89		94	99	107	115	123
			0.030	19	25	33	42	45	54	59		63	69	75	83	91
			0.035	16	21	27	35	37	45	51		55	60	66	73	82
			0.040	13	18	23	30	32	40	45		48	53	59	66	75
0.001	1 in 1,000	5.280	0.010	94	113	131	150	155	171	179		184	190	197	235	212
			0.012	73	89	105	122	127	142	149		154	160	167	175	182
			0.020	36	45	56	68	71	83	89		93	99	105	113	120
			0.030	21	27	34	42	45	54	59		63	68	74	81	89
			0.035	17	22	28	35	38	46	51		54	59	65	72	79
			0.040	14	18	24	30	33	40	45		48	52	58	65	72
0.01	1 in 100	52.8	0.010	95	114	133	151	156	171	179		184	190	196	204	210
			0.012	74	90	107	123	128	142	149		154	160	166	173	181
			0.020	36	46	57	68	72	83	89		93	99	105	112	119
			0.030	21	27	35	43	45	55	59		63	68	74	80	87
			0.035	17	22	29	35	38	46	51		55	59	65	71	78
			0.040	14	19	24	31	33	40	45		48	52	58	64	71

* Reprinted by permission from "Handbook of Hydraulics" by Horace W. King, published by McGraw-Hill Book Company, Inc.

TABLE 5.3. TWO-THIRDS POWERS OF NUMBERS
(For use in Manning's formula)

Number	.00	.01	.02	.03	.04	.05	.06	.07	.08	.09
.0	.000	.046	.074	.097	.117	.136	.153	.170	.186	.201
.1	.215	.229	.243	.256	.269	.282	.295	.307	.319	.331
.2	.342	.353	.364	.375	.386	.397	.407	.418	.428	.438
.3	.448	.458	.468	.477	.487	.497	.506	.515	.525	.534
.4	.543	.552	.561	.570	.578	.587	.596	.604	.613	.622
.5	.630	.638	.647	.655	.663	.671	.679	.687	.695	.703
.6	.711	.719	.727	.735	.743	.750	.758	.765	.773	.781
.7	.788	.796	.803	.811	.818	.825	.832	.840	.847	.855
.8	.862	.869	.876	.883	.890	.897	.904	.911	.918	.925
.9	.932	.939	.946	.953	.960	.966	.973	.980	.987	.993
1.0	1.000	1.007	1.013	1.020	1.027	1.033	1.040	1.046	1.053	1.059
1.1	1.065	1.072	1.078	1.085	1.091	1.097	1.104	1.110	1.117	1.123
1.2	1.129	1.136	1.142	1.148	1.154	1.160	1.167	1.173	1.179	1.185
1.3	1.191	1.197	1.203	1.209	1.215	1.221	1.227	1.233	1.239	1.245
1.4	1.251	1.257	1.263	1.269	1.275	1.281	1.287	1.293	1.299	1.305
1.5	1.310	1.316	1.322	1.328	1.334	1.339	1.345	1.351	1.357	1.362
1.6	1.368	1.374	1.379	1.385	1.391	1.396	1.402	1.408	1.413	1.419
1.7	1.424	1.430	1.436	1.441	1.447	1.452	1.458	1.463	1.469	1.474
1.8	1.480	1.485	1.491	1.496	1.502	1.507	1.513	1.518	1.523	1.529
1.9	1.534	1.539	1.545	1.550	1.556	1.561	1.566	1.571	1.577	1.582
2.0	1.587	1.593	1.598	1.603	1.608	1.613	1.619	1.624	1.629	1.634
2.1	1.639	1.645	1.650	1.655	1.660	1.665	1.671	1.676	1.681	1.686
2.2	1.691	1.697	1.702	1.707	1.712	1.717	1.722	1.727	1.732	1.737
2.3	1.742	1.747	1.752	1.757	1.762	1.767	1.772	1.777	1.782	1.787
2.4	1.792	1.797	1.802	1.807	1.812	1.817	1.822	1.827	1.832	1.837
2.5	1.842	1.847	1.852	1.857	1.862	1.867	1.871	1.876	1.881	1.886
2.6	1.891	1.896	1.900	1.905	1.910	1.915	1.920	1.925	1.929	1.934
2.7	1.939	1.944	1.949	1.953	1.958	1.963	1.968	1.972	1.977	1.982
2.8	1.987	1.992	1.996	2.001	2.006	2.010	2.015	2.020	2.024	2.029
2.9	2.034	2.038	2.043	2.048	2.052	2.057	2.062	2.066	2.071	2.075
3.0	2.080	2.085	2.089	2.094	2.099	2.103	2.108	2.112	2.117	2.122
3.1	2.126	2.131	2.135	2.140	2.144	2.149	2.153	2.158	2.163	2.167
3.2	2.172	2.176	2.180	2.185	2.190	2.194	2.199	2.203	2.208	2.212
3.3	2.217	2.221	2.226	2.230	2.234	2.239	2.243	2.248	2.252	2.257
3.4	2.261	2.265	2.270	2.274	2.279	2.283	2.288	2.292	2.296	2.301
3.5	2.305	2.310	2.314	2.318	2.323	2.327	2.331	2.336	2.340	2.345
3.6	2.349	2.353	2.358	2.362	2.366	2.371	2.375	2.379	2.384	2.388
3.7	2.392	2.397	2.401	2.405	2.409	2.414	2.418	2.422	2.427	2.431
3.8	2.435	2.439	2.444	2.448	2.452	2.457	2.461	2.465	2.469	2.474
3.9	2.478	2.482	2.486	2.490	2.495	2.499	2.503	2.507	2.511	2.516
4.0	2.520	2.524	2.528	2.532	2.537	2.541	2.545	2.549	2.553	2.558
4.1	2.562	2.566	2.570	2.574	2.579	2.583	2.587	2.591	2.595	2.599
4.2	2.603	2.607	2.611	2.616	2.620	2.624	2.628	2.632	2.636	2.640
4.3	2.644	2.648	2.653	2.757	2.661	2.665	2.669	2.673	2.677	2.681
4.4	2.685	2.689	2.693	2.698	2.702	2.706	2.710	2.714	2.718	2.722
4.5	2.726	2.730	2.734	2.738	2.742	2.746	2.750	2.754	2.758	2.762
4.6	2.766	2.770	2.774	2.778	2.782	2.786	2.790	2.794	2.798	2.802
4.7	2.806	2.810	2.814	2.818	2.822	2.826	2.830	2.834	2.838	2.842
4.8	2.846	2.850	2.854	2.858	2.862	2.865	2.869	2.873	2.877	2.881
4.9	2.885	2.889	2.893	2.897	2.901	2.904	2.908	2.912	2.916	2.920

TABLE 5.4. SQUARE ROOTS OF DECIMAL NUMBERS
(For use in Manning's formula)

Number	. −0	. −1	. −2	. −3	. −4	. −5	. −6	. −7	. −8	. −9
.00001	.003162	.003317	.003464	.003606	.003742	.003873	.004000	.004123	.004243	.004359
.00002	.004472	.004583	.004690	.004796	.004899	.005000	.005099	.005196	.005292	.005385
.00003	.005477	.005568	.005657	.005745	.005831	.005916	.006000	.006083	.006164	.006245
.00004	.006325	.006403	.006481	.006557	.006633	.006708	.006782	.006856	.006928	.007000
.00005	.007071	.007141	.007211	.007280	.007348	.007416	.007483	.007550	.007616	.007681
.00006	.007746	.007810	.007874	.007937	.008000	.008062	.008124	.008185	.008246	.008307
.00007	.008367	.008426	.008585	.008544	.008602	.008660	.008718	.008775	.008832	.008888
.00008	.008944	.009000	.009055	.009110	.009165	.009220	.009274	.009327	.009381	.009434
.00009	.009487	.009539	.009592	.009644	.009695	.009747	.009798	.009849	.009899	.009950
.00010	.010000	.010050	.010100	.010149	.010198	.010247	.010296	.010344	.010392	.010440
.0001	.01000	.01049	.01095	.01140	.01183	.01225	.01265	.01304	.01342	.01378
.0002	.01414	.01449	.01483	.01517	.01549	.01581	.01612	.01643	.01673	.01703
.0003	.01732	.01761	.01789	.01827	.01844	.01871	.01897	.01924	.01949	.01975
.0004	.02000	.02025	.02049	.02074	.02098	.02121	.02145	.02168	.02191	.02214
.0005	.02236	.02258	.02280	.02302	.02324	.02345	.02366	.02387	.02408	.02429
.0006	.02449	.02470	.02490	.02510	.02530	.02550	.02569	.02588	.02608	.02627
.0007	.02646	.02665	.02683	.02702	.02720	.02739	.02757	.02775	.02793	.02811
.0008	.02828	.02846	.02864	.02881	.02898	.02915	.02933	.02950	.02966	.02983
.0009	.03000	.03017	.03033	.03050	.03066	.03082	.03098	.03114	.03130	.03146
.0010	.03162	.03178	.03194	.03209	.03225	.03240	.03256	.03271	.03286	.03302
.001	.03162	.03317	.03464	.03606	.03742	.03873	.04000	.04123	.04243	.04359
.002	.04472	.04583	.04690	.04796	.04899	.05000	.05009	.05196	.05292	.05385
.003	.05477	.05568	.05657	.05745	.05831	.05916	.06000	.06083	.06164	.06245
.004	.06325	.06403	.06481	.06557	.06633	.06708	.06782	.06856	.06928	.07000
.005	.07071	.07141	.07211	.07280	.07348	.07416	.07483	.07550	.07616	.07681
.006	.07756	.07810	.07874	.07937	.08000	.08062	.08124	.08185	.08246	.08307
.007	.08367	.08426	.08485	.08544	.08602	.08660	.08718	.08775	.08832	.08888
.008	.08944	.09000	.09055	.09110	.09165	.09220	.09274	.09327	.09381	.09434
.009	.09487	.09539	.09592	.09644	.09695	.09747	.09798	.09849	.09899	.09950
.010	.10000	.10050	.10100	.10149	.10198	.10247	.10296	.10344	.10392	.10440
.01	.1000	.1049	.1095	.1140	.1183	.1225	.1265	.1304	.1342	.1378
.02	.1414	.1449	.1483	.1517	.1549	.1581	.1612	.1643	.1673	.1703
.03	.1732	.1761	.1789	.1817	.1844	.1871	.1897	.1924	.1949	.1975
.04	.2000	.2025	.2049	.2074	.2098	.2121	.2145	.2168	.2191	.2214
.05	.2236	.2258	.2280	.2302	.2324	.2345	.2366	.2387	.2408	.2429
.06	.2449	.2470	.2490	.2510	.2530	.2550	.2569	.2588	.2608	.2627
.07	.2646	.2665	.2683	.2702	.2720	.2739	.2757	.2775	.2793	.2811
.08	.2828	.2846	.2864	.2881	.2898	.2915	.2933	.2950	.2966	.2983
.09	.3000	.3017	.3033	.3050	.3066	.3082	.3098	.3114	.3130	.3146
.10	.3162	.3178	.3194	.3209	.3225	.3240	.3256	.3271	.3286	.3302

Many other discharge formulas have been presented from time to time by eminent engineers, but the two just given have been the most generally adopted through the past century. However, extensive research in the past 25 years has resulted in some modifications in the use of both the Chézy-Kutter and the Manning formula as well as in the development of some new formulas better suited to the design of certain special types of channels and conduits. In this connection the work of Ramser [111] and Scobey [126] and [127] is especially noteworthy (see Chap. 7, Design of Open Ditches).

MEASURING DEVICES

The use of formulas (9) and (10) to determine the actual discharge of a stream involves making various measurements not always immediately and readily obtainable, such as depth of stream, shape of cross section, and slope of stream surface and bed, and these measurements must be followed by various computations for r, a, and the formula itself. However there are various mechanical devices and methods directly available in many cases, such as weirs, venturi flumes, and current meters.

TABLE 5.5. $1.486/n'$
(For use in Manning's formula)

n'	$1.486/n'$	n'	$1.486/n'$	n'	$1.486/n'$	n'	$1.486/n'$
0.008	185.75	0.016	92.88	0.030	49.53	0.065	22.86
0.009	165.11	0.017	87.41	0.033	45.03	0.070	21.23
0.010	148.60	0.018	82.56	0.035	42.45	0.075	19.81
0.011	135.09	0.020	74.30	0.040	37.15	0.080	18.58
0.012	123.83	0.022	67.55	0.045	33.02	0.090	16.51
0.013	114.31	0.0225	66.04	0.050	29.72	0.100	14.86
0.014	106.15	0.025	59.44	0.055	27.02	0.125	11.89
0.015	99.07	0.0275	54.04	0.060	24.77	0.150	9.91

5.4. Weirs. In its simplest form a weir is merely a thin flat metal plate set vertically, at right angles to the line of flow in a stream, and having cut in it a notch of definite geometrical form, as, for example, a rectangle, a trapezoid, or a triangle. There are various types of weirs, all illustrated and fully discussed in almost any standard book on irrigation. However, as the use for weirs in the drainage field is limited and as for reasons stated below this use is largely confined to the Cipolletti trapezoidal weir, only that type is discussed in this book. In all weir formulas L is the length of the weir crest (the horizontal edge of the weir notch over which the water flows) in feet and H is the head on the weir crest (the depth of the water on the crest) in feet.

5.5. Cipolletti Trapezoidal Weir. The formula for discharge over the Cipolletti trapezoidal weir, shown in Fig. 5.2, is

$$q = 3.367LH^{3/2} \tag{11}$$

Weirs are preferable to most other water-measuring devices. This type of weir has the disadvantages of weirs in general as it collects considerable sediment and requires considerable drop at the weir section. However, its advantages exceed those of other types as it is simple to construct and install and is suitable for measurement of a wide range of discharges from relatively small to relatively large. It should be noted that its side slopes have a rise of 4 feet to 1 foot horizontal run.

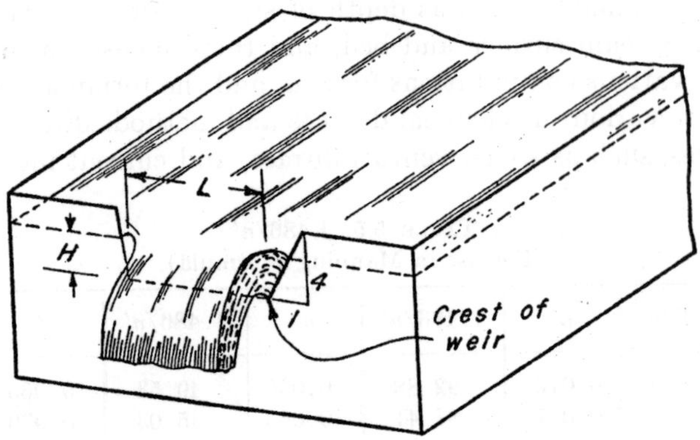

FIG. 5.2. Cipolletti trapezoidal weir.

5.6. Rules for Setting Weirs. Strict observance of the following rules is essential for good results:

1. Set weir at lower end of long pool, of width and depth to assure velocity of approach not over 0.5 foot per second—practically still water.
2. Set center line of weir box parallel with line of flow.
3. Set face of weir vertical, perpendicular transversely to flow line.
4. Be sure of straight, sharp crest, set absolutely level.
5. Set crest about $3H$ above bottom of pool and ends at least $2H$ from sides of pool.
6. Place gage or weir scale in weir pond or box far enough upstream (4 or 5 feet) to be free of the drawdown curve over the crest, best to one side, in practically still water, with zero of scale absolutely level with the weir crest.
7. For measurement of head use carpenter's rule or equivalent scale, fastened vertically to gage post—zero of scale exactly level with weir crest.
8. Set weir crest high enough to leave air space under freely falling water. If crest becomes submerged by water below it, measurements will be unreliable unless complicated corrections are made.

TABLE 5.6. DISCHARGE OVER CIPOLLETTI TRAPEZOIDAL WEIRS BY THE FORMULA $q = 3.367LH^{3/2}$, CUBIC FEET PER SECOND

Head H		Length of weir, ft									
		1.0	1.5	2.0	2.5	3.0	4.0	5.0	6.0	7.0	8.0
Ft	In.	q values									
0.05	$0\frac{5}{8}$	0.038	0.06	0.08	0.09	0.11	0.15	0.19	0.23	0.26	0.30
0.10	$1\frac{3}{16}$	0.107	0.16	0.21	0.27	0.32	0.43	0.53	0.64	0.75	0.85
0.15	$1\frac{13}{16}$	0.196	0.29	0.39	0.49	0.59	0.78	0.98	1.17	1.37	1.56
0.20	$2\frac{3}{8}$	0.301	0.45	0.60	0.75	0.90	1.20	1.51	1.81	2.11	2.41
0.25	3	0.421	0.63	0.84	1.05	1.26	1.68	2.10	2.53	2.95	3.37
0.30	$3\frac{5}{8}$	0.553	0.83	1.11	1.38	1.66	2.21	2.77	3.32	3.87	4.43
0.35	$4\frac{3}{16}$	0.697	1.05	1.39	1.74	2.09	2.79	3.49	4.18	4.88	5.58
0.40	$4\frac{13}{16}$	0.852	1.28	1.70	2.13	2.56	3.41	4.26	5.11	5.96	6.81
0.45	$5\frac{3}{8}$	1.016	1.52	2.03	2.54	3.05	4.07	5.08	6.10	7.11	8.13
0.50	6	1.190	1.79	2.38	2.97	3.57	4.76	5.95	7.14	8.33	9.52
0.55	$6\frac{5}{8}$	1.373	2.06	2.75	3.43	4.12	5.49	6.87	8.24	9.61	10.99
0.60	$7\frac{3}{16}$	1.565	2.35	3.13	3.91	4.69	6.26	7.82	9.39	10.95	12.52
0.65	$7\frac{13}{16}$	1.764	2.65	3.53	4.41	5.29	7.06	8.82	10.59	12.35	14.11
0.70	$8\frac{3}{8}$	1.972	2.96	3.94	4.93	5.92	7.89	9.86	11.83	13.80	15.77
0.75	9	2.187	3.28	4.37	5.47	6.56	8.75	10.93	13.12	15.31	17.49
0.80	$9\frac{5}{8}$	2.409	3.61	4.82	6.02	7.23	9.64	12.05	14.45	16.86	19.27
0.85	$10\frac{3}{16}$	2.638	3.96	5.28	6.59	7.92	10.55	13.19	15.83	18.47	21.11
0.90	$10\frac{13}{16}$	2.875	4.31	5.75	7.19	8.62	11.50	14.37	17.25	20.12	23.00
0.95	$11\frac{3}{8}$	3.117	4.68	6.23	7.79	9.35	12.47	15.59	18.70	21.82	24.94
1.00	12	3.367	5.05	6.73	8.42	10.10	13.47	16.83	20.20	23.57	26.93
1.10	$13\frac{3}{16}$	3.884	5.83	7.77	9.71	11.65	15.54	19.42	23.30	27.19	31.07
1.20	$14\frac{3}{8}$	4.426	6.64	8.85	11.06	13.28	17.70	22.13	26.55	30.98	35.40
1.30	$15\frac{5}{8}$	4.990	7.48	9.98	12.47	14.97	19.96	24.95	29.94	34.93	39.92
1.40	$16\frac{13}{16}$	5.577	8.36	11.15	13.94	16.73	22.31	27.88	33.46	39.04	44.62
1.50	18	6.185	9.28	12.37	15.46	18.56	24.74	30.92	37.11	43.30	49.48
1.60	$19\frac{3}{16}$	6.814	10.22	13.63	17.03	20.44	27.25	34.07	40.88	47.70	54.51
1.70	$20\frac{3}{8}$	7.462	11.19	14.92	18.65	22.39	29.85	37.31	44.77	52.24	59.70
1.80	$21\frac{5}{8}$	8.130	12.20	16.26	20.32	24.39	32.52	40.65	48.78	56.91	65.04
1.90	$22\frac{13}{16}$	8.817	13.22	17.63	22.04	26.45	35.27	44.09	52.90	61.72	70.54
2.00	24	9.522	14.28	19.04	23.80	28.57	38.09	47.61	57.13	66.66	76.18

* Adapted by permission from "Handbook of Hydraulics," by Horace W. King, published by McGraw-Hill Book Company, Inc.

9. The head of water should not exceed one-third the length of the crest, nor should it be less than 2 inches.

10. Riprap or otherwise protect the ditch bed just below the weir against scour by the falling water.

5.7. The Parshall Measuring Flume. This device, formerly called the *improved venturi flume* but more recently renamed the *Parshall measuring* *flume* after its inventor, Ralph L. Parshall, is fully described and discussed by him in Refs. 101 and 102. In the opinion of many engineers the Parshall measuring flume meets more fully than any other known device the objections to all types of weirs just discussed. On this account it has even been stated by some engineers that it is likely, within the next 25 years, to replace all other forms of weirs.[1] This seems likely to be true for projects under close engineering supervision. In other cases it may not be so.

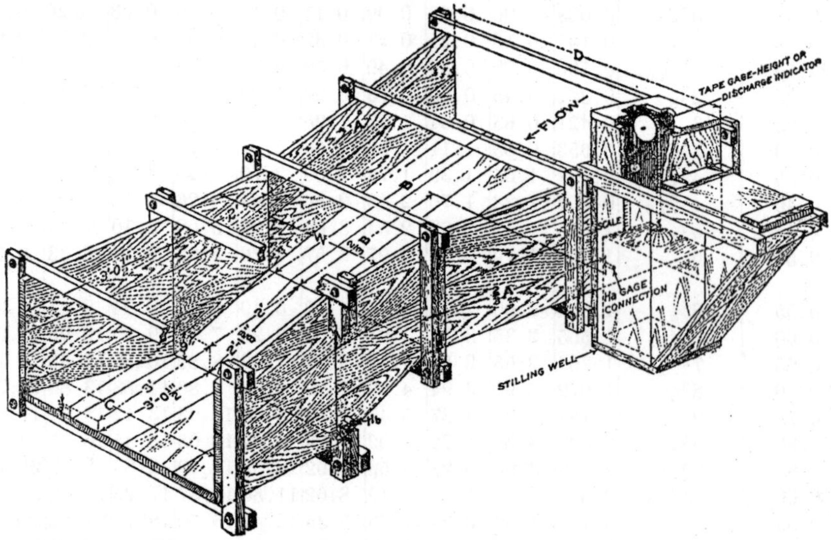

FIG. 5.3. Improved venturi (Parshall) measuring flume equipped with stilling well and indicating-tape device. (*Colo. Agr. Expt. Sta. Bul.* 336.)

The following is a condensed version of a discussion by Clyde [29] of the advantages and use of the Parshall measuring flume: Figure 5.3 shows the improved venturi flume and stilling well with float indicating tape. Briefly, the improved venturi flume consists of a box which may be made of wood, concrete, or metal, a level floor with a converging section, and vertical side walls. At the end of the converging section the floor slopes downward 9 inches in 2 feet. The outlet section is 3 feet long and diverges. The floor of the outlet rises 6 inches in 3 feet. The lower end of the outlet is 3 inches lower than the crest. Figure 5.3 and Table 5.7 show the standard dimensions of the improved venturi flume. The head is measured at a point $\frac{2}{3}B$ (Fig. 5.3) from the crest.

[1] Other engineers and technical men, however, think this is overstating the case, for, they say, its construction and setting are too complicated for use by the average farmer or land operator, who usually much prefers either the Cipolletti or the triangular weir for farm use because it is easily made, easily set, and readily transported while available tables and mechanical scales make their use quite as simple.

To insure conditions of free flow, the head H_b should not be greater than $0.7H_a$, both referred to the level of the crest as datum. If the depth H_b is greater than $0.7H_a$, the discharge for a given head H_a is decreased and a correction must be made for submergence.

The Improved Venturi Flume possesses many advantages over the other types of measuring devices. It requires only a small amount of head; it is not affected by velocity of approach; it is self-cleaning; it has a relatively large range of capacity; it is simple to construct and easily installed; the discharge is determined by the measurement of but one head; and it lends itself to the attachment of automatic registers for the securing of continuous records. . . .[1]

TABLE 5.7.* STANDARD DIMENSIONS AND CAPACITIES OF IMPROVED VENTURI FLUME
(Letters refer to Fig. 5.3)

Crest length W, ft	Dimensions, ft and in.					
	A	$\frac{2}{3}A$	B	$\frac{2}{3}B$	C	D
1	4′6″	3′0″	4′ 4⅞″	2′11¼″	2	2′ 9¼″
2	5′0″	3′4″	4′10⅞″	3′ 3¼″	3	3′11½″
3	5′6″	3′8″	5′ 4¾″	3′ 7⅛″	4	5′ 1⅞″
4	6′0″	4′0″	5′10⅝″	3′11⅛″	5	6′ 4¼″
5	6′6″	4′4″	6′ 4½″	4′ 3″	6	7′ 6⅝″
6	7′0″	4′8″	6′10⅜″	4′ 6⅞″	7	8′ 9″
7	7′6″	5′0″	7′ 4¼″	4′10⅞″	8	9′11⅜″
8	8′0″	5′4″	7′10⅛″	5′ 2¾″	9	11′ 1¾″
10	9′0″	6′0″	8′ 9⅞″	5′10⅝″	11	13′ 6⅜″

* Table 9 of *Utah Agr. Expt. Sta. Cir.* 77 or Table 1 of *Colo. Expt. Sta. Bul.* 423.

The general formula for free discharge through the improved venturi flume, as given by Parshall [101], is as follows:

$$q = 4WH_a^{1.522W^{0.026}} \qquad (12)$$

in which W is the width of throat in feet and H_a is the upper head in feet, observed at a point distant upstream from the crest two-thirds the length of the converging section.

For convenience in the determination of discharges from the various devices discussed, the discharge tables (Tables 5.6 to 5.8) are given.

5.8. The Current Meter. The current meter, one well-known type of which is shown in Fig. 5.4, makes possible another rather simple method of measuring stream flow particularly adapted to large streams. Details of the design of the current meter are immaterial to this discussion. If such a discussion is desired, the reader will find it in the catalogues of the various makers. The meter is calibrated by moving it through still water at known rates of speed and tabulating the number of revolutions per

[1] From Clyde [29].

TABLE 5.8.* FREE-FLOW DISCHARGE THROUGH IMPROVED VENTURI FLUMES, BY THE FORMULA $q = 4WH_a^{1.522W^{0.026}}$, CUBIC FEET PER SECOND

Upper head *Ha*		Throat widths, ft								
Ft	In.	1	2	3	4	5	6	7	8	10
		q values								
0.20	2 3/8	0.35	0.66	0.97	1.26					
0.25	3	0.49	0.93	1.37	1.80	2.22	2.63			
0.30	3 5/8	0.64	1.24	1.82	2.39	2.96	3.52	4.08	4.62	
0.35	4 3/16	0.80	1.57	2.32	3.06	3.78	4.50	5.22	5.93	
0.40	4 13/16	0.99	1.93	2.86	3.77	4.68	5.57	6.46	7.34	9.10
0.45	5 3/8	1.19	2.32	3.44	4.54	5.63	6.72	7.80	8.87	11.00
0.50	6	1.39	2.73	4.05	5.36	6.66	7.94	9.23	10.51	13.05
0.55	6 9/16	1.62	3.17	4.70	6.23	7.74	9.25	10.76	12.24	15.22
0.60	7 3/16	1.84	3.62	5.39	7.15	8.89	10.63	12.36	14.08	17.52
0.65	7 13/16	2.08	4.11	6.12	8.11	10.10	12.08	14.05	16.01	19.94
0.70	8 3/8	2.33	4.60	6.86	9.11	11.36	13.59	15.82	18.04	22.48
0.75	9	2.58	5.12	7.65	10.16	12.67	15.17	17.66	20.14	25.13
0.80	9 9/16	2.85	5.66	8.46	11.25	14.04	16.81	19.59	22.36	27.89
0.85	10 3/16	3.12	6.22	9.30	12.38	15.46	18.52	21.58	24.64	30.76
0.90	10 13/16	3.41	6.80	10.17	13.55	16.92	20.29	23.66	27.02	33.74
0.95	11 3/8	3.70	7.39	11.07	14.76	18.44	22.12	25.79	29.47	36.82
1.00	12	4.00	8.00	12.00	16.00	20.00	24.00	28.00	32.00	40.00
1.05	12 5/8	4.31	8.63	12.96	17.28	21.61	25.94	30.28	34.61	43.28
1.10	13 3/16	4.62	9.27	13.93	18.60	23.26	27.94	32.62	37.30	46.66
1.15	13 13/16	4.94	9.94	14.94	19.94	24.96	30.00	35.02	40.06	50.13
1.20	14 3/8	5.28	10.61	15.96	21.33	26.71	32.10	37.50	42.89	53.71
1.25	15	5.62	11.31	17.02	22.75	28.50	34.26	40.02	45.80	57.37
1.30	15 5/8	5.96	12.01	18.10	24.21	30.33	36.47	42.62	48.78	61.12
1.35	16 3/16	6.32	12.74	19.20	25.69	32.20	38.74	45.26	51.84	64.96
1.40	16 13/16	6.68	13.48	20.32	27.21	34.11	41.05	47.99	54.95	68.90
1.45	17 3/8	7.04	14.23	21.27	28.76	36.06	43.42	50.76	58.14	79.92
1.50	18	7.41	15.00	22.64	30.34	38.06	45.82	53.59	61.40	77.02
1.55	18 5/8	7.80	15.78	23.84	31.95	40.09	48.28	56.48	64.71	81.21
1.60	19 3/16	8.18	16.58	25.05	33.59	42.17	50.79	59.42	68.10	85.49
1.65	19 13/16	8.57	17.38	26.29	35.26	44.28	53.34	62.42	71.56	89.85
1.70	20 3/8	8.97	18.21	27.55	36.96	46.43	55.95	65.48	75.08	94.29
1.75	21	9.38	19.04	28.82	38.69	48.61	58.60	68.59	78.66	98.81
1.80	21 5/8	9.79	19.90	30.13	40.45	50.83	61.29	71.75	82.29	103.4
1.85	22 3/16	10.20	20.76	31.45	42.24	53.09	64.01	74.98	86.00	108.1
1.90	22 13/16	10.62	21.63	32.79	44.05	55.39	66.81	78.24	89.76	112.9
1.95	23 3/8	11.06	22.53	34.14	45.90	57.72	69.63	81.57	93.59	117.7
2.00	24	11.49	23.43	35.53	47.77	60.08	72.50	84.94	97.48	122.6
2.05	24 5/8	11.93	24.34	36.94	49.67	62.48	75.42	88.37	101.4	127.6
2.10	25 3/16	12.37	25.27	38.35	51.59	64.92	78.37	91.84	105.4	132.7
2.15	25 13/16	12.82	26.20	39.79	53.54	67.39	81.36	95.37	109.5	137.8
2.20	26 3/8	13.28	27.15	41.25	55.52	69.90	84.41	98.94	113.6	143.0
2.25	27	13.74	28.12	42.73	57.52	72.43	87.49	102.6	117.8	148.3
2.30	27 5/8	14.21	29.09	44.22	59.56	75.01	90.61	106.2	122.0	153.7
2.35	28 3/16	14.68	30.08	45.74	61.61	77.61	93.77	110.0	126.3	159.1
2.40	28 13/16	15.16	31.09	47.27	63.69	80.25	96.97	113.7	130.7	164.6
2.45	29 3/8	15.64	32.10	48.82	65.80	82.92	100.2	117.6	135.1	170.2
2.50	30	16.13	33.11	50.39	67.93	85.62	103.5	121.4	139.5	175.8

* Adapted from *Utah Agr. Expt. Sta. Cir.* 77, "Measurement of Irrigation Water."

second for each speed. These meters are carefully rated at the factory where they are made, and their accuracy may be checked at any time at the rating station operated by the Bureau of Standards, Washington, D.C. There are two types of current meters—the electric current meter and the acoustic current meter. The former communicates to the operator through a telephone receiver, by means of an electric current, the number of times the current wheel revolves. The latter uses a tapping

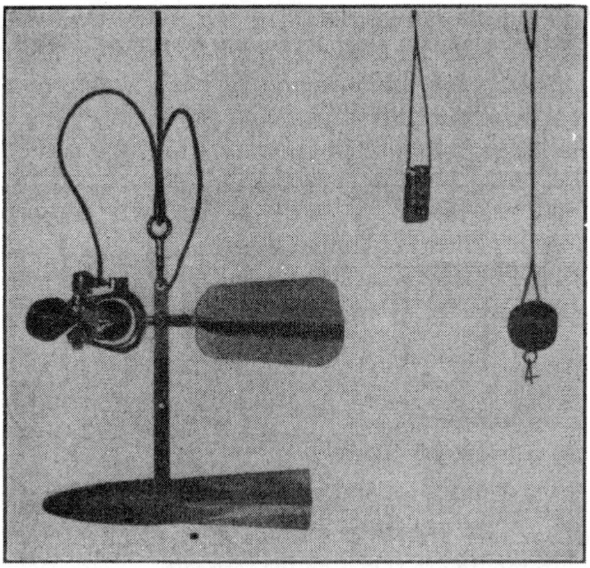

FIG. 5.4. Gurley current meter.

device the sound of which is conveyed to the ear of the operator through the handle of the current meter and a rubber tube. The electric meter is more generally used than the other type.

A rating table accompanies each instrument. The velocity at any point in a stream is obtained by holding the meter still at that point in the stream, noting the number of revolutions per second and reading from the rating table the corresponding velocity of the stream. If the average velocity of a stream is to be determined from a single reading or from a limited number of readings, it is of interest to know that experience has shown that, in streams up to 1.5 feet deep, the closely approximate velocity for the entire stream is that found at about 0.6 of the depth. For streams exceeding 1.5 feet depth, the approximate average velocity is close to an average of the readings at 0.2 and 0.8 of the depth. As a rule, more reliable results may be obtained, however, by taking readings at a number of fixed key points in a surveyed cross section, as indicated in the following discussion, or by moving the meter slowly, at a uniform rate from top to bottom of stream at the center or at several selected points. This method is difficult for high velocities. Where circumstances permit, the most satisfactory method is as follows: Select a straight stretch of channel where shape, depth, and width of channel are as uniform as possible and where the gradient is such that the flow is never turbulent. The exact shape of the cross section of the stream is found by a survey at a point near the lower end of the selected stretch, usually where there is a permanent bridge which may be used as a base line both for the survey and for the elevations of all future measurements.

Figure 5.5 shows a stream cross section of this type. The depth measurements d_1, d_2, etc., are taken at intervals b_1, b_2, etc., of 2 to 20 feet, depending on the width of the stream and the shape of the channel bottom. The discharge is computed separately for each subsection, a_1, a_2,

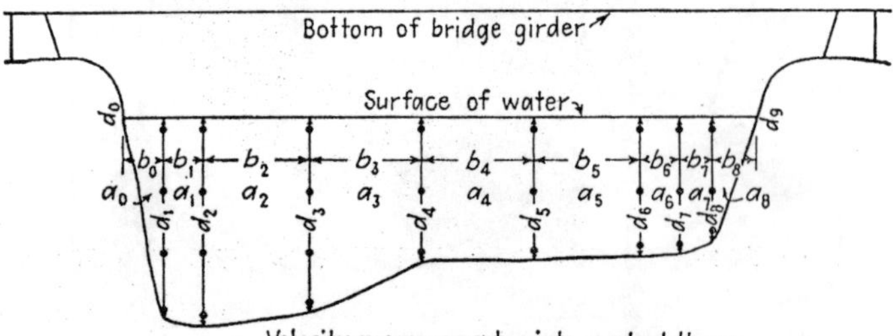

FIG. 5.5. Typical river-bed cross section.

etc. The end subsections (a_0 and a_8 in the figure) are triangles; all others are trapezoids. Areas are either measured by planimeter or computed. Velocity measurements are taken *on the verticals.* Velocities at d_0 and d_9 are considered zero. Those in the end sections are taken as the arithmetic averages on d_1 and d_8, respectively, and those in each other subsection as averages of all velocities measured on the vertical boundaries of the subsection, for example, in the figure, the discharge through subsection a_2, or $q_2 = a_2v_2 = \frac{1}{2}b_2(d_2 + d_3) \times$ (average of v readings on d_2 and d_3). The total discharge for the stream is the sum of the discharges through all the subsections, or

$$Q = q_0 + q_1 + q_2 + q_3 + \cdots + q_8 = \Sigma q \tag{13}$$

This method is used universally by the Federal government throughout the country on large rivers. At permanent gaging stations the cross section is determined and plotted on cross-section paper to a definite scale, and the areas are determined for various stages of water and uniformly tabulated for future use. Corresponding velocities are also measured with a current meter, and capacity or discharge tables for the given stream are computed from the measured results. Then all that has to be done at any time is to observe the stage of water against the gage board always permanently set at the gaging station, refer to the table, and read the discharge of the stream.

5.9. The Float Method. The *float* method consists of determining the velocity of the surface water in the open channel by means of a float. This float may be a chip or a piece of wood of any kind. In using this method it is necessary to select a section of the channel which is straight and smooth. Place marks of some kind, such as poles or stakes on the ditch bank, at some distance apart, say 100 feet. Place the float on the stream above one of these marks, and determine the length of time it takes it to travel the distance between the two marks. This should be repeated a number of times and the average taken, in order to get a result as accurate as possible. The velocity should be converted into feet per second. The mean velocity of a stream varies from 0.8 to 0.95 of the surface velocity, the smaller percentage being used on shallow streams. Wind blowing up or down a stream will decrease or increase the surface velocity. The mean velocity times the average cross-sectional area (in square feet) of the channel, between the two points, will give the discharge. This method, at best, is more or less crude, and the results obtained from its use are inaccurate. It can be used, however, under conditions where approximations are satisfactory for the purpose at hand.

PROBLEMS

1. (*a*) What is the equivalent of 1 cfs in gallons per minute? In acre-inches per day? In inches of rainfall over 1 acre per hr and per day?

2. Relative to weirs,

(*a*) Must the velocity of the water through the weir notch be known in order to find q? Explain.

(*b*) Compute q over a Cipolletti weir when the length of crest is 3 ft and the head on the weir measured at a point 5 ft upstream from the weir is 8 in. Check your results against a table.

3. Compute q for a venturi flume where the throat is 3 ft wide and the head is 2 ft. Check against a table.

4. How many times does doubling the head on a rectangular or trapezoidal weir increase the discharge? Demonstrate.

5. By use of a current meter the following mean velocities in feet per second were obtained in the cross section of a ditch at the various points shown: $A = 0.00$, $B = 2.46$, $C = 2.53$, $D = 3.01$, $E = 2.84$, $F = 2.60$, $G = 0.00$. Depth soundings in feet were as follows: $A = 0$, $B = 4.2$, $C = 5.4$, $D = 5.8$, $E = 4.9$, $F = 3.2$, $G = 0$. The distance between any two consecutive stations is 4 ft, except between F and G, which is 4.3 ft. What is the discharge of the ditch?

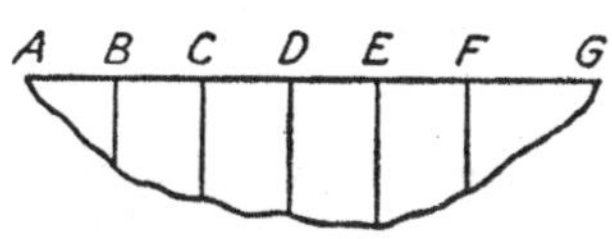

6. Find the area of cross section in the accompanying figure. Using the table given below, find the average velocity of total stream, the average surface velocity, and the average peripheral velocity. Then determine at what proportion of the depth a horizontal line of a few readings would give a reliable average velocity for the whole stream. How does your result in this respect

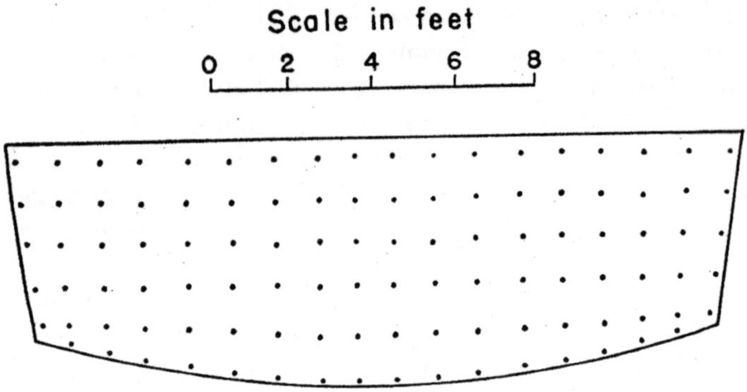

compare with accepted experience and practice? Compute the ratio of the surface velocity to the average velocity and the ratio of the peripheral velocity to the average velocity. Determine the per cent of error introduced by using the surface velocity; the peripheral velocity.

Referring to the sketch and reading left to right on each line of dots beginning with the top line, the observed velocities by current meter in feet per second are:

Top line	Second line	Third line	Fourth line	Fifth line	Bottom line
2.099	2.210	2.326	2.158	2.003	2.035
2.042	2.511	2.550	2.522	2.321	2.175
2.561	2.698	2.653	2.566	2.367	2.323
2.604	2.752	2.744	2.616	2.438	2.444
2.692	2.780	2.857	2.761	2.673	2.394
2.754	2.868	2.906	2.917	2.684	2.401
2.827	2.955	2.969	2.972	2.783	2.343
2.841	2.960	3.039	2.944	2.854	2.463
2.944	3.039	3.091	2.980	2.767	2.465
2.923	3.022	3.053	2.963	2.652	2.412
2.901	2.992	2.952	2.860	2.574	2.396
2.754	2.917	2.892	2.862	2.428	2.210
2.710	2.812	2.818	2.698	2.363	2.246
2.672	2.763	2.607	2.544	2.223	2.116
2.555	2.769	2.637	2.511	2.200	2.105
2.547	2.619	2.533	2.351	2.152	1.980
2.402	2.449	2.353	2.300	2.096	
2.096	2.233	2.320	2.233	2.003	

7. Assuming $s = 0.0005$ and $n = 0.03$ and using the area found in Prob. 6, compute v by the formula $v = C\sqrt{rs}$, taking C from the table. Compare the result with the average velocity obtained in Prob. 6. Remember that $r = a/p$.

CHAPTER 6

MAJOR TYPES OF DRAINS AND DRAINAGE PROBLEMS[1]

Some areas, relatively limited in extent compared to the total of agricultural lands, have *natural* drainage in one of two ways: (1) through permeable substrata, usually complete, needing no artificial drains; or (2) through surface runoff, usually needing to be supplemented by artificial drains.

The larger part of the agricultural area must be provided with *artificial* drains which are of two main types: (1) open channels both artificial and natural, and (2) underdrains.

OPEN CHANNELS: TYPES

Under the especially favorable but rare conditions of thoroughly permeable soil and subsoil and sufficient natural slope, open channels may provide ample drainage, both surface and subsurface, for limited types of crops such as the more water-loving grasses, including reed-canary grass, alta fescue, and meadow foxtail mixed frequently with certain of the legumes, as bird's foot trefoil, and some other crops than grasses, such as rice. In most cases, however, open ditches alone do not provide a complete drainage system but fulfill an equally important function in receiving and carrying off excess surface water and acting as outlets for needed underdrainage systems.

6.1. Natural Channels. All stream beds are drainageways, even if dry part of the year, usually being the outlets to natural flow or artificial drainage from adjacent lands. Often the smaller channels, especially, are the outlet ditches for areas through which they flow, but, particularly through flat lands, these stream beds often require straightening and deepening in order to lower the water table sufficiently for farming. A good example of this type is the lower Pudding River in western Oregon.

6.2. Outlet Ditches. Outlet ditches are large open ditches whose primary function is to collect from large areas all free surface flow resulting from any cause, and the discharge from tile drainage systems, and to carry it to final outlets at low enough elevations and of sufficient capacity to

[1] For discussion of drainage design, location, construction methods, and costs, see Chaps. 7 to 10, 12, and 15.

receive the total without overflow. If deep enough through open soils and subsoils and on ample slope, such ditches also provide limited underdrainage (see Sec. 6.8).

Except where the bulk of the precipitation comes in short, intense downpours, causing heavy flash flooding, a ditch with a 4-foot base, minimum 4-foot depth, and 1½ to 1 side slopes is usually ample to carry off the excess surface water; but this depth is wholly inadequate to serve as an unimpeded outlet for a tile drainage system. For this purpose the ditch must be deep enough freely to receive the flow from the tile drains entirely above the normal flow in the ditch itself. Therefore, on flat areas or through moderately uniform topography, such a ditch must be about twice as deep as one which has to carry only the ordinary surface flow. That is, simply to secure sufficient depth without need for the extra capacity, an outlet ditch for tile drains usually involves between three and four times the amount of excavation needed for the ordinary surface ditch. This fact and their peculiar function place such ditches in a class by themselves (see Fig. 10.1).

Other types of outlets for tile drainage systems than open ditches are possible as discussed under Design of Tile Drainage in Chap. 10.

Outlet ditches are usually but not always of a public character, affecting several landowners and are, therefore, generally provided for by statutory processes (see Chap. 16).

6.3. Intercepting Ditches. Where high land discharges a considerable quantity of surface water onto the low land, it is often possible to cut off this flow at or near the foot of the hill and carry it away to some nearby outlet without permitting it to spread out over the low land. The excavated material is usually thrown up to form a low dike on the lower side, thus increasing the capacity of the ditch. The dike and sides of the ditch are smoothed off and sown to permanent grass. As the capacity of the ditch is usually required only occasionally and for short periods, the crop will not be killed and no land is wasted by the ditch.

6.4. Field Drainageways. By field drainageways are meant broad shallow V-shaped channels with side slopes sufficiently flat to be farmed across without bridging. Such ditches are a distinct asset on flat lands infested with shallow landlocked pockets tending to become permanent ponds (see Fig. 7.1). The pond may readily be eliminated through use of these V-shaped channels, connecting the pockets in a chain and leading on a nonerosive gradient to and outletting in roadside or other deep existing ditches. They eliminate the need for deep lateral ditches that cut up the fields and that can be crossed only on bridges, and they help to prevent overloading of tile drains passing under the pockets. They prevent standing water with the result that the fields can be cultivated soon after rains, and they are in no way a hindrance to field operations. Some operators consider it an advantage to keep them in permanent sod for a width of 15 or 20 feet.

6.5. Dead Furrows. What may be considered a simple modification of field drainageways are regularly spaced parallel dead furrows leading across a field on a nonerosive gradient to a regular drainage outlet channel. Such a system of dead furrows is a main feature of the system of field management known as bedding, which is discussed in detail under drainage problems in the latter part of this chapter (see Fig. 6.4).

SUBSURFACE DRAINS

6.6. Purpose of Subsurface Drainage. The purpose of subsurface drainage is to lower the free water level within the root zone to as near the average rooting depth of crops to be grown as is possible and to hold it as nearly constant at that level as possible. The required depth at which the drains must be placed to accomplish this result, which is chiefly governed by the character of soil and subsoil, type of crops to be raised, and climatic and rainfall characteristics, is discussed in Chap. 10 on Location and Design of Underdrains. The artificial means used to accomplish subsurface drainage are open ditches, pole drains, stone drains, wooden box drains, and tile drains.

6.7. Natural Subsurface Drainage. Where the soil deposition is such that there is a layer of sand or gravel 2 to 5 or 6 feet below the surface layer of rich topsoil there is seldom any need of artificial subdrainage. Areas where such natural subdrainage exists are, however, relatively rare, and the effectiveness of such drainage is largely dependent on the permeability of the surface soil and the existence of a natural outlet for the sand and gravel strata.

6.8. Open Ditches. Limited areas where open ditches of suitable depth effectively relieve excessive wetness within the root zone are to be found in many agricultural regions and somewhat noticeably so in the United States west of the continental divide. The following are good illustrations.

In the Kittitas Valley near Ellensburg in eastern Washington, a common condition is a water-bearing gravel layer under pressure, just below the normal blanket of surface silt and clay loam soils and resting upon a lower impervious stratum. This water-bearing stratum keeps the surface layer excessively wet until relieved by artificial means. On the flatter slopes, ditches across the slope, extending through this water-bearing layer, remove the excess water to an extent that gives to the surface soil relief from excess wetness for a distance of as much as ¼ mile upslope and ½ mile downslope from the ditch.

In the lower Columbia Valley near Longview in western Washington, where widely varying types of surface soil over permeable subsoils prevail, open ditches whose bottoms are well below the average root zone of local crops lower a high water table sufficiently to permit growth of all good pasture grasses.

In the Imperial Valley in southern California, where soils of a light, pervious character are prevalent, widely spaced open ditches of a depth well through the root zone often give good and complete drainage for all the wide variety of crops raised in that region.

In the majority of agricultural areas, however, open ditches cannot be depended upon for subsurface drainage. This fact is generally true of the northern Mississippi Basin. The studies in northern Minnesota by Averill and McGrew [8], for example, have shown conclusively that subsurface drainage is seldom appreciable in the less pervious soils as far as 50 feet from the ditch (see Fig. 6.13). It is obvious that open-ditch laterals 100 feet apart in farm fields are economically and operationally prohibitive. For the most part, open-ditch drainage must be supplemented by some form of covered subsurface drains.

6.9. Pole Drains. Before the advent of tile, farmers often dug ditches 2 or 3 feet deep and 1 foot wide in which were placed three poles. These poles were arranged in the shape of a triangle. The ditches were then filled up. They were given an outlet into an open ditch. The water would seep in around the poles and flow along them to the open ditch. This method is more or less temporary, because as soon as the poles decay the earth falls in and stops up the drain. The outlay of money is small, however, because the poles can, as a rule, be cut on the farm and the ditch be dug by the farm hands during a slack time. This method is still practiced to some extent in some regions, but as it is poor economy, it is not advocated except where the tile itself or the money with which to buy it is not available. In such a case it may be a justifiable temporary step. The digging of the trench is the most costly part of the enterprise. This is just as true when tile is installed, but once properly installed, drains of good tile are practically permanent.

6.10. Stone Drains. Drainage by means of stones is accomplished in a way somewhat similar to that by means of poles. If large flat stones are available, they are placed at the sides of the ditch with one over the top, thus forming a small channel which is more durable than the poles. If flat stones are not to be had, small *irregular* stones are thrown into the ditch to the depth of 6 or 8 inches. The water flows through the interstices to the outlet. In many parts of the country no stones are available for such use and, even if there were, the latter, especially, of the two methods is not advocated because the interstices become clogged with silt, stopping the water flow and constituting a serious interference with the installation of new drains.

6.11. Box Drains. Box drains, usually made of low-grade, locally available lumber, free of shakes and rot, are used extensively in many areas where tile is hard to get or is considered too costly. They are probably the best cheap substitute for tile and give effective results. They are usually made rectangular in cross section of four boards per section (see Fig. 6.1) two 4-inch and two 6-inch boards, or even four 6-inch boards for the smaller sizes corresponding roughly to 4- and 5-inch tile. Some prefer to leave the bottom entirely open, with slats regularly spaced to keep the sides from spreading. This type is satisfactory if the bottom of the trench is in fairly firm soil.

If sound lumber is used, such drains will last a number of years in any kind of soil especially where they stay continually wet. In a true peat

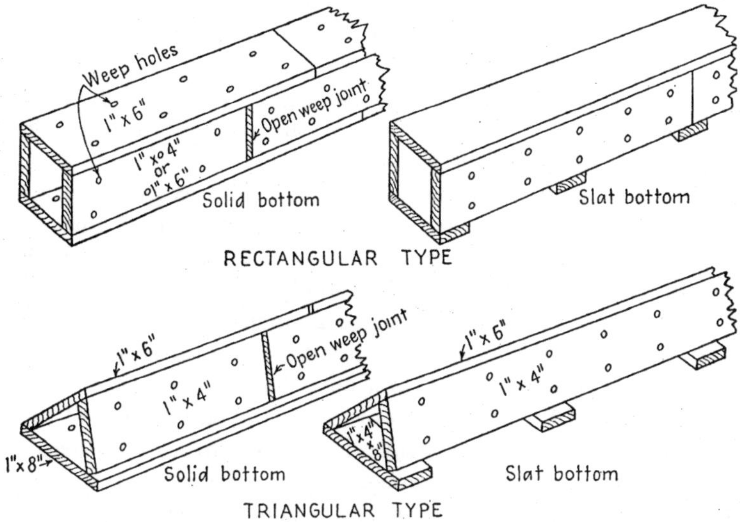

FIG. 6.1. Box drains approximating 4- and 5-in. tile in capacity.
NOTE: Space ½- or ¾-in. seep holes about 1 ft horizontally and 2 in. vertically.

soil, in which they are often used, they will last almost indefinitely. Treating the lumber would give a similar long life in mineral soils. If larger sizes are used, cross slats across the top should be placed to reinforce it against splitting under pressure of the fill. Some prefer a triangular cross section with three boards or with two side boards and bottom slats, but they are slightly more difficult to make and install. In general, box drains are easy to make and to lay to grade.

6.12. Tile Drains. Tile drains of either clay or concrete are practically permanent if the tile are of good quality and carefully installed. (See Chap. 11 for further details on quality and permanence of drain tile.) Seconds in bell-and-spigot glazed shale sewer tile and in concrete lap-joint irrigation pipe are sometimes available, at a price advantage, for use in farm drains. They usually possess all the qualifications of a good drain tile, which makes them a good investment, but the purchaser should make sure of that by insisting that they pass the standard American Society for Testing Materials specifications for good drain tile reprinted in Chap. 11. Many miles of both these kinds of pipe are now in satisfactory use in drainage systems.

6.13. Advantages and Disadvantages of Open Ditches and Tile Drains. *Open Ditches.* The only advantage of open ditches over tile drains is that great capacity is possible in an open ditch at far less cost than that of a tile drain of equal capacity. The disadvantages of open ditches are marked. They take up much land area even when spread spoil banks permit cropping right up to the ditch bank. They cut up the fields, greatly increasing the difficulties and cost of farm operations because they can be crossed only at bridges. They need constant maintenance, which is usually neglected until they are rendered ineffective by mud banks or vegetal growth by which cleaning is made so difficult as often to cost as much as the excavation of the original ditch. They are unsightly and often a menace to livestock.

Usually, on the modern farm, open ditches, except the V-shaped drainageways, have little place as a permanent type of improvement. Even on the farm, however, open ditches as main drains are usually more economical, over a period of years, beyond the point where the required diameter of the tile main would exceed 24 inches.

Tile Drains. About the only disadvantage of a tile drain as compared with an open ditch is its greater first cost than an open ditch of equal or even greater capacity. The most notable of the advantages of tile drains over open ditches are the following: They take up no land area. They require practically no maintenance. They offer no interference with field operations. They are permanent if good tile is used and properly installed. They grow more efficient year by year. They are not unsightly but, on the contrary, eliminate unsightly spots. Mains of comparatively small size are cheaper than open ditches.

SOME SPECIAL TYPES OF DRAINS

6.14. Mole Drains. A mole drain is simply a small, round channel drawn through the subsoil by tractor power at a limited depth by a cartridge-like lug attached behind the steel frame of a special type of deep-tillage machine called a mole ditcher (see Fig. 6.2). Regulation of depth, avoidance of backfall, and the maintaining of a definite gradient is extremely difficult—often impossible—except on flat lands or those of uniform topography, where the drains are made to follow uniform surface gradients set out with a level. Success with mole drains is possible only in soils having strong coherence when compressed by the lug making the opening. Previous to the early 1940's they were hardly conceded to be

Fig. 6.2. Mole ditcher and drain. *Above:* Side view of ditcher. Note hydraulic cylinder mounted on the beam, by means of which operator regulates depth of mole on uneven ground. (*USDA Farmers' Bul.* 2046. *Courtesy of John G. Sutton.*) *Below:* Cross section of the mole drain. (*Courtesy of Killefer Manufacturing Company.*)

drains at all, being regarded merely as short aerating channels that might, under favorable conditions of gradient and character of soil, aid drainage by accumulating soil water and carrying it more quickly to the tile drains.

Nevertheless, mole drainage has been extensively used with reputed success in some foreign countries where soils prevail of greater coherence than is found in most of the soils in the United States. According to Olsen [98] it is widely used with satisfaction in Great Britain. A similar experience is claimed for it in New Zealand. In the United States, in 1928, Robey [119] in Michigan and, in 1943, Barr and Turnbull [10] in Louisiana reported effective use of mole drainage in some areas; both installation and replacement costs are low. In 1951, Fred H. Larson of the Soil Conservation Service staff of the Pacific Region reported that farmers in the Beaver Drainage District near Clatskanie, Oreg., in the lower Columbia River Valley, use mole drainage effectively to aid underdrains in the flat areas of peat and muck soils stratified with clay-loam inwash. They run the drains in checkerboard pattern at 20-foot spacing and 22 to 32 inches deep. The gradients are set out with a level. Original cost as well as replacement cost in case of failure are both low.

At best the life of mole drains is short, however, and their successful use is possible only on limited areas embracing soils with strong cohesion under compression. They are conceded to have two good characteristics—low cost and some influence in aerating the soil. In spite of the favorable reports just mentioned it is the firm opinion of some authorities in this country that mole drainage is generally ill-suited for extensive use in the United States. They warn against reliance on foreign data as a guide to its use here, advise great caution in its adoption, and advise using it only after careful soil examination and minor tests of its behavior indicate its suitability in any given region and the probability that serious erosion will not result.

An interesting development from the research side is a study in the use of small, perforated flexible tubes in mole drains, reported by Glenn O. Schwab [133] at Iowa State College in 1951. To any one interested this contribution should prove worthwhile reading (see also Ref. 134).

6.15. Vertical Drains. In many localities scattered over the United States it is common practice to drain sinkholes, field pockets, and other low land having no gravity outlet into wells which penetrate into a water-absorbing stratum some distance below the ground surface which has natural outlet lower than the area to be drained. Such drains naturally fall into two classes: shallow vertical drains and deep-well vertical drains.

6.16. Shallow Vertical Drains. Shallow vertical drains are those which outlet into nonwaterlogged strata of coarse sand or gravel 20 feet or less below the surface and extend, laterally, to the banks of neighboring streams or other bodies of water into which they discharge through numerous streamlets or by heavy seepage. Notably in Wisconsin, in parts of southern Minnesota, and in other states it is possible easily to secure vertical outlets of this type under comparatively thin layers of clay or other relatively impervious soils. Such a well is easily and cheaply dug by hand with a post auger, preferably of the Iwan type with readily extensible handle of ¾-inch gas pipe (see Fig. 6.3), and as easily cased with

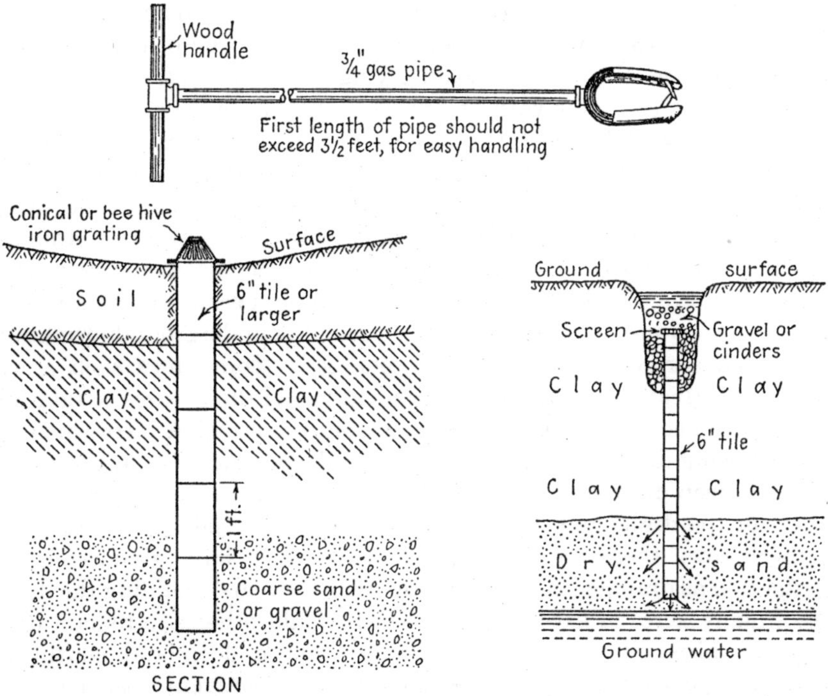

FIG. 6.3. Shallow vertical drain and equipment for digging it. *Above:* Post auger of the Iwan type. *Below:* Left, drain capped with iron grating. Right, drain capped with screen and gravel blanket.

drain tile and capped with a conical grate or with a screen and a blanket of gravel (see Fig. 6.3).

However, such a shallow vertical drain is always a gamble and cannot be recommended as a means of permanent relief as there is no sure means of determining whether the pervious stratum into which it is dug has any outlet. If there is no outlet, the life of the drain will be definitely limited by the capacity of the pervious deposit. In any case it is soon apt to clog with the silt that will be washed into it through the drain.

Early in the twentieth century, vertical drains were investigated in nine Lake and upper Mississippi Valley states, by Fred Shafer, then senior drainage engineer of the U.S. Department of Agriculture. These investigations showed 100 per cent failure, within a short time after their installation, of the many shallow vertical drains investigated, due to the silting just mentioned. Nevertheless, if a stratum of coarse sand or gravel lies within 10 or 15 feet of the surface of a field pocket too deep to obtain an economical outlet for a tile drain, it may be feasible as a means of temporary relief from surface flooding to install a shallow vertical drain that may eventually result in conditions that will later make possible economical installation of a tile drain. The cost will be so little as to be worth trying. Up to 1944 two cases were on record in Minnesota where such an experiment took place, the vertical drain giving good service for several years and being replaced later with an extensive tile drainage system.

6.17. Deep-well Vertical Drains. A deep-well vertical drain is just what the name indicates—a drain into deep rock fissures or into a regular drilled and cased deep well. The investigation by Shafer, above reported, showed that a deep-well vertical drain is likely, from a purely drainage standpoint, to be permanently effective. Nevertheless, at the start, even a deep drainage well is more or less of a gamble, because it is impossible to secure exact and complete data concerning the underground formations into which surface water will be discharged. Sometimes such a well functions perfectly for a few weeks or months and then ceases to take any more water because there is no escape from the stratum into which the water is being drained. Still, in southern Minnesota in 1944, there were two such deep drainage wells, each of which had furnished a satisfactory outlet for the drainage from a large farm for a long period of years. In northern Iowa there are several such wells which have given complete satisfaction as drainage outlets for more than 30 years, enabling the drainage of many acres of excellent land which would otherwise remain unreclaimed. Most of these wells discharge into a deep and continuing stratum of pervious rocks, with an iron casing to prevent obstruction by caving, extending within 4 or 5 feet of the surface. Others extend through a deep stratum of limestone into a formation furnishing good outlet.

The warning is given, however, that such deep-cased wells usually tap permanent sources of water for livestock, domestic, or community use. Hence the feeling is common that such an outlet for farm drainage *should never be permitted,* as pollution of the water supply is sure to happen sooner or later with the resultant liability of disease epidemics. On this account the practice of using deep wells as outlets for farm drainage is, therefore, forbidden by law or, as in Minnesota, by regulations of official boards of health, which have the force of law. Vertical drainage is not recommended for general use. In some parts of the irrigated West, however, drainage by means of wells has proved the only satisfactory solution to the drainage problem (see Chap. 13 on Drainage of Irrigated Lands).

DRAINAGE PROBLEMS

Agricultural drainage problems fall naturally into three main classes according to what may be considered effective drainage:

Class I. Purely *surface* drainage problems, involving only the rapid removal of surface water.

Class II. Combined *surface and subsurface* drainage problems, involving both removal of surface water and the reducing of excessive wetness in the soil itself within the root zone of crops.

Class III. Purely *subsurface* drainage problems, involving only reduction of excessive wetness in the soil itself. Class III problems are not discussed under a separate heading because they are identical with those encountered in the second portion of Class II.

All three classes generally have to cope with two conditions: (1) entirely stagnant water or water with no appreciable movement either over the surface or through the soil, without artificial aid, and (2) water which moves over broad areas either over the surface or through the soil or both.

CLASS I

Class I largely but not wholly embraces problems of a pioneer or transitional character, surface drainage being almost universally a first step in drying the surface and making possible the breaking and first cropping of virgin lands, but contemplating future complete drainage which eventually involves underdrainage as a final essential for high production of most crops and especially such high-value crops as small fruits and vegetables.

Another type of problem strictly within Class I is the region where soil and subsoil are so impermeable that underdrainage is nearly or wholly ineffective, such, for instance, as the clays, clay loams, and tighter silty-clay loams. The essential requirement here is to remove free water, from whatever source, quickly from the surface, before it has a chance to so thoroughly wet the surface soil as to prevent farming operations for a long period after rains. Considerable areas of this character occur in most agricultural regions. A good illustration is the wide area of Dayton soils lying between the Willamette and the Calapooya rivers in western Oregon. A common method of handling the drainage in such areas is what is known as *bedding*.

Bedding. The method of bedding consists in dividing the land into a series of strips usually 65 to 100 feet wide, bounded by dead furrows running in the general direction of the main field slope on a definite nonerosive gradient and outletting in a natural channel or district ditch which is the drainage outlet for the region. The surface of the strip is worked up into a ridge whose center line, midway between the dead furrows, is about 1 foot higher than the land surface at the edges of the bounding furrows (see Fig. 6.4). This gives sufficient lateral slope to carry the surface water to the dead furrows which, in turn, discharge the water into the outlet channel at the edge of the field. Most natural streams build up a ridge on either bank appreciably higher than the land surface back from

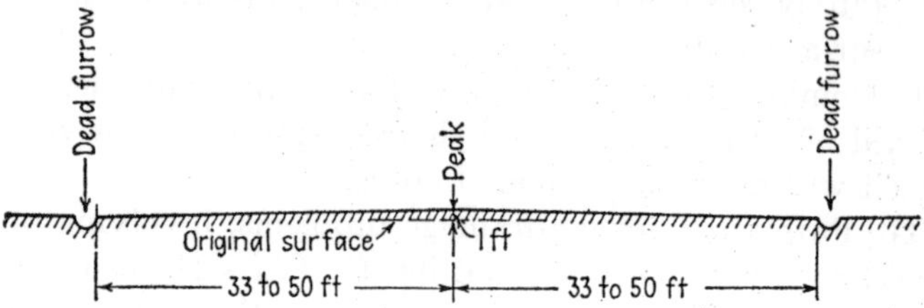

FIG. 6.4. Typical cross section of bedding strip.

the creek. Care should be taken to cut the bedding dead furrows through these ridges on the general gradient of the furrows so that the water discharges into the outlet channel without impediment. This method is widely practiced in the United States.

CLASS II

Underdrainage is an absolute essential for many crops, especially such high-value crops as small fruit and vegetables and some other crops.

FIG. 6.5. Land too flat for natural drainage. *Above:* Flooding and drowning out on an undrained field. *Below:* Rye on tile-drained field similar to the one shown above. These two views were taken within a few minutes of each other from positions *A* and *B*, Fig. 6.6, facing in general directions indicated by the arrows

Hence Class II is the most universal class of drainage problems. It includes all the following types.

6.18. Lands Too Flat for Natural Drainage. There exist, widely scattered over the agricultural regions of the world, considerable areas of land, at least partly agricultural, with little or no slope, where the water from precipitation, not removed by infiltration and deep seepage, lies

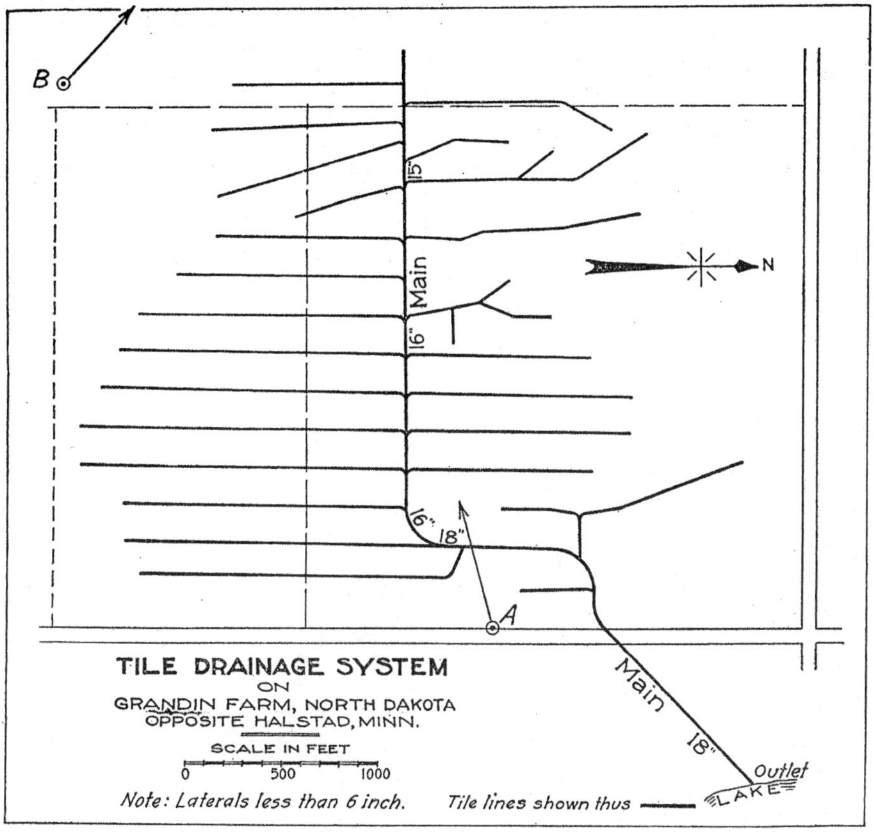

FIG. 6.6. Map of farm lands, Valley of the Red River of the North, on the Grandin Farm on the North Dakota side of the river, just opposite Halstad, Norman County, Minnesota. The parallel tile lines 200 ft apart are the minimum essential for these lands.

stagnant until it has evaporated or moves so slowly as to have the same ill effects as stagnant water (see Fig. 6.5). Many of these areas are extremely fertile, potentially, and produce excellent crops after they are drained. The drainage system for such an area is usually based upon either an open ditch or a large tile main located along the line of the natural drainageway or of the greatest natural depression of the surface and made just deep enough to provide good outlet for the laterals. The small field laterals, usually tile, are laid at depths and spacings discussed in Chap. 10 (see Fig. 6.6). The main must empty into some larger

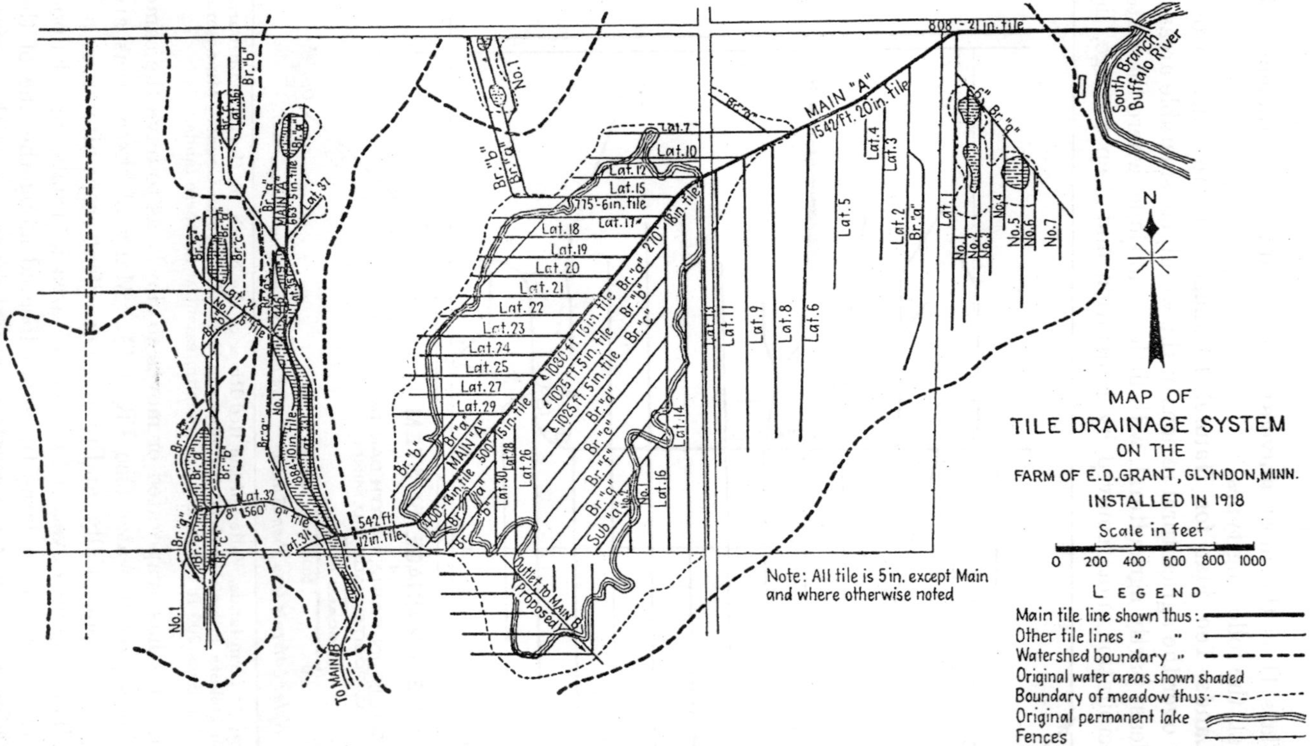

FIG. 6.7. Map of about 240 acres of a farm in Clay County, Minnesota, separated from its natural drainage outlet, the Buffalo River, by a higher-lying impervious ridge. Where the summit of the ridge crosses the tile line about 300 ft back from the river, it was necessary to lay the tile about 11 ft deep to secure an outlet.

channel or body of water, either by gravity flow or by pumping from a sump or reservoir (see Fig. 10.2).

6.19. Lands Separated from Their Natural Outlet Streams by High Impervious Ridges. If too flat for natural drainage, such lands must be treated like the preceding case, but the mains must be cut through the ridges to the outlet streams (see Fig. 6.7). Where flood stages of the outlet streams may be high enough to submerge the mouth of the main drain and cause backflow and consequent flooding of the low lands behind

Fig. 6.8. Typical backwater gate. (*Courtesy of California Corrugated Culvert Company.*)

the ridge, an automatic drainage gate is needed (see Fig. 6.8). An alternative method is to outlet the tile mains into a sump or reservoir on the land side of the ridge and install an automatic pumping plant to lift the water from the interior drainage over the ridge (see Fig. 10.2).

6.20. Ponded Areas. The two types of problems just discussed are usually complicated by the existence of shallow pockets of widely varying size, and depths of a few inches to 2 or 3 feet below the general level, scattered irregularly over the entire area. In these pockets the remnants of the surface water settle and produce the condition commonly known as *ponding*, for which there is usually no natural outlet. Hence, if not artificially drained, these ponds tend to become permanent and spoil the fields for effective and economical farming. They should be handled in the manner earlier described under Field Drainageways, Sec. 6.4. In addition, field inlets, such as illustrated in Fig. 10.22, into the tile laterals passing under the pockets, where such laterals exist, are an effective aid in the elimination of ponding.

6.21. Marsh Pockets and Running Sloughs. Marsh pockets and running sloughs partially hemmed in by subsoils of low permeability and with too flat a slope or too shallow an outlet to drain naturally commonly exist on farms practically everywhere. The best drainage design for these is a double main arrangement, or, better still, a main tile through the center of the pocket or slough and tile laterals around or above the sides at just the right depth to intercept the seepage water from the fields before it breaks through to the surface can be used (see Figs. 1.2, 1.3, and 6.9).

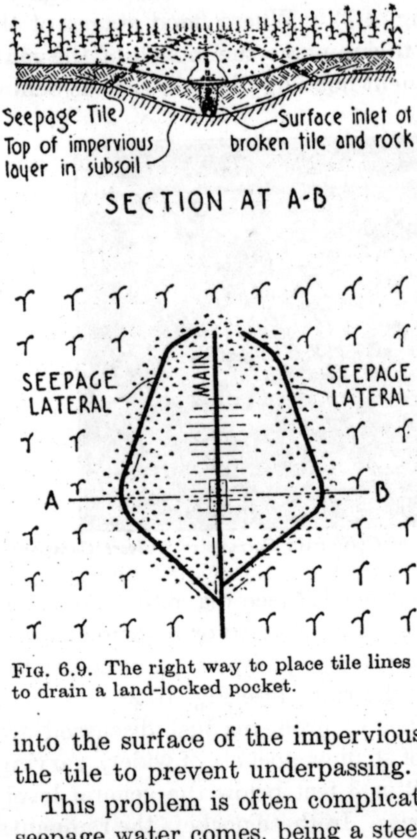

Fig. 6.9. The right way to place tile lines to drain a land-locked pocket.

6.22. Seeped Land Drainage. Where bluffs and slopes or low lands at their base are wet, owing to seepage from higher pervious and water-bearing strata or from irrigation canals, the cure is indicated in Fig. 6.10. Tile lines should be run along the foot of the slope a little above the wet area, at just the right depth to catch the seepage water at the top of the layer of impervious subsoil over which it runs, before it breaks through onto the low area. The low land may have good natural drainage in itself. Hence, cutting off the seepage from the lower land by a single tile line may be all that is needed. If one line does not catch all the water, a second line a few feet from the first and parallel to it usually will. These seepage or interceptor lines of tile should be sunk into the surface of the impervious layer about one-half the diameter of the tile to prevent underpassing.

This problem is often complicated by the higher land, from which the seepage water comes, being a steeply sloping agricultural area, itself wet from seepage over its entire expanse. Here a parallel series of interception drains, up the entire breadth of the slope and leading to a common outlet, will be needed to drain the entire hillside. For determination of spacing for such a parallel system see Chap. 10 on Location and Design of Underdrains.

A combination of the two last-named types is frequently found in rolling country dotted with marsh pockets, running sloughs, and field watercourses. In such a case the mains and submains should follow the lines of the natural surface drainage (see Fig. 1.2).

6.23. Overflowed Lands. Overflowed lands are low-lying lands along rivers which are overflowed practically every spring or during extra-heavy rainfall or rapid snow melt in the mountains by the river rising over its normal banks or tidal marshes overflowed daily by the rise of the ocean tides. In such cases, the needed first step is the construction of levees along the river banks or the ocean shore, broad and high enough to contain the highest floods. The lands behind the levees must be drained in the manner already described for lands separated from their outlet streams by high impervious ridges. During flood stages the ordinary drainage from the fields must be caught in reservoirs on the land side of the levees.

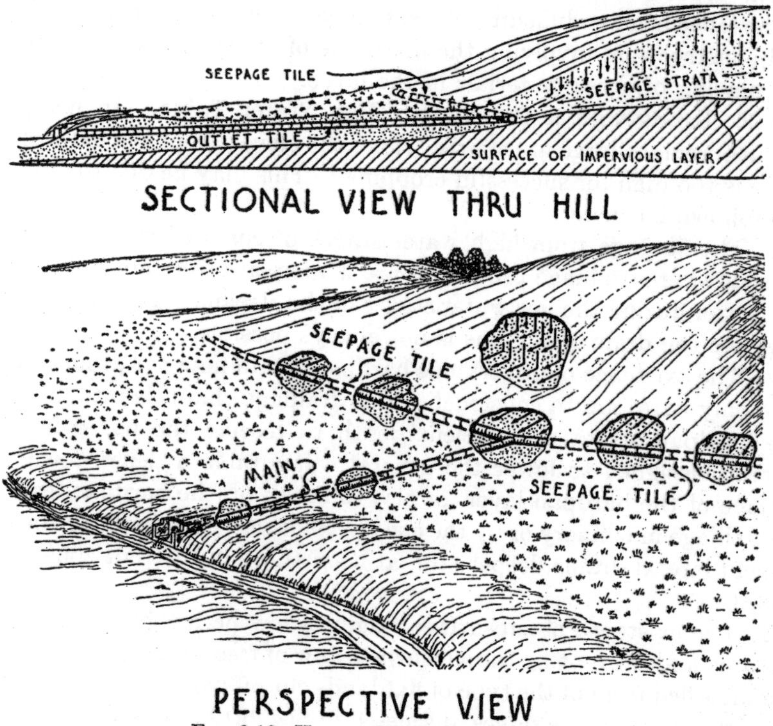

FIG. 6.10. How to drain a seepage area.

If the capacity of the reservoirs is overtaxed during high-water periods, an automatic pumping plant is used to lift the interior drainage water over the levee and empty it into the river or the sea.

Tidal marshes have saline soils. The excess salts must be removed in the manner outlined for drainage of alkali lands in Chap. 13.

6.24. Hydrostatic Pressure. Holding back the ocean tides and confining the waters of streams between levees at high-water stages at once pose the problem of seepage due to hydrostatic pressure through or under the levees and the constant danger of a break-through of water en masse and serious flooding of the lowlands the levees were intended to protect. Among the best-known evidences of this seepage are the so-called "boils" that frequently occur in the lands behind the levees, where the seepage water, under hydraulic pressure from the stream, breaks through the relatively thin shell of surface soil and boils out over the surface like a spring. A temporary method of control of these boils is to build a "chimney" of sand bags around them to the height of the surface of the flood water in the stream; but this is purely an emergency measure and offers no permanent solution.

The problem of prevention of such seepage and of the eventual failure of the levee is most difficult. It is, to a considerable extent, one of design and construction, involving the materials of which the levee is built and the placing of these materials in the levee.

6.25. Lands with a Permanent High Water Table. In many agricultural regions large areas exist where the elevation of the natural water table is too high for successful cropping. This may be due to any one of the following causes:

1. The seepage from high water stages of confined streams, just discussed, where drainage provisions are inadequate to remove all the seepage water during the low water stages of the stream. The only cure is to improve the drainage. This is not always possible.

2. The existence of impervious hardpan strata near the ground surface. The only cures are to construct outlet drainage channels into or through the hardpan strata of sufficient depth to receive the water from field laterals designed to lower the general water table at least down to the surface of the hardpan layer, or else to break up the hardpan layer with deep tillage machinery or by the use of explosives, thus enabling the laying of field lateral drains deeper still, at least through the root zone of crops raised.

3. The existence of subsurface ridges of low permeability surrounding the area and preventing the subsurface flow of free water. This condition is a modified form of the type of flat lands cut off from their outlet streams by higher impervious ridges, discussed in Sec. 6.19. It should be treated in the manner that has been outlined for that type.

4. Lack of sufficient depth in regional outlet channels freely to receive water at all times from field lateral drains designed to give complete drainage for crops raised in the area. This is a common condition in flat areas, whether narrow valley floors or wide prairies. One excellent illustration of many is that of East Muddy Creek in the central part of the Willamette River Valley above and to the east of Albany in western Oregon. This creek is the drainage outlet for a large area of flat land. It quickly rises to or over bankfull stage from every considerable rain and, owing to a crooked alignment and low gradient, the flood stages reduce very slowly. The present depth of the channel is none too adequate as an outlet for tile drainage even at low-water stages of the stream. The only sure cure for such a condition is to increase the capacity of the creek channel by

FIG. 6.11. Comparative cross sections of black spruce trees of about equal age, from adjacent drained and undrained areas in the swamp forest region of northern Minnesota

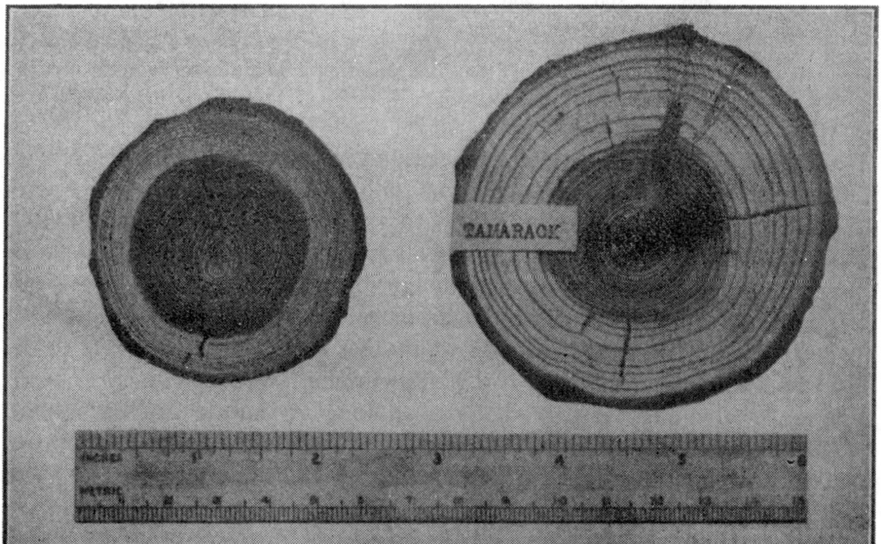

FIG. 6.12. Comparative cross sections of tamarack trees of about equal age from similar areas and locations to those referred to in Fig. 6.11 (cross section at right from drained area).

straightening and deepening it to the point where it will always be effective as an outlet for required lateral drainage.

6.26. Forest Drainage. There has been some complaint from time to time in the public press in the Lake states that so much drainage has been injurious to forest growth. This claim is not borne out by experience in Europe and by investigations in this country. In 1928, Averill and McGrew [8] made a comprehensive study of the effect of drainage on swamp forests in 26 widely scattered areas in six representative counties

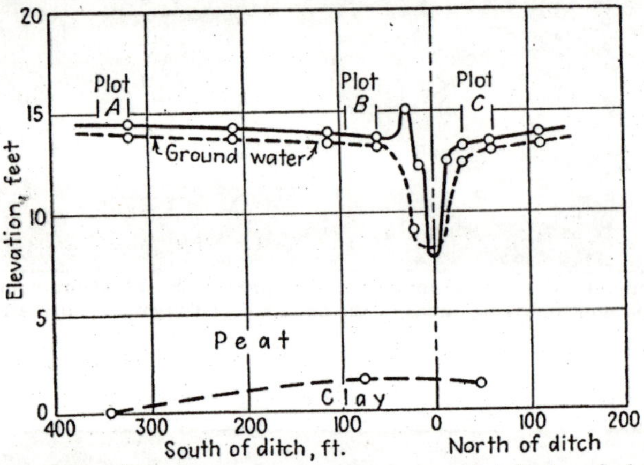

FIG. 6.13. Typical cross section of a northern Minnesota outlet ditch and vicinity, showing rapid rise of water table laterally from the ditch; location a peat-swamp forest in Aitkin County.

of northern Minnesota. So far as drainage is directly concerned, their findings may be briefly summarized as follows:

No evidence was found of injury to trees from thorough drainage. Confirming long experience in northern Europe in forest drainage, it was found that the greatest need of drainage, by trees, is to keep the water changing rather than necessarily to accomplish thorough and deep drainage. It should be noted, however, that trees are naturally deep rooters, and the studies by Averill and McGrew, further sustained by European experience, also clearly showed that, far from having an injurious effect on tree growth, deep drainage enables firmer and deeper rooting and both volume and quality of growth are much greater on drained than on undrained areas (see Figs. 6.11 to 6.13 taken from the report cited).

6.27. Eroded Lands. The control of soil erosion by water is a complex problem involving the use of both surface and subsurface drainage methods closely coupled with soil science and agronomic procedures. It is, therefore, beyond the scope of this book except as concerned with details of drainage fully dealt with herein under specific headings.

6.28. Peat and Muck Lands. Peat and muck land drainage is also an exceedingly complex problem of such size and importance that an entire chapter is devoted to it (see Chap. 14 on The Soil Moisture Control Problem on Peat and Muck Lands).

6.29. Irrigated Lands and the Alkali Problem. Good underdrainage is an absolutely essential accompaniment of irrigation. It, like that of peat and muck lands, is a problem of such importance, size, and complexity that an entire chapter is devoted to it (see Chap. 13).

CHAPTER 7

DESIGN OF OPEN DITCHES

FACTORS IN DESIGN

The following factors enter into the design of any drainage system:

1. Purpose of the drainage system.
2. Outlet: depth and capacity.
3. Ruling gradients obtainable.
4. Character of soil and subsoil.
5. Character of the local rainfall and source of seepage, if any.
6. Watershed: size, shape, topography, and vegetal cover.

These factors must be determined, primarily, by field surveys.

DRAINAGE SURVEYS

In the presentation of this section it is assumed that the reader is trained in the principles and practices of plane surveying. Careful surveying is essential for good results in drainage.

Three surveys are theoretically required. These are the reconnaissance survey, the preliminary survey, and the construction survey. On single farm drainage projects of comparatively small area the last two surveys may often be combined in one.

On the first two of these surveys, aerial photographs are particularly valuable as they show the relative locations of the dry, wet, and flooded areas more plainly and exactly than can be shown in any other way, as is indicated by Fig. 7.1.

7.1. Reconnaissance Survey. The first thing to be done in a survey for both open ditches and tile drains is to go over the ground on foot or in some suitable conveyance and get, from such inspection, as complete an idea as possible of the drainage problem. The following facts should be noted for both open-ditch and tile-drain design:

1. Topography of the watershed area.
2. Size and extent of the area to be drained.
3. Possible locations of the drainage outlet.
4. General character of soil throughout drainage area.
5. Amount of cultivated and uncultivated lands in the general vicinity.
6. Kind and grade of crops.
7. Number, extent, condition, and size of natural waterways.
8. Kind of survey that it will be necessary to make.

In addition to this, the engineer should collect every county, private, state, and government map which will give him any data to assist in the solution of the problem at hand.

The care with which the reconnaissance survey is made will have considerable influence on the other surveys and will determine the amount of time and money that can be expended for any and all of them. It is important that the engineer, after the reconnaissance, have a clear picture

FIG. 7.1. Horrible example of need for auxiliary surface drainage on fertile fields well served by tile. Scene in Palo Alto County, Iowa, May 5, 1951. (*Courtesy of Des Moines Register and Tribune.*)

in his mind of the size and extent and general feasibility of the drainage problem.

7.2. Preliminary Survey. After the reconnaissance survey, the drainage engineer should proceed to obtain the facts enumerated above as accurately as time and conditions will permit. This step is called the *preliminary survey*. The size and character of the entire project will determine the kind of preliminary survey; the necessary survey for a small area, for instance, would be different than for a large one, while that called for where the land was in a narrow valley would not be the same as for a flat swamp. The surveys, especially for public projects, should comply with the state laws, however—whatever they may be.

The preliminary survey will result in a map of the drainage area which will show some topography, the approximate location of the ditch or ditches, and sufficient information to design the ditch. The various items on which information is to be accumulated are the following:

1. Area and shape of watershed.
2. Rainfall data for area and amount of underground seepage.
3. Topographical map of sufficient area to locate ditches.
4. Type and condition of subsoil.
5. Ownership and value of land.
6. Amount of cultivated and uncultivated land in the project.
7. Location and gaging of natural waterways.
8. Natural storage basins.
9. Capacity of possible outlet channels and their depth below the land to be drained.
10. Profile levels along center lines of mains and submains.

Location of Open Ditches. There is no fixed rule or set of rules that will enable one to locate every ditch, as each location presents an individual problem which must be solved by taking into consideration the conditions involved. This calls for original study and the use not only of the principles of surface drainage but also of a great deal of common sense. Parsons [103] gives the following general points to be kept in mind in the location of open ditches:

1. The value to stream flow of straight courses and gradual curves.
2. The desirability of locating drains along property lines wherever possible.
3. The damage to farms of ditches so located as to cut off small inaccessible corners, or extending across them at an angle or in irregular directions.
4. The economy of following natural depressions.
5. The injurious effect of unstable soils upon ditch construction and maintenance.

In flat country it is often possible to run a ditch straight for miles, along established survey lines, to the outlet stream regardless of direction. However, in northern latitudes, where heavy snowfall often drifts and fills open ditches, long east and west stretches of large ditches should be avoided and they should be made to follow, as nearly as compatible with topography and slope, a general northerly and southerly direction. This is because east and west ditches do not receive the full melting power of the spring sun until so late that they are often still obstructed with ice and snow until after the flooded conditions of the spring breakup are past. Their capacity is thus greatly reduced during a critical period, often resulting in damaging floods over adjacent fields. On the other hand, north and south ditches receive the full melting power of the mounting spring sun and open up quickly, thus obviating floods.

Laterals should always be turned downstream a short distance back from the junction, at an angle with the main not in excess of 60 degrees. Otherwise velocity of flow in the main channel will be checked, thus causing undue silting and possible erosion of the far bank of the main by the flow from the lateral.

7.3. Construction Survey. The construction or location survey is that in which the work is staked out on the ground exactly as it is to be constructed. The map and profiles made in the preliminary survey are the guide. Together, they are the basis for the design. On them the locations of the ditches (main, submains, and tile lines) and the grade lines have been drawn. The construction survey can be made only after the design is completed; hence discussion of it will be taken up in Chaps. 8 and 12 covering construction of drains.

As indicated briefly, earlier, on a farm-drainage project it is usually possible to combine the preliminary and construction surveys if the area involved is relatively small—80 to 320 acres or, in rare cases of rather uniform topography, even a full section of 640 acres—with some pauses in the field work to do the office work on the design.

7.4. Equipment for Drainage Surveys. The essentials in surveying equipment for drainage surveys are an engineer's level (preferably equipped with stadia hairs and a reliable compass), a level rod with markings lending themselves readily to stadia readings, 100-foot steel tape, small ax, extensible soil auger with wrenches, blue or black lumber crayon, marker stakes, and on large jobs such as a public ditch and on smaller jobs in rugged topography, a stadia transit. On such jobs, a plane table with full-circle compass and telescopic alidade is often a convenient timesaver enabling completion of the map in the field as the survey progresses. On small single-farm jobs where the topography is not very rugged, the equipment shown in Fig. 7.2 has proved an effective timesaver and the transit will not be needed.

Special Light Equipment for Farm-drainage Surveys. Such equipment consists of the following items: a light, reliable engineer's or architect's wye level with good lenses and equipped with stadia hairs; a self-reading level rod with markings lending themselves readily to stadia readings; a light traverse board (preferably of a size that will hold a map of a standard section, 640 acres, on a scale of 200 feet to the inch); an open-sight alidade; placement pins with knobbed heads; an engineer's 6-inch scale (flat with 10 and 40 scale usually most convenient); a full-circle reliable compass, about 3 inches in diameter, that can be securely fastened to a corner of the traverse board; a boy-scout ax; a 100-foot steel tape with reel; a blue lumber crayon; an extensible soil auger, with wrenches; a good ruby eraser containing a little grit; a 3H and HB drawing pencil, and a pencil-sharpening strip; a pocket scratch pad; an engineer's field book; stakes (half lath for markers, ¾ by 1½ by 6 inches for hubs). Most of these items are shown in Fig. 7.2. A hand level of any standard type is also a convenience in securing approximate elevations on nearby desirable points of secondary importance.

FIG. 7.2. **Most convenient instruments for use in making a farm drainage survey:** (1) lumber crayon; (2) small ax; (3) 100-ft steel tape (in reel); (4) architect's wye level, with tripod, compass, and stadia hairs; (5) level rod good for stadia reading; (6) traverse board; (7) open-sight alidade; (8) placement needles; (9) full-circle compass; (10) engineer's 6-in. scale; (11) soil auger and wrenches; (12) level case; (13) canvas and cord.

ANALYSIS OF LEADING DESIGN FACTORS

7.5. Purpose of the Drainage System. In planning a drainage system it is necessary to know, first of all, what the land is to be used for, as this largely fixes the extent of drainage required, hence is a leading factor in the economy of the design. Powers and Teeter [110] and the engineers of the U.S. Department of Agriculture Soil Conservation Service agree with the authors in recognizing three major classes of drainage according to the degree of drainage required. These classes are listed as *A*, *B*, and *C* drainage, described as follows:

A Drainage. This class includes the most complete drainage that it is reasonably possible to secure, a system that will quickly relieve excess wetness in the root zone, resulting from mean maximum rains. It is concerned with lands of unusually high productive capability, when well drained, which, under intensive agriculture, will give a high return in truck crops, small fruits, and other special crops. In this case complete and continuously effective tile and auxiliary surface drainage is a first essential, often fully justified, even at exceedingly high first cost.

B Drainage. This class will remove surface water and eliminate ponding within 24 to 48 hours after rain ceases and will begin to remove excess wetness from the root zone soon after heavy rainfall starts. It is concerned chiefly with those widely distributed large areas which, when fairly well drained, are suited to general farming of potentially high production capability for standard farm crops, including small grains, hay crops, and most forage and fiber crops. The drainage requirements of such lands are, in their early development, systematic provision for quick removal of surface water and elimination of ponding. These should be followed progressively, as returns from the land justify the expense and give promise of increase in returns, by gradual increments of tile drainage which will relieve the root zone of excess wetness within a reasonable time. The rate of drainage must be sufficiently consistent with the normal development of the crops to prevent damage and to permit deeper rooting as the season advances. Development of such a system may be by progressive steps over a period of 10 years or more. The construction should follow a carefully worked out progressive and comprehensive plan, beginning at the outlet and branching out upstream to include units drained by specific submains and their laterals. The earlier steps in this development, having to do with surface drainage, lie in the class known as *C* drainage.

C Drainage. This class is mainly open-channel drainage for removal of surface water and elimination of ponding soon after a rain stops. It is chiefly concerned with lands where raising of livestock and feeds for livestock is the principal objective. The drainage required is of a character that will stimulate good pasturage, hay crops, and those rougher forage crops fairly tolerant of wet soil conditions. Only rarely is tile drainage justified and then only to a limited extent to meet some special local problem.

In the agricultural development of virgin lands with potentially good production capability, the normal order of drainage development starts with class *C*, advances gradually into class *B* and, in selected areas of good promise, through class *B* into class *A*.

7.6. Required Capacity of the Ditch. The required capacity of a given open ditch will depend mainly (1) on whether it is to be an outlet for tile drains, to furnish underdrainage in lieu of tile drains, or merely a surface-flow carrier and (2) on the rate of water removal required by the class of drainage selected. Hence, in fixing the needed capacity, not only must the total water it must carry off resulting from any given storm period be known but, more important, the maximum time allowable for removal of the free water without damage to the crops must be determined. This will vary widely—2 to 4 hours for sensitive truck crops such as potatoes at the tuber-setting stage, 2 days for corn and small grains, 4 days to a week for the general run of the more water-tolerant grass crops.

7.7. The Drainage Coefficient *D*. The foregoing makes it clear that the required *rate* of drainage is the key factor in establishing the needed capacity of a ditch. This rate, expressed as the depth in inches of water that is drained off from a given area in 24 hours, or as the cubic feet per second drained off per acre, is the *drainage coefficient D*. It can be determined in several ways according to the local sources of information available.

1. The maximum total water to be removed by the ditch from a single storm is the total rainfall from the heaviest known rainstorm or series of rainstorms covering the entire area to be drained, through a number of consecutive days, diminished by evaporation, transpiration, and deep seepage, through the total time allowed for removal of the free water. If the ditch is merely a carrier of surface water, there must also be subtracted from the total rainfall from the given storm the amount taken up by the soil surface through infiltration (see Chap. 3) and its eventual removal through underdrainage. For example, if the total rainfall from the given storm is 6 inches, with 4 days allowed for removal of the free water, during which time evapo-transpiration removes ¾ inch, deep seepage ¼ inch, and tile drainage 1 inch, the surface water to be directly removed by the ditch is 4 inches, the required daily rate of removal being 1 inch. This 1 inch is *D*.

2. If all records needed by the foregoing method are not available, *D* may be determined by use of Ramser's coefficient *C* in the rational formula, Chap. 4.

3. Still another method is to measure the runoff corresponding to the maximum rate of rainfall from the total watershed through the outlet stream. By this method for the larger streams, where there are regular gaging stations, the average value may be taken from the government records for the given time of year.

The unit of measure most commonly used for the drainage coefficient is inches of depth removed in 24 hours. For the most part this unit will be used in this book. Another unit used in many water-consumption computations is cubic feet per second per acre for a fixed period of time.

The result from the first method (1) will be in inches of depth for the entire period. This value divided by the number of days allowed for completing the drainage will give the required *D* value. The result in each of the last two methods will be in cubic feet per second for the whole area. This result, divided by the area of the watershed in square miles, will give the drainage coefficient expressed as cubic feet per second per square mile, or, divided by the area of the watershed in acres, as cubic feet per second per acre.

7.8. Conversion of D to Cubic Feet per Second per Acre. Discharge formulas are expressed in terms of cubic feet per second, Q for discharge from the total area, q for the discharge per acre. Hence D values, to be usable in drainage design, must be converted to equivalent cubic feet per second per acre.

Note the following illustrative example, using an area of 1 acre and D as 1 inch as a basis:

1 acre = 43,560 sq ft (water 1 ft deep = 43,560 cu ft per acre)
1 acre-inch = 43,560/12 = 3,630 cu ft
24 hr = 86,400 sec
q = cfs per acre = 3,630/86,400 = 0.0420 cfs per acre

Table 7.1 gives the value of q for several different values of D.

TABLE 7.1. VALUES OF q FOR VARIOUS DRAINAGE COEFFICIENTS

Drainage Coefficient D, Inches per 24 Hours	*q, Cubic Feet per Second per Acre*
1	0.0420
15/16	0.0394
7/8	0.0367
13/16	0.0341
3/4	0.0315
11/16	0.0289
5/8	0.0262
9/16	0.0236
1/2	0.0210
7/16	0.0184
3/8	0.0157
5/16	0.0131
1/4	0.0105
3/16	0.0079
1/8	0.0052
1/16	0.0026

7.9. Selection of the Drainage Coefficient. For small drainage areas, a small error in the drainage coefficient would make very little difference one way or the other. It would be best, of course, to have the error on the side of safety by making it too large rather than too small, as this would affect the size of the ditch very little and not materially increase the cost. In large projects, however, an error in the drainage coefficient of even a small amount would mean a financial loss to the district building the ditch. It is necessary, therefore, that a drainage engineer be employed on large projects in order that the drainage coefficient may be selected by one who has had considerable experience.

Where there is no record of Q for the outlet stream for the given watershed, it will be well to compute D, if possible, by all three methods just discussed, as a check.

7.10. Range of D Values. The range of D values for open ditches varies from ¼ to 4 inches in extreme cases, according to topography, width of area normal to the ditch, slopes of approach, surface cover, and infiltration rate of the surface soil. For tile drainage the variation will usually range from ⅛ to ¾ inch or possibly 1 inch in extreme cases. D for tile drainage is more fully discussed in Chap. 10.

7.11. Sutton's Curves. In some cases on large watersheds or regions involving several watersheds where engineers of state or Federal agencies

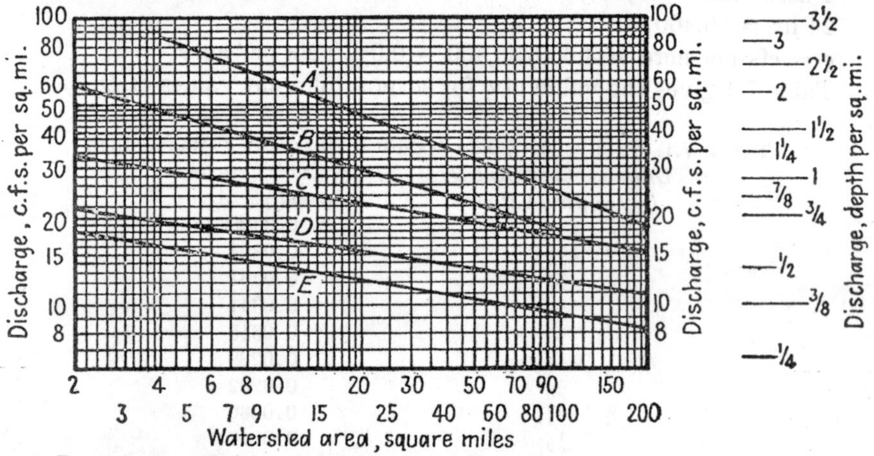

A. For good protection from overflow (not maximum runoff).
B. For excellent drainage.
C. For very good agricultural drainage in Ohio, Indiana, Illinois, Iowa, and northern Missouri; for good agricultural drainage in Kentucky and southern Missouri.
D. For fair agricultural drainage in Ohio, Indiana, Illinois, Iowa, and northern Missouri. (This curve is also accepted by the Minnesota State Division of Engineering and Water Conservation and by the Drainage Office of the Minnesota Agricultural Experimental Station, for the central and southern sections of Minnesota.)
E. Standard recommended jointly to the Soil Conservation Service by the two above-named Minnesota state agencies, for use in the Valley of the Red River of the North.

FIG. 7.3. Runoff curves for the central and northern Mississippi Basin. *A, B, C,* and *D* standards established by U.S. Department of Agriculture; Soil Conservation Service for use by Central District Drainage Camps. (*Courtesy of Soil Conservation Service and Agricultural Engineering.*)

have long been familiar with the regimen of the outlet streams, they have worked out runoff curves giving stream discharges in terms of cubic feet per second per square mile, based on size and character of watershed and behavior of the outlet stream. The use of these curves is recommended in drainage and flood-control design problems for specified regions. A case of their use in the work of the Soil Conservation Service is presented by Sutton [146]. The practical features of this case as regards runoff rates are summarized in Fig. 7.3, which recommends values of D for several regions and specified purposes in the central and northern Mississippi Basin, with one curve added for the basin of the Red River of the North, recommended jointly by the engineers of the Minnesota State Department of Drainage and Water Conservation and of the State Agricultural Experiment Station. These curves, the result of long and careful study by engineers of high standing, are a reliable guide in the selection of proper values of D for open ditches in most cases of comparatively flat watersheds generally comparable to those from which the curves were drawn.

7.12. The Outlet. The outlet is a prime factor in drainage design. The proper value of D having been determined, the next step is to determine whether the regional[1] outlet channel has sufficient capacity to obviate overflow of the adjacent lands. In other words, is the flood stage of the regional outlet channel enough below the land to be drained to receive the water from the main ditch at all times. On flat lands where the boundaries between watersheds are often rather indefinite, there is possibly more than one stream which might serve well as a regional outlet for the given ditch. It must then be determined in which watershed the ditch really lies, and which of the several streams will be the best and least expensive outlet. There may also be several points on the outlet channel where the ditch might discharge its water. It must then be decided which of these locations will provide the best outlet under all conditions. It may be that the normal regional outlet is already carrying its capacity load before the new ditch enters. This may be due to a too shallow channel or to fouling of the stream by silting or by the growth of brush. In this case the channel should be cleaned and probably deepened and otherwise enlarged to provide ample capacity. Such increase in flood flow is almost sure to affect the drainability of lands farther downstream detrimentally, thus giving rise to serious legal and economic problems with the affected downstream landowners. All such matters should be looked into in the study of the outlet problem.

Depth to the water surface in the outlet channel at normal flow should be enough to permit the water surface in the new ditch at normal flow to be slightly below the average rooting depth of most field crops—not less than 4 feet for the best results. Where the ditch is to serve as an outlet for tile drains, the depth must be much greater—at least a foot below the mouth of the tile main (see Chap. 10). Lack of adequate outlet practically assures the ultimate failure of the drainage system. One of the disastrous results from lack of an adequate outlet channel is clearly illustrated by the experience of the Minnesota Agricultural Experiment Station on its Coon Creek Peat Experimental Fields. The final drainage outlet for the entire Coon Creek Basin of many thousand acres is Coon Creek itself. The bed of this creek was improved a year or two before the installation of tile drains on the Peat Experimental Fields. Even after the improvement of the creek channel, its capacity was insufficient to

[1] By regional outlet channel is meant the creek, river, or other natural channel carrying the drainage from an entire large watershed or region.

carry off the water from sudden heavy floods over the entire basin, because of failure to cut through a solid rock ridge across the creek bed a half mile downstream from the experimental fields. This ridge, acting as a dam, greatly reduced the effective gradient of the creek for miles upstream.

The tile drains had been installed on the bog in 1919 and during that season and the spring of 1920 had seemed to be functioning effectively; but a very heavy rainfall early in June of that year flooded the entire basin. The inability of the creek channel to remove this flood quickly resulted in the complete inundation of the tiled experimental fields (see Fig. 7.4) for about 10 days following. As a result, the colloidal material

FIG. 7.4. Effects of an inadequate drainage outlet channel: Minnesota Peat Experimental Fields at Coon Creek, Anoka County, after the flood of June, 1920.

NOTE: These fields were completely tiled out in 1919, but the utter inadequacy of Coon Creek as an outlet completely clogged the tile-drainage system which was practically ineffective for the entire season of 1920 because of puddling of the joints.

in the peat puddled the soil throughout the root zone and sealed the joints of the tile drains to such an extent that no adequate drainage was secured for the rest of that season, and except for the tougher grasses, the crops on the experimental fields were a total failure. Not until well into the summer of 1921 did normal drainage conditions again obtain, although inspection revealed that the tile lines were entirely free of internal obstructions and their gradients were unimpaired.

Detailed essentials in the design of open-ditch outlets and of tile outlets differ appreciably. Outlets for tile drainage systems are discussed in Chap. 10.

7.13. Ruling Gradients Obtainable. Other influences being equal, the steeper the drainage gradient, the faster the water will run. Hence it is apparent that the amount of water a drain of a given size will carry depends largely on the rate of fall of the drain. This must not, however, be interpreted to mean that a drain of a given size with a fall of $\frac{1}{10}$ of a foot in 100 feet will carry just half as much water as a drain of the same size with twice as rapid a fall. Best practice limits gradients on open ditches, as far as physically and economically feasible, to those which will produce neither erosive nor silting velocities. This theoretically ideal velocity for any given type of soil is usually called the *safe velocity*. The ideal is never attained in practice because of the wide variations in soil character and condition in the stream bed, in stream stages at different stages, and various other factors that affect velocity and the general behavior of the stream at different points and times. Keeping the gradient of an open ditch as uniform as possible is good practice, especially on the main, but it must often be adjusted to meet the requirements of the laterals. This circumstance often causes considerable variation in the gradient of the main. A trial working grade line, apparently supported by the information available from the surveys, should be laid on the preliminary profile, but details of design may uncover difficulties not at first apparent and force such major changes in the trial grade line as to require further field study to enable perfecting of the design. The selection of a suitable gradient for a given open ditch is greatly influenced and largely determined, in good practice, by the following factors: rate of drop of the natural surface along the ditch, character of the soil and subsoil through which the ditch runs, depth of normal water surface in the channel of the outlet stream below the adjacent land surface, depth of normal maximum flow of water in the ditch, shape of the ditch cross section, friction of the flowing water against the bed of the channel, and velocity of flow in the ditch (the most effective factor, itself dependent on the adjustment of all the other factors just above listed).

It is impossible to set any hard and fast rules for gradients to be used in open ditches as all elements that enter into design of any given ditch must be adjusted to result, as nearly as possible, in a safe velocity, with especial attention to depth of flow. The designer will do well to keep his maximum velocity below 4 feet per second because it is common experience that velocities as high as 4 feet per second are apt to cause serious erosion on unprotected ditch slopes even in dense clay soils. This precaution is especially timely because, on a rising stage in a stream, it is probable that a velocity considerably higher than the design velocity will develop, especially in the surface. Under these conditions, serious erosion of the ditch bed may occur.

7.14. Safe Velocities of Flow in Open Ditches. Fortier and Scobey, as a result of estimates by a number of irrigation engineers of wide experience, suggest permissible canal velocities shown in Table 7.2. For ditches of average large size experience has shown that velocity must be held within a range of 2 to 4 feet per second, and the proper gradients will usually lie between 2 and 5 feet per mile, with an average of about 3 feet. Sometimes the fall of the natural surface in the general direction of flow is so great that the ditch gradient cannot safely follow parallel to it. In such a case either a longer circuitous route must be worked out, or grade reducers must be introduced at strategic intervals to reduce the ditch gradient to a safe value.

TABLE 7.2.* PERMISSIBLE CANAL VELOCITIES

Original material excavated for canal	Velocity, feet per second, after aging, of canals carrying—		
	Clear water, no detritus	Water transporting colloidal silts	Water transporting noncolloidal silts, sands, gravels, or rock fragments
(1)	(2)	(3)	(4)
Fine sand (noncolloidal)	1.50	2.50	1.50
Sandy loam (noncolloidal)	1.75	2.50	2.00
Silt loam (noncolloidal)	2.00	3.00	2.00
Alluvial silts when noncolloidal	2.00	3.50	2.00
Ordinary firm loam	2.50	3.50	2.25
Volcanic ash	2.50	3.50	2.00
Fine gravel	2.50	5.00	3.75
Stiff clay (very colloidal)	3.75	5.00	3.00
Graded, loam to cobbles (noncolloidal)	3.75	5.00	5.00
Alluvial silts when colloidal	3.75	5.00	3.00
Graded, silts to cobbles (noncolloidal)	4.00	5.50	5.00
Coarse gravel (noncolloidal)	4.00	6.00	6.50
Cobbles and shingles	5.00	5.50	6.50
Shales and hardpans			

* Reprinted by permission from "Permissible Canal Velocities" by Fortier and Scobey, *Trans. A.S.C.E.*, Vol. 89, p. 940, 1926.

7.15. Grade Reducers in Open Ditches. A grade reducer in an open ditch is, in plan and effect, a broad-crested trapezoidal weir whose crest lies in the base of the ditch at the upper level. It discharges into a silting basin whose bottom is usually 2 feet or more below the bottom of the ditch at the lower level, with the top of the overflow wall in the base of the ditch at the lower level. Figure 7.5 shows the plan of a drop based on the investigations of Morris and Johnson [92] and approved and widely used by the Soil Conservation Service in the Pacific Region. The symbols used in the figure are defined in the legend.

Grade reducers in open ditches are relatively expensive and, as may be seen from Fig. 7.5, are of complicated design. They should be avoided when some other safe alternative can be worked out.

7.16. Transition at Change of Ditch Width. Wherever for any cause, as for example a sudden increase in the local volume of flow or because of the entry of a lateral, a change in width of ditch base is required, the change should not be abrupt but by gradual transition. If the total change does not exceed 4 feet, with the low velocities that usually obtain in drainage ditches the total transition need not exceed 100 feet in length, all in one section. If the change is greater than 4 feet, a longer transition is needed. In that case a good practical rule to follow is to make the transition of such length as to allow an additional width of 2 feet for each 100 feet of length. For example, with the exact point of total change at station 100 with an increase from a 6-foot width to a 12-foot width, the transition would be 300 feet long. It should be cared for in the setting of the slope stakes. For example, at and above station 103, 6 feet; at station 102, 8 feet; at station 101, 10 feet; and at station 100 and below, 12 feet.

7.17. Unavoidable Excessively Low Gradients. Several possible conditions occur where it is practically impossible to secure a nonsilting gradient in an open ditch. Notable among these are the following:

1. Where large watersheds have only a slight natural fall, possibly of only a few inches per mile and are barred from their outlet channels by a ridge of relatively impervious material.
2. Where the elevation of the bed of the regional outlet stream is too high to permit the depth desired in the main ditch of the drainage system.
3. Where subsoil or other subsurface conditions may be such as practically to prohibit the desired depth of main ditch. (For example, either a sand stratum or a rock ridge of considerable extent near enough to the surface to make a shallower ditch economically desirable or even imperative.)

In such cases as those listed, the flatter gradient with its almost certain tendency to heavy silt deposit and the frequent need for cleaning of the ditch at high cost is often the only reasonable solution. The only alternative is a cumulative deepening of the ditch, proceeding downstream to the outlet, there providing a storage reservoir to receive the drainage flow, and an automatic pumping plant to lift the water into the regional outlet channel. This solution cannot be applied to the third case mentioned above. Provided the normal flood flow in the regional outlet channel is at a sufficiently low elevation to receive the discharge from such a main, it is physically possible to solve the first case, without pumping, by use of a large tile main, cumulatively deepened; but the tile main is apt to prove prohibitive in first cost.

7.18. Influence of Change to Flatter Grade. The restrictions imposed on velocity of flow in unlined channels also limit it to streamline flow.

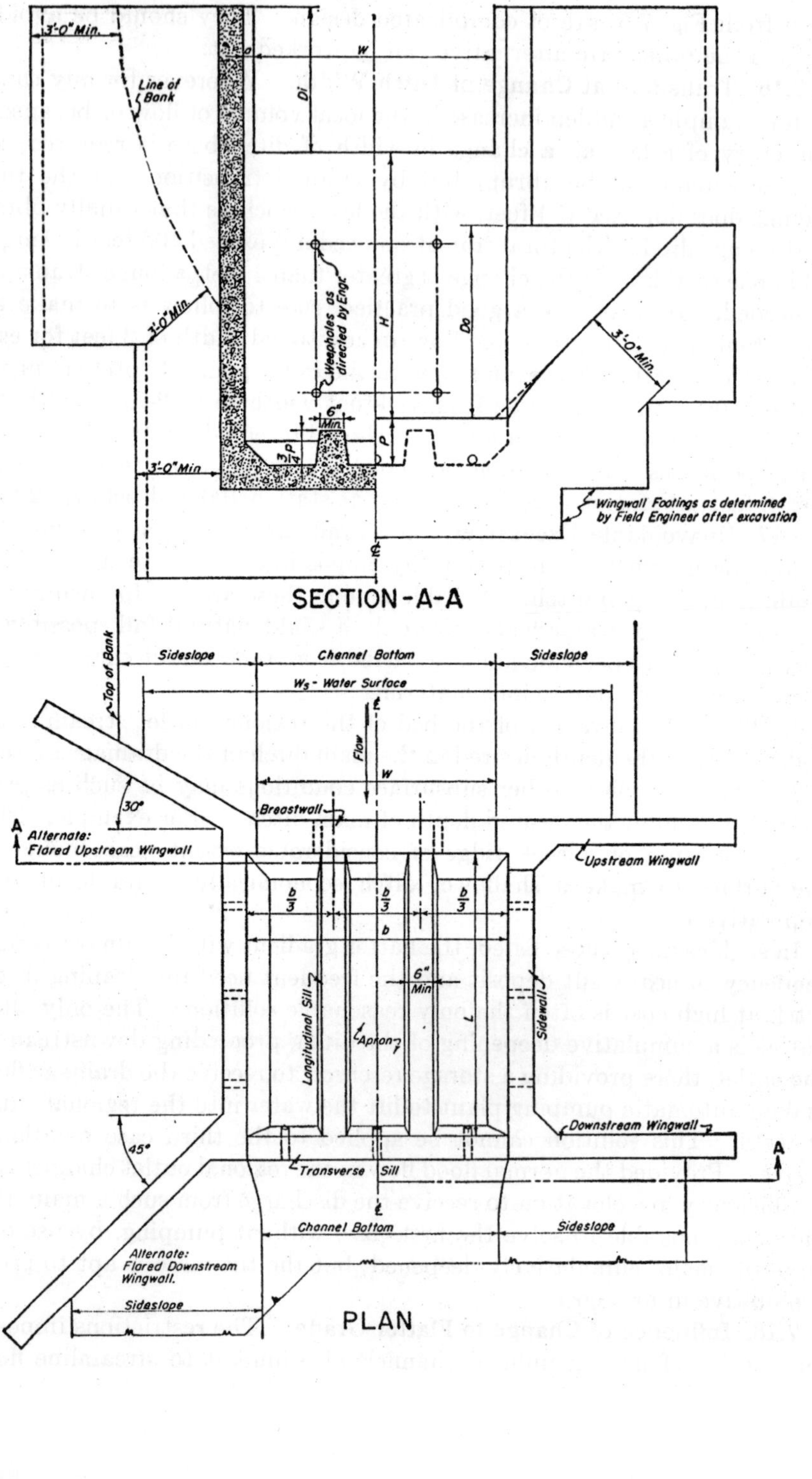
3'-0" Min.
Line of Bank
W
Di
Weepholes as directed by Engr.
H
Do
3'-0" Min.
3'-0" Min.
6" Min.
P
3/4 P
3'-0" Min.
Wingwall Footings as determined by Field Engineer after excavation
SECTION-A-A
Top of Bank
Sideslope
Chennel Bottom
Sideslope
W_S - Water Surface
Flow
W
30°
Breastwall
A
Alternate: Flared Upstream Wingwall
Upstream Wingwall
b/3
b/3
b/3
b
6" Min
Longitudinal Sill
Apron
L
Sidewall
45°
Downstream Wingwall
A
Transverse Sill
Channel Bottom
Sideslope
Alternate: Flared Downstream Wingwall.
Sideslope
PLAN

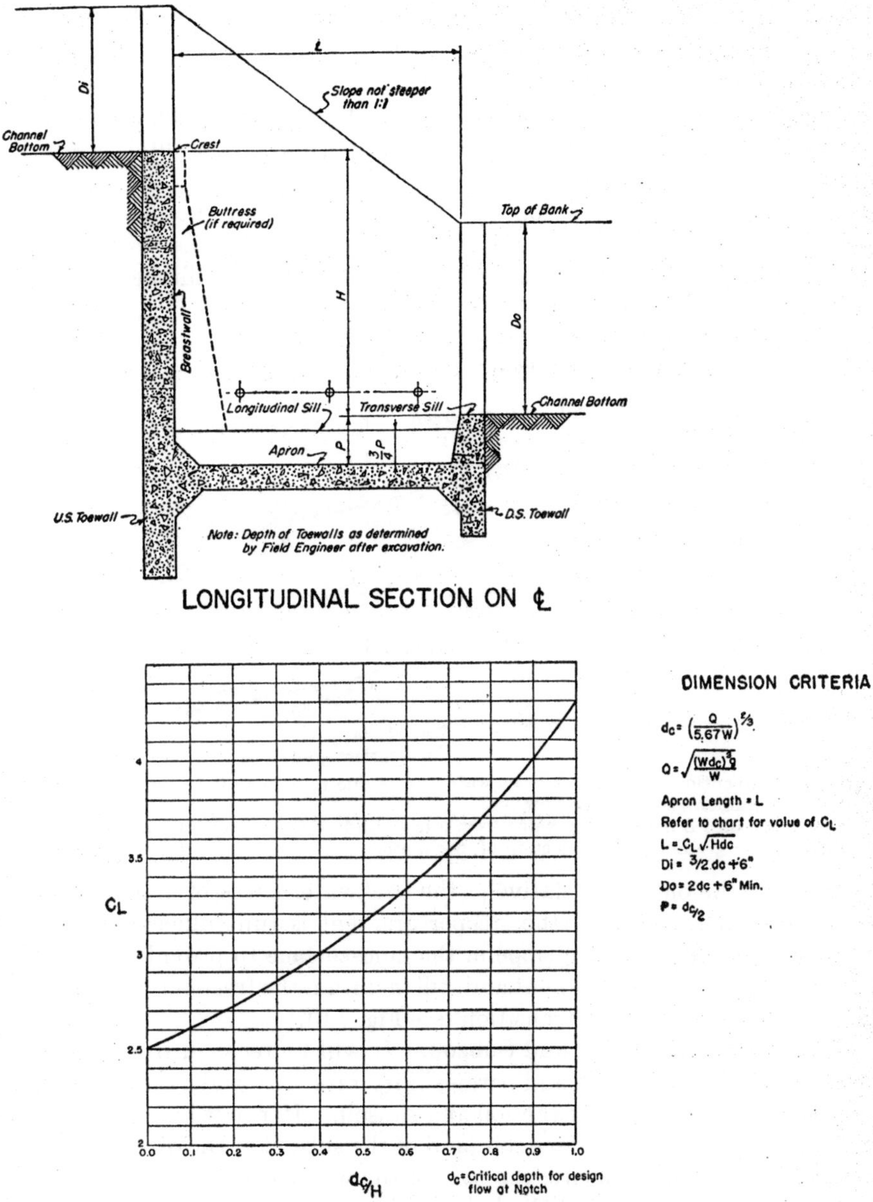

Fig. 7.5. Standard drop structure. (*Courtesy of Soil Conservation Service, Pacific Region.*) Q = discharge, cfs; W = width of crest, ft; H = height from floor of stilling basin to crest of notch, ft; L = apron length, ft; d_c = critical depth, ft; P = depth of stilling basin, ft; C_L is an empirical coefficient of apron length; D_1 = depth of notch, ft; D_0 = depth of downstream walled channel, ft.

Hence, ordinarily, the hydraulic jump will not occur in such channels except where drops are used; but backwater curves occur at marked changes from a mild gradient to a milder gradient. The backwater curve extends wholly *upstream* from the point in the stream surface vertically above the point of change of grade in the bottom of the ditch channel, according to Woodward and Posey [163]. At this point, the width of ditch bottom and the side slopes of the ditch remaining the same on both gradients, the depth of flow may, in practice, be taken as the depth of normal flow on the flatter gradient. In earth drainage ditches this fact

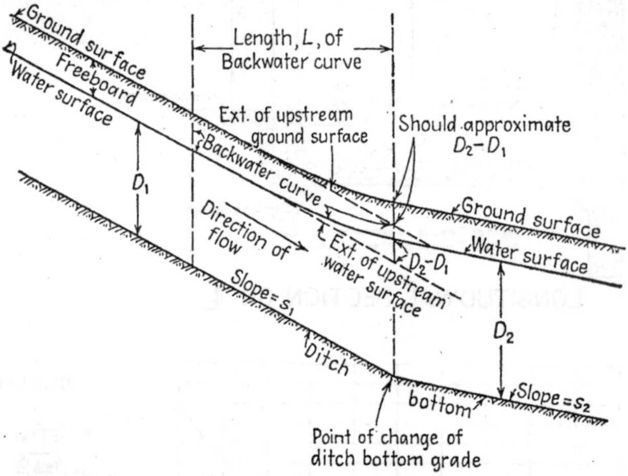

FIG. 7.6. Conventional diagram of backwater curve due to change from one flat gradient to a flatter gradient (streamline flow).

NOTE: Distance L may be any value from a few hundred feet to several miles, according to difference in gradients, cross section of ditch, etc.

is apt to be the only one relative to the backwater curve of major interest to the designer, provided the steeper gradient is run sufficiently downstream from the change of slope in the land surface to provide the depth required by the flatter gradient and still maintain the freeboard considered necessary throughout the transition (see Fig. 7.6).[1]

7.19. Character of Soil and Subsoil. It will be recalled from Chap. 3 that both infiltration rate and permeability rate vary widely according to character and condition of the soil and subsoil. Hence it is clear that the infiltration rate of the surface soil (and of the subsoil in the ditch channel) determine, in large measure, the part of the water from rainfall and from melting snow that an open ditch must have the capacity to handle, that is, what the capacities of the given open ditches and their spacing must be. Further, as the permeability rate through the soil and subsoil largely determines the rate at which subsurface flow reaches the tile drains, it is

[1] If the designer has further interest in the backwater curve he is referred to Refs. 67 and 163.

a leading factor in determining the required size and spacing of the tile drains in the given system. No proper design for drainage can be worked out without giving particular attention to all the foregoing and following relations. The formulas used in design all show the factors determining velocity of flow to be the rate of slope or simply the slope s of the ditch, already discussed, the shape of cross section of the ditch, represented by the hydraulic radius r, and the roughness factor n appearing in both the Chézy-Kutter and the Manning formulas, all of which are more or less influenced by the nature of the soil.

7.20. Shape of Cross Section of Open Ditches. It has just been shown that the shape of cross section of an open ditch is closely related to both the velocity of flow and the character and behavior of the soil and subsoil through which the ditch runs. Theoretically, the most efficient cross section for an open ditch is a semicircle, as the wetted perimeter p would always be a minimum and its hydraulic radius a maximum. However, a semicircular cross section is impractical from both a construction and a maintenance standpoint. Hence it is necessary to use a trapezoidal form, which closely approximates the advantages of the semicircular form and is readily constructed and maintained. That particular trapezoidal cross section should be sought which will give the largest possible hydraulic radius compatible with the depth made necessary by natural surface conditions and with safe velocity for the given soil and subsoil through which the ditch runs. In the ideal trapezoidal form for any given case, a definite theoretical relationship exists between the angle of the side slopes with the base, depth of flow, and width of base. Standard works on hydraulics show by the calculus that the theoretically most economical cross section under favorable structural conditions is

$$b = 2d \tan \frac{\theta}{2} \tag{14}$$

in which b is the width of base, d is the flow depth, and θ is the angle of the side slopes with the horizontal. However, local topographic, soil, crop, and economic conditions and certain practical construction and maintenance limitations usually make necessary considerable departure from the relationship shown in Eq. (14), which can, therefore, serve only as a general guide.

The late E. R. Jones held that, in general, there is no proper place for an open ditch with a bottom width less than 4 feet and a maximum flow depth less than 6 feet. No mention is made of slope. It will be seen that, where the side slopes are $1\frac{1}{3}$ to 1, Jones's minimum ditch practically follows Eq. (14) because where $b = 4$ and $d = 6$ the equation evaluates to $b = 0.67d$, very nearly making $b = 4.0$.

There are two other practical reasons why, in the majority of cases, the minimum base width of a ditch should be 4 feet or more. A narrower ditch of appreciable depth is difficult of economical construction with almost any equipment commonly used for ditch digging. A narrower ditch will silt up or otherwise clog more rapidly than will one 4 feet wide or more and will need far greater, more frequent, and more difficult maintenance. If a narrower ditch must be used (as local or temporary conditions frequently require), it must usually be dug either by hand or by special machinery apt to be of a cumbersome character, resulting in higher unit cost. Furthermore the minimum ditch suggested by Jones is conservative as only under rather unusual conditions will a ditch not over 6 feet deep be an effective outlet for tile drains.

7.21. Side Slopes. The side slopes of an open ditch should, under no circumstances, be steeper than the angle of repose of the type of material through which the ditch runs. With practically all types of soil the angle of repose is different in air than it is under water. Table 7.3 gives the approximate angle of repose, expressed in terms of horizontal run per foot of rise, for the major soil types. Solid rock will stand vertical if desired; loose laminated rock ledges will usually stand at ¼ to ½ to 1; and many tough fibrous peats will stand vertical or nearly so for long periods, but in time they tend to weather to slopes varying from about ½ to 1 to ¾ to 1.

Table 7.3. Repose Slopes of the Major Classes of Loose Soils

Kind of soil	Approximate angle of repose		
	In air	Under water	Notes
Sand	2 or 2½ to 1	3 or 4 to 1	According to fineness
Average agricultural loams	1½ or 1¾ to 1	2 or 2½ to 1	
Clay	1 or 1¼ to 1	1¼ or 1½ to 1	

7.22. Spoil Bank and Berms. Up to the early 1930's it was common practice to make the original side slopes conform closely to those shown in the left-hand column of Table 7.3. Most of the open drainage ditches in the United States have been dug since the beginning of the twentieth century. On the majority of them, little or no maintenance has been done at any time with the result that few are any longer functioning effectively. Deterioration has been hastened in many instances in the past from piling the excavated material in high, irregular spoil banks at inadequate distances back from the edge of the ditch bank. This space is called the *berm*. The spoil banks, thus placed with an inadequate berm, have imposed a heavy overload on the normal ditch bank, causing failure thereof with bulging of the banks and caving of large masses of the natural soil of the bank into the ditch. Thus the flow of the water is obstructed and eventually stopped altogether. This condition has been aggravated, too, by severe erosion of the spoil banks from heavy rainstorms, thus washing large quantities of soil back into the ditch.

Proper width of berm depends, in any case, on the stability of the soil through which the ditch is dug and on the weight of the spoil banks per square foot of the area they occupy. Parsons [103] states: "As the line of cleavage of the wedges of soil which tend to cave from the sides of any trench is always of curved shape, if the spoil banks were deposited outside the surface extremity of this cleavage line their sliding effect would be entirely prevented." He suggests using 100 pounds per cubic foot of wet soil as an average in computing the weight of waste bank from a given size of ditch (see also Secs. 8.5 and 9.7).

Extended experience since about 1920 has shown the error of using too steep side slopes on the ditch, and heavy concentrated spoil banks, and inadequate berms, all three of which have greatly increased the maintenance requirement, which, too, has been greatly neglected. The result has been a strong trend toward side slopes flat enough not only to obviate side erosion and slippage but to make it possible for the slopes to be cropped above the normal flow line to a continuous grass cover which can be cut with a hay mower. On this account it has become common practice to make the side slopes of drainage ditches as flat as 2, 3, or 4 to 1 except in the denser clay loam and clay soils where 1½ to 1 and even 1 to 1 side slopes are still commonly used. The flatter slopes add considerably to first cost, but this is soon overcome by the increased efficiency of the ditch and the great decrease in maintenance cost. Another advantage of the flatter slopes is the greatly increased storage capacity per foot of rise at flood stages, thus preventing overflow and the flooding of adjacent fields. On pumping jobs where electrical pumping must be off-peak this is important.

7.23. Spreading the Spoil Banks. Since about 1920, common practice has been to grade the spoil banks down to a height of 1.5 to 2 feet at the brink of the ditch banks and to spread the material over the adjacent fields in a continually thinner layer until it merges in the natural surface at a distance of a few rods from the ditch. This practice enables the ground to be cropped right up to the brink of the ditch and, at the same time, eliminates the weed-catching and weed-breeding nuisance created by the old-type high rough spoil banks. It also distributes the load on the ground surface near the ditch and thus largely removes the danger of the spoil banks causing slips and cave-ins on the ditch slopes.

Some of the older authorities object to the flatter slopes and to the cropping of the same on the ground that high water is apt to kill the vegetation, that in many soils flat slopes erode worse than steep slopes, and that it is difficult to secure proper maintenance. Some also object to spreading the spoil banks on the ground that the great mass of the spoil-bank material is far less fertile than the usual topsoil of the field and that, thereby, the cropping capability of the area covered by the spread spoil banks is greatly reduced.

Experience since about 1920 and particularly since about 1930 does not support these contentions. As far back as 1915 both of these practices were used with success and satisfaction by farmers in some areas in

FIG. 7.7. Open ditch with spoil banks spread.

southern Minnesota (see Fig. 7.7). In 1944, Lewis A. Jones, Chief of the Division of Water Conservation and Drainage of the U.S. Department of Agriculture Soil Conservation Service, stated that these practices were being used in the ditch rehabilitation work of the Service in cooperation with county officers, ditch boards, and farmers of the central Mississippi Valley states. In a recent letter Jones further says:

The Soil Conservation Service in rehabilitating old ditches and constructing new ditches is still recommending 2 to 1 slopes in most soils. . . . The leveling of the spoil banks has proven very attractive to land owners and in many cases they are insisting on such work being done. . . .

There are many instances where the fertility of the area over which the waste banks have been leveled is not as high, but the farmers insist they can build up the fertility in a short time and they can use their land up to the edge of the ditch. The practice materially lowers the cost of maintenance on ditches.

This last statement is strongly supported by Dr. F. J. Alway, retired Chief in Soils at University Farm, St. Paul, Minn., and by his successor, Dr. C. O. Rost, both of whom maintain that the spread spoil banks can be restored to full fertility in 2 years time or less by proper fertilizer treatment and soil management. These two authorities have proved their statement in Minnesota, and they have told of doing the same thing in Nebraska, years ago.

7.24. The Roughness Factor. In Table 5.1, page 115, are given the values of the roughness factor n customarily used in design with both the Chézy-Kutter and the Manning formulas, but the following values of n, resulting from more recent investigations, are recommended for use in open-ditch design.

Ramser's Recommendations for Earth Channels (especially adapted to the Middle West, East, and South). As the result of extended study on existing open channels under widely varying conditions throughout the South and the Middle West, Ramser [111] recommends values of n in Kutter's and Manning's formulas as follows:

1. $n = 0.025$ in design of large channels, rolling country, high velocity, and sufficient low-water flow to prevent rapid vegetal growth; slick silt lining the wetted perimeter; maintenance.[1]

2*a*. $n = 0.030$ in design of large channels, in rolling country, with sufficient low-water flow to prevent rapid vegetal growth; moderate erosion; maintenance.

2*b*. $n = 0.030$ in design of large channels, flat country, with fairly large low-water flow; no appreciable erosion; annual clearing.

3*a*. $n = 0.035$ in design of small channels, in rolling country, with small low-water flow; erosion causing some irregularities; maintenance.

3*b*. $n = 0.035$ in design of channels, flat country, insufficient low-water flow to prevent rapid vegetal growth in lower part of channel; annual clearing.

3*c*. $n = 0.035$ in design of large channels, high velocity and large low-water flow; rapid erosion causing large irregularities; no vegetation.

4. $n = 0.040$ in design of channels, flat country, very fertile, loamy soil conducive to rapid vegetal growth; very small low-water flow or dry in summer; annual clearing.

Scobey's Recommendations for Earth Channels (especially adapted to mountain and western conditions). From a somewhat different and more detailed viewpoint, Scobey [127] gives the following recommendations for earth channels:

1. $n = 0.016$ for excellent condition of channels; low velocity; slick silt deposit or natural material smooth when wet; minimum influence of

[1] By maintenance is meant that the ditch is kept in good condition by cleaning out growth and obstruction as often as necessary, generally every 2 or 3 years.

vegetation at edges; water free of aquatic growth; no bends or sharp curves.

2. $n = 0.018$ for conditions intermediate between types 1 and 3; volcanic-ash soils with no vegetation; or large canals in very fine silt.

3. $n = 0.020$ for well-constructed canals in firm earth or fine-packed gravel; velocities permitting silting in of the gravel; banks clean cut, free of disturbing vegetation; alignment reasonably straight. Very large canals of type 4 may take $n = 0.020$.

4. $n = 0.0225$ for average well-constructed canals with eventually medium-smooth bottom; graded level; grass on edges; average alignment or silt at both sides and a few scattered stones in the middle; hardpan in good condition, clay and lava-ash soils. This is also a good value where retarding influence of moss, dense grass near edges, or scattered cobbles begin to show, for small head ditches serving only one or two farms, and in canals in hilly terrain, with upper bank largely in cut and subject to clogging by debris rolling from the hillside. (This eliminates Scobey's type 5.)

6. $n = 0.030$ for canals subject to heavy aquatic growths, having irregular banks with dense rootlets and bottom rough from rock fragments or erosion or complete covering of coarse gravel not finely bedded. Suited to earth conglomerate and large floodways well maintained.

7. $n = 0.035$ for canals about half choked with moss or channels that will not receive continuous maintenance.

8. $n = 0.040$ for canals badly choked by vegetative growth, large cobblestones or boulders, approaching in character a mountain stream with clean loose cobbles. Floodways poorly maintained require values of 0.050 up.

7.25. Character of the Local Rainfall. After ridges have been cut through and outlet mains installed in a given area, so that the impounded water has had a chance to drain out, the water that a drainage system must be designed to carry away is usually only that from rain or snow. Average and maximum annual rainfall have very little to do with determination of the proper size of a drain. (At this point Secs. 4.6 to 4.33 should be reviewed.) The important objective is that the drain be able to carry off, quickly, both the free surface water and the excess water in the soil, the total of which is covered by what has earlier been defined as runoff. Hence the drainage engineer should design his main on the basis of the intense rainstorm or series of closely consecutive rainstorms of a frequency and total duration of runoff characteristic of the region (see Fig. 4.2). The convenient time unit of capacity will usually be found to be a 24-hour day, but in special cases it may be found to be much smaller according to the tolerance of excessive wetness by the local crops.

Records of rainfall for considerable areas may be obtained from any regular weather station of the U.S. Weather Bureau in or near the area to be drained. However, the engineer should also draw upon his own knowledge and reports of residents regarding amount of local rainfall in the heaviest known short storms in a single day or in a series of successive days of rain.

7.26. Watershed: Size, Shape, and Topography. The area, topography, and shape of the watershed from which a main drain is to carry off the water is one of the most important factors in determining the size of the drain. With a given gradient and the same class of surface conditions and soil throughout the watershed, the total amount of runoff to be taken care of will be proportional to some fractional power of the area as shown by the curves of Fig. 7.3. For areas of less than 2 square miles where other conditions are the same, 160 acres would require a main drain of about double the capacity required for 80 acres. The surface runoff resulting from a given storm for a long, narrow area of steep, rocky, or clay land practically bare of vegetation may be a raging torrent for a very short time, after which it will quickly subside. The channel required to carry off such a flow safely and quickly would be many times as large as that required for an equal area of practically flat or sandy or peat land. Such land may absorb a considerable portion of the water as it falls and release the residue slowly, owing to the very low rate of natural fall, possibly extending the discharge over a long period of time. The whole area of the watershed should be taken into consideration. The stem of the funnel has to pass all the water that falls within its rim. The main drain is that stem.

ILLUSTRATIVE PROBLEM IN OPEN-DITCH DESIGN

Scope of the Problem. The problem is to design the main open drainage ditch of a given system which has been located in the field, the profile levels having been run and the profile drawn (see Fig. 7.8). The general conditions to be met are as follows: The ditch is 4 miles long. The water to be carried off is wholly that from the local rainfall on the watershed. Investigation has shown that, at maximum discharge, the drainage coefficient should be 2 in. of depth in 24 hr or 0.084 cfs per acre of watershed. The direct discharge is from the area directly drained by the main and its minor laterals, and the increase in the discharge is proportional to the distance from station 211 to station 158, from station 158 to station 106, from station 106 to station 53, and from station 53 to station 0. In addition, on the downstream side of each of the following stations a submain enters the main at the same elevation of surface water and same velocity as the main: station 158, station 106, and station 53. At the outlet, at maximum discharge, the elevation of the water surface in the main ditch and its velocity are the same as in the outlet channel. The general quantities to be dealt with in design are shown in detail in the accompanying table.

Mile	Station	Approx. fall of surface, ft per mile	Drainage area, acres			Required capacity, cfs		
			Directly tributary to main	Tributary to submains	Total	Main	Submains	Total
	211		1,200		1,200	101		101
4		4.5						
	158		2,400	800	3,200	202	67	269
3		3.0						
	106		4,000	500	4,500	336	42	378
2		2.0						
	53		5,500	500	6,000	462	42	504
1		0.5						
	0		7,000		7,000	588		588

As far as compatible with good design it is desired to maintain a closely approximate freeboard in the regular excavation, of 1.5 ft. The plan calls for the spoil banks to be spread at or near the brink of the ditch slopes to a depth not to exceed 18 in. and to gradually thin out, laterally, on the surface of the fields to a distance of 50 to 200 ft on each side, according to the amount of excavation per unit length of the ditch.

The elevation of the water surface at flood stages in the outlet channel was found to be 1,000.00. This is taken as the original control level from which to start design of the main. In order to maintain the freeboard desired and to parallel the natural ground surface as nearly as practicable, study revealed the desirability of making the elevations given opposite the following stations practically fixed points on the final hydraulic grade line determined upon:

Station 0 + 00: 1,000.00
Station 53 + 00: 1,000.50
Station 106 + 00: 1,002.50
Station 158 + 00: 1,005.40
Station 211 + 00: 1,009.90

However, complications interfering with the best design later indicated the wisdom of slight changes in these elevations at stations 158 and 211, as appears in the course of the discussion.

The design in this problem is based on Manning's formula. Two to one side slopes are used. Study of the outlet conditions indicated that it would be desirable to limit the flow depth in the main at the outlet to 10 ft, the ditch to be designed for bad order, with a roughness factor n of

TABLE 7.4. AREAS OF CROSS SECTION OF FLOW DEPTHS, HYDRAULIC RADII AND TWO-THIRDS POWER OF HYDRAULIC RADII FOR OPEN DITCHES

Depth, ft	Bottom width 3 ft												Depth, ft
	1½ to 1 slopes			2 to 1 slopes			3 to 1 slopes			4 to 1 slopes			
	a	r	$r^{2/3}$	a	r	$r^{2/3}$	a	r	$r^{2/3}$	a	r	$r^{2/3}$	
2.0	12.00	1.18	1.12	14.00	1.17	1.11	18.00	1.15	1.10	22.00	1.13	1.09	2.0
2.2	13.86	1.27	1.17	16.28	1.27	1.17	21.12	1.25	1.16	25.96	1.23	1.15	2.2
2.4	15.84	1.36	1.23	18.72	1.36	1.23	24.48	1.35	1.22	30.24	1.33	1.21	2.4
2.6	17.94	1.45	1.28	21.32	1.46	1.29	28.08	1.45	1.28	34.84	1.43	1.27	2.6
2.8	20.16	1.54	1.33	24.08	1.55	1.34	31.92	1.54	1.33	39.76	1.53	1.33	2.8
3.0	22.50	1.63	1.39	27.00	1.64	1.39	36.00	1.64	1.39	45.00	1.62	1.38	3.0
3.2	24.96	1.72	1.44	30.08	1.73	1.44	40.32	1.74	1.45	50.56	1.72	1.44	3.2
3.4	27.54	1.81	1.49	33.32	1.83	1.50	44.88	1.83	1.50	56.44	1.82	1.49	3.4
3.6	30.24	1.89	1.53	36.72	1.92	1.55	49.68	1.93	1.55	62.64	1.92	1.55	3.6
3.8	33.06	1.98	1.58	40.28	2.01	1.59	54.72	2.03	1.60	69.16	2.02	1.60	3.8
4.0	36.00	2.07	1.62	44.00	2.10	1.64	60.00	2.12	1.65	76.00	2.11	1.65	4.0
4.5	43.88	2.29	1.74	54.00	2.33	1.76	74.25	2.36	1.77	94.50	2.36	1.77	4.5
5.0	52.50	2.50	1.84	65.0	2.56	1.87	90.00	2.60	1.89	115.0	2.60	1.89	5.0
5.5	61.88	2.71	1.94	77.0	2.79	1.98	107.25	2.84	2.01	137.5	2.85	2.01	5.5
6.0	72.00	2.93	2.05	90.0	3.01	2.09	126.00	3.08	2.12	162.0	3.09	2.12	6.0
6.5	82.88	3.14	2.14	104.0	3.24	2.19	146.25	3.32	2.23	188.5	3.33	2.23	6.5
7.0	94.50	3.35	2.24	119.0	3.46	2.29	168.00	3.56	2.33	217.0	3.58	2.34	7.0
7.5	106.88	3.56	2.33	135.0	3.69	2.39	191.25	3.79	2.43	247.5	3.82	2.44	7.5
8.0	120.00	3.77	2.42	152.0	3.91	2.48	216.00	4.03	2.53	280.0	4.06	2.55	8.0
8.5	133.88	3.98	2.51	170.0	4.14	2.58	242.25	4.27	2.63	314.5	4.31	2.65	8.5
9.0	148.50	4.19	2.60	189.0	4.36	2.67	270.00	4.51	2.73	351.0	4.55	2.75	9.0
9.5	163.88	4.41	2.69	209.0	4.59	2.76	299.25	4.75	2.83	389.5	4.79	2.84	9.5
10.0	180.00	4.62	2.77	230.0	4.81	2.85	330.00	4.98	2.92	430.0	5.04	2.94	10.0
10.5	196.88	4.83	2.86	252.0	5.04	2.94	362.25	5.22	3.01	472.5	5.28	3.03	10.5
11.0	214.50	5.04	2.94	275.0	5.26	3.02	396.00	5.46	3.10	517.0	5.52	3.12	11.0
11.5	232.88	5.25	3.02	299.0	5.48	3.11	431.25	5.70	3.19	563.5	5.76	3.21	11.5
12.0	252.00	5.45	3.10	324.0	5.71	3.20	468.00	5.94	3.28	612.0	6.01	3.31	12.0

TABLE 7.4. AREAS OF CROSS SECTION OF FLOW DEPTHS, HYDRAULIC RADII AND TWO-THIRDS POWER OF HYDRAULIC RADII FOR OPEN DITCHES. (*Continued*)

Depth, ft	Bottom width 4 ft												Depth, ft
	1½ to 1 slopes			2 to 1 slopes			3 to 1 slopes			4 to 1 slopes			
	a	r	$r^{2/3}$	a	r	$r^{2/3}$	a	r	$r^{2/3}$	a	r	$r^{2/3}$	
2.0	14.00	1.25	1.16	16.00	1.23	1.15	20.00	1.20	1.13	24.00	1.17	1.11	2.0
2.2	16.06	1.35	1.22	18.48	1.33	1.21	23.32	1.30	1.19	28.16	1.27	1.17	2.2
2.4	18.24	1.44	1.28	21.12	1.43	1.27	26.88	1.40	1.25	32.64	1.37	1.23	2.4
2.6	20.54	1.54	1.33	23.92	1.53	1.33	30.68	1.50	1.31	37.44	1.47	1.29	2.6
2.8	22.96	1.63	1.39	26.88	1.63	1.39	34.72	1.60	1.37	42.56	1.57	1.35	2.8
3.0	25.50	1.72	1.44	30.00	1.72	1.44	39.00	1.70	1.42	48.00	1.67	1.41	3.0
3.2	28.16	1.81	1.49	33.28	1.81	1.49	43.52	1.80	1.48	53.76	1.77	1.46	3.2
3.4	30.94	1.90	1.53	36.72	1.91	1.54	48.28	1.89	1.53	59.84	1.87	1.52	3.4
3.6	33.84	1.99	1.58	40.32	2.00	1.59	53.28	1.99	1.58	66.24	1.97	1.57	3.6
3.8	36.86	2.08	1.63	44.08	2.10	1.64	58.52	2.09	1.63	72.96	2.07	1.62	3.8
4.0	40.00	2.17	1.68	48.00	2.19	1.69	64.00	2.19	1.69	80.00	2.16	1.67	4.0
4.5	48.38	2.40	1.79	58.50	2.42	1.80	78.75	2.43	1.81	99.00	2.41	1.80	4.5
5.0	57.50	2.61	1.90	70.0	2.65	1.92	95.00	2.67	1.93	120.0	2.65	1.92	5.0
5.5	67.38	2.83	2.00	82.5	2.88	2.02	112.75	2.91	2.04	143.0	2.90	2.03	5.5
6.0	78.00	3.04	2.10	96.0	3.11	2.13	132.00	3.15	2.15	168.0	3.14	2.14	6.0
6.5	89.38	3.26	2.20	110.5	3.34	2.23	152.75	3.39	2.26	195.0	3.39	2.26	6.5
7.0	101.50	3.47	2.29	126.0	3.56	2.33	175.00	3.63	2.36	224.0	3.63	2.36	7.0
7.5	114.40	3.68	2.38	142.5	3.79	2.43	198.75	3.87	2.47	255.0	3.88	2.47	7.5
8.0	128.0	3.90	2.48	160.0	4.02	2.53	224.00	4.11	2.57	288.0	4.12	2.57	8.0
8.5	142.4	4.11	2.57	178.5	4.24	2.62	250.75	4.34	2.66	323.0	4.36	2.67	8.5
9.0	157.5	4.32	2.65	198.0	4.47	2.71	279.00	4.58	2.76	360.0	4.61	2.77	9.0
9.5	173.4	4.53	2.74	218.5	4.69	2.80	308.75	4.82	2.85	399.0	4.85	2.87	9.5
10.0	190.0	4.74	2.82	240.0	4.92	2.89	340.00	5.06	2.95	440.0	5.09	2.96	10.0
10.5	207.4	4.95	2.90	262.5	5.14	2.98	372.75	5.30	3.04	483.0	5.34	3.06	10.5
11.0	225.5	5.16	2.99	286.0	5.37	3.07	407.00	5.54	3.13	528.0	5.58	3.15	11.0
11.5	244.4	5.37	3.07	310.5	5.59	3.15	442.75	5.77	3.22	575.0	5.82	3.24	11.5
12.0	264.0	5.58	3.15	336.0	5.82	3.24	480.00	6.01	3.31	624.0	6.07	3.33	12.0

TABLE 7.4. AREAS OF CROSS SECTION OF FLOW DEPTHS, HYDRAULIC RADII AND TWO-THIRDS POWER OF HYDRAULIC RADII FOR OPEN DITCHES. (Continued)

Depth, ft	Bottom width 6 ft												Depth, ft
	1½ to 1 slopes			2 to 1 slopes			3 to 1 slopes			4 to 1 slopes			
	a	r	$r^{2/3}$	a	r	$r^{2/3}$	a	r	$r^{2/3}$	a	r	$r^{2/3}$	
2.0	18.00	1.36	1.23	20.00	1.34	1.22	24.00	1.29	1.19	28.00	1.25	1.16	2.0
2.2	20.46	1.47	1.29	22.88	1.44	1.28	27.72	1.39	1.25	32.56	1.35	1.22	2.2
2.4	23.04	1.57	1.35	25.92	1.55	1.34	31.68	1.50	1.31	37.44	1.45	1.28	2.4
2.6	25.74	1.67	1.41	29.12	1.65	1.40	35.88	1.60	1.37	42.64	1.56	1.35	2.6
2.8	28.56	1.77	1.46	32.48	1.75	1.45	40.32	1.70	1.42	48.16	1.66	1.40	2.8
3.0	31.50	1.87	1.52	36.00	1.85	1.51	45.00	1.80	1.48	54.00	1.76	1.46	3.0
3.2	34.56	1.97	1.57	39.68	1.95	1.56	49.92	1.90	1.53	60.16	1.86	1.51	3.2
3.4	37.74	2.07	1.62	43.52	2.05	1.61	55.08	2.00	1.59	66.64	1.96	1.57	3.4
3.6	41.04	2.16	1.67	47.52	2.15	1.67	60.48	2.10	1.64	73.44	2.06	1.62	3.6
3.8	44.46	2.26	1.72	51.68	2.25	1.72	66.12	2.20	1.69	80.56	2.16	1.67	3.8
4.0	48.00	2.35	1.77	56.00	2.34	1.76	72.00	2.30	1.74	88.0	2.26	1.72	4.0
4.5	57.38	2.58	1.88	67.50	2.58	1.88	87.75	2.55	1.87	108.0	2.51	1.85	4.5
5.0	67.50	2.81	1.99	80.0	2.82	2.00	105.00	2.79	1.98	130.0	2.75	1.96	5.0
5.5	78.38	3.03	2.09	93.5	3.05	2.10	123.75	3.04	2.10	154.0	3.00	2.08	5.5
6.0	90.00	3.26	2.20	108.0	3.28	2.21	144.00	3.28	2.21	180.0	3.25	2.19	6.0
6.5	102.4	3.48	2.30	123.5	3.52	2.31	165.75	3.52	2.31	208.0	3.49	2.30	6.5
7.0	115.5	3.70	2.39	140.0	3.75	2.41	189.00	3.76	2.42	238.0	3.74	2.41	7.0
7.5	129.4	3.92	2.49	157.5	3.98	2.51	213.75	4.00	2.52	270.0	3.98	2.51	7.5
8.0	144.0	4.13	2.57	176.0	4.21	2.61	240.00	4.24	2.62	304.0	4.23	2.62	8.0
8.5	159.4	4.35	2.67	195.5	4.43	2.70	267.75	4.48	2.72	340.0	4.47	2.71	8.5
9.0	175.5	4.56	2.75	216.0	4.66	2.79	297.00	4.72	2.81	378.0	4.72	2.81	9.0
9.5	192.4	4.78	2.84	237.5	4.89	2.88	327.75	4.96	2.91	418.0	4.96	2.91	9.5
10.0	210.0	4.99	2.92	260.0	5.12	2.97	360.00	5.20	3.00	460.0	5.20	3.00	10.0
10.5	228.4	5.21	3.01	283.5	5.35	3.06	393.75	5.44	3.09	504.0	5.45	3.10	10.5
11.0	247.5	5.42	3.09	308.0	5.57	3.14	429.00	5.68	3.18	550.0	5.69	3.19	11.0
11.5	267.4	5.63	3.17	333.5	5.80	3.23	465.75	5.92	3.27	598.0	5.93	3.28	11.5
12.0	288.0	5.85	3.25	360.0	6.02	3.31	504.00	6.16	3.36	648.0	6.18	3.37	12.0

TABLE 7.4. AREAS OF CROSS SECTION OF FLOW DEPTHS, HYDRAULIC RADII AND TWO-THIRDS POWER OF HYDRAULIC RADII FOR OPEN DITCHES. (*Continued*)

Depth, ft	Bottom width 8 ft												Depth, ft
	1½ to 1 slopes			2 to 1 slopes			3 to 1 slopes			4 to 1 slopes			
	a	r	$r^{2/3}$	a	r	$r^{2/3}$	a	r	$r^{2/3}$	a	r	$r^{2/3}$	
2.0	22.00	1.45	1.28	24.00	1.42	1.26	28.00	1.36	1.23	32.00	1.31	1.20	2.0
2.2	24.86	1.56	1.35	27.28	1.53	1.33	32.12	1.47	1.29	36.96	1.41	1.26	2.2
2.4	27.84	1.67	1.41	30.72	1.64	1.39	36.48	1.57	1.35	42.24	1.52	1.32	2.4
2.6	30.94	1.78	1.47	34.32	1.75	1.45	41.08	1.68	1.41	47.84	1.63	1.39	2.6
2.8	34.16	1.89	1.53	38.08	1.85	1.51	45.92	1.79	1.47	53.76	1.73	1.44	2.8
3.0	37.50	1.99	1.58	42.00	1.96	1.57	51.00	1.89	1.53	60.00	1.83	1.50	3.0
3.2	40.96	2.10	1.64	46.08	2.06	1.62	56.32	2.00	1.59	66.56	1.94	1.56	3.2
3.4	44.54	2.20	1.69	50.32	2.17	1.68	61.88	2.10	1.64	73.44	2.04	1.61	3.4
3.6	48.24	2.30	1.74	54.72	2.27	1.73	67.68	2.20	1.69	80.64	2.14	1.66	3.6
3.8	52.06	2.40	1.79	59.28	2.37	1.78	73.72	2.30	1.74	88.16	2.24	1.71	3.8
4.0	56.00	2.50	1.84	64.00	2.47	1.83	80.00	2.40	1.79	96.00	2.34	1.76	4.0
4.5	66.38	2.74	1.96	76.50	2.72	1.95	96.75	2.66	1.92	117.0	2.60	1.89	4.5
5.0	77.50	2.98	2.07	90.0	2.96	2.06	115.00	2.90	2.03	140.0	2.85	2.01	5.0
5.5	89.38	3.21	2.18	104.5	3.20	2.17	134.75	3.15	2.15	165.0	3.09	2.12	5.5
6.0	102.00	3.44	2.28	120.0	3.44	2.28	156.00	3.40	2.26	192.0	3.34	2.23	6.0
6.5	115.4	3.67	2.38	136.5	3.68	2.38	178.75	3.64	2.37	221.0	3.59	2.35	6.5
7.0	129.5	3.90	2.48	154.0	3.91	2.48	203.00	3.89	2.47	252.0	3.84	2.45	7.0
7.5	144.4	4.12	2.57	172.5	4.15	2.58	228.75	4.13	2.57	285.0	4.08	2.55	7.5
8.0	160.0	4.34	2.66	192.0	4.38	2.68	256.00	4.37	2.67	320.0	4.33	2.66	8.0
8.5	176.4	4.56	2.75	212.5	4.61	2.77	284.75	4.61	2.77	357.0	4.57	2.75	8.5
9.0	193.5	4.78	2.84	234.0	4.84	2.86	315.00	4.86	2.87	396.0	4.82	2.85	9.0
9.5	211.4	5.00	2.92	256.5	5.07	2.95	346.75	5.10	2.96	437.0	5.06	2.95	9.5
10.0	230.0	5.22	3.01	280.0	5.30	3.04	380.00	5.34	3.06	480.0	5.31	3.04	10.0
10.5	249.4	5.44	3.09	304.5	5.53	3.13	414.75	5.58	3.15	525.0	5.55	3.14	10.5
11.0	269.5	5.65	3.17	330.0	5.76	3.21	451.00	5.82	3.24	572.0	5.80	3.23	11.0
11.5	290.4	5.87	3.25	356.5	5.99	3.30	488.75	6.06	3.32	621.0	6.04	3.32	11.5
12.0	312.0	6.09	3.34	384.0	6.22	3.38	528.00	6.30	3.41	672.0	6.29	3.41	12.0

TABLE 7.4. AREAS OF CROSS SECTION OF FLOW DEPTHS, HYDRAULIC RADII AND TWO-THIRDS POWER OF HYDRAULIC RADII FOR OPEN DITCHES. (*Continued*)

Depth, ft	Bottom width 10 ft												Depth, ft
	1½ to 1 slopes			2 to 1 slopes			3 to 1 slopes			4 to 1 slopes			
	a	r	$r^{2/3}$	a	r	$r^{2/3}$	a	r	$r^{2/3}$	a	r	$r^{2/3}$	
2.0	26.00	1.51	1.32	28.00	1.48	1.30	32.00	1.41	1.26	36.00	1.36	1.23	2.0
2.2	29.26	1.63	1.39	31.68	1.60	1.37	36.52	1.53	1.33	41.36	1.47	1.29	2.2
2.4	32.64	1.75	1.45	35.52	1.71	1.43	41.28	1.64	1.39	47.04	1.58	1.36	2.4
2.6	36.14	1.87	1.52	39.52	1.83	1.50	46.28	1.75	1.45	53.04	1.69	1.42	2.6
2.8	39.76	1.98	1.58	43.68	1.94	1.56	51.52	1.86	1.51	59.36	1.79	1.47	2.8
3.0	43.50	2.09	1.63	48.00	2.05	1.61	57.00	1.97	1.57	66.00	1.90	1.53	3.0
3.2	47.36	2.20	1.69	52.48	2.16	1.67	62.72	2.08	1.63	72.96	2.01	1.59	3.2
3.4	51.34	2.31	1.75	57.12	2.26	1.72	68.68	2.18	1.68	80.24	2.11	1.65	3.4
3.6	55.44	2.41	1.80	61.92	2.37	1.78	74.88	2.29	1.74	87.84	2.21	1.70	3.6
3.8	59.66	2.52	1.85	66.88	2.48	1.83	81.32	2.39	1.79	95.76	2.32	1.75	3.8
4.0	64.00	2.62	1.90	72.0	2.58	1.88	88.00	2.49	1.84	104.0	2.42	1.80	4.0
4.5	75.38	2.88	2.02	85.5	2.83	2.00	105.75	2.75	1.96	126.0	2.68	1.93	4.5
5.0	87.50	3.12	2.14	100.0	3.09	2.12	125.00	3.00	2.08	150.0	2.93	2.05	5.0
5.5	100.40	3.36	2.24	115.5	3.33	2.23	145.75	3.26	2.20	176.0	3.18	2.16	5.5
6.0	114.00	3.60	2.35	132.0	3.58	2.34	168.00	3.51	2.31	204.0	3.43	2.27	6.0
6.5	128.4	3.84	2.45	149.5	3.82	2.44	191.75	3.75	2.41	234.0	3.68	2.38	6.5
7.0	143.5	4.07	2.55	168.0	4.06	2.55	217.00	4.00	2.52	266.0	3.93	2.49	7.0
7.5	159.4	4.30	2.64	187.5	4.30	2.64	243.75	4.25	2.62	300.0	4.18	2.60	7.5
8.0	176.0	4.53	2.74	208.0	4.54	2.74	272.00	4.49	2.72	336.0	4.43	2.70	8.0
8.5	193.4	4.76	2.83	229.5	4.77	2.83	301.75	4.74	2.82	374.0	4.67	2.79	8.5
9.0	211.5	4.98	2.92	252.0	5.01	2.93	333.00	4.98	2.92	414.0	4.92	2.89	9.0
9.5	230.4	5.21	3.01	275.5	5.24	3.02	365.75	5.22	3.01	456.0	5.17	2.99	9.5
10.0	250.0	5.43	3.09	300.0	5.47	3.11	400.00	5.46	3.10	500.0	5.41	3.08	10.0
10.5	270.4	5.65	3.17	325.5	5.71	3.20	435.75	5.71	3.20	546.0	5.66	3.18	10.5
11.0	291.5	5.87	3.25	352.0	5.94	3.28	473.00	5.95	3.28	594.0	5.90	3.27	11.0
11.5	313.4	6.09	3.34	379.5	6.17	3.36	511.75	6.19	3.37	644.0	6.15	3.36	11.5
12.0	336.0	6.31	3.42	408.0	6.40	3.45	552.00	6.43	3.46	696.0	6.39	3.44	12.0

TABLE 7.4. AREAS OF CROSS SECTION OF FLOW DEPTHS, HYDRAULIC RADII AND TWO-THIRDS POWER OF HYDRAULIC RADII FOR OPEN DITCHES. (*Continued*)

Depth, ft	Bottom width 12 ft												Depth, ft
	1½ to 1 slopes			2 to 1 slopes			3 to 1 slopes			4 to 1 slopes			
	a	r	$r^{2/3}$	a	r	$r^{2/3}$	a	r	$r^{2/3}$	a	r	$r^{2/3}$	
2.0	30.00	1.56	1.35	32.00	1.53	1.33	36.00	1.46	1.29	40.00	1.40	1.25	2.0
2.2	33.66	1.69	1.42	36.08	1.65	1.40	40.92	1.58	1.36	45.76	1.52	1.32	2.2
2.4	37.44	1.81	1.49	40.32	1.77	1.46	46.08	1.70	1.42	51.84	1.63	1.39	2.4
2.6	41.34	1.93	1.55	44.72	1.89	1.53	51.48	1.81	1.49	58.24	1.74	1.45	2.6
2.8	45.36	2.05	1.61	49.28	2.01	1.59	57.12	1.92	1.55	64.96	1.85	1.51	2.8
3.0	49.50	2.17	1.68	54.00	2.12	1.65	63.00	2.03	1.60	72.00	1.96	1.57	3.0
3.2	53.76	2.28	1.73	58.88	2.24	1.71	69.12	2.15	1.67	79.36	2.07	1.62	3.2
3.4	58.14	2.40	1.79	63.92	2.35	1.77	75.48	2.25	1.72	87.04	2.17	1.68	3.4
3.6	62.64	2.51	1.85	69.12	2.46	1.82	82.08	2.36	1.77	95.04	2.28	1.73	3.6
3.8	67.26	2.62	1.90	74.48	2.57	1.88	88.92	2.47	1.83	103.36	2.39	1.79	3.8
4.0	72.00	2.73	1.95	80.0	2.67	1.93	96.00	2.58	1.88	112.0	2.49	1.84	4.0
4.5	84.38	2.99	2.08	94.5	2.94	2.05	114.75	2.84	2.01	135.0	2.75	1.96	4.5
5.0	97.50	3.25	2.19	110.0	3.20	2.17	135.00	3.10	2.13	160.0	3.01	2.09	5.0
5.5	111.38	3.50	2.31	126.5	3.45	2.28	156.75	3.35	2.24	187.0	3.26	2.20	5.5
6.0	126.00	3.75	2.41	144.0	3.70	2.39	180.00	3.61	2.35	216.0	3.52	2.31	6.0
6.5	141.38	3.99	2.52	162.5	3.95	2.50	204.75	3.86	2.46	247.0	3.77	2.42	6.5
7.0	157.50	4.23	2.62	182.0	4.20	2.60	231.00	4.11	2.57	280.0	4.02	2.53	7.0
7.5	174.38	4.47	2.71	202.5	4.44	2.70	258.75	4.36	2.67	315.0	4.27	2.63	7.5
8.0	192.00	4.70	2.81	224.0	4.68	2.80	288.00	4.60	2.77	352.0	4.52	2.73	8.0
8.5	201.38	4.93	2.90	246.5	4.92	2.89	318.75	4.85	2.87	391.0	4.77	2.83	8.5
9.0	229.50	5.16	2.99	270.0	5.16	2.99	351.00	5.10	2.96	432.0	5.01	2.93	9.0
9.5	249.38	5.39	3.07	294.5	5.40	3.08	384.75	5.34	3.06	475.0	5.26	3.02	9.5
10.0	270.00	5.62	3.16	320.0	5.63	3.17	420.00	5.58	3.15	520.0	5.51	3.12	10.0
10.5	291.38	5.84	3.24	346.5	5.87	3.25	456.75	5.83	3.24	567.0	5.75	3.21	10.5
11.0	313.50	6.07	3.33	374.0	6.10	3.34	495.00	6.07	3.33	616.0	6.00	3.30	11.0
11.5	336.40	6.29	3.41	402.5	6.34	3.43	534.75	6.31	3.42	667.0	6.25	3.39	11.5
12.0	360.00	6.51	3.49	432.0	6.57	3.51	576.00	6.56	3.50	720.0	6.49	3.48	12.0

Checking, as in mile 1, the rectilinear character of the hydraulic grade line, the average depth of 8.4 at the center of the mile checks within less than 0.05 ft of being a straight line. This small difference may be due to uneven refinements in computation all along the line and should be ignored.

The foregoing design makes the elevation of the bottom of the ditch at the upper end of the second mile 1,002.5 − 8.1, or 994.40.

Design of the Third Mile. The elevations of the hydraulic grade line at the upper and lower ends of mile 3 are tentatively fixed by the original conditions of the problem at 1,005.40 and 1,002.50, respectively, which also fixes the hydraulic gradient, assuming the hydraulic grade line to be a straight line throughout the mile, at 2.9 ft per mile or $s = 0.0005577$, whence $\sqrt{s} = 0.02362$.

Following the same method of analysis as was used in mile 2 and assuming that the elevation of the bottom of the ditch at the junction of miles 2 and 3 is fixed by the foregoing design at 994.40 at a depth of 8.1 ft, the next step, as was done in mile 2, is to seek the proper base width on the above stated hydraulic gradient for mile 3 to fit this fixed elevation and depth of ditch. It appears that this proper theoretical width is less than 3.0 ft, a base of 2.9 ft being required for the depth of 8.1 ft. This is not a good or practical ditch. Good practice suggests the use of at least the previously approved minimum base width of 4.0 ft. However, the Q value required at the lower end of mile 3 is 336 cfs which requires a depth of only 7.8 ft for a 4-ft base on the given s value of 0.0005577. This would leave a jump in the ditch bottom of 8.1 − 7.8 or 0.3 ft which can be taken care of in one of three ways: (1) by a concrete drop which is relatively too expensive considering the smallness of the drop and is never considered desirable if it can be avoided; (2) by calculation of the normal backwater curve which is unreasonably cumbersome and time-consuming from both a design and a construction standpoint for any advantage gained by its use and is, therefore, not recommended; or (3) by determining a new rectilinear hydraulic grade line for the mile, on a 4-ft base width, that would meet the depth requirement of 8.1 ft at the lower end of the mile and, at the same time, meet the original elevation of the hydraulic grade line at the upper end of the mile fairly closely.

With the depth and base width known the area is also known, leaving $\sqrt{s}$ the only unknown factor in the equation for Q, which then reads

$$Q = 336 = 163.7 \times 37.15 \times 2.55 \times \sqrt{s}$$

Whence

$$\sqrt{s} = \frac{336}{163.7 \times 37.15 \times 2.55} = 0.02167$$

Then

$$s = (0.02167)^2 = 0.0004695$$

Starting with the elevation of the hydraulic grade line at the junction of miles 2 and 3 at 1,002.50, the above new s value for mile 3 will give the elevation of the hydraulic grade line at station 158 as 1,004.94 which is 0.46 ft below the elevation of 1,005.40 tentatively fixed in the originally stated conditions of the problem, and this change will gradually increase the freeboard in the length of the third mile from the desired 1.5 to 2.0 ft at the upper end of the mile. Nevertheless this is the best design to meet the situation and is on the safe side as to velocity of flow. It is, therefore, recommended.

Then, at the upper end of mile 3 the problem is simply to determine, by the usual method, the depth of ditch of 4-ft base width required on the new hydraulic gradient of 0.0004695 to carry the required Q of 269 cfs. This depth is found to be 7.4 ft, making the elevation of the bottom of the ditch at this point 997.54.

As on the preceding mile, the average of the depth at the two ends of mile 3 checks within 0.05 ft, the rectilinear character of the hydraulic grade line.

Design of the Fourth Mile. The elevations of the hydraulic grade line on this mile were tentatively fixed by the original conditions of the problem at stations 158 and 211 at 1,005.40 and 1,009.90, respectively. This also fixed the hydraulic gradient at 4.5 ft for the mile or $s = 0.000849$ and the $\sqrt{s} = 0.02913$. It develops that the same general condition exists here in the relations between mile 3 and mile 4 as was found between miles 2 and 3 in that designing a continuous bottom for the two miles would call for a width of bottom on mile 4 of less than 3 ft—an undesirable and impractical result. The better way, therefore, is to continue the minimum 4-ft bottom through mile 4, although with the hydraulic grade line as fixed by the original terms of the problem the same kind of complication arises as occurred between miles 2 and 3 of a drop at the junction of the two miles, which must be eliminated by some kind of allowable adjustment.

By the usual method the required depths at the upper and lower ends of mile 4 with a 4-ft base, and on the fixed s value, were found to be 4.3 and 5.8 ft, respectively, with the usual close check on the rectilinear character of the hydraulic grade line for the mile. The change in the elevation of the hydraulic grade line at the junction of miles 3 and 4, made above in the design of mile 3, necessitates a further change either in the gradient of mile 4 from the 4.5 ft per mile, above listed, to 4.96 ft per mile, with a corresponding lowering of the elevation of the hydraulic grade at the upper end of mile 4 from 1,009.90 to 1,009.44, thus making the final position of the hydraulic grade line parallel to and 0.46 ft lower than its original tentative position. The latter is the more desirable plan as it does not affect the velocity of flow over the original. It is, therefore adopted. These changes make the corresponding elevations of the ditch bottom at the lower and upper ends of the mile to be 999.14 and 1,005.14, respectively. The result is still a 1.6-ft jump in the ditch bottom at the junction of miles 3 and 4 at, but on the upper side of, station 158. Good design in view of probable future improvements in drainage requires that this situation be handled as follows: If this entire ditch is to serve as an outlet for tile drains at ordinary stages of flow, the 1.6-ft drop in the ditch bottom should be eliminated by making the ditch that much deeper throughout mile 4. This is what is done, making the depths at the lower and upper ends of the fourth mile 7.4 and 5.9 ft and the ditch bottom elevations 997.54 and 1,003.54, respectively, resulting in a much better outlet for tile drains. Why this is so is made entirely clear in Chap. 10 on Location and Design of Underdrains.

The design of the entire ditch can be summarized as shown in Table 7.5.

7.27. Backwater Curves in Earth Drainage Ditches Negligible. In any such problem as the foregoing, each change, proceeding downstream, from one given mild gradient to a milder gradient actually results in a backwater curve extending upstream from the station of the change of grade in the ditch bottom. However, as the allowable velocity of flow in earth drainage channels never reaches the "critical" (such as to cause a hydraulic jump), the resultant flow in them is always streamline flow. The backwater curves formed are relatively short, and the departure of the actual water surface through them from the rectilinear hydraulic grade line from station of change of ditch grade to station of next change of ditch grade is so slight as to be negligible in design and construction practice. The use of rectilinear hydraulic grade lines as is done in the foregoing problem is, therefore, common practice because it greatly simplifies the work of design, and any small error thus introduced is on the side of safety as to capacity.

7.28. Simplified Design for Open Farm Ditches. The Regional Office of Region III of the U.S. Department of Agriculture Soil Conservation Service [141] supplies for the use of its farm planners a simplified procedure for design of limited open ditches on farms by employing the curves shown in Fig. 7.9. The curves in this figure are applicable only to flat watershed areas having average slopes less than 25 feet per mile. The watershed area is to be determined above each section of ditch for which capacity is to be computed. For lands in need of drainage with somewhat steeper slopes than 25 feet per mile the next higher runoff curve may be used to give substantially equivalent protection.

7.29. Explanation of Curves. The equivalent *M* curve was developed to give fair agricultural drainage in the valley of the Red River of the North in western Minnesota. Each of the other equivalent curves (*D*, *C*, and *B*) are expected to give fair to good, good to very good, and very good

TABLE 7.5. SUMMARY OF DESIGN

Mile	Station	Approx. fall, ft per mile	Drainage area, acres			Required capacity Q, cfs			Elevations			s	Gradient of ditch bottoms, %	b, ft	d, ft
			Main	Laterals	Total	Main	Laterals	Total	Ground surface	Hydraulic grade line	Bottom of ditch				
	211		1,200		1,200	101		101	1,011.5	1,009.44	1003.54			4.0	5.9
4		4.5										0.0008490	0.11320		
	158		2,400	800	3,200	202	67	269	1,007.0	1,004.94	997.54			4.0	7.4
3		3.0										0.0004695	0.06038		
	106		4,000	500	4,500	336	42	378	1,004.0	1,002.50	994.40			4.0/8.0	8.1
2		2.0										0.0003774	0.04906		
	53		5,500	500	6,000	462	42	504	1,002.0	1,000.50	991.8			8.0/30.0	8.7
1		0.5										0.0000943	0.02264		
	0		7,000		7,000	588		588	1,001.5	1,000.00	990.6			30.0	9.4

to excellent drainage, respectively. While these curves were developed for use in Region III of the Soil Conservation Service, they are considered to apply, within the drainage ranges shown, anywhere in the United States. The greater rainfall experienced in the South, East, and Far West is usually offset to a great extent by flatter watershed slopes. It is

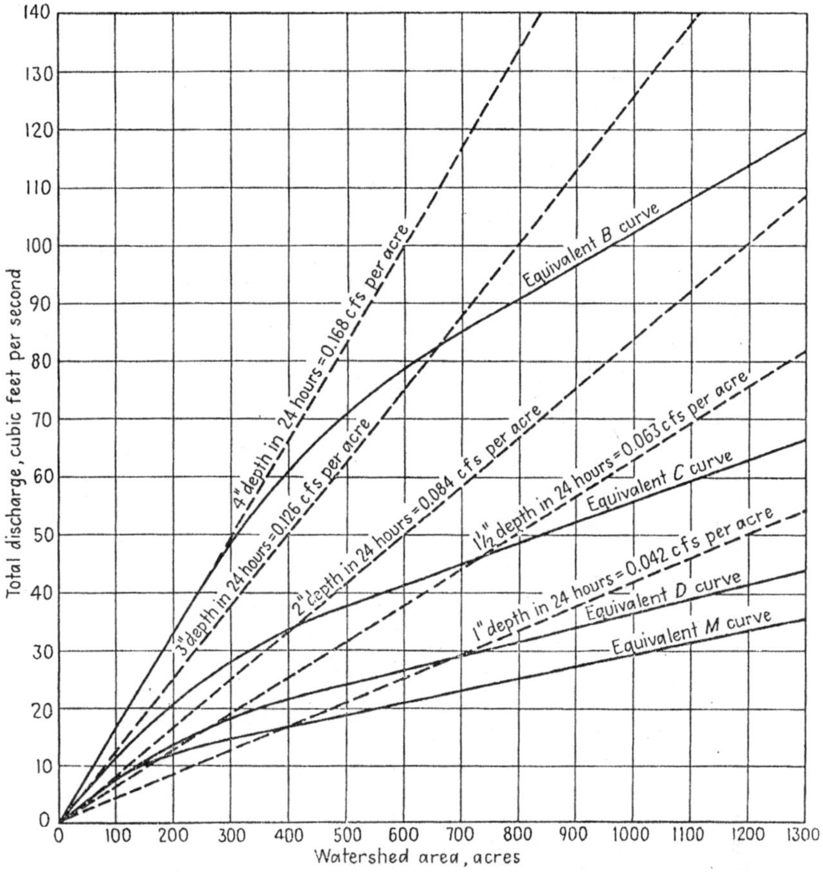

FIG. 7.9. Drainage runoff curves for open ditch design in flat watershed areas of less than 2 square miles. (Fig. 84, p. 147, "Engineering Handbook for Farm Planners—Upper Mississippi Valley Region.") (*Courtesy of Soil Conservation Service, Region III, Milwaukee, Wis.*)

to be noted that these curves apply primarily to areas of less than 2 square miles and are intended for use only in the design of farm ditches. The main and major lateral ditches of large drainage districts, especially in the South, have been and are designed with very much less capacity per unit area of watershed in accordance with the favorite runoff formula of the engineer in charge.

TABLE 7.6.* CUBIC FEET PER SECOND HANDLED BY DRAINAGE DITCH

Side slopes		2 to 1										3 to 1										
Bottom width, ft	% slope† / Ft depth	0.02	0.04	0.05	0.06	0.08	0.10	0.20	0.30	0.40	Top width, ft	0.02	0.04	0.05	0.06	0.08	0.10	0.20	0.30	0.40	0.50	Top width, ft
4	1.5	5	8	9	9	11	12	17	21	24	10	7	9	10	11	13	15	20	25	29	32	13
6	1.5	7	10	12	13	15	16	23	29	33	12	8	12	13	15	17	19	26	33	38		15
8	1.5	9	13	15	16	19	21	30	37		14	10	15	17	18	20	23	33	40	46		17
10	1.5	11	16	18	20	23	26	36	44		16	12	17	19	21	25	27	39	48	55		19
4	2.0	10	14	15	17	19	22	31	38		12	12	17	19	21	24	27	38	46			16
6	2.0	13	18	20	22	26	29	40	50		14	15	21	24	26	30	33	47	58			18
8	2.0	16	23	25	28	32	36	50			16	18	26	29	31	36	40	57	70			20
10	2.0	19	27	30	33	38	43	61			18	21	30	34	37	42	47	67				22
4	2.5	15	22	24	27	31	34	49			14	19	27	30	33	39	42	61				19
6	2.5	20	28	31	34	40	44	62			16	24	33	37	41	47	53	75				21
8	2.5	24	34	38	42	49	54	77			18	28	40	45	49	56	64	89				23
10	2.5	29	41	45	50	58	65	91			20	33	46	52	57	66	73	104				25
4	3.0	23	32	36	39	45	51	72			16	29	41	46	51	59	65	93				22
6	3.0	28	40	45	49	57	64	90			18	35	50	55	61	70	78	111				24
8	3.0	34	49	55	60	69	77				20	41	58	65	71	82	92					26
10	3.0	41	58	64	71	81	91				22	47	67	75	82	94	105					28

TABLE 7.6.* CUBIC FEET PER SECOND HANDLED BY DRAINAGE DITCH. (*Continued*)

Side slopes		4 to 1											6 to 1										
Bottom width, ft	% slope† / Ft depth	0.02	0.04	0.05	0.06	0.08	0.10	0.20	0.30	0.40	0.50	Top width, ft	0.02	0.04	0.05	0.06	0.08	0.10	0.20	0.30	0.40	0.50	Top width, ft
0	1.5	4	5	6	7	8	9	12	15	17	19	12	6	8	9	10	12	13	18	22	26	29	18
4	1.5	7	11	12	13	15	17	23	29	33	37	16	9	13	15	16	19	21	30	36	42	47	22
6	1.5	9	13	15	16	19	21	30	36	42		18	11	16	18	20	23	25	36	44	50	56	24
8	1.5	11	16	18	20	22	25	36	43	50		20	13	19	21	23	26	29	42	51	59		26
0	2.0	8	12	13	14	16	18	26	32	37		16	12	18	20	22	25	28	40	48	56		24
4	2.0	14	20	22	24	28	31	44	54			20	18	26	29	32	36	41	58	71			28
6	2.0	17	24	27	29	34	38	54	66			22	21	30	33	36	42	47	67	82			30
8	2.0	20	28	32	35	40	45	64	78			24	24	34	38	42	48	54	77	94			32
0	2.5	15	21	24	26	30	33	47	58			20	23	32	36	39	45	51	72	88			30
4	2.5	23	33	37	40	47	52	74				24	31	44	49	54	62	69	98	120			34
6	2.5	28	39	44	48	55	62	87				26	35	50	56	61	70	79	111				36
8	2.5	32	45	50	55	64	72	101				28	40	56	63	68	79	89	125				38
0	3.0	24	34	39	42	49	54	77				24	37	52	58	64	73	83	117				36
4	3.0	36	50	56	61	71	79	112				28	48	68	76	83	96	108	152				40
6	3.0	41	58	65	72	83	92	131				30	53	76	85	93	107	120	170				42
8	3.0	47	67	74	82	94	106	149				32	59	84	94	103	119	133	188				44

NOTE: Maximum velocity 2.5 fps; $n = 0.04$.

* Courtesy of the Soil Conservation Service, Region III, Milwaukee 3, Wis. (being Fig. 85, p. 149, "Engineering Handbook for Farm Planners—Upper Mississippi Valley Region.").

† Slope refers to ground surface slope.

7.30. Use of Curves. For standard farm crops the size and grade of the ditch, where levees are not involved, should be such that the runoff water will be removed from the ground surface in a 24-hour period during and following an ordinary rain. The curves in Fig. 7.9 provide a simple means for design of such ditches.

First select a curve applicable to the area and to the degree of drainage desired. This is done through location and description of the watershed relative to slope (feet per mile). Then, if the area of the watershed in acres is known, the runoff to be used in design can be obtained from the figure. For some truck crops such as onions, potatoes, or celery the crop will be damaged if covered for 24 hours, and a curve of higher value than those normally indicated should be selected. The required size of ditch for field drainageways (see Chap. 6) can be determined from Table 7.6. Since the ditch-size table covers only ditches up to 3 feet deep it is obvious that for other types of ditches at least 1 or 2 feet of freeboard should be provided in using them since the minimum recommended depth of ditch exceeds 4 feet, possibly by as much as 4 extra feet. This is especially true where the ditch is to serve as an outlet for tile as tile will not function effectively while the water in the ditch is at a stage which submerges the tile outlet.

ILLUSTRATIVE EXAMPLE

Watershed, 500 acres; location, northern Iowa; land slope, less than 25 feet per mile. The C curve in Fig. 7.9 should be used. This curve gives the runoff Q for a 500-acre watershed as 38 cfs. Let it be assumed that 2 to 1 side slopes will be used and that the fall of the ditch will be 0.06 per cent, the ditch to be used as a tile outlet. Table 7.6, in the 2 to 1 side slope section, shows that the minimum base width of 4 ft and a depth of 3 ft will handle the water with 1 cfs spare capacity (spare depth), but to serve effectively as a tile outlet for 4- to 6-in. tile laterals the ditch should be made 6 ft deep.

PROBLEMS

1. A ditch discharges 20.01 cfs and drains an area of 46 sq miles. What is the drainage coefficient?

2. A ditch discharges 14.21 cfs and drains an area of 3½ sq miles. What is the drainage coefficient?

3. The drainage coefficient of an area of 5 sq miles is ½ in. What is the discharge of the ditch draining this area?

4. The drainage coefficient of an area of 13 sq miles is ⅜ in. What is the discharge of the ditch draining this area?

5. Compute in complete detail:

(*a*) The equivalent of q in cubic feet per second per acre, to the nearest fourth decimal, of a drainage coefficient D of ½ in. in 24 hr.

(*b*) The equivalent D in inches of rainfall, to the nearest 1/16 in., of a runoff equal to 0.033 cfs per acre; to 0.116 cfs per acre.

(*c*) The total runoff Q per second from an area of 160 acres under the conditions of 5(*a*); from an area of 900 acres.

6. A piece of land 2 miles square must be drained by a ditch which has a fall of 1½ ft per mile; 1 in. per 24 hr is to be removed. The soil is sandy. Design the ditch.

7. Design a ditch to drain 25,000 acres of sandy loam land. The fall is 1 ft per mile and the drainage coefficient ⅛ in.

8. Determine the size of ditch at the outlet to fit the following conditions: Area of watershed 5,000 acres; fall per mile, 3 ft; drainage coefficient, ¾ in.; side slopes of ditch, 4 to 1; condition of ditch, fair; extra depth for safety, 1 ft.

9. Design a ditch to fit the following conditions, there being no natural limitations on allowable depth at the junction with the outlet channel: Watershed at head of ditch, 1,000 acres; at end of first mile, 2,500 acres; at end of second mile, 4,000 acres; at center of third mile, 5,000 acres; at end of third mile (outlet), 6,000 acres. Fall of natural surface per mile at head and end of first mile, 2½ ft; throughout second mile, 4 ft; throughout first half of third mile, 3 ft; throughout second half of third mile, 2 ft. Drainage coefficient 1½ in., side slopes of ditch, 2 to 1, condition of ditch, poor, 10% to be added to the theoretical value of Q.

10. What are the most important considerations in ditch design, and which of these considerations, when carried out in construction, will promote maintenance?

11. What two major steps are involved in the design procedure of all structures carrying water?

CHAPTER 8

OPEN-DITCH CONSTRUCTION

THE CONSTRUCTION SURVEY

With the location and design of the open ditches completed by means of the reconnaissance and preliminary surveys, the next step is to lay out the work on the ground as a basis for more exact estimates of cost for contractors' bids and as a guide to construction. This process is called the *construction survey.* The first move in the construction survey is to check the chainage and profile levels along the center lines of the ditches. If not done in the preliminary survey, a series of bench marks at about half-mile intervals in convenient locations should be set along the right-of-way of the ditch. These should be carefully checked when rerunning the profile levels before any cross-sectioning is done. In the preliminary survey the ditches were located by angle courses. The next step is to replace the sharp angles by circular curves and to follow that by setting the slope stakes at each station and at the necessary plusses between stations by the process known as cross-sectioning.

8.1. Curves in Open Ditches. Changes in direction of open ditches should never be by abrupt angles, but tangents should be joined by circular curves. Some authorities feel that such curves should be spiraled, as is done on main railroad tracks, to make the change of direction gradual and obviate erosion on the outer bank of the ditch. However, in drainage ditches, the velocity of flow is generally so low as not to cause erosion. In the rare cases where it seems likely to occur, a far simpler way than spiraling the curves is the following: Beginning back along the approach tangent about 100 feet from the beginning of the curve, gradually flatten the outer slope of the ditch, so that when the beginning of the curve is reached it will be about 3 to 1, if the general slope of the ditch banks is steeper than that (say 1½ or 2 to 1), and carry the flatter slope the entire length of the curve, easing it back to the steeper slope at the downstream end of the curve by the reversal of the action taken at the upstream end. In this way a water cushion is formed in the danger zone, acting as a buffer that prevents the flowing water from exerting any serious erosive force directly against the ditch slope. Experience by various engineers has shown that this practice is fully effective in the prevention of ditch-bank erosion on curves. Where the standard bank slope along the entire ditch is 3 to 1 or flatter, no such additional flattening will be likely to be needed.

The sharpness of the curve to use in any given case should be decided, on the judgment of the engineer, by local soil and topographic conditions, relative position of ownership boundaries, and the attitude of affected landowners. Some authorities maintain that keeping between the limits of a 4-degree and a 20-degree curve is desirable. (The degree of a curve is expressed as the angle subtended at the curve center by a chord 100 feet long at the circumference. The radius of a 1-degree curve is 5,730 feet. For curves in earth channels it is close enough to assume in any case that the radius of a given curve equals 5,730 divided by the degree of the given curve: the radius of a 2-degree curve is 2,865 feet; of a 6-degree curve, 955 feet; of a 10-degree curve, 573 feet; etc.)

All beginnings and ends of curves should be tied in by diagonal cross-reference lines intersecting in each case at the given point and with transit hubs set on the ditch right-of-way, well out beyond the lines of the slope stakes.

8.2. Cross-sectioning. For small or deep but narrow ditches on comparatively flat land with a top width not exceeding 3 or 4 feet (see Sec. 7.20), a single line of marker and hub stakes, such as is set in staking tile drains (see Fig. 10.6) along one side of the proposed ditch, is all that is needed if a list of top widths and cuts at the various stations is provided the workman. For larger ditches, except where the land surface is under water, slope stakes are set on both sides of the center at every full station and at such plusses as variations in ground surface along the line of the ditch seem to indicate as necessary. Usually cross sections needed at +25, +50, or +80 can as well be taken at +27, +54, and +81, which, being respectively, one, two, and three times the number of cubic feet in a cubic yard, facilitate figuring volumes.

The process of cross-sectioning and of computing end areas and volumes is fully covered in all standard surveying manuals; hence it need not be repeated here.

8.3. Hub Stakes at Full Station Slope Stakes. It is excellent practice, at the time of cross-sectioning, to set a firm hub, driven flush with the ground surface, by each of the full station slope stakes along one side of the ditch, preferably the lower side if there be such (see Fig. 8.1). These hubs will greatly aid the grading foreman, and save his time and that of the engineer, if used by the foreman in checking his own grade in the manner indicated in Fig. 8.2.

8.4. Protection of Stakes in Position. It is one of the chief responsibilities of the grading foreman to protect all slope stakes and hubs, in position, before grading begins, in order that a negligible number of them will be disturbed during construction.

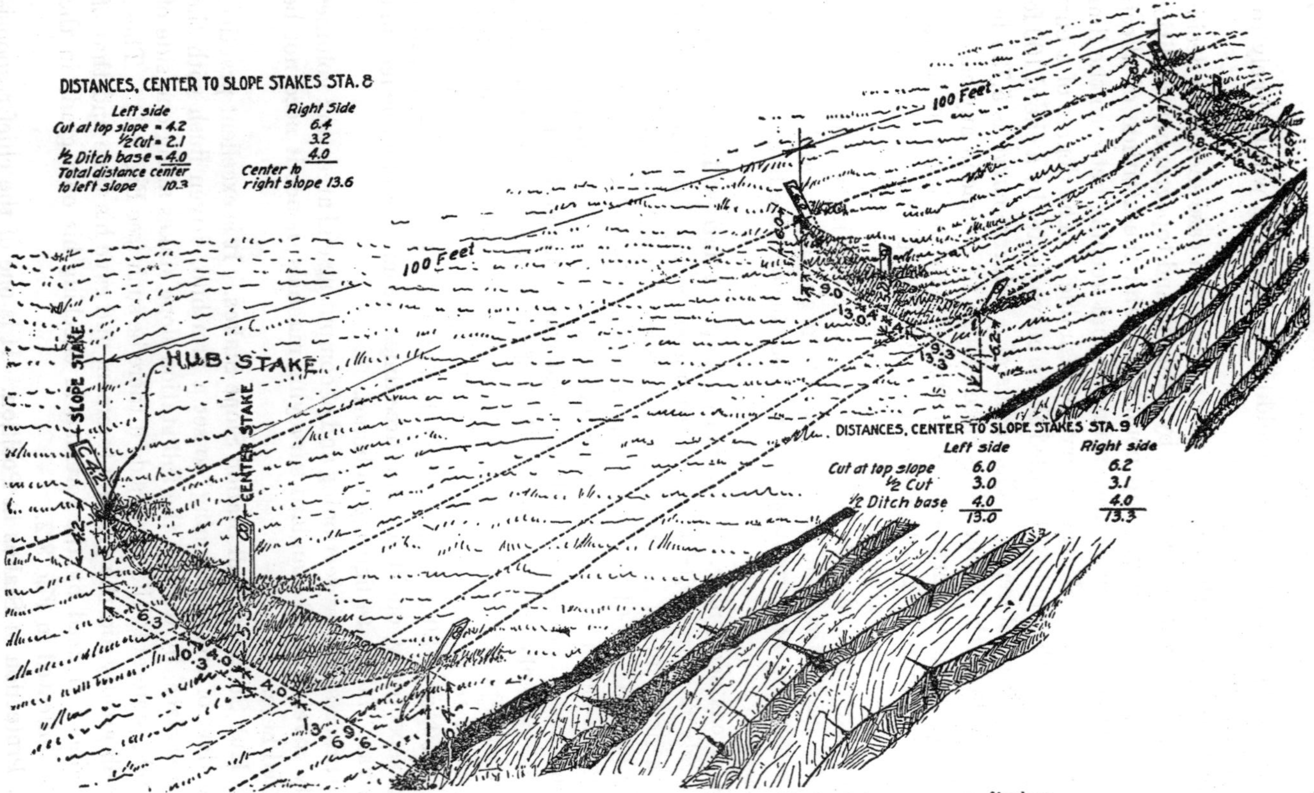

FIG. 8.1. Method of setting and marking slope and hub stakes on open ditches.

8.5. Berms and Waste Banks. In ditch construction, as stated in Sec. 7.22, the excavated material is piled in spoil banks on the natural ground surface at the sides of the ditch, allowing what is intended to be sufficient width of berm to avoid causing failure and caving of the ditch banks (see Secs. 7.22 and 7.23).

The edge of the berm next to the ditch should be clean-cut, leaving no overhang of sod or other clods to break away and obstruct the ditch, and the berm should always be so wide that there is no danger of any of the waste rolling back into and clogging the ditch.

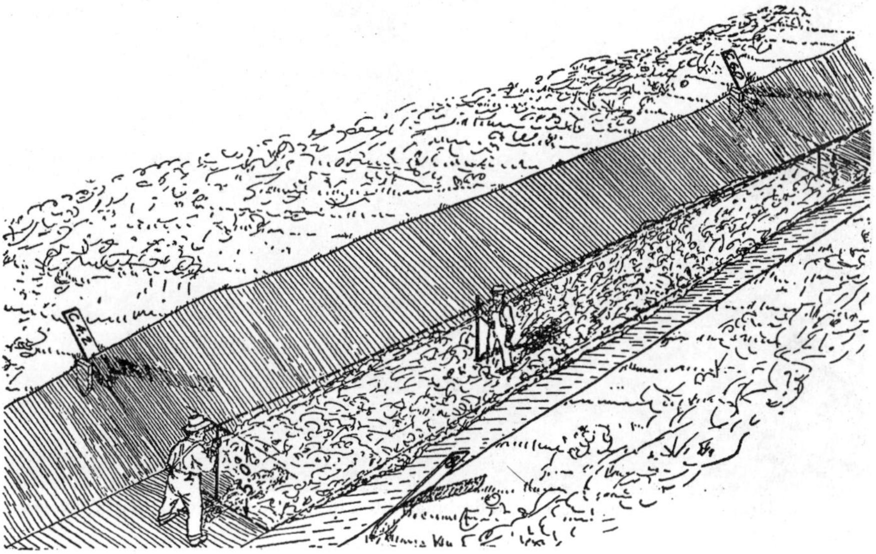

FIG. 8.2. Method of testing grade of large ditch between stakes, where exact levels have been taken on hub stakes, set along one line of slope stakes.

Even on very small hand-excavated ditches the berm should be at least 4 feet wide. On large, deep ditches or in soft marshes it should be 8 to 12 feet wide. Pickels [104] urges a 15-foot berm for ditches with bottom widths up to 14 feet. The width of the berm for the different reaches of a ditch should be given in the specifications, and Pickels further warns that no ditching contract should be let, regardless of the size of his bid, to any contractor who does not have available for the work equipment capable of meeting the specifications in full. A tough permanent sod should be maintained on the berm.

It is often desirable, where a ditch traverses a low area or borders a highway, railway, or small natural stream, to place all the waste on one side only of the ditch. This serves as a dike increasing the capacity of the ditch to more fully protect the low land against overflow and the highway and railway from washout from excessively high stages of water;

or it may serve as a road embankment (see Fig. 8.3). Where the usual practice of depositing the waste equally on both sides of the ditch is followed, modern good ditching practice provides that it shall be spread out at practically uniform depth of 1.5 to 2 feet along and close to the edge of the ditch and gradually thin out until it merges with the natural surface at a distance of several rods from the ditch as discussed in Sec. 7.23,

FIG. 8.3. Dragline enlarging Big Bayou drainage ditch in Arkansas. (*Courtesy of Bucyrus-Erie.*)

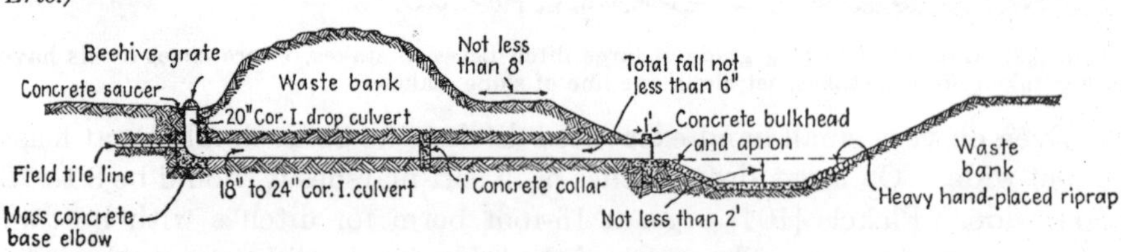

FIG. 8.4. Inlet for surface flood behind spoil bank of open ditch.

Fig. 7.7. In any case the waste banks should be smoothed off sufficiently, before becoming dry and hard-baked, to permit seeding and subsequent easy mowing of the vegetal cover. Otherwise, besides being unsightly, they become catchers and breeders of weeds.

Spoil banks obstruct surface runoff from adjacent lands; hence at strategic points, usually in natural depressions which they cross, openings should be made through them and maintained to provide for the passage of surface runoff into the ditch. Mere ditchlike openings with high overfall into the ditch tend to result in heavy erosion, development of gullies, and fouling of the ditch. Hence a better type of runway for the surface water accumulating behind the waste banks is shown in Fig. 8.4, which is self-explanatory. The culvert in this structure is not set so low as the bottom of the ditch. The discharge end of the culvert pipe in this structure should be supported by a concrete bulkhead provided with a suitable spillway into the ditch to prevent undue erosion and undermining.

8.6. Junctions of Open-ditch Laterals with Mains and Submains. As a rule the normal bottom gradient of a lateral ditch or of a submain does not directly merge with that of the main or submain at the junction. The

FIG. 8.5. Grassed inlet in branch ditch at junction with main ditch. *(Courtesy of Soil Conservation Service by L. A. Jones.)*

result is an overfall from the lateral into the main. This is undesirable and results in a deep pool being washed in the bed of the main ditch with the formation of a silt bar in its channel immediately downstream from the pool. The normal way to obviate this trouble in construction is to excavate an easement in the bottom grade of the lateral to meet the grade of the main and to protect this easement with sod or riprap. The Division of Water Conservation and Drainage of the Soil Conservation Service has carried out elaborate experiments with this type of easement and the Division Chief, Lewis A. Jones (since retired), has reported on these experiments as follows:[1]

> We . . . have found that by digging the lateral ditch back to about the center of the waste bank at a grade level to the outlet ditch and then sloping the bottom of the inlet, on about 6:1 to 10:1 grade, to the bottom of the lateral and establishing a good growth of grass, silt deposits from the lateral are largely controlled. (See Fig. 8.5.)

[1] Personal letter to the authors.

This procedure is a reasonable one, and the results such as reported by Jones are to be expected. The method is recommended, and the necessary excavation should be provided for in setting the slope stakes.

CONSTRUCTION METHODS AND EQUIPMENT

The digging of open ditches may be done in a number of different ways: by hand, V-drag, team and scraper, dynamite, road or terracer graders, dredges, and dragline excavators. Each of these has its proper place according to the kind of job and the character of the material to be moved.

FIG. 8.6. V-drag grader much used for digging shallow open ditches. (*Univ. Minn. Agr. Expt. Sta. Bul.* 217.)

8.7. Ditching by Hand. In this age of practical mechanical development, there is little justification for ditching by hand with pick and shovel except where the ditch is so small or where stumps, rocks, or other impeding elements are so numerous that no excavating machines can be used effectively. Ditches with a bottom width much under 3 feet, a depth not in excess of 4 feet, and a maximum top width of about 6 feet can often best be dug by hand but always at a relatively high cost, varying widely according to the material to be handled and the price of labor.

FIG. 8.7. The grader shown in Fig. 8.6 in action. (*Univ. Minn. Agr. Expt. Sta. Bul.* 217.)

Hand ditching is most efficiently done by men experienced in this kind of work, with tools such as are used in tiling by hand (see Fig. 12.1). The digger should be free to cast the material to either side of the ditch, leaving a 4-foot berm.

8.8. Ditching by V-drag. The V-drag (built in the shape of a V) can be built of wood at home, but that made commercially of steel is a more effective implement. For a temporary V-shaped ditch up to 3 feet deep, in the bottom of which, after bottoming by hand to exact grade, tile may eventually be laid, the adjustable drag grader (see Figs. 8.6 and 8.7) is an effective machine in skilled hands. It is also useful in construction of the broad, shallow V-shaped drainageways described in Chap. 6. However, such ditches are more effectively made with the road or terracer graders, discussed later.

8.9. Team and Scraper Work. Where animal power is readily available and neither rock nor water is present in such quantity as to make use of such power impractical, for ditches of base width 3 feet and upward and depths up to about 8 feet, scrapers are effective ditching equipment according to type as described below. However, in scraper work the soil must usually be loosened first in successive layers by plowing. Side slopes should be finished by hand with a mattock in undisturbed natural soil.

8.10. Digging Ditches with Scrapers. For many small ditching jobs on farms some kind of inexpensive scraper is still a handy tool. Any

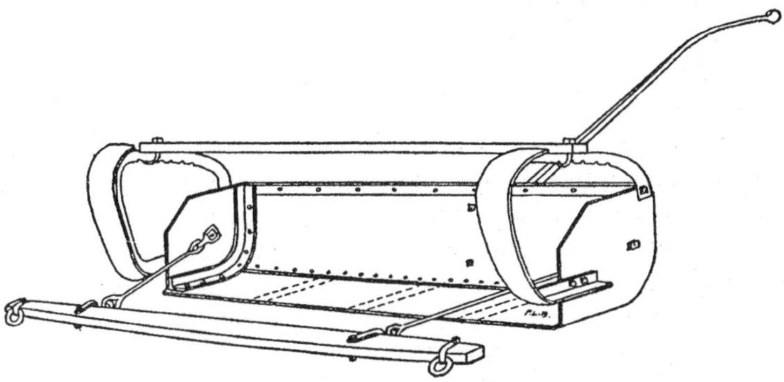

FIG. 8.8. Fresno scraper.

hardware or implement dealer can supply full information and prices so that only brief discussion is needed here.

The slip or scoop scraper is of small capacity, usually holding 3 to 7 cubic feet. These scrapers are well adapted to small jobs where the earth is moved only a short distance. The Fresno scraper (Fig. 8.8) is a wide shallow scraper with two runners extending in front and a long handle behind. This scraper slides over the ground in the same manner as the slip scraper. It may be had in widths varying from 3½ to 5 feet and capacities from 8 to 12 cubic feet. The advantages of this scraper are that it can be filled and emptied by the driver with greater ease than the slip scraper, and the earth, as it is emptied, can be distributed uniformly over a given area.

The wheel scraper consists of a large slip scraper, mounted on wheels, to which a tongue is attached. This scraper is provided with levers for raising and lowering the scoop to load and empty it. It has a capacity of 9 to 16 cubic feet. Where the earth is to be moved over 100 feet and the soil is dry, it is desirable to use this type of scraper as it is more economical in long hauls. Should the soil to be moved be stony and full of roots, the slip scraper is best, as it moves over the ground better. In ditching, the hauls are hardly ever long, except in such cases as building levees or dams across streams. The cost of moving earth with this type of scraper varies with the distance the earth is to be hauled.

8.11. Ditching with Explosives. Ditching with explosives has a number of advantages in temporary work. In emergency cases, such as where a temporary ditch must be had in a limited time to relieve a threatening flood situation, where emergency cleaning is called for, where pilot ditches are needed to firm the surface where extremely wet conditions leave the ground too soft for ditching by mechanical methods, or where the presence of many large stones or stumps makes mechanical methods impractical, temporary ditching with explosives is a quick and relatively inexpensive step. But, in the opinion of many drainage engineers, excavating *permanent* ditches with explosives is seldom if ever justified because the explosives never leave a finished job, leaving behind in the channel quantities of loose material due to three causes: (1) the sides and bottom of the channel which must be left in a firm and undisturbed natural state in a finished ditch are badly shaken by the explosive; (2) much of the excavated material falls directly back into the channel, following the explosion; and (3) the sides must be sloped by hand in undisturbed soil thus augmenting the residual loose material. All this loose material must be promptly removed by mechanical means, or it will quickly foul the ditch in action.

Ditching with explosives has one definite advantage in that it widely distributes most of the material thus excavated, over the adjacent field surface, thus eliminating much of the work of spreading spoil banks. This advantage is relatively small, economically considered, because when done at the right time mechanical spreading of the spoil banks adds but a small percentage to the total cost of the ditch and makes a more uniform job.

In any case the use of explosives for excavating ditches less than 18 inches deep and less than 3 feet wide is not generally recommended.

Where ditching is to be done with explosives, it being a highly specialized technique, it is recommended that the services of an explosives engineer be secured to direct the work and that the publications of the Atlas Powder Company [7], E. I. du Pont de Nemours and Company [39,40], and of the Institute of Makers of Explosives [148] be consulted.

DITCHING WITH HEAVY EQUIPMENT

8.12. Light Graders Drawn by Either Animal or Mechanical Power. This type, which is illustrated in Figs. 8.9 and 8.10, but made with various modifications, may be drawn by a good two-horse team or by a light

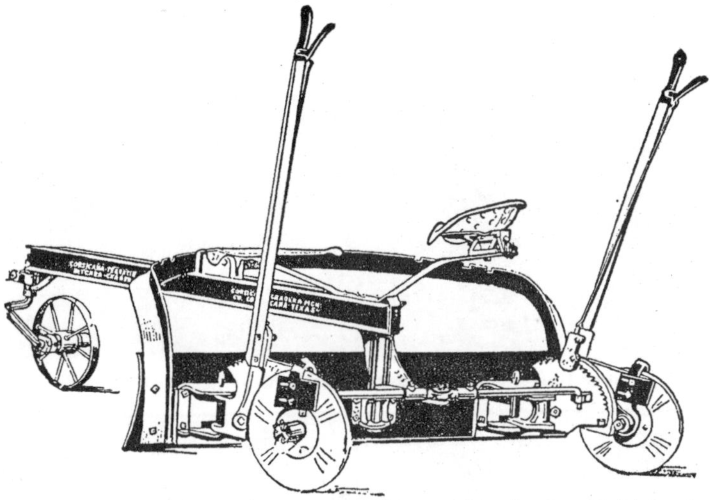

Fig. 8.9. Corsicana terracer or ditcher, used either with animal or light tractor power. (*Courtesy of Corsicana Ditcher and Grader Company.*)

Fig. 8.10. Illustrating use of Corsicana grader, shown in Fig. 8.9, shaping grassed waterway in Iowa.

tractor. These graders are justifiable equipment for the individual farmer and are thus extensively owned and used in making V-shaped ditches up to a depth of about 3 feet.

8.13. Heavy Motor Graders. Except under conditions especially adapted to team and scraper work or light graders of the type described

FIG. 8.11. Heavy 10-ft blade terracer grader drawn by heavy track-laying tractor.

FIG. 8.12. Heavy-duty motor grader: an excellent machine for digging broad V-shaped drainageways on farm fields. (*Courtesy of Austin-Western Co.*)

in the foregoing paragraph, and for depths of ditch ranging from about 3 to 6 feet, the most economical methods of ditch excavation are by means of some type of heavy excavating machine operated by mechanical power. The best machine of this class of equipment for constructing any V-shaped channel is the heavy-duty motor grader, either drawn by a heavy track-laying tractor or propelled by a self-contained power unit (see Figs. 8.11 and 8.12). Ownership by the individual farmer of these heavy machines is not generally justified, but they are often owned by counties or by soil-conservation districts from whom they may be rented at a nominal rate. They require firm ground on which to work and will not operate satisfactorily on soft wet ground.

8.14. The "Gradall" Telescoping Boom Ditcher. This ditcher (Figs. 8.13 and 9.10) must rest on dry firm ground but has given good satisfaction under district ownership in both digging and cleaning farm ditches of

FIG. 8.13. "Gradall" ditching machine at work in Virginia. It is owned by Thomas Jefferson Soil Conservation District and is rented to farmers for $12 an hour. Note that two men are required to operate it. (*USDA Farmers' Bul.* 2046, *courtesy of John G. Sutton.*)

moderate size. It requires two men to operate, but according to Sutton [147], it will either dig or clean out ditches of considerable width and depth and is available to farmers of some soil-conservation districts at relatively low cost.

8.15. Dredges. During the earlier years of the twentieth century, when large drainage operations first received great impetus in the United States, various types of large dipper dredges such as the floating and walking dredges and the clamshell and orange-peel bucket were developed and widely used. Since about 1920 most of these older types of dredges, except the clamshell and orange-peel buckets, have ceased to be extensively used, although still available for stream improvement, levee construction, and ditching in heavily flooded marshlands.

Strictly speaking, a dredge is a power excavator equipped with floating supports. If the same machinery were mounted on "caterpillars," rollers, or some other form of land support, it would be called a *land excavator* as opposed to the floating dredge (see Fig. 8.14). It serves the purpose of this book best, however, to call all such machines simply *dredges*, which implies that the actual digging is done under water.

From a practical standpoint it is necessary that the engineer know in advance what kind of machinery is to be used in construction and the limitations of such machinery, so that he may design the ditches to suit.

FIG. 8.14. A typical power shovel with ¾-cu-yd bucket. If mounted on floating supports or a spud steadied barge, it would have all the essential parts of a floating-dipper dredge. (*Courtesy of Osgood-General.*)

For instance, if a floating dredge is to be used, the ditches must all be so interconnected that the dredge may float from one to the other.

Of the many kinds of dredges used for digging ditches the principal ones mainly used nowadays naturally divide into two groups: *floating-dipper dredges* and *hydraulic* or *suction dredges*.

8.16. The Floating-dipper Dredge. The floating-dipper dredge (see Fig. 8.15) consists of a boat on which is mounted a complete power unit together with the necessary machinery for operating a dipper to do the digging. This dredge is a steam shovel placed on a boat. In its smaller sizes it is used for digging ditches for drainage and irrigation purposes. This dredge works downstream whenever possible, as in this way it is assured of sufficient water to float it without building dams across the ditch. In the relatively near past this machine has been used more than any other type for ditching, in spite of the fact that it cannot do smooth work. One objection to it is that, as it works downstream, quantities of soft mud, which do not always wash out later, are likely to accumulate behind it.

FIG. 8.15. Typical dipper dredge dumping into scow alongside. Note the vertical bottom spuds. When equipped with inclined bank spuds, this type of dredge was used extensively in the pioneer days of drainage ditch construction. *(Courtesy of Bucyrus-Erie.)*

8.17. The Hydraulic Dredge. The hydraulic dredge consists of a boat equipped with power plant and pumps (see Figs. 8.16 and 8.17). The material to be moved must be soft (unless a special cutter head is used) and contain considerable water. It is sucked up through a large pipe by means of a pump and discharged through a pipe at some distance. This dredge is used more often for cleaning out channels than for digging ditches because the material should be such as is easily moved (see Fig. 8.18). Its principal use is probably in the government service in improvement of rivers and harbors.

Of late years various types of dragline excavators have been developed that have largely replaced the older types of dredges on most ditching work.

8.18. Dragline Excavators. Dragline excavators have been modified and improved until they are now considered the most effective general-purpose machines for any open-ditch drainage work. They differ from dredges in that they always work on the land to be excavated or alongside of it. The operating machinery is similar to that of a power-shovel dredge, but it is mounted on a sort of tractor frame equipped with caterpillar treads and is self-propelled in addition to operating the ditching mechanism (see Fig. 8.19). Where the ground is exceptionally soft, mats are used which enable draglines to operate anywhere a man can walk (see Fig. 8.20). The dragline works successfully in practically any soil condition and does equally well in original ditch excavation and in cleanout work if the length of boom and size of bucket are selected to fit the given

FIG. 8.16. Hydraulic dredge at work on the lower Willamette River opposite Swan Island, Portland, Oreg. (*Courtesy of U.S. Army Engineers, Portland District.*)

FIG. 8.17. Discharge pipe from the dredge shown in Fig. 8.16. (*Courtesy of U.S. Army Engineers, Portland District.*)

job. These machines range in size from those with a ⅜-cubic yard bucket with a 25- to 30-foot boom to those with a 5-cubic yard bucket with a 100-foot boom. Some dragline machines are now made in such a way that the buckets are interchangeable as between the regular dragline bucket, the clamshell, and the orange-peel. The illustrations in Chaps. 8 and 9 show draglines of various sizes operating under a variety of local conditions and types of job (see especially Figs. 8.3, 8.21, 8.22, 9.5, and 9.6).

8.19. Dippers and Buckets. The bucket on most dredges is of the dipper type, but several kinds of buckets are often used for the same

FIG. 8.18. Clamshell dredge "Alabama" excavating drainage canal near Charenton, La. (*"Photograph March, 1939, by Corps of Engineers, U.S. Army, New Orleans District"; reproduced by the courtesy and permission of the District Engineer.*)

NOTE: Although the dredge shown is a clamshell dredge, the local condition and type of material to be moved also clearly illustrate those well suited to hydraulic dredge work.

general purpose, namely, the clamshell, orange-peel, and dragline bucket and the trench hoe.

The dredge dippers are of rectangular shape, having an open top and a hinged bottom (see Figs. 8.14 and 8.23). They are loaded by scooping the earth in at the top and are unloaded by tripping a latch on the bottom, allowing it to fly open. The lip of the dipper which does the scraping is provided with a number of heavy teeth to prevent rapid wear on the bucket itself. These teeth are generally removable so that they can be easily replaced or taken off and sharpened. The dippers usually range in capacity from ¾ to 15 cubic yards. Wherever stumps and rock are

FIG. 8.19. Dragline excavating new drainage ditch on Marymoor farm a mile southeast of Redmond, Wash., in King County Soil Conservation district. Water in ditch is at level of Samamish River. (*Courtesy of Soil Conservation Service, Portland, Oreg.*)

FIG. 8.20. Close-up of the wide treads used on a ¾-cu-yd dragline for stable underfooting on soft ground. (*Courtesy of Osgood-General.*)

encountered, the dipper dredge with reinforcing teeth should be used (see Fig. 8.23).

Clamshell buckets are shaped and operated like a clamshell. They are built light and, therefore, are not adapted to heavy work; but as they close tightly they are used in fine or soft dry material. They are built in capacities of 1 to 5 cubic yards (see Fig. 8.24).

The orange-peel bucket is shaped like one-half of an orange. It opens

FIG. 8.21. Dragline at work in the Florida Everglades. (*Courtesy of Bucyrus-Erie.*)

FIG. 8.22. Scene depicting typical conditions near outlet of a main drainage ditch in the South. Dragline appears to be working from both sides, using crossovers to be excavated later. (*Courtesy of Koehring Company.*)

FIG. 8.23. Power shovel working on project in Oklahoma showing bucket in act of dumping. (*Courtesy of Harnischfeger Corporation.*)

FIG. 8.24. One and one-half cubic yard clamshell excavating sand and gravel from the Mississippi River bed near St. Louis, Mo. (*Courtesy of Osgood-General.*)

at the end by dividing into quarters or thirds. These buckets are dependent on their weight and leverage for their digging power, hence are not adapted to very hard digging. They are built in capacities of 1 to 10 cubic yards.

The dragline bucket is rectangular in shape, similar to the dipper, but

FIG. 8.25. Dragline bucket in act of dumping while excavating drainage ditch in Michigan. (*Courtesy of Bucyrus-Erie.*)

FIG. 8.26. One-half cubic yard trench-hoe bucket. Machine is designed with gooseneck boom. (*Courtesy of Osgood-General.*)

is operated differently. It is dragged toward the operator along the place to be excavated and is thus filled (see Fig. 8.19). It is unloaded by tipping it over the point to receive the load (see Fig. 8.25). It can be used for heavier digging than the clamshell or orange-peel and is very widely used in ditching. It is now built in sizes of $\frac{3}{8}$ to 5 cubic yards capacity.

The trench-hoe bucket is of the form of an inverted dipper and is used largely for digging narrow trenches by drawing it toward the operator like a hoe. The smaller sizes only are needed for tile trenching, etc. (see Figs. 8.26 and 12.14 to 12.16).

Ditches of large size and great length are dug most cheaply with dredges. About the smallest ditch that can be dug with a floating dredge is one with a 12-foot bottom width and a 5-foot depth. The cost of digging with this type of dredge under favorable conditions is usually a minimum for such work and 40 to 50 per cent higher for unfavorable conditions.

It is hard to make uniform side slopes with floating dredges. With a land excavator it is easier. For flat slopes, scrapers or a dragline should be used to finish the side slopes, or, in extreme cases, they may be finished by hand.

8.20. Spreading the Waste Banks. After being allowed to drain and dry out just enough to reduce slipperiness and make the material

FIG. 8.27. Heavy cable-operated bulldozer at work: Front view, Keating SCD. (*Courtesy of Soil Conservation Service, Portland, Oreg.*)

FIG. 8.28. Same machine as Fig. 8.27. Close-up front view.

sufficiently firm to bear the weight of heavy equipment, the best machines for spreading the waste banks are the terracer earlier discussed and the bulldozer (see Figs. 8.27 and 8.28).

8.21. Finishing the Side Slopes of Ditches. Regardless of the method by which a ditch is excavated, special care must be exercised to see that the excavating equipment does not gouge below the surface of the side slopes. Then the slopes should be finished by hand with a mattock or other similar tool made for the purpose or with a scraper or dragline excavator. This kind of treatment is good insurance against sloughing and caving of the sides of the ditch and thus reduces maintenance cost (see Sec. 9.3). It is best practice, also, to sow the slopes above the normal flow line to some quick-growing, tough, sod-forming grass as soon as the ditch is completed. This provides protection against weathering and erosion of slope material, which, when loosened, rolls or is washed into the bottom of the ditch, augmenting the silting and even forming mud bars that obstruct the flow.

8.22. Bridges and Culverts. Substantial structures over the ditches should be provided at all road crossings. Creosoted wooden bridges with strips of sheet iron, bolted along the treads of traffic, have proved satisfactory in many locations; but maintenance costs will be lower if either steel bridges or culverts of concrete or of asphalt-coated, asbestos-bonded corrugated iron are used. The abutments and floors of bridges must be carefully inspected at least twice a year, and necessary repairs should be made at once.

CHAPTER 9

OPEN-DITCH MAINTENANCE

9.1. Importance of Timely Maintenance. It is an almost invariable unfortunate practice to neglect open drainage ditches, once they are dug; this occurs in spite of the fact that the drainage statutes in a number of states provide the machinery for securing the necessary funds and make it mandatory on ditch supervisors and boards to carry out the necessary maintenance works. Such statutes are largely ignored. This situation is, to say the least, unfortunate. It is well understood that earth construction begins to deteriorate rapidly as soon as it is complete and that, due to weathering from wind, rain, etc., without immediate and constant maintenance, destruction soon ensues. It is a fact well known among engineers and many farmers that even a large drainage ditch, without maintenance, will often be obliterated in 10 years time. Furthermore, it is usually more costly to redig it than was the first construction. Maintenance of ditches neglected for only a year or two is always expensive and often difficult to accomplish.

The problem of ditch maintenance resolves itself into two important sets of operations: (1) how to prevent deterioration entirely, or keep it as small as possible; and (2) how to repair the damage well and economically after it has occurred and restore the ditch to its original condition.

DETERIORATION PREVENTION AND CURE

Silt Control

Anything that prevents silt from entering ditches or that tends to carry it through without being dropped by the water comes under this head. Engineers have profited by past mistakes in this respect and now insist on many procedures that were not formerly considered necessary.

9.2. Soil Washing. If, by proper land use, cultivation, terracing, or other methods, soil washing from ditch watersheds can be kept under control, much of the silt evil will be eliminated, because silt will not have a chance to get into the ditch (see Fig. 9.1).

9.3. Diversion Ditches. The greatest trouble from silting generally occurs where water from upland ditches with steep slopes is emptied into flat ditches in bottom land. The velocity of flow is decreased in the flat ditches to such an extent that the silt settles out rapidly and necessitates almost constant maintenance to keep them open. Sometimes, the trouble is minimized by the use of diversion ditches, running around the lower edge of the high ground, which collect the hill water, highly charged with sediment, and carry it for some distance to a favorable point, where it is emptied into a drainage channel; but the diversion ditch itself needs to be cleaned out occasionally, and its burden of silt generally causes trouble farther downstream.

FIG. 9.1. Silt eroded by a single rain from an uncontoured field in western Iowa.

9.4. Settling Basins. Another method that has proved effective, where the topography permits its use, is to collect the hill water in stilling basins and allow the silt to settle out before the water enters bottom ditches. Frequently, natural depressions and hollows can be utilized for this purpose, but unless there is another into which the water can be turned when one is filled with sediment, they too must be cleaned out at regular intervals.

9.5. Flooding. Still another, and the most sensible way of all, is to treat the silt as a friend rather than an enemy and utilize it to build up and enrich low bottom lands. This method was originated by Roy N. Towl, a consulting engineer of Omaha, Neb., and it has been used by him with much success on one of his farms near Tekamah, Nebr., since 1917. Rich silt has been deposited at an appreciable rate per year, and the land, formerly sandy and level, has now assumed a considerable surface slope away from and at right angles to the overflow ditch. Water is caused to spread out in a slowly moving thin sheet which does not injure growing crops and does not prevent cultivation a short time after it recedes. Good crops are thus produced at the same time that the land is being continually enriched and built up.

Figure 9.2 illustrates clearly the action that takes place. The central or overflow ditch is made just large enough to take care of a normal flow. Any excess delivered to it from the larger nonoverflow hill ditch passes through the culvert in the upper dam and is prevented from entering the main drainage ditch by the lower dam. The water continues to rise until it slowly overflows the banks and travels laterally in both directions, depositing a thin film of silt as it goes. By the time it reaches the lateral collecting ditches, 1,000 feet away, practically all the silt has been dropped, and the water is delivered to the main drainage ditch in a clarified condi-

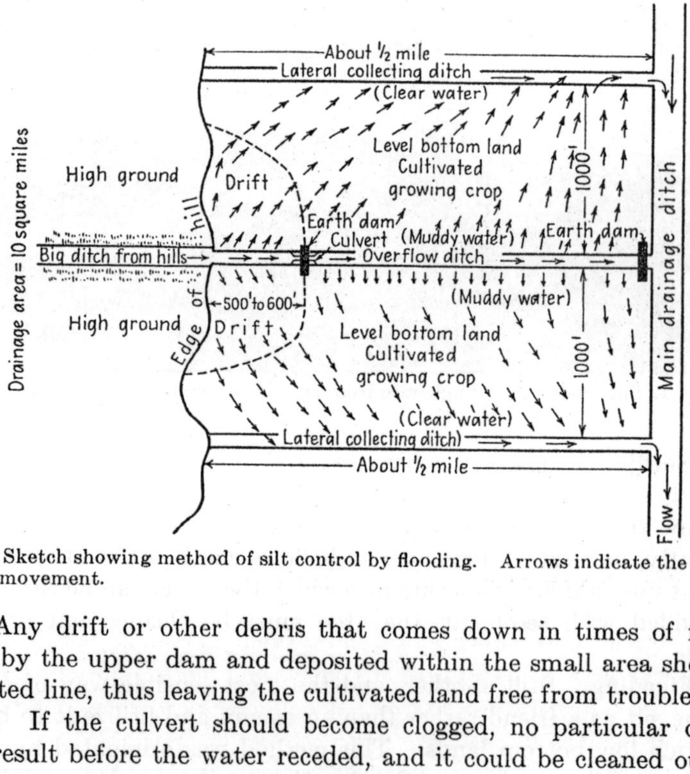

FIG. 9.2. Sketch showing method of silt control by flooding. Arrows indicate the direction of water movement.

tion. Any drift or other debris that comes down in times of flood is caught by the upper dam and deposited within the small area shown by the dotted line, thus leaving the cultivated land free from trouble in this respect. If the culvert should become clogged, no particular damage would result before the water receded, and it could be cleaned out.

The layout shown in the sketch is suited to take care of the runoff from a watershed of 10 square miles. A larger watershed would require proportionately more land for the water to spread over. Great care is necessary to see that the water thins out uniformly and travels slowly. The bed and banks of the overflow ditch build up at an equal rate so that, in a few years, the ditch flows along the crest of a ridge which may have a lateral slope of as much as 1 foot per hundred.

9.6. Caving Banks. There are two common causes for caving banks—either the grade of the ditch is excessive, resulting in erosive velocities, or the side slopes have not been designed to fit the character of the local soil.

If there is only a moderate excess grade, caving can usually be prevented by use of proper side slopes planted to grass which can be mowed.

Fig. 9.3. Loose rock wing dam or jetty to prevent erosion of outer bank of a creek on a curve through a field at high-water stage. (*Courtesy of Soil Conservation Service, Pacific Region.*)

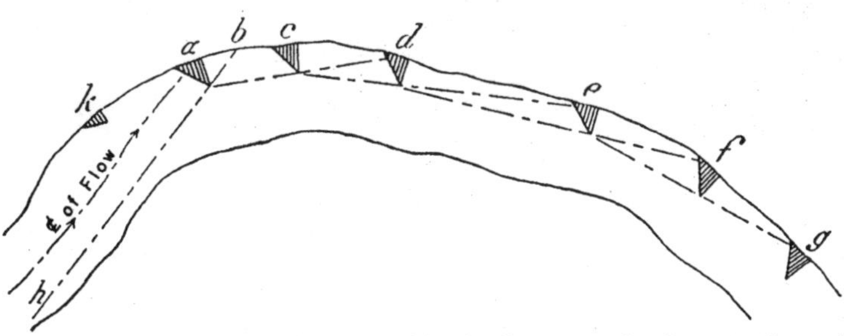

Fig. 9.4. Method of determining location of jetties for stream-bank protection. Point *a*, the location of the first jetty, is the intersection of the flow line and the eroding bank; jetty *c* is located by drawing the line *hb* parallel to the flow line and across the toe of jetty *a*; *ac* is twice *ab*; jetty *d* is located by projecting a line across the toe of jetties *a* and *c*; the remaining jetties are located in the same manner as jetty *d*; supplementary jetty *k* is located a distance *ac* upstream from jetty *a* and should be approximately one-half regular size. (*From USDA Farmers' Bul.* 1813.)

At critical points, usually near curves, heavier protective devices may be needed. The most satisfactory although not the cheapest protection has usually proved to be heavy hand-placed riprap or brush revetments. Loose rock jetties (see Fig. 9.3) may also be used, spaced at about stream width with a length about one-fourth the spacing (see Fig. 9.4). But deflection devices should be placed with great care not to overdo current deflection lest the problem simply be transferred to the opposite downstream bank.

9.7. Side Slopes. Taking advantage of moderately high grades to construct the ditches much smaller than necessary, depending on erosion to enlarge them, is dangerous practice *which should never be permitted* as the eroded material, deposited downstream, may cause low-flow meanders and consequent caving. In any case, heavy erosion, once started, is hard

FIG. 9.5. Dragline flattening side slopes of a main outlet ditch in the Middle West. (*Courtesy of Koehring Company.*)

to check at any point and will usually ruin any chance of having eventually stabilized side slopes.

It is difficult to secure side slopes which will be absolutely permanent in any kind of soil, and soil conditions often vary widely and unexpectedly on different reaches of the same ditch, a condition which makes the slope-ratio problem wholly indeterminate, technically, and dependent on the experience and judgment of the engineer (see Fig. 9.5). It is practically impossible to dig flatly sloping banks with floating dredges, and the erstwhile common practice of the dredge operators of digging to the specified top width but with steep side slopes and greater than specified depth, leaving erosion and weathering to finish the sloping, cannot be justified. Allowed in the past, even by good engineers, on the theory that the banks would soon cave and slide and finally take the designed form of ditch cross section, *this practice has almost invariably ended in disaster.* Experience has shown that, usually, it is wise to go to extra trouble and expense to construct the ditches in the first place with smooth side slopes in undisturbed soil, flat enough to give reasonable assurance of permanence.

In the *National Reclamation Magazine* of Apr. 1, 1925, Jacob A. Harmon of the Elliott-Harmon Engineering Company, in an article based on an experience in the lower Mississippi Valley, tells substantially as follows what is likely to happen under the "cut-and-slide" plan just discussed.

The job, a reconstruction project, took place in the summer of 1919 in southeastern Missouri. In the original construction there had been careless digging (it is inferred, on the cut-and-slide plan) with resulting serious deterioration. The redigging of the system of ditches was done for several miles with the side slopes carefully finished, under strict inspection, to a continuous 1 to 1 slope. The finished job looked fine and seemed to be standing up well. However, about March, 1921, some of the ditches were failing seriously. Within 6 weeks after the rainy season started, on some of the ditches 15 to 20 feet wide, with carefully cut 1 to 1 side slopes and an original depth ranging from 10 to 15 feet, serious caving had resulted in filling the ditches clear across, from 3 to 6 feet above the excavated grade. It was decided that, for stability, the side slopes must be much flatter. A 3 to 1 slope was adopted and carefully adhered to in spite of the marked increase in excavation. This has proved continuously satisfactory. The slopes were planted to Bermuda grass to protect them against washing from heavy rainfall. The farmers can now mow the grass on this slope with a power mower. Harmon wished to emphasize the point that if the ditches are so dug that the farmers cannot use their normal farm equipment for maintenance they deteriorate until they must be redug at great expense; and that by the use of flat slopes, even at 15 to 20 per cent additional excavation in the original construction, followed by farmer maintenance, appreciable economic savings to many of the drainage districts are effected. The soil in this region is an exceedingly fine black clay, commonly known as *gumbo* or *buckshot.* When wet it loses all its cohesive character and melts away, sliding in almost horizontal layers along lines of cleavage. A material similar in behavior is common in the prairie river valleys of the northern Middle West in the United States and in Canada. This is in sharp contrast to the behavior of the more common coarser clays in the upper Mississippi Valley and in many other parts of the United States. These clays break along nearly *vertical* lines and roll or slump down into the ditch bottom. They frequently stand up fairly well on side slopes of 1 to 1 but are of uncertain behavior. On all soil types here mentioned much flatter slopes than even 1½ to 1 are recommended.

On a large construction job in the Little River Drainage District, also in southeastern Missouri, the Morgan Engineering Company used side slopes of 1½ to 1 at ordinary depths, and on deep ditches, 2 to 1 with apparent success. But these slopes are too steep to mow with an ordinary mower.

In the lower Mississippi Valley and other similar wide river valleys throughout the United States, much of the soil is alluvial silt deposited over fine water-bearing sand. In such cases, if ditches of required depth cannot be secured without penetrating the sand layer, the work must proceed very cautiously and slowly. The sand must be excavated in thin layers and sufficient time allowed between the removal of consecutive

FIG. 9.6. Three-quarter cubic yard dragline engaged in cleanout work on a well-maintained drainage ditch.

layers for the surplus water to drain away and permit the sand to settle and become firmer. Otherwise it will flow into the ditch in large quantities, undermining the surface layers of silt which then drop into the cavities thus formed.

As stated in Sec. 8.5, Berms and Waste Banks, ditch slopes sometimes fail by bulging out, owing to the weight of the waste banks deposited with too narrow a berm to fit the local soil character and conditions. Once started, the only way to check this action effectively is to distribute the load by spreading the waste banks as discussed in Sec. 7.23.

9.8. Equipment for Removal of Silt Deposit. In any open ditch there will always be more or less fouling by silt deposits in time. These alone are serious and when combined with layers of vegetable growth year after year they rapidly fill and obliterate the ditch or greatly reduce its capacity unless systematic removal of the silt is done at frequent intervals. In very small ditches, silt deposits are still sometimes removed by hand by men skilled in handling the ditching spade and shovel. However, such work is onerous, uneconomical, and slow. Besides, skilled workmen of this type have become almost nonexistent. Various excavating machines are now available for effective and economical ditch-cleaning work (see Fig. 9.6).

Floating dredges cannot be used for such work unless the depth of mud exceeds 2 feet, because there would not be enough water to enable them

FIG. 9.7. The English dragline for digging and cleaning farm ditches discussed by Olsen. The bucket is tapered to fit the ditch section. The side boom facilitates working from one side of the ditch. (*Courtesy of Soil Conservation Service, by L. A. Jones.*)

to float. Such dredges usually dig downstream and hold water against the slopes for long periods of time. This softens the slopes and may cause them to cave badly when the water goes out with a rush.

Light dipper excavators, known as straddle dredges, powered by internal-combustion engines and mounted on platforms which straddle the ditch, are admirably adapted for such work in ditches of moderate top width but are no longer commercially available. They are propelled on flanged wheels and track sections or on rollers or skids, and they dig upstream.

On ditches of medium to large size the most popular excavator for cleanout work is the standard dragline equipped with a bucket of ½ to 1 cubic yard capacity, as the smaller sizes are still found more effective for such work than are the large buckets often used in ditch construction. They move along on either berm on track-laying treads or on "walkers," on mats if needed (see Fig. 9.6). They are capable of excavating the bottom cleanly and of cutting smooth side slopes at any desired angle. They can do this work with ease up to 125-foot bottom width. Figures 8.21 to 8.24 show draglines operating on cleanout jobs of various kinds.

FIG. 9.8. A close-up of the tapered bucket on the ditcher shown in Fig. 9.7. (*Courtesy of Soil Conservation Service, by L. A. Jones.*)

In exceptional cases where it is necessary to spread out the excavated material at a considerable distance from the ditch, machines operating on the hydraulic principle have been used to advantage.

For small ditches of narrow base and top and of no great depth, Olsen [98] tells of a dragline widely used in England for cleanout work, equipped with a side boom and a special small bucket of ¼ to ½ cubic yard capacity of a shape calculated closely to fit the cross section of small ditches (see Figs. 9.7 and 9.8). L. A. Jones reported by letter in 1944 that Olsen was then developing, in Louisiana, an attachment of this kind for small draglines used in this country and stated that "small draglinese quipped with this attachment can perform clean-

FIG. 9.9. Bucket-ladder excavator for open ditches, especially designed for cleanout jobs. Powered by a standard diesel engine and fully equipped with hydraulic controls. With adjustable extension units it can handle ditches up to 30 ft wide and 7 ft deep at a speed of 50 buckets per minute. (*Courtesy of American Steel Dredge Company.*)

out work on small ditches very effectively, especially in areas such as the sugar cane region of southern Louisiana where there is little or no tile drainage and the fields are drained by small ditches."

There are available modifications of the bucket-ladder and bucket-wheel tile-trenching machines, shown in Figs. 12.10 to 12.13, that are suited to cleaning open ditches of moderate size, but neither is intended for original excavation of open ditches. Both machines must rest on firm dry ground. Each type has its own special features. The bucket-ladder type (see Fig. 9.9) is the faster worker and will clean a much larger ditch than will the bucket-wheel type but, being less rigid than the latter,

FIG. 9.10. Specially designed machine being demonstrated near Blackville, S.C. The boom maintains uniform ditch banks. This excavator is truck-mounted and may be readily transported for small jobs. (*USDA Farmers' Bul.* 2047, *courtesy of John G. Sutton.*)

will not carry as smooth a grade. It is especially suited to both widening and cleanout work even where the excavation is under water. The bucket-wheel type is merely the machine shown in Fig. 12.11, additionally equipped with wing shovels for sloping the sides of the ditch. In the hands of a skillful operator it does the smoother job of the two but is not suited to work in a wide or a very deep ditch and will not work under water or in very wet ground. Either of these machines costs several thousand dollars, and the maintenance cost is high. The Gradall telescoping boom ditcher (see Figs. 8.13 and 9.10) is in much the same class.

9.9. Explosives in Ditch Cleaning. Where the volume of silt to be removed does not seem to justify employing an excavating machine for the job, explosives are sometimes used to remove the deposit, which they will do, but such action is subject to all the disadvantages listed in Sec. 8.11 as accompanying original excavation of ditches with explosives. If used, the operation should be carried out under the close supervision of an experienced man, with caution not to overdo it.

9.10. Ditch Deterioration Due to Vegetation. Rank growth of grass, weeds, and brush in and alongside ditches and the existence of water weeds in the channel constitute two of the most serious agencies of ditch deterioration. These by themselves are serious, but they also aggravate silt deposition and together make a mass of obstructive deposit difficult and expensive to remove unless systematic work of this kind is done regularly every spring and fall. Figure 9.11 clearly illustrates the result

FIG. 9.11. A large public outlet ditch in the valley of the Red River of the North fouled by several years growth of grass and wild-sown grain.

of neglect to clear out grass, weeds, and brush at least annually, in the dry part of the fall, when they may usually be burned out. However, fire should never be used for this purpose on peat bogs. For methods of treatment see Sec. 9.13 and Fig. 9.12, and Sec. 9.17 and Fig. 9.15.

Some types of weeds, such as the Canada and Russian thistle, the mature growth or seeds of which are especially bad if allowed to develop unchecked, quickly clog and obliterate even large ditches.

Prevention is always better than cure. This is especially true in handling the vegetation problem in ditches. Such vegetation constantly appears in spite of the greatest care to eliminate it. The essential need is to destroy it or drastically check its growth while it is still young and tender before it matures its spores or seeds. Various methods and means for its control have proved effective according to the type of vegetation. The problem is so important that a thorough search for its solution has been made since about 1940 by the Bureau of Reclamation of the U.S.

Department of Interior in cooperation with several of the services of the U.S. Department of Agriculture. This study is still active. The following discussion is based on the findings of the study as reported in various papers of the two departments and contributions from various other individuals and agencies.[1]

9.11. Types of Objectionable Vegetation—How Spread. For convenience in dealing with this problem, weeds are considered in two classes, *ditch-bank weeds* (land weeds) and *channel weeds* (water weeds), both of which are a serious problem in drainage and irrigation-ditch maintenance, but in drainage the land weeds constitute the major problem. There are many troublesome species in both major classes; but only a few of the main offenders will be referred to here. They are spread by various agencies, all of which must be considered in activities for their control and elimination. Leading among these agencies are wind and water, both of which are great weed and weed-seed carriers; animals, including man himself, by seed and even plants or parts of plants adhering to clothing, car fenders, and other equipment; and animal droppings.

9.12. Methods of Control and Eradication of Weeds. Most vegetable growth is easiest to control and eradicate while it is still young and tender, but it is often necessary to deal with it at all stages of growth. There are two general systems of control, *mechanical* and *chemical.*

As the mechanical methods of control differ widely between land weeds and water weeds the two types of weeds are here treated separately under mechanical control; but, as in chemical control the chemicals used are mostly the same or in the same general group for the two types of weeds, chemical controls of the two types are here treated under one heading.

MECHANICAL CONTROL

Land Weeds

Land weeds, especially in their mature stages, seriously interfere with drainage operation and maintenance in three specific ways: (1) by mass growth they prevent thorough inspection; (2) by breaking over and falling into the water or by breaking loose and being rolled by wind into the

[1] "Control of Weeds on Irrigation Systems," multigraphed pamphlet of the U.S. Department of Interior Bureau of Reclamation, Branch of O. and M., July, 1946. "Aquatic Weed Control," by B. B. Boyer, published by Cloroben Corp., March, 1946. "Controlling Submerged Water Weeds on Irrigation Systems with Aromatic Solvents," a mimeographed Joint Report by the U.S. Department of Interior and U.S. Department of Agriculture, April, 1949. "Submerged Aquatic Weed Control in Irrigation Channels," by V. F. Burns, a contribution from the Western Weed Conference, Denver, Colo., Jan. 30, 31, and Feb. 1, 1950, an official paper of the U.S. Department of Agriculture, Agricultural Research Administration, Bureau of Plant Industry, Soils and Agricultural Engineering, and the Washington Agricultural Experiment Station, cooperating.

water they catch floating weeds and debris, thus slowing the water flow and causing silt in suspension to deposit and form mudbanks and dams in the channel; and (3) by increasing the weed problem on adjacent crop land. Rodents are more plentiful in badly weed-infested strips along ditches than where the weeds are kept under control. Bindweed, quackgrass, whitetop, burdock, and both Canada and Russian thistle are leading offenders among the herbaceous land weeds.

The first step in mechanical control of land weeds along ditches is to spread or smooth the waste banks as earlier discussed, in order that the

FIG. 9.12. This ditch was constructed by dragline in 1945. Maintenance has been too long delayed. (*USDA Farmers' Bul.* 2047, *courtesy of John G. Sutton.*)

mowing and cultural operations may readily be carried right up to the ditch. Flat side slopes make possible extension of such operations down to or near the permanent water line. The chief mechanical methods of control are by *mowing*, *pasturing*, and *burning*, according to the exact conditions to be met.

9.13. Mowing. In their immature stage the best method of land-weed control is frequent mowing—at least twice during the growing season. Where the objectionable herbaceous growths tend toward maturity before they are effectively attacked, mowing several times a season is an effective control procedure. Flat side slopes, 3 or 4 to 1, make it possible to use a horse-drawn mower. Many farm wheel tractors now have a sickle bar for mowing weeds that can be operated from the ditch berm on slopes as steep or even steeper than 1½ to 1, and many of the makers of farm machinery now make a mower especially designed to operate from the ditch berm and mow ditch slopes of practically any steepness (see Fig.

9.12). Sickle-bar attachments of similar adaptability are also made to be attached to and operated by almost any type of self-propelled power unit. Such a device is now in continuous use by highway departments.

There must be sufficient width of unobstructed berm immediately adjacent to the ditch to serve as a road on which to operate such equipment. All stumps and woody growths much in excess of ½ inch in thickness and all large rocks as well as gullies must be removed beforehand from the path of the sickle bar. Mowing the banks at least twice a season is strongly advocated (see Fig. 9.13).

FIG. 9.13. Well-vegetated shallow field drain in Virginia that can be maintained with equipment ordinarily found on the farm. (*USDA Farmers' Bul.* 2047, *courtesy of John G. Sutton.*)

9.14. Pasturing. In the South, where there is little or no dormant season for plant growth, where rainfall is high and vegetative growth is rapid and luxuriant, more frequent and possibly continuous attention to ditch maintenance is required. Writing of conditions in Louisiana, where the annual rainfall is 50 to 60 inches, Watson [158] advocates keeping ditches in condition by grazing (see Fig. 1.7). He cites cases where the luxuriant growth has decreased the capacity of an unpastured ditch 90 per cent in 2 years, and others where, under similar conditions, the ditch has become choked and filled up in 5 years time, requiring reexcavation, whereas pastured ditches were kept continuously in good condition.

Up to the early 1930's, permitting livestock in open ditches did not find favor in the central and northern United States, but since the Federal drainage rehabilitation campaign of 1935, flatter grassed slopes have come into general use, with the idea of permitting controlled pasturage of the ditches for maintenance purposes, and this practice is now widely and effectively used (see Fig. 9.14).

Controlled pasturing of ditch slopes kills weeds, exposes jungles of noxious weeds which may then be eliminated by more drastic treatment, and results in reducing rodent colonies. However, certain preparatory measures are essential, including the removal of poisonous weeds such as water hemlock and whorled milkweed and fencing the ditch right-of-way which is ordinarily a district responsibility. In some districts the individual farmers set their own fences some distance back from the ditch

FIG. 9.14. A four-section water gate across a drainage ditch in Indiana that keeps animals in but lets trash through. Running water swings the gate forward, and debris floats on down the channel. Can be built in three sections for smaller ditches. (*USDA Farmers' Bul.* 2047, *courtesy of John G. Sutton.*)

to enlarge the pasturing strip. The district fences are thus eliminated. Rotation pasturing of large areas traversed by the ditch has much the same effect and helps check the spread of weed infestation to adjoining crop lands. Cattle guards across the ditchbank roads at area boundaries are much more satisfactory than the inconvenient farm gates in the fence lines.

Classes of Stock to Pasture. All types of pasturing stock except hogs are used, the hog being excepted because of his rooting habits which could cause difficulties in ditch maintenance (see Fig. 1.7). If the soil of the ditch slope puddles or roughens badly when wet, under the trampling of heavy stock, this can be met by confining pasturing of such areas to sheep and goats.

9.15. Blanket of Pasture Crops Desirable. Once the weed situation is under practical control good pasture grasses sown on the strip will help to keep it so. Grasses should be chosen whose natural habitat is similar to the given locality in essential plant-growth characteristics. In northern areas *Bromus inermis* and crested wheatgrass are considered among the best, but redtop, rye grass, orchard grass, and buffalo grass are also suitable. In the South and West the pasture crops recommended are buffalo grass, curley mesquite, vine mesquite, Mediterranean salt grass, and kikuyu grass.

9.16. Procedure with Mature Land Weeds. If ditchbank weeds have been allowed to reach a fully matured stage or even, as in numerous cases

FIG. 9.15. An oil-burning flame thrower drawn by farm tractor and used for burning weeds in small drainage ditch. (*USDA Farmers' Bul.* 2047, *courtesy of John G. Sutton.*)

on record, several seasons have elapsed without any attempt to get rid of them, the situation calls for a much more drastic treatment than any thus far discussed. It is apt to involve heavy work with a bulldozer or other heavy equipment, whose function is the same, followed by assembling, piling, and burning. This is especially true in dealing with such pests as Russian thistle which, driven by the wind, often piles up in great windrows along waste banks and in ditch channels. The latter check and cause to be deposited both wind- and water-borne silt, and the weeds and silt combined often completely fill sections of large ditches in a single season. In this case fire will not solve the problem, but heavy excavating machinery must be called into play at great expense.

9.17. Burning. With other types of mature weed growth still anchored to the soil, burning, either in prairie-fire style or, usually better, with the

FIG. 9.16. Chemical spraying to control vegetation in a drainage ditch in the Middle West. The hose is long enough to permit operator to walk in bottom of ditch if needed. (*USDA Farmers' Bul.* 2047, *courtesy of John G. Sutton.*)

mechanical types of burners used along railroads and highways may be the only effective treatment (see Fig. 9.15). Burning is widely used with various forms of burners the description of which, with their action, is outside the scope of this book but is given in the references earlier cited. Searing alone rather than complete burning of the still anchored and growing weeds has proved more effective in many instances. However, *on peat areas, burning should never be resorted to for weed control.*

FIG. 9.17. Windrow of full-grown willow trees permitted to take root and achieve maturity in the botton of a public ditch in southern Minnesota.

9.18. Spraying. Spraying with chemicals is one of the most effective methods of land-weed control. Effective sprays are discussed in considerable detail in Sec. 9.20. Spraying equipment in common use along railroads and highways is equally effective in ditch maintenance (see Fig. 9.16).

9.19. Woody Growths. Brush and young trees should never be permitted to develop in or alongside a ditch from one season to the next. The results of failure to remove brush while still young and small are well illustrated in Fig. 9.17 which shows only one of a number of such cases that have been allowed to develop in the northern Mississippi Valley. The elimination of this condition required heavy work for an entire summer season with explosives and heavy mechanical equipment, all of which could have been avoided had the willows been cut and grubbed out years before while still only small brush.

Water Weeds

The water weeds with which one has to deal in drainage considerations are of three main types: (1) *floating*, not anchored in the soil at all but loose in the water, such as water lettuce and water hyacinth; (2) *emergent*, rooted in the soil of the ditch bottom, a part of the plant growing below the water surface, such as the cattail, tule, and water cress; and (3) *submerged*, grown entirely below the water surface, such as pondweed, rooted, and coontail, not rooted. These types, while more serious in irrigation channels, also give much trouble in drainage ditches in some regions. Both mechanical and chemical methods of control apply to water weeds also.

Mechanical control of water weeds is quite different from such control of land weeds. Rooted types may be removed by special forms of dragline machines. Such removal may have to be frequent and is apt to be expensive. Cattails, cut several times a season, under water, weaken and die. Where there is a perceptible current, long, pliable band saws, weighted to hug the bottom and drawn back and forth diagonally while moving upstream, have been used to advantage. Men are stationed downstream with pitchforks to catch the floating stems and cast them out.

Another common method of cutting *emergent* weeds loose is to drag a heavy chain, such as a ship anchor chain, up and down the channel by means of a tractor attached to each end and traveling on the two banks. Weeds, once loose, float to the top and may be removed from the stream by various methods.

Another device which has been used extensively in England and also in India, Egypt, and Australia is described as follows:[1]

An English device for clearing rushes, water-lilies, weeds and other growth from drainage and irrigation ditches, rivers, lakes and ornamental waters consists

[1] *Eng. News-Record,* Vol. 89, No. 4.

of power driven cutters operated from a small motor boat of 12-inch draft (see Fig. 9.18).

Each cutter consists of a steel blade, V-shaped in plan with a wavy or scalloped cutting edge and having the apex attached to a long rod. There is a rod on each side of the boat, the upper end being attached to a rocker arm and also to a link on a lifting lever by which the cutter is raised and lowered as desired. The rocker mechanism is driven from the launch engine and, as the boat moves forward, the cutters are given a rapid oscillating motion, so that they engage the weeds at high velocity and strike them with a diagonal cut.

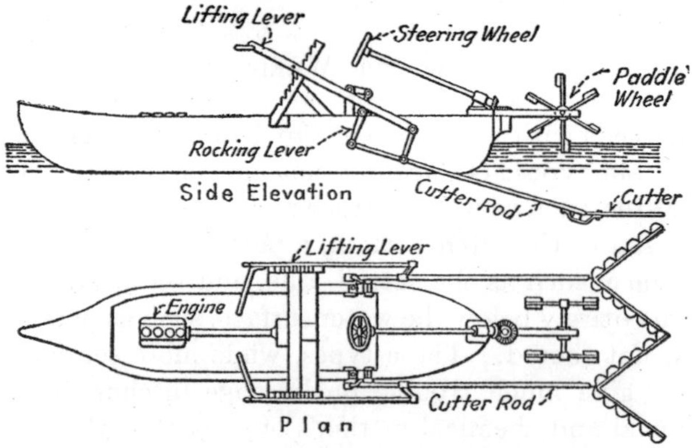

FIG. 9.18. Power-driven weed cutter on motor boat.

In shallow water the cutters trail on the bottom, being provided with spring shoes to keep them out of the mud and to guide them over logs or other obstructions. In deeper water they are set usually to cut at a depth of 4 to 6 feet. It is stated that, in ordinary service, the boat will clear as much as 4 acres in an hour. A small double paddle wheel is used for propulsion as it will not be clogged with the weeds. This wheel serves also as a rudder, being carried by a pivoted frame which has a radial movement in a horizontal plane. For small ditches a rowboat or a scow pulled by tow lines may be fitted with a single, hand-power cutter, the oscillating mechanism of which is operated by a crank handle.

CHEMICAL CONTROL

All the mechanical methods for control of water weeds involve heavy, onerous work and in most cases do not permanently eliminate the weeds. Studies that have been going on for many years have resulted in the development of chemical treatments which kill the weeds and leave them in a condition to be permanently disposed of either by disintegration or by physical removal. Water weeds killed by chemicals usually sink to the bottom. If disintegration is not complete and the accumulation becomes so great that it tends to cause suspended silt to precipitate and build up appreciable layers that gradually reduce the capacity of the ditch, a dragline will have to be employed from time to time.

Where willows and other water-loving softwood trees get started in the ditch, they must be cut while still small; but very soon after cutting, young shoots sprout from the old stumps and grow rapidly, thus necessitating recutting within a short time. In the attempt to prevent such sprouting, O. C. Kulicka, a drainage engineer of Greenville, Miss., experimented with various chemicals which were intended to destroy all the live tissue of the stumps and roots. He found that a solution of granulated caustic soda and white arsenic, 10 pounds of each dissolved in 8 gallons of water, proved most effective. After the mixture cools, 2 more gallons of water are added, and the compound is applied to fresh cuts and splits in the stumps by means of a homemade mop. The rate of progress for a crew of eight men was from ½ to 3 miles per day, working on a ditch of 125 feet base width, with a very dense growth. On smaller ditches, only three or four men were used.

Great care must be taken not to inhale or swallow any of the dust or poisonous fumes arising while the preparation is being made, and all stock should be kept away for at least 10 days from ditches undergoing this treatment. Vinegar, or some other acid, acts as an antidote in case of accident.

9.20. Limited List of Control Chemicals. The extended studies by Federal agencies above referred to have resulted in the production of various chemicals which accomplish the results desired in control of water weeds, and many of them are also used effectively on land weeds. Complete listings and descriptions of these chemicals, their action and directions for their use are beyond the scope of this book, but the following list of some of the best known and most effective, as of 1952, seems justified. For further information interested parties are referred to the reports earlier cited and to their authors and sponsors, and to *USDA Farmers' Bulletin* 2047.

2,4-*D*. A form of highly organic acetic acid or, as used, an ethyl acetate, still in the experimental stage but known to be destructive to a wide range of undesirable vegetation. Spray should be as near the ground as possible to avoid drifting to and injuring neighboring crop plants many of which are very susceptible to it. To be effective it must be applied at an air temperature of about 70 degrees Fahrenheit, to vegetation in its vigorously growing stage. Some time is required for complete action.

Sinox. A highly organic sodium compound. Applied when plants are in seedling stage. Range of action more limited than that of 2,4-D. Works well on many broadleaved plants but also attacks certain grasses, hence should be used with caution and only in special cases. Consult cited references for guidance.

Ammonium Sulfamate. Name indicates chemical nature. Especially effective in controlling such growths as willows and poison ivy.

Sodium Chlorate. Applied to soil surface, acts as sterilizer. (Action lasts 2 to 5 years.) Best applied in the fall in the dry form. Should be well raked into the soil. May also be used effectively as a spray to check quite a list of weeds.

Boron Compounds. These are also soil sterilizers. Often used effectively in combination with sodium chlorate.

Carbon Bisulfide. Also kills roots of plants. More expensive and costs more to apply than the chlorates. Requires moist but not excessively moist soil. Is inflammable and should be used under the same precautions as apply to handling gasoline.

Copper Sulfate. Used effectively in weak solutions in waters infested with algae and other tender floating growths.

Sodium Arsenite. Very weak solutions will kill coontail and various species of pondweed; but is very poisonous and on that account not usually recommended.

Benoclor. A chlorinated hydrocarbon, a solvent of natural rubber. Dissolves chlorophyll rather than poisons tissues. Used largely as an into-the-water spray to control water weeds but is also a good herbicide on land weeds as a spray. What is known as Benoclor 3-C is considered the best form. Requires special equipment to apply in the water, a very small amount of water and a very low velocity of flow—½ to 1 foot per second. Final and full value not yet known, but field of influence among vegetable tissues is very wide. Cost is heavy, but results are good enough to justify use. Without strong toxic effect on animals such as sheep and cattle but of little danger to them anyway because the taste is very objectionable to them. *Fish* will not survive benoclor treatment. Weeds sink to the bottom about 24 hours after treatment.

MISCELLANEOUS ITEMS IN OPEN-DITCH MAINTENANCE

All outlets of relief culverts through waste banks and of tile drains emptying into the ditch and all bridge abutments, pier foundations, ice breakers, and other structures liable to become clogged or fouled should be inspected each spring and fall, and all foreign obstructions of every character cleared away. Such work is usually miscellaneous hand labor that must be done carefully and thoroughly.

9.21. Enlarging Outlets. Experience has frequently indicated the desirability of increasing the capacity of outlet ditches at the same time that the cleanout work is being done (see Figs. 8.3, 8.20, 8.22, and 9.6). This is sometimes accomplished by deepening; however, if there is danger of encountering quicksand at greater depths, the increased capacity had better be secured by widening. This generally consists of removing the slope and spoil bank on one side and is often tedious work on account of buried logs and stumps.

A novel method, used by the Morgan Engineering Company in their work in Missouri, was to carry the runoff from distant areas in one or more "high-flow-line" ditches where the water was confined between tight spoil banks. Lateral drainage of land in the vicinity was taken care of by parallel "low-flow-line" ditches.

9.22. Civilian Conservation Corps Drainage Camps. In 1935 it was decided to assign 45 Civilian Conservation Corps Camps to the former U.S. Bureau of Agricultural Engineering, now a part of the Bureau of Plant Industry, Soils and Agricultural Engineering, to be used for maintenance work in organized drainage districts. This has provided much-needed impetus for drainage maintenance and has resulted in important benefits to the public by providing better outlets for all kinds of private drains. The actual work itself, however, has been strictly limited to public projects duly organized under the various state laws.

The first camp was located at Bancroft, Kossuth County, Iowa, and the following record of work accomplished by this one camp during its first 3 years is cited as typical.

Clearing: 3,210,536 square yards requiring 46,456 man-days.

Excavation: 2,225,993 cubic yards requiring 15,475 man-days.

Tile work: 64,343 linear feet requiring 9,674 man-days.

The commercial value of this work is estimated to be $614,711.

PROBLEMS

1. What two leading types of vegetation cause serious trouble in open-ditch maintenance? Name three prominent species of plant life under each type.

2. What are the leading methods of control of vegetation in open ditches:

(*a*) For land weeds?

(*b*) For water weeds?

3. What precautions are necessary in the use in open ditches for controlling weeds, of

(*a*) Burning?

(*b*) Pasturing?

4. Name four leading types of excavators giving good service in open-ditch maintenance and describe their leading differences and limitations. Which has the widest scope?

5. A ditch 1 mile long traverses flat land on a grade of 0.3 ft per mile and carries surface drainage from higher ground. The ditch dimensions are: side slopes, 1 to 1; depth 6 ft; bottom width, 10 ft.

A storm occurs at a time when the discharge rate is 10 cfs.

Five hours after the first increase occurs the ditch is flowing full and remains at bankful stage for 5 hr, after which the flow decreases during a period of 10 additional hours to the original rate of 10 cfs.

The dry weight of silt contained in each cubic foot of inflow water is 3 lb which is reduced by sedimentation to 0.5 lb at the outlet.

Take arithmetical averages during the rising and falling stages as sufficiently accurate for computing total discharge and assume 50 lb of dry silt equivalent to a volume of 1 cu ft when wet.

If the silt is uniformly distributed lengthwise throughout the ditch,

(*a*) How deep is the deposit resulting from this one storm?

(*b*) By what percentage is the discharge capacity of the ditch reduced?

CHAPTER 10

LOCATION AND DESIGN OF UNDERDRAINS

The nature and function of underdrains has been explained in Chap. 6. It has been indicated in Chap. 3 that permeability and infiltration rates are primary considerations, and in what way. The factors that enter into drainage design have been fully discussed in Chap. 7. These should all be reviewed at this point, and frequent reference to them will be found desirable, with special attention to soil characteristics and soil-water relationships.

Although various items in the design of underdrainage are of a purely office character after the survey, in general the survey and design are interdependent and must largely be considered together. Furthermore, with the exception of extensive and intricate public projects, the preliminary and location or construction surveys can usually be combined, thereby saving time and expense.

The first essential step in the location and design of underdrains is a reconnaissance survey of the drainage area (see Chap. 7).

10.1. Outlet. The very fact of a contemplated underdrainage system automatically causes the project to fall in the class of *A* drainage. That point settled, as was shown to be the case in open-ditch design, the factor of first importance in design of underdrainage is the *outlet*. A tile drainage system usually outlets directly into an open channel. An artificial channel is usually the best as it is built for that purpose, although a natural channel will serve fairly well, and if available should be used. The location of the tile outlet depends on the presence or absence of natural drainageways across the area to be drained. The tile mains usually follow such drainageways. The elevation of the surface of normal flow in any prospective outlet channel relative to the elevation of the land to be drained must be determined. Location of high-water marks is also helpful in determining whether the capacity of the proposed outlet channel is sufficient to prevent overflow of the adjacent farm land.

10.2. Depth of Outlet Channel. As has been pointed out in Chap. 6, outlet ditches for tile drainage must be deep enough to permit free discharge from the tile main above the normal flow stage in the outlet channel. In spite of their great depth such channels serve as direct drainage channels only to a limited extent. An outlet ditch for tile drainage in flat country should not be less than 8 feet deep unless it runs through an area or depression lying appreciably lower than the land to be drained by the tile. Figure 10.1 shows why. From the figure it is seen that, on a uniform 0.10 per cent gradient, to maintain it on flat land the depth must be increased from 3.0 feet at points *b*, *d*, *f*, or *h* to 6.9 feet, an additional 3.9 feet. The increase of size of tile from 5-inch tile at the head of the system to the 18-inch tile required at the outlet, because of gradually

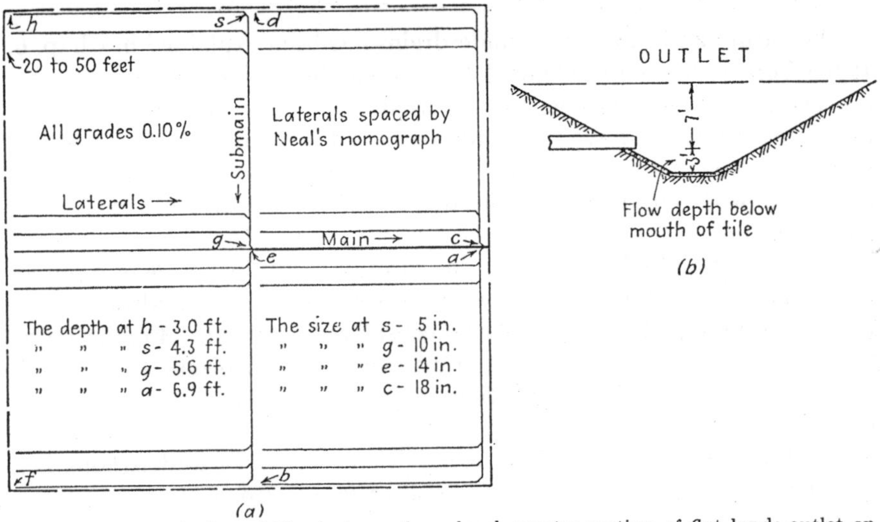

FIG. 10.1. Formal plan of tile drainage for a legal quarter section of flat land; outlet an open ditch along one side of the tract. (*a*) Formal plan of the drainage system. (*b*) Cross section of open ditch and tile outlet into same, showing minimum depth of ditch required.

increasing acreage, amounts to the difference in the outside sizes of 1.3 feet. The required depth of outlet channel, for best practice is as follows:

Primary depth of laterals at head of system	3.0 feet
Increase in depth to maintain gradient	3.9 feet
Increase in depth called for by increase in size from the head to the outlet	1.3 feet
Free flow depth	3.0 feet
Total	11.2 feet

Or call it 11.0 feet.

This is an unusual situation; yet it may occur. Seldom does the natural surface slope of the land itself furnish sufficient fall for the tile in all directions. Even where such a fortunate condition exists, a depth of over 7 feet is needed if a free flow depth of 3 feet in the outlet ditch is retained. If the natural slope of the land gave half the needed fall and only 2 feet free flow in the outlet ditch were called for, the minimum depth of outlet ditch must still be 8 feet. Various adjustments are usually possible, such as using an 0.05 per cent gradient on the cross lines of the submains and on the main, but this change necessitates further increase in tile size to 16 inch at *e* and 22 inch at the outlet, thus partially offsetting the advantage of the flatter gradient. Rarely do any adjustments reduce the minimum requirement for depth of outlet ditch below 8 feet, unless the ditch itself lies in a swale whose bed is appreciably below the general level of the land to be drained. Hence even under favorable conditions a depth of 8 feet is the minimum satisfactory for an open outlet channel on flat lands.

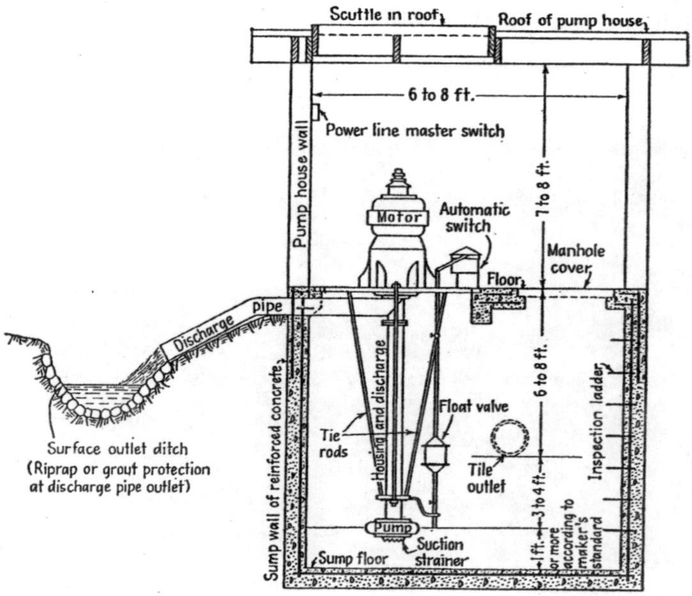

FIG. 10.2. Diagrammatic sketch of pumping outlet for a tile-drainage system.

10.3. Pumping Outlets for Tile Drains. Frequently farming communities find it difficult to secure badly needed outlet ditches for tile drainage. Occasionally, where local conditions are favorable, a farmer with electric power available can use a pumping outlet. Pumping outlets for drainage are extensively used in conjunction with irrigation in the irrigated West, in Michigan, and to some extent in Minnesota, as well as frequently in the central and lower Mississippi Valley where drainage alone is the problem.

The essential features, shown diagrammatically in Fig. 10.2 in their proper relative positions, are as follows (see also Fig. 10.3):

1. A sump or cistern 6 to 8 feet square with reinforced concrete walls, bottom, and cover, and of sufficient depth to receive the water from the tile main, to permit setting the pump 3 feet or more below the mouth of the main and a foot or more above the bottom of the sump. (The sump

should be located on the higher ground at the margin of the low area to prevent danger of flooding of pump house and motor.)

2. A vertical centrifugal pump, electrically driven, that will easily remove the maximum runoff from the main as fast as it arrives.

3. An electric motor of sufficient horsepower to operate the pump easily under maximum load.

4. A float valve connecting with the master switch of the motor, adjustable to enable it to start the motor automatically at any desired height of water in the sump and, when it drops below this point, to throw the switch and shut off the motor.

FIG. 10.3. Drainage outlet secured by use of small pumping plant and rural electric power in Iowa. Note weeds in undrained field beyond fence. 1949 photograph.

5. A discharge pipe which should be of ample capacity to conduct the maximum flow of the tile-drainags system and discharge it into an existing surface channel or road ditch leading to the final outlet ditch or basin. The outer end of the discharge pipe should be turned sharply downward, outside the pump house, as far as the ground surface leading to the ditch will allow, to give as much siphon action as possible. Heavy riprap or a grout apron should be provided in the surface ditch at the pipe outlet to prevent erosion.

6. A pump house of any suitable building material and equipped either with a large scuttle in the roof directly over the motor or with a sliding roof, to permit of drawing the pump for repairs. (The walls of the sump serve as a foundation for the pump house, whose floor is the cover of the sump.)

7. A manhole and inspection ladder to the bottom of the sump.

Some General Principles of Design. Specifications for general use, of size and capacity, cannot be given as each case is an individual problem, the design for which should be carefully worked out.

Economy calls for ample capacity to carry an overload for short periods during unusual floods and to provide for further extensions of the drainage system not originally contemplated.

Where winter operation is required, special provisions against freezing must be made.

Size and capacity of pump and of motor are largely determined by capacity of the drainage main, degree of drainage required, and the

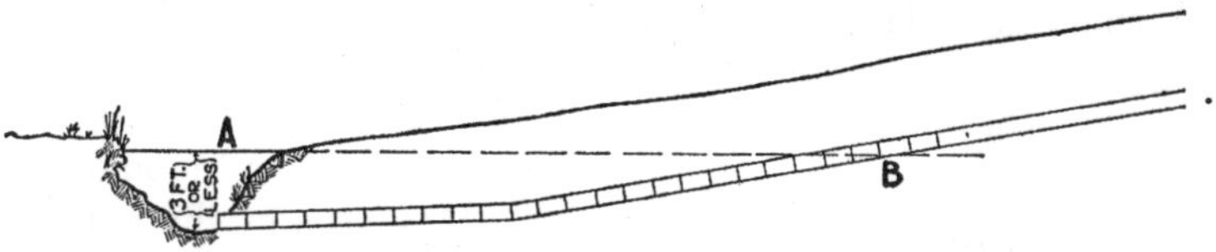

A POOR OUTLET FOR TILE DRAINAGE. IT LACKS DEPTH AND CAPACITY. AT PERIODS OF FLOOD STAGE THERE WILL BE NO DRAINAGE BELOW THE LINE AB. SO CALLED FORCED OUTLETS ARE NOT OUTLETS. TILE SHOULD HAVE ENDED AT B AND A SMALL OPEN DITCH CUT FROM A TO B.

FIG. 10.4. Diagram of a "forced" or submerged outlet, showing its utter worthlessness.

necessary lift; but a 3-inch pump and a 5-horsepower motor are about the minimum sizes desirable even where the theoretically required capacity may be less. Motor and pump are built as a combination unit by several nationally known makers. It is possible to get multiple-unit pumps on which the successive units cut in at different stages of water in the sump. Such a pump is apt to be desirable where the rainfall is usually moderate but where short periods of intense floods also occur at infrequent intervals. However, the extra cost of such pumps is relatively high.

Cost. First cost of such a plant is considerable. It may be different in every case but, for ordinary farm areas, will usually run from $1,000 to $3,000. However, it will frequently pay for itself in 1 to 3 years. Operating and maintenance costs will be small. The normal life of such a plant is 25 to 30 years.

Advantages. The pumping plant gives the farmer an independent drainage outlet, and by proper adjustment of the automatic feature, he can control the water level in the drained area to a considerable extent and, in dry years, retain some gravitational water against deficient rainfall in those low areas which yield the best crops.

10.4. Submerged Outlets. Permanently submerged or so-called "forced" outlets are not really outlets. Figure 10.4 shows their uselessness. There is no outlet for the drainage water flowing through the tile below point *B*, and the portion of the tile line from point *B* to point *A* is of no effect. An open ditch from *A* to *B* should have been dug in place of the tile which is useless and, therefore, a needless expense.

10.5. Stakes. An outlet having been definitely located, the location-construction survey should next be made. For this work both marker and hub stakes will be needed. These should be of material that will withstand hard usage. The hub stakes should be so long that, when driven nearly flush with the ground surface without burring, they will not be settled by trampling livestock. In ordinary tile-drainage work on the farm a good-quality pine lath will serve effectively—one-half a 4-foot lath for marker stake and one-quarter lath or less for hub stake. On public projects the stakes should be of sound ¾-inch material.

10.6. Classes of Tile Lines. Four classes of tile lines may occur in a tile drainage system, commonly designated as follows:

1. A *main* is a line of large tile that carries the discharge from submains and laterals branching out from it. It must be deep enough to permit free discharge into it from its branches. The zero of the main is at the outlet. A large drainage area may naturally divide into two or more tiling systems, each served by its own separate and independent main. These multiple mains are usually designated as Main A, Main B, etc.

2. A *submain* is a smaller branch main serving a definite section of the drainage area served by the main, and discharging into the main. The zero of a submain is its junction with the main. Submains are usually designated by number, number 1 being the one nearest the outlet.

3. A *lateral* is a single line which collects the water from the soil and empties it into a main or submain. The zero of a lateral is its junction with the main or submain. The laterals branching out from any given main or submain are usually designated by lower-case letters beginning with a for the one nearest the outlet of its main or submain.

4. A *branch* is a line collecting the water from the soil and discharging it into a lateral. The zero of the branch is the junction with the lateral. A branch is usually labeled branch x of the lateral.

To illustrate, a given line may be designated as branch 1 of lateral e of submain 1 of the main (see Fig. 10.5). This system, carried into all records, correlates all phases of the work and tends to prevent errors by anyone working on the project.

LAYING OUT TILE DRAINS

In laying out tile drains the best and customary method is as follows: Starting at the outlet of the main, the line is measured with a 100-foot steel tape; a marker stake, with the station number on the front face, and a hub stake are placed at each station or 100 feet on straight lines and every 10 to 50 feet on curves according to the sharpness of the curve. The station numbering begins with zero at the outlet and proceeds consecutively upstream. The method of marking stations and junctions of lines is shown on the type map, Fig. 10.5. The marker stakes are marked

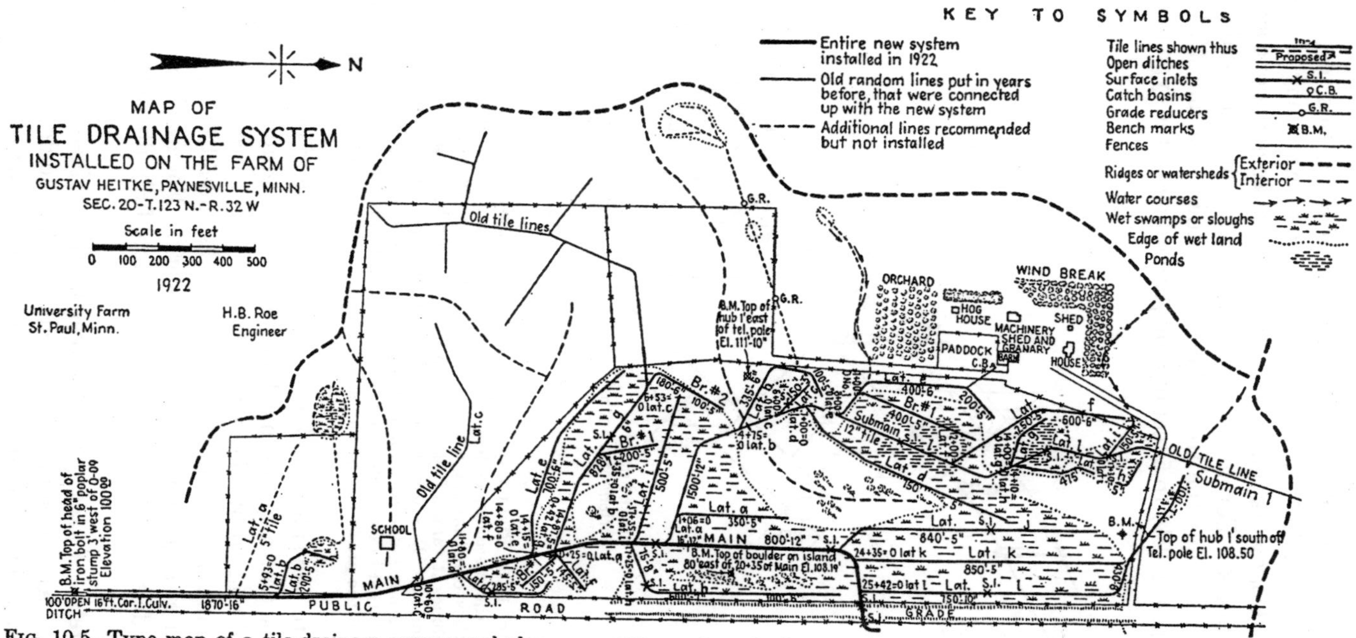

Fig. 10.5. Type map of a tile-drainage survey and plan on a 120-acre farm in Stearns County, Minnesota. Note: 18 + 30 of main = 0 of submain 1.

in the same way as the corresponding stations on this map, and the marker stakes and hub stakes are placed as shown along line HH' in Fig. 10.6, for *hand* work, and the trench is dug close along one side of the line of hub stakes. For *machine* work the marker and hub stakes are set in the same way; but the trench is actually dug along line MM' in Fig. 10.6, a distance A ($2\frac{1}{2}$ to 5 feet or more, according to the type of machine

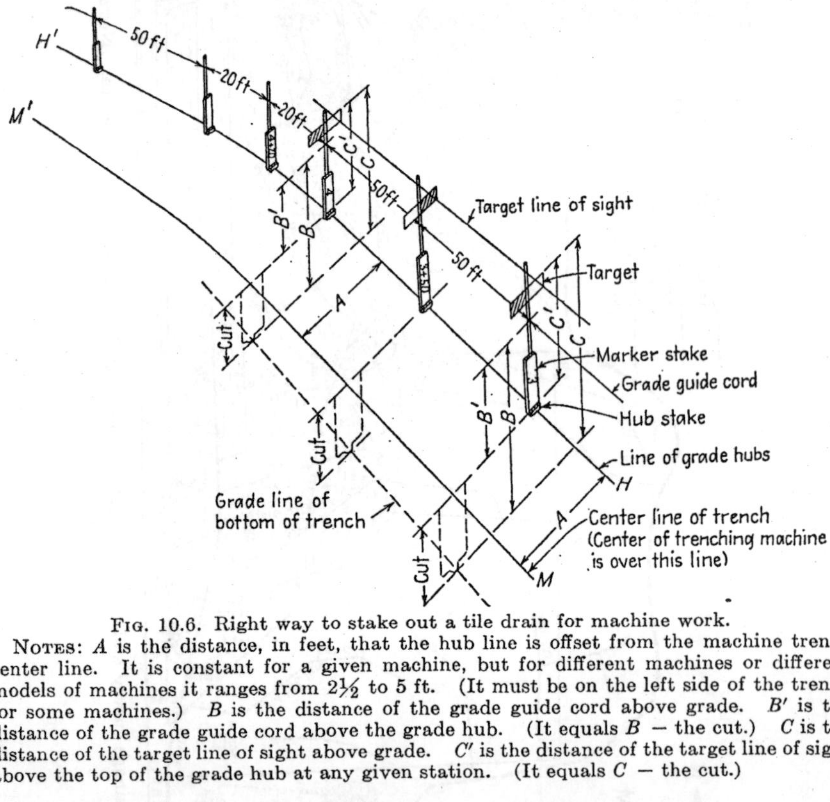

FIG. 10.6. Right way to stake out a tile drain for machine work.

NOTES: A is the distance, in feet, that the hub line is offset from the machine trench center line. It is constant for a given machine, but for different machines or different models of machines it ranges from $2\frac{1}{2}$ to 5 ft. (It must be on the left side of the trench for some machines.) B is the distance of the grade guide cord above grade. B' is the distance of the grade guide cord above the grade hub. (It equals B − the cut.) C is the distance of the target line of sight above grade. C' is the distance of the target line of sight above the top of the grade hub at any given station. (It equals C − the cut.)

to be used) to one side, usually the left, of the line of hub stakes to allow for the truck width of the machine, to prevent grade hubs from being ridden down and to fit the method for carrying correct grade employed on the machine used. On this account hubs should be placed every 50 feet even on straight lines on machine work. For machine methods of carrying grade see the more detailed and more fully illustrated discussion in Chap. 12.

10.7. Levels. Exact levels are taken by the engineer on the tops of the hub stakes. At the same time, bench marks should be set in readily accessible spots over the drainage area at intervals of $\frac{1}{8}$ to $\frac{1}{4}$ mile on farm projects and about every $\frac{1}{2}$ mile on public projects. A concise description of the character and location of each bench mark should be entered in the level notes opposite the recorded elevation. The bench mark and its elevation should be plainly marked in the field, either on a large heavy witness stake or on a fence post or telephone pole alongside the bench mark. A piece of ½- or ¾-inch gas pipe, driven with a sledge to a firm bearing and its top left projecting about an inch above the ground surface, makes an excellent field bench mark. A list of these bench marks, with description, location, and elevation of each, should be a definite item among the filed records of the project (see Table 10.1). From the levels on the hubs the grades on the tile lines are determined and the cuts from tops of hubs to finished grade at bottom of trenches computed. The cuts are tabulated by line and station, points of change of grade and rates of grade being also listed (see Table 10.2), and a copy is given to the tiler to guide him in setting the targets. Some engineers mark the cuts at each station and plus on the marker stake, but it is better practice to furnish the tiler a table of cuts for each line. There are

TABLE 10.1. SAMPLE LIST OF PERMANENT BENCH MARKS

Elevation, Ft	*Description*
100.00..	Top of head of iron bolt in poplar stump 3 ft west of main statin 0 − 09
108.19..	Top of boulder on island 80 ft east of main station 20 + 35
108.50..	Top of hub 1 ft south of telephone pole 50 ft north of end of lateral *j* of the main
111.10..	Top of hub on east side of telephone pole 55 ft left of station 1 + 65 of lateral *b* of submain 1

other methods than those outlined here, but the one given here has been found to be generally the most efficient. Laying tile to water grade is not recommended.

The locations of all special features such as surface inlets, catch basins, and junctions of one tile line with another should be shown by special stakes plainly marked to show what they stand for; for example, 8 + 65 of lateral $d = 0$ of branch 2. The end stake of any line should be so marked; for example, 7 + 60, end of lateral *b*.

In the work of laying out the tile lines, a topographic map is not an essential except on some large and intricate projects, or on public projects where required by law. With a little experience and initiative a good engineer can flag out each main and submain ahead of the chainmen by inspection. After the levels are run and the determination of grades on the various lines is under way, it may occasionally be necessary to make minor revisions in location of some lines, but these are relatively infrequent. Seepage lines on or at the base of slopes or around the edges of marsh pockets should be located by the use of the soil auger accompanied by the level. In the light of comparatively recent developments in methods for determining proper spacings between parallel lines on flat areas or on low uniform slopes, such systems should be worked out only after locations of mains and submains are completed and after thorough examination of soil and subsoil and the determination of such pertinent

TABLE 10.2. SAMPLE TABLE OF LEVELS, GRADES, AND CUTS ON TILE LINES FOR DRAINAGE SYSTEM

On Farm of Gustav Heitke, Paynesville, Stearns County, Minnesota

Station No.	Elevations, ft		Gradient, per cent	Cut, top of hub to grade, ft	Remarks
	Tops of hubs	Grade			
14	108.00	100.08	+0.12	7.92	
14 + 15 = 0 of lateral *e*	107.64	100.10		7.54	
14 + 80 = 0 of lateral *f*	105.25	100.18		5.07	
14 + 87.5 = 0 of lateral *g*	104.74	100.19		4.55	
15	105.40	100.20		5.20	
15 + 30	106.59	100.24		6.35	
16	106.24	100.32		5.92	
16 + 10	106.15	100.33		5.82	
16 + 20	106.00	100.34		5.66	
16 + 90	105.76	100.43		5.53	
17 + 25 = 0 of lateral *h*					
17 + 35 = 0 of lateral *i*					
17 + 50	105.87	100.50		5.37	
18	105.79	100.56		5.23	
18 + 30 = 0 of submain 1	106.02	100.60		5.42	
19	106.09	100.68 for 16 in.		5.41	Make top of 12-in. tile inside flush with top of 16-in. tile inside
		101.05 for 12 in.		5.04	

characteristics as moisture equivalent, plastic limits, or permeability rates, either by direct test or by comparison with charts of known existing projects in the same region. Worked out through these agencies by methods later discussed, the parallel laterals can then be laid on the record mensuration map of the project before they are staked out in the field.

10.8. Staking Curves on Tile Lines. Turns in tile lines should never be by sharp angle bends except at junctions. They should be by approximately circular curves of such length, where reasonably possible, that the openings on the outside of the curve at the joints between the individual tiles shall not exceed about ⅛ inch. Otherwise either the ends of the tiles on the curves must be trimmed to give close joints or the joints must be well protected by tar paper or tile "bats." The lining out of curves should not be left to the tiler, but the engineer should consider it part of his job. Ordinarily it is not necessary to run out even long curves on such work with a transit as there are shorter methods by chain that are quicker and accurate enough for the purpose. Many engineers, on short curves of small radius, simply line out the steel tape on a curve smoothed up by eye, but this method is not especially desirable with the following simple and more accurate method available. The method, as here presented, was carefully worked out at the Minnesota station and used effectively in hundreds of cases. It is recommended for general use.

Simple Method for Staking Out a Circular Curve. This method depends on the facts that, for short circular curves, (1) the middle ordinates for any two arcs vary nearly as the lengths of the arcs and (2) the middle ordinate and the external secant are nearly equal.

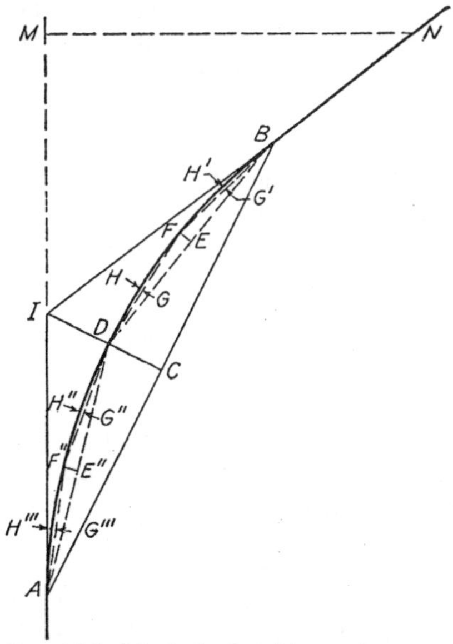

FIG. 10.7. Method of staking out a practically true circular curve on a tile line.

NOTE: After the length of the curve LC has been selected from Table 10.4 and points A and B are set, it will be accurate enough on tile drain lines to take $CD = \frac{1}{2}CI$, FE and $F'E' = \frac{1}{4}CD$, such lines as $HG = \frac{1}{4}FE$, and so on.

The slight variation by this method from a true circular curve gives the effect of a crude spiral at each end of the curve, thus making it of the character desired. Referring to Fig. 10.7, one should proceed as follows: from the physical conditions of the problem select the desired length of curve from Table 10.4 and then proceed according to Fig. 10.7 with the aid of Tables 10.3 and 10.4.

After the line of the curve is secured, it should be rechained, setting stakes at the desired regular stations and plusses. If the end of the curve comes at an odd plus, it can usually, without serious injury to the alignment, be pushed ahead or drawn back to the nearest 5 feet; that is, if the end of the curve comes, say, at 7 + 63, it can be pushed ahead along the forward tangent to 7 + 65, or if it comes at 7 + 61, it can be pulled back along the forward tangent to 7 + 60.

TABLE 10.3. FOR CD

Angle, MIN, deg	IN	MN, ft	MI, ft	AI or BI, ft	CD, ft
0–30	Always 100 ft	0–50	100–86.6	½ of curve length	½ of CI
31–45		51.5–70.7	85.7–70.7	55 % of curve length	49 % of CI
46–60		71.9–86.6	69.5–50	55 % of curve length	48 % of CI
61–75		87.5–96.6	48.5–25.9	60 % of curve length	47 % of CI
76–90		97.0–100	24.2–0.0	⅔ of curve length	43 % of CI

TABLE 10.4. CURVE LENGTHS AND RADII, TO NEAREST FOOT, ALONG INNER SIDE OF TILE, FOR JOINT OPENINGS OF ⅛ AND ¼ INCH AT OUTER SIDE OF TILE*

Total angle of curve		For ⅛-in. joint openings							For ¼-in. joint openings						
			15°	30°	45°	60°	75°	90°		15°	30°	45°	60°	75°	90°
Diameter of tile															
Inner, in.	Approx. outer, ft	R†	LC†	LC	LC	LC	LC	LC	R	LC	LC	LC	LC	LC	LC
4	0.4	40	10	20	30	40	50	60	19	5	10	15	20	25	30
5	0.5	50	13	25	38	50	63	75	24	6	12	18	25	31	38
6	0.6	59	15	30	45	60	75	90	29	8	15	23	30	38	45
7	0.7	68	18	35	53	71	88	105	34	9	18	26	35	44	53
8	0.8	77	20	40	60	80	100	120	38	10	20	30	40	50	60
9	0.9	87	23	45	68	90	113	136	43	11	23	34	45	57	68
10	1.0	96	25	50	75	101	126	151	48	13	25	38	50	63	75
12	1.2	115	30	60	91	121	151	181	58	15	30	45	60	75	90
12	1.2	231	60	120	181	242	302	362	115	30	60	90	120	150	180
14	1.4	269	70	141	211	281	352	422	134	35	70	105	140	175	210
15	1.5	288	75	151	226	302	377	452	144	38	75	113	151	189	226
16	1.6	308	80	161	241	322	402	483	154	40	80	121	161	201	241
18	1.8	346	90	181	271	362	452	543	173	45	90	136	181	226	271
20	2.0	385	101	201	302	402	503	603	192	50	100	151	201	251	302
22	2.3	423	111	221	332	443	554	664	211	55	111	166	221	276	332
24	2.4	462	121	241	363	483	603	724	230	60	121	182	242	303	363

* Individual tiles are considered as true length as commercially listed, that is, as 12 in. for 4- to 12-in. tile and 24 in. for 12- to 24-in. tile.

† LC = length of curve, R = radius.

Where a rate of curvature is required that would naturally cause gaps at the joints of more than ¼ in., the ends of the tile should be trimmed.

10.9. Gradients for Tile Drains. The working grade line of a tile drain, during the process of laying, is the line of the bottom of the tile exteriors as they lie in the bottom of the trench. As earlier indicated, the levels taken on the tops of the hub stakes along the tile lines enable profiles of these lines to be drawn, and it is best to draw them of the mains and submains. On these profiles the grade lines of the bottoms of the tiles can be drawn at such depths as will fit the required depths of the laterals at their junctions with the mains and submains. The longer the reach on a given gradient, the better. These grade lines should follow closely the general fall of the land surface unless this is less than the allowable minimum grade for tile lines. For the general run of *laterals*, profiles are not necessary as, with a little practice, the engineer can establish the grades by inspection of the levels. On the laterals, while scrupulously avoiding backfall, frequent changes in gradient more closely to follow the fall in the land surface and thus save much needless digging are allowable and common practice with many engineers because the capacity of the lateral tile is almost invariably far in excess of the need; hence the loss of head at these frequent changes of gradient is negligible in practice.

Some authorities hold that minimum gradients on laterals of small tile (3, 4, and 5 inch) should be 0.2 to 0.3 per cent in most agricultural soils. However, experience in the Middle West and the West, especially in the northern Mississippi Valley and in Oregon and Washington, indicates that such minimum gradients are far in excess of long-tried effective minima. If this were not so the tile drainage of millions of acres would have been impossible. A commonly used effective gradient on tile lines of all classes and sizes, in the regions just mentioned, is 0.1 per cent and, on occasional laterals of limited length (1,000 feet or less), a perfectly flat gradient has proved effective when there was a good outlet for the given line, as the head furnished by the tile running full—even as small as 4 inch—has been sufficient to produce the necessary flow if the silt burden of the drain water is light.

It has been suggested that tile gradients be determined between a minimum velocity of ⅔ foot per second and a maximum velocity of 1⅓ feet per second, but with a 4-inch tile and a gradient of 0.1 per cent, the average velocity of flow is ¼ foot per second. Hence, general experience would indicate that, if velocity of flow is to govern selection of gradients, ¼ foot per second would be a more satisfactory minimum than ⅔ foot per second. Furthermore, as 4- and 5-inch lateral tile rarely flow full under spacings called for by soil character and structure, velocity of flow may safely be considered a practically negligible factor in determining gradients of laterals. It has generally been so considered with satisfactory results. A gradient of 0.5 per cent is a safe maximum to use under most soil conditions, avoiding the common disadvantages of excessively steep gradients. Still, in many soils for short distances, gradients as high as 1.00 per cent can be used safely where sufficient care is exercised in installation of the drains. But, in soils easily cut by running water, churning at the joints is almost sure to result in early undermining of the tile and ruin of the drain where the gradient is in excess of 1.00 per cent, and it sometimes happens with gradients of 1.00 per cent or somewhat less. Therefore on those short stretches of tile where the natural fall is exces-

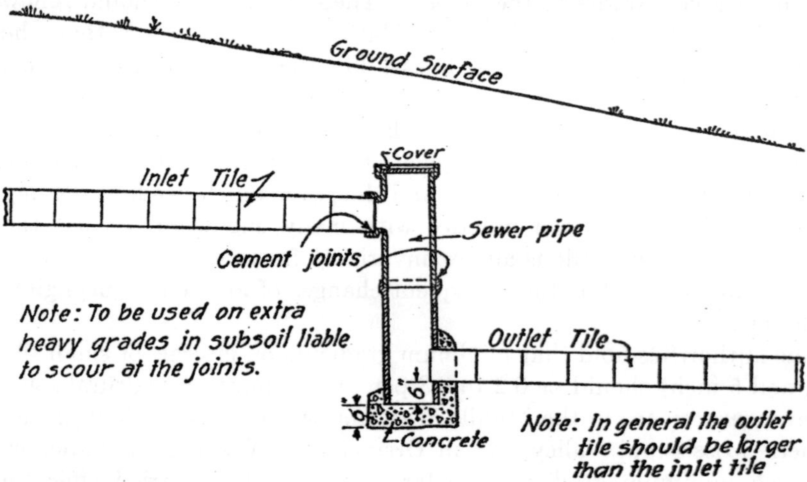

FIG. 10.8. Grade reducer for use on excessively steep natural grades along tile lines.

sive, one of the three following methods should be used to reduce the danger of undermining:

1. Locate the tile lines across the heavy drop by an indirect route that will enable use of a safe gradient.

2. Introduce grade reducers at intervals in the tile line (see Fig. 10.8). They have not always proved satisfactory and sooner or later give trouble. If they are used, it will be well to run the tile on the flatter grade obtaining just above such stretch of excessive fall, to a point where its depth has become the minimum allowable, and there introduce the grade reducer, at the bottom of the vertical fall in which a flatter grade can be again started. The base of the grade reducer should be a thick plate of good concrete, well tamped to prevent any chance of the falling water washing it out. Grade reducers should be readily accessible from the land surface. They should be inspected frequently and cleaned out after each heavy storm to prevent clogging of the tile lines. They should be used only when special conditions indicate them as the best solution to the given problem. As a rule the use of either sewer tile or corrugated metal pipe, as discussed next, is the better practice.

3. Throughout the section of steep gradient, use either bell-and-spigot sewer tile and cement the joints tightly or else long sections of corrugated metal pipe with bolted connecting bands. The few joints can be made watertight by the use of a mastic substance under the bands.

On the other hand, if gradients are so low as to reduce velocity of flow below a self-cleaning value, in open soils the water often seeps away before the drains can receive it. In such cases it leaves behind its silt burden which, in time, clogs the tile. Even in tight soils, the sluggish flow permits a small amount of sediment to settle and, in course of time, clog the tile. Therefore, except under special circumstances for a short distance, gradients appreciably less than 0.08 per cent should be avoided on tile laterals. However, as earlier stated, even a level gradient on straight single lateral tile drains of 1,000 feet or less in length will function properly when the drop into the submain or main is clear and unobstructed by water flowing in them. This is not, however, good practice for very crooked lines. Where short lengths of gradients less than 0.08 per cent seem necessary, extra care in installation, with ample free drop into the mains and submains, and elimination of branches, curves, and bends, is necessary.

10.10. Mensuration Map. A major part of the location-construction survey is to make, on a convenient scale, a complete mensuration map of the tract to be drained. For a single average-size farm, a scale of 200 feet to 1 inch is satisfactory. For small areas a scale of 100 feet to 1 inch may be best. On public drainage districts, 400 or 800 feet to 1 inch or even smaller may be best practice. This map should show the exact location of all the drain lines, whether tile or open ditch; size of tile and linear feet of each size on each separate line; junction points of each branch line with its main and submain; location of outlet, catch basins, surface inlets, grade reducers, permanent bench marks, boundaries of orchards and other permanent fields, permanent fences, buildings, and other artificial features within the limits of the area to be drained; all ridges, and particularly the ridge bounding the watershed; all natural watercourses, whether dry or wet; and the approximate boundaries of all swamps and wet land areas and permanent meadow. Such a map of a limited farm drainage project is shown in Fig. 10.5. It may sometimes be necessary to make a contour map of a large district for use in design, and on public work it is properly often required by law, but it is not necessary on most single farm drainage projects and is of little use to the average farmer. The mapping can be done along with the leveling. The most effective mapping equipment on large projects is a standard plane table, full-circle compass, open-sight alidade, engineer's level with stadia hairs, and a self-reading level rod so marked as to serve as a stadia rod. On farm jobs up to 1,000 acres a light traverse board of a size to hold a map of a section and a half on a scale of

200 feet to the inch and a good architect's level with stadia hairs will effectively replace the standard plane table and engineer's level. This lighter equipment makes possible rapid work of approved accuracy.

10.11. Lengths of Tile Lines. The maximum allowable length of a line of tile of a given size may theoretically be determined as shown by Etcheverry [45] on the basis of the required value of the drainage coefficient D, the width of land strips to be drained by the given tile, and its carrying capacity Q in cubic feet per second, according to the gradient used, on the assumption that the tile flows full at the lower end of the reach. However, while this method will be satisfactory and is followed in practice for the large sizes of tile needed on mains and submains, on field laterals it tends to crowd the capacity of the line, which will be narrowly limited by the practical quality of construction work, on which account the smaller sizes of tile, the 2 and 3 inch, have dropped out of use, and even 4 inch tends to do the same.

Elliott [41] was about the first among the earlier writers in this field to give definite information on allowable minimum gradients and maximum allowable lengths of tile lines of different sizes, and these figures were very helpful in the earlier development of tile drainage, but more recent research and long practical experience have shown that somewhat higher gradients and shorter laterals, especially for the smaller sizes of tile, give more reliable results under average soil conditions and quality of construction work obtainable. A summary of Elliott's figures will be found in Table 5 of Ref. 41.

Most tile-drainage installation is now done by machine, and such work, by and large, will not and probably cannot be done with the accuracy required where the very small tile of yesteryear were used. Extremely skillful handwork was essential for success with it. A vast amount of tiling is being done in the Pacific Northwest. It is all machine work because it can be done much more rapidly and cheaply than by hand and, furthermore, capable men experienced in tiling by hand are no longer readily available.

The best modern practice in tile drainage tends to limit the length of single field laterals, or even those with one or two short branches, to 1,000 or 1,200 linear feet, although local conditions of topography, soil, and general plan of the system may sometimes justify the occasional use of laterals up to 1,500 or 2,000 feet in length under unusually favorable circumstances or to avoid other adjustments that may be relatively more expensive.

10.12. Drainage Coefficient *D* for Tile Drainage. The maximum amount of water planned to be taken off from an area by tile drains in 24 hours must depend on the amount that can move through the soil and subsoil to the drains in that time. This largely depends on the two factors of rainfall intensity and soil density. It cannot reliably be stated which of these two is the more influential, and it probably varies from one to the other in different regions. Pickels [104] points out that the soil cannot absorb the water from an intense rain as fast as it falls and that the larger part of it flows away over the surface, whereas were the same amount of water precipitated much more slowly most of it would be taken up by the soil and carried off through the tile drains at a rate dependent upon the normal permeability of the soil and subsoil. This general fact is well supported by the results of investigations by Schlick [131,132] on specific watersheds in Iowa.

Schlick [131] calls attention also to the evident fact that D tends to vary inversely, on flat lands, with the spacing of tile laterals. He maintains that the value of D, established by early investigators at $\frac{1}{4}$ inch for average sandy-loam and silt-loam soils, should probably be raised to $\frac{5}{16}$ to $\frac{3}{8}$ inch and that "when the drains are provided with inlets for admitting surface water (even) this rate of runoff should be increased." For laterals 4 feet deep and 100 feet apart without surface inlets he recommends that D be given a value of $\frac{5}{16}$ to $\frac{3}{8}$ inch and that, for a spacing of 50 feet, this be increased to $\frac{1}{2}$ inch or more.

In a later investigation of closer spacing in Iowa, Kirkham and Frevert [60] experimented with tile lines $3\frac{1}{4}$ feet deep and spaced 24, 30, 36, and 48 feet apart. The soil was described as a clayey sand with enough clay to clog the sand pores but not enough to make good soil structure. There were no cracks, worm holes, or other soil channels through which water could seep. They found that the 24-foot spacing lowered the water table midway between the drains four times as much as the 48-foot spacing and produced twice as much *drained soil* for each 100 linear feet of tile. Intermediate spacings showed proportionately intermediate results.

Powers and Teeter [110] state that, with a 40-inch rainfall, on large fields without surface inlets, a D value of $\frac{1}{3}$ inch is suitable, apparently as a general average, but that, in smaller fields with occasional inlets, a D value of $\frac{1}{2}$ inch is recommended.

The engineers of the Minnesota Agricultural Experiment Station found, through the experience of 25 years, that for the lacustrine and most of the glacial soils a D value of $\frac{3}{8}$ inch and for the residual and the denser glacial clay loams a value of $\frac{1}{4}$ inch usually gave the most satisfactory results on lateral systems, while, in general, $\frac{1}{2}$ inch for mains and submains was desirable, often caring for surface inlets as well.

10.13. Annual Rainfall as a Guide. Some authorities suggest the annual rainfall as a guide, allocating values to the entire areas of groups of several states as follows: $\frac{1}{4}$ inch for areas with 30-inch rainfall or less, $\frac{5}{16}$ to $\frac{3}{8}$ inch for areas with a 30- to 40-inch rainfall, $\frac{1}{2}$ inch for areas with a 40- to 50-inch rainfall, and $\frac{3}{4}$ inch or higher for areas with rainfall of 50 to 60 inches. In view of the broad allocations and the wide variations in both rainfall and soil conditions over each group of states, it seems that these values could be considered only as broad general trends that must often be modified within any given group by local conditions. It appears that a modification thereof according to the following practice submitted by F. H. Larson of the Soil Conservation Service staff in the Pacific Region would be more reliable and satisfactory. Larson has stated that it has been his observation in the Northwest that D values of approximately 1 per cent of the local annual rainfall seem to be about right, judged by results. This suggestion seems to check closely with the above recommendations for Iowa, Minnesota, and Oregon.

No part of the foregoing discussion applies to drainage of irrigated lands, which is discussed in a subsequent chapter.

DESIGN OF TILE DRAINAGE

10.14. Types of Tile-drainage Systems. There are really only three main types of tile-drainage systems according to the arrangement of the various lines in the systems. These are, as illustrated in Fig. 10.9, the *natural system*, the *cutoff* or *interception system*, and the *parallel system*. Most authors list several other systems, but these are only minor modifications of one of those shown in Fig. 10.9. The most prominent of these modifications are two parallel systems known as the *gridiron* and the *herringbone* systems, as shown in Fig. 10.10. The herringbone system is a modification of the gridiron system.

The Natural System. This system is called for where the land needing drainage is a relatively long narrow swamp or low wet spot with various arms reaching out among the adjacent knolls and ridges from which the low area receives and holds the surface runoff. The main tile runs up through the approximate center of the low area, and branches run out from this main to tap the various arms of the swamp. The tile lines are readily flagged out by eye, without map or instrumental aids except the steel tape, ax, and stakes. Where the main swamp or some of its arms are broad, it may be necessary to run in some extra laterals or even a small parallel system covering the broad part and outletting in the main or submain.

The Cutoff System. This system, also known as the *interception system*, is one with the drains so placed as to intercept seepage from either surrounding or adjoining higher land before it can break out on the surface and keep the low land wet (see Secs. 6.21 and 6.22 and Figs. 6.9 and 6.10).

The Parallel System. This system is applicable to flat lands or those of low uniform slope. The particular form of the parallel system for the given part of the drainage area which it fits depends on whether all parallel lines lie on one side of the main or whether they are distributed in secondary systems on both sides according to the directions of natural slopes in the various areas which they individually serve. The herring-bone system will work on both sides, each lateral feeding directly into the main. A given gridiron system connects with the main only on one side, but there may be several relatively small gridiron systems on both sides,

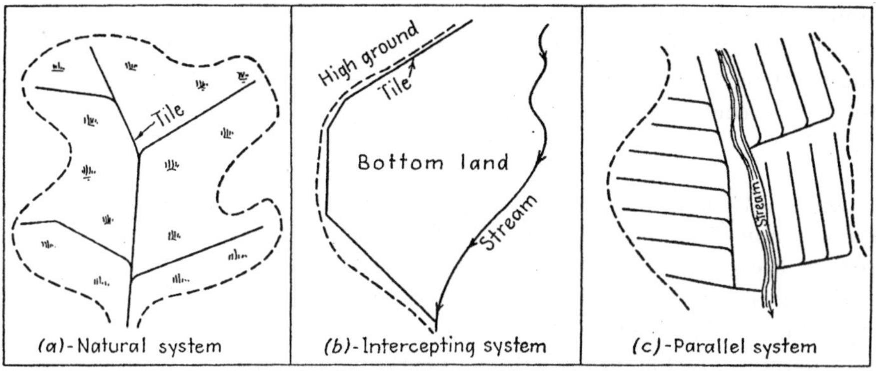

FIG. 10.9. Types of tile-drainage systems.

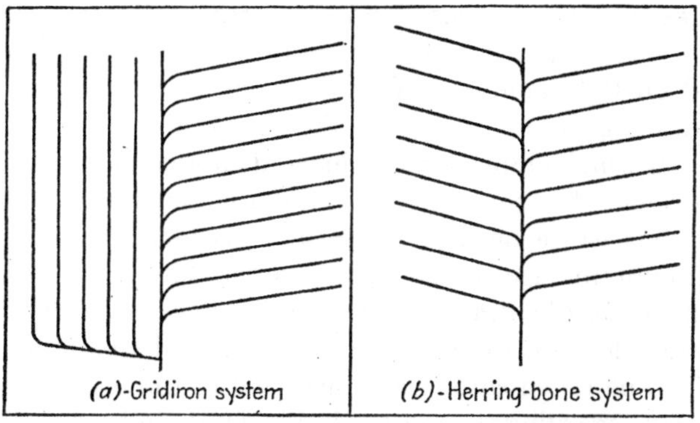

FIG. 10.10. Two much-used types of parallel systems.

each independent of each other and each joining the main at separate advantageous points. This type of system is usually the most economical of all and especially for the larger sizes of tile used on the submains. Double drained areas are least with this system. Figure 10.1 shows the advantages of the gridiron system on flat areas of considerable size.

10.15. Sizes of Tile Available. The size of a tile is understood to mean its internal diameter in inches. A full listing of standard sizes is given in Chap. 11, but for convenience it is stated here that all sizes from 4 to 42 inches are available by 2-inch increments and in some instances by 3-inch increments above 12 inches. Some makers also offer sizes of 45, 48, 54, and 60 inches, usually only on special order.

10.16. Proper Sizes of Tile for Laterals. *Never use a smaller tile than 4 inch* in drainage work. Three-inch tile are sometimes available, and they and even 2 inch have been used in farm drainage work, but the practice is discontinued because these small tile are harder to lay properly, as their small size makes imperative, and they quickly fill with silt, making it necessary either to redig and lay them anew or to clean them out, which is a difficult and expensive job. There is a marked tendency to discontinue the use of 4-inch tile. The use of 5-inch tile as a minimum is rapidly growing in favor. Many farmers prefer not to use any tile smaller than 6 inch, but they are larger than needed on well-designed and carefully constructed laterals.

DESIGN OF MAINS AND SUBMAINS

Especially on fairly large tile drainage systems the mains and submains do only a negligible part of the direct draining of the adjacent wet soil but receive the great bulk of the water they carry direct from branch conduits —the laterals. Hence their design is not directly affected by soil conditions but, as has been earlier stated, is a problem of pure hydraulics of flow in pipes, the mains and submains serving primarily as carrying sewers for the lateral system.

10.17. Sizes of Tile on Mains and Submains. Determination of the size of tile needed on mains and submains is a problem in hydraulic design based on the value of D used and on the area from which each carries the drainage water. Never take the combined capacities of the lateral tile lines which empty into mains and submains to determine their size. It is evident that this method would give much larger tile than are needed and thus cause needless expense. Six-inch tile is usually considered the minimum size to use on mains and submains.

The design of a tile drainage system involves a knowledge of the hydraulic principles governing flow in underground conduits, estimation of the runoff to be taken care of and selection of the proper tile capacity to fit, a decision regarding the most suitable depth and spacing of laterals, and the arrangement of the drains into systems that will best suit the ground and economically provide thorough drainage. Economical locating systems for tile and the most suitable depth and spacing of laterals for various soil types, are discussed in a later section. The other design factors will now be taken up in the order named.

10.18. Flow of Water in Tile Drains. The laws governing flow of water in pipes have long been the subject of research by investigators of many nations, with the result that scientific knowledge concerning them is well advanced. Their application to the special case of flow in tile drains, however, is attended by certain distinctions that should always be kept in mind. The main characteristic that distinguishes tile drains from other conduits is the fact that water enters at every joint all along the line (for small tile at 1-foot intervals). To be conservative and practical, tile are always considered as flowing full (the maximum discharge occurs when flow depth is about 95 per cent of the diameter) under the motivating force of gravity, and the velocity of flow is taken to be the average of the rate of movement of all the filaments within the tile. Actually, this velocity is not constant but varies from a maximum near the center of the tile to a minimum at the circumference and diverges from a motion of pure translation by the formation of cross currents and eddies, which occur at protruding edges and at other obstructions. In this respect, however, flow formulas are no different from other engineering formulas that are based on theoretical assumptions, and when modified by the introduction of an empirical coefficient, determinable by test, the error resulting from their use becomes negligible.

10.19. Factors Governing Discharge. Of the factors governing the discharge from a tile drain the following three, already defined in Chap. 5, are the most important: (1) the slope s on which the tile is laid or the effective head or *gradient;* (2) the resistance or "skin friction" developed by the water in sliding along the walls of the tile, expressed as the "roughness coefficient," n in the Manning formula; and (3) the shape of the free cross section of the drain which, for drain tile, is always assumed to be circular and a measure of the hydraulic efficiency. It is represented by the hydraulic radius r which, in a circular tile flowing either full or just half full, is one-fourth the internal diameter.

10.20. Velocity Formulas. Several formulas are in common use among drainage engineers for computing the velocity of water flowing in tile drains. Each one has its strong advocates, notwithstanding the fact that all such formulas were originally developed for continuous and uninterrupted flow. It has been found necessary to modify them materially to make them fit drain-tile conditions. The Manning formula (see Chap. 5) is the simplest and most easily used of the generally recognized velocity formulas and, as modified in the following paragraph, will be used exclusively in this book, with certain changes in the value of n in tile drainage design.

10.21. The U.S. Department of Agriculture Formula. In 1920, Yarnell and Woodward [166], after a number of experiments extending over several years, in a government laboratory at Arlington, Va., developed a formula which is generally considered to be especially adapted for use with drain tile. This formula, commonly known as the U.S. Department of Agriculture formula, is really what the Manning formula reduces to with an n value of approximately 0.011. The formula is

$$v = 138r^{2/3}s^{1/2} \tag{15}$$

where v = mean velocity, fps
s = hydraulic gradient or slope h/l in which
h = fall, ft, between points considered and
l = length, ft, between points considered
r = mean hydraulic radius a/p in which
a = internal area of cross section of the tile, sq. ft, and
p = wetted perimeter or internal circumference of the tile, ft

Now

$$Q = av$$

and

$$A = \frac{Q}{q}$$

where Q = discharge of the tile, cfs
q = discharge, cfs per acre
A = area drained by the tile, acres

10.22. Solution by Chart. The solution of the U.S. Department of Agriculture formula can readily be accomplished by the use of logarithms,[1] but with the assistance of a graph such as the one shown in Fig. 10.11, it is not necessary to perform any arithmetical computations. This chart has been prepared by computing the discharge for the various sizes of tile, on different grades, and then plotting the results on a logarithmic base. The diagonal lines marked with the size or diameter of tile represent the capacity of the tile when flowing full. The velocity of flow is indicated by the broken diagonal lines running approximately at right angles to the tile-capacity lines. The numerals across the bottom of the graph represent the slope or fall in feet per 100 feet, and those across the top show the same values in inches per 100 feet.

10.23. How to Use the Chart. To find the capacity of any tile line from 4 to 48 inches, follow up (or down, if expressed in inches) the vertical line representing its slope in feet per 100 feet until it intersects the diagonal line for the size of tile being considered. From this intersection, trace a horizontal line to the *left* and read the discharge in cubic feet per second from the scale at the left edge of the graph. A horizontal line extended from the point of intersection toward the *right* will show the number of acres that this particular tile will drain, with any drainage coefficient from ¼ to 1 inch.

To make this operation clear, assume a 12-inch tile, laid on a grade of 0.20 foot per 100 feet (nearly 2½ inches per 100 feet). It is desired to determine how many acres this main will serve if the proper drainage coefficient is ⅜ inch. Follow up the vertical line representing a slope of 0.20 foot per 100 feet until it intersects the diagonal line marked 12 inches,

[1] Or Tables 5.3 and 5.4 giving two-thirds powers of r and one-half powers of s (square roots) may be used.

and then follow a horizontal line to the scale at the left, where it is found that this tile has a discharge capacity of 1.9 cubic feet per second. By following the same horizontal line to the right side of the diagram, it is found in the column over "D (runoff) = ⅜ inch" that this drain would

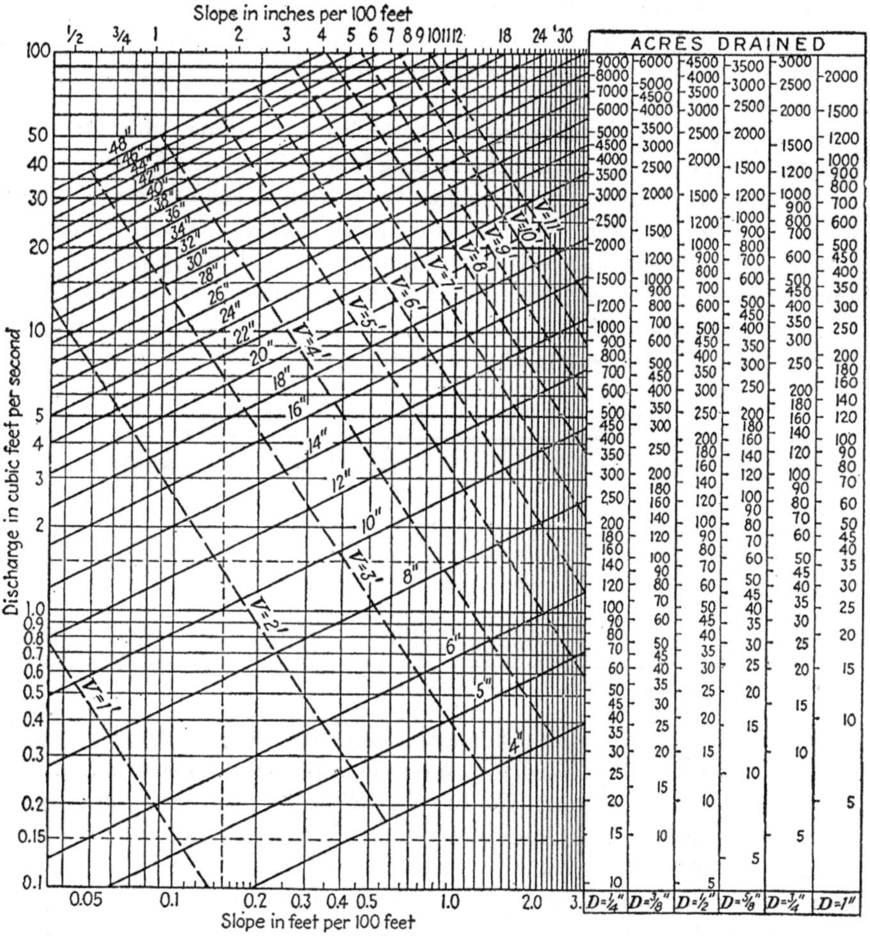

FIG. 10.11. Discharge curves for drain tile. Based on U.S. Department of Agriculture formula: $V = 138r^{2/3}s^{1/2}$, D = drainage coefficient. (*From USDA Bul.* 854.)

take care of the discharge from 120 acres. If the drainage coefficient were ¼ inch, this same tile would drain an area of 180 acres. The velocity of flow in the tile, as read from the diagram, would be nearly 2.5 feet per second.

It is thus seen that this diagram saves a large amount of work in making laborious calculations, and its use gives results that are accurate enough for all practical purposes. In practice, the problem is generally presented the other way around, for it is usually desired to know what size of tile is necessary to drain a given area with the slope or fall shown on the profile for that drain. After selecting the proper drainage coefficient, a horizontal line (real or imaginary) is drawn from the number of acres to be drained on the right scale toward the left until it intersects the vertical line representing the slope in feet per 100 feet. The size of tile, as represented by the diagonal line next above the point of intersection, is the proper size to use.

10.24. The U.S. Department of Agriculture Formula with Other *n* Values. It seems logical that the leading factors influencing the selection of a formula for determining the capacity of drain tile should be the value of the friction factor *n*, which is intimately involved, either visibly or understood, in the Chézy-Kutter, Manning, and U.S. Department of Agriculture formulas, and with the ease of manipulation of the formula in practical problems.

King [67] gives the values of *n* for drain tile shown in Table 10.5.

TABLE 10.5. VALUES OF *n* FOR DRAIN TILE

Kind of pipe	Value of *n* according to condition of the tile line			
	Best	Good	Fair	Bad
Common clay drain tile..............	0.011	0.012*	0.014*	0.017
Concrete pipe........................	0.012	0.013	0.015*	0.016

* Values commonly used in designing.

Studies by Scobey [126] indicate that, in any given class of tile, the resulting value of *n* tends markedly to increase with the diameter of the tile. The studies by Yarnell and Woodward [166] out of which they developed the U.S. Department of Agriculture formula show a similar indication; yet that formula is based on a fixed value for *n* of 0.01077, which is clearly seen to be suitable only for relatively small sizes of tile in drains laid with great care and in excellent condition—a situation seldom, if ever, attained in general practice. Pickels [104], therefore, seems justified when he says that he considers the *v* values given by the U.S. Department of Agriculture formula too large. However, when he recommends that such values be reduced 10 per cent for submains and 15 per cent for laterals, the evidence of the studies by Scobey and by Yarnell and Woodward, above referred to, and the following view by Etcheverry [45] seem to reverse the order of the correction percentages suggested by Pickels. Etcheverry states in part as follows: "The value of *n* should vary with the diameter of the conduit. The comparison of formulas presented further indicates that a value of n equal to 0.011 is suitable for tiles 6 inches or less in diameter, 0.012 is suitable for tile 8 to 12 inches in diameter, and 0.013 is suitable for tile 14 to 24 inches in diameter." On the basis of this statement one should use $n = 0.014$ for tile 26 to 36 inches in diameter.

In 1924 the Minnesota station published tables of capacities of drain tile prepared by the late John T. Stewart, computed by the Chézy-Kutter formula with n held constant at 0.013. These tables have been used with all sizes of tile for the past 20 years with entire satisfaction. It does not seem desirable from any standpoint to use a formula that crowds the limit of safe capacity.

The late Samuel H. McCrory once stated that there seemed still to be a considerable difference of opinion among engineers as to the proper value of n to use in tile drainage design, that it is probable that, in general practice, exactness of installations would vary considerably from the experimental installations in the Yarnell-Woodward experiments. Hence the determination of the proper value of n to use would rest largely with the judgment of the engineer according to local conditions that must be met. The logic in this statement seems evident. Therefore, when it is desired to use another value of n than the 0.01077 of the U.S. Department of Agriculture formula, the chart in Fig. 10.11 may still be used to advantage because, by multiplying the Q and A values obtained by use of the chart in Fig. 10.11 by the proper coefficient from Table 10.6 according to the value of n it is desired to use, the proper values of Q and A will result. For it is evident that, other factors remaining constant, the values of Q and A will vary inversely as n.

TABLE 10.6. REDUCTION FACTORS FOR USE OF OTHER VALUES OF n IN THE U.S. DEPARTMENT OF AGRICULTURE FORMULA

n	Reduction factor	Reduction, per cent	n	Reduction factor	Reduction, per cent
0.012	0.897	10	0.015	0.718	28
0.013	0.829	17	0.016	0.673	33
0.014	0.769	23	0.017	0.634	37

10.25. Depth of Mains and Submains. The depths of mains and submains should be just sufficient to permit free flow from all laterals and branches into the top portion of their cross section.

DESIGN OF LATERAL SYSTEMS

Hydraulic design of laterals is unnecessary as the capacity of clean 4- and 5-inch laterals is nearly always in excess of requirements. The laterals receive the water direct from the soil and are seldom loaded to capacity. Their proper design is, therefore, largely governed by the topography of the area, the normal rooting depth and, consequently, the depth of the drainage required by the crops to be raised and the character of the soil and subsoil. Their design, therefore, consists in determining what is the optimum depth and where is the best location (what distance apart on flat lands) at which they must be placed to function most effectively.

Rolling farm lands do not present much of a problem in design of tile drainage as they have fair natural drainage except in the depressions. In such cases only a skeleton natural system is needed, with the submains and laterals following the natural surface drainageways and made to pass beneath the lowest points in the depressions. There is little or no exact knowledge of proper depth for tile in skeleton systems except that based on experience and the normal rooting depth of crops grown in the given area.

It is doubtful if depths of laterals appreciably less than 3½ feet, except for short distances under deep pockets, can be justified where general farming is practiced and a considerable number of crops with widely differing rooting habits are raised on the same areas in rotation.

Fred H. Larson of the Soil Conservation Service staff, Pacific Region, suggests determination of required capacity and depth of tile drains in natural systems by the same methods by which they are determined in parallel systems on flat lands, discussed in Sec. 10.34. In other words, if spacing for the soil in question calls for 100 feet if it were in a parallel system, then the required tile capacity for a random line would be based on an acreage obtained by using a width of 100 feet times the length of the tile line. This also applies where the tile is used to intercept side-hill seepage, unless there is definite evidence to the contrary.

Flat lands with poor natural drainage require a thorough investigation in order to design a tile drainage system therefor with proper depth and spacing. Through a series of years vigorous crops, especially the deep-rooting crops, tend to improve subdrainage conditions by slight opening up of the soil, thus making it more responsive to the influence of tile drains. However, the chief influence of crops upon the proper design of tile drainage systems is the fact that different types of plants vary widely, not only in their normal rooting depths but also in their tolerance for excess water in the soil; hence drainage does not have to be so deep for the shallower rooting crop types as it does for the deeper rooting ones or those more sensitive to excess wetness in the soil.

10.26. Crop Type as an Index of Optimum Drainage Requirement. Thus far information on drainage requirements according to crop type is based almost wholly on observation and practical experience as to the susceptibility of specific crops to damage from excessive wetness in the soil (see Manson [75]) and as to the normal rooting depths of different crops. In this latter connection Table 10.7, compiled from studies by numerous reliable authorities, will be found a helpful guide in tile-drainage design.

10.27. Field and Truck Crops. Table 10.7 shows the rooting depths of crops ordinarily grown by farmers. Where the land is to be used continuously to produce certain groups of crops, such as small grains, hay crops, corn, potatoes, and the commoner truck crops, an average depth of drainage can be selected which will not put the ground-water table hopelessly beyond reach of the shallower rooters and, at the same time, will provide for the deepest rooters at least average conditions for their optimum development. In most humid regions a wide range rotation, such as above suggested, seems to be taken care of by putting the tile laterals at a depth of 3½ to 4 feet with the proper spacing to fit the soil and subsoil.

On the other hand, if a tract is to be given over to only one or two specialized crops, the sensible thing to do may be to install a depth of drainage particularly suited to the rooting habits of these special crops. Certified seed potatoes are a good example of a specialized crop. When it became necessary to give the soil on a particular field a rest from potatoes and build it up again with legumes, the 3- to 3½-foot drainage for potatoes would serve very well for a single-year grain crop to be followed by red clover to be plowed under.

10.28. Orchards and Bush Fruits. Orchards ordinarily root deep—up to about 8 feet with the bulk of the roots in the first 4 to 6 feet. Rows of trees will vary from 20 to 50 feet apart, depending on kind of tree, topography, soil, and climate. For the wider spacings in heavy soils, frequently a tile drain is laid between each two rows of trees at depths of 6 to 8 feet. When the rows are close, the tile may be laid in every second or third space and at depths to suit the kind of tree and character of the soil.

Bush fruits vary greatly in depth of rooting, but 4 feet is about the maximum limit of rooting depth, while the dense portion of the roots mostly lies within the first 24 to 30 inches. With raspberries and others of the same family, the bulk of the roots lie in the first 6 to 12 inches, with a few main roots reaching much deeper. The rows of bush fruits are much closer together than those of trees in an orchard, and the bushes are much closer in the row. Therefore drainage for such will usually follow the general rule of spacing and depth that governs in the case of fields where annuals are grown. However, on a permanent bush fruit field, it is desirable to vary the spacing sufficiently from standard to bring the tile lines mid-way between any two rows, never directly under the bushes, and to fit the depth of the drains to the rooting habits of the given type of bush.

TABLE 10.7. ROOT PENETRATION FOR CROPS (AVERAGE LOAM SOIL CONDITIONS)
(Compiled from various reliable sources)

Crop	Depth of dense root zone, in.	Zone of moderate branching, in.	Extreme depth of tap or main roots, in.	Approximate optimum depth for tile drains, ft	Approximate* average optimum depth for tile drains for each class of crop, ft
Field Crops					
Small grains:					
Wheat	12	12–36	48–60	4 –4½	
Rye	20	20–30	60–90	3 –5	3½–4
Oats	18–24	24–30	48–60	3 –4	
Barley	24–30	30–42	54–78	3½–4½	
Flax	12–18	18–24	30	2½–3	
Forage crops:					3 –3½
Sunflower	12	12–36	108	3½–5	
Sugar beets	.....	12–48	60–72	4½–6	3½–5
Field corn	18–36	36–54	60–72	4 –5	
Hay and grass crops:					
Alfalfa	36	36–72	72–108	5 –6	5
Clover	24–48	.....	48–96	5 –6	
Brome grass	.....	.....	66–78	5 –6	
Kentucky bluegrass	18–20	.....	60–84	3 –4½	
Redtop	22	.....	40	3 –4	2½–3½
Timothy	15	.....	36	2½–3	
Vegetable Crops					
Beets	20	20–50	60–120	3 –5	
Carrots	36	.....	90	3 –7	
Onions	24	24–36	60	3½–5	
Cabbage	24	24–36	36–90	3½–5	3½–5
Cauliflower	12–20	20–36	36–54	3½–5	
Lettuce	8–12	12–48	90	4 –5	
Chard	18	.....	72	3 –5	
Tomatoes	24	24–36	48	3½–4	
Peppers	12–18	18–36	48	3½–4	
Sweet corn	24	24–36	48	3½–4	3½–4
Cucumbers	12	12–36	36–48	3½–4	
Muskmelons	8–12	12–36	36–48	3½–4	
Beans	12	12–30	36	2½–3½	
Peas	12	12–30	36	2½–3½	
Radishes	6	.....	24	2 –2½	2½–3½
Spinach	10	10–24	42	2 –3	
Potatoes	10–12	12–40	40	3 –3½	
Small Fruits					
Strawberries	12–18	.....	36	....	2

* General averages run from 3½ to 5 feet. In average mineral soils the deeper of these, or 5 feet, is in no danger of causing overdrainage.

INFLUENCE OF THE SOIL FACTOR IN TILE DRAINAGE

10.29. Ground-water Table. In all soils requiring drainage, there exists at, or just below, the ground surface a zone of saturated soil which is completely filled with water. The elevation to which water will rise in test wells in such soil delineates the surface of gravitational water, which is variously spoken of as the *ground-water table, plane of saturation,* or simply *surface of ground water.* Since the roots of most field crops will not penetrate below the ground-water table, and since the various crops normally produce root systems of different lengths, it is the work of the drainage engineer not only permanently to lower the plane of saturation but

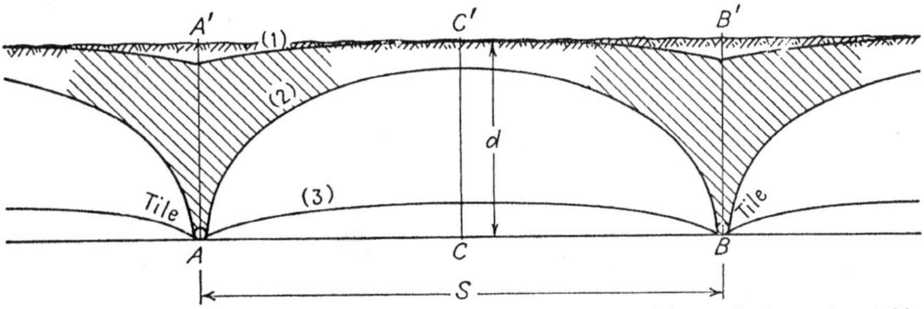

FIG. 10.12. Conventional sketch showing several typical positions of the water table between two parallel tile drains.

also to hold it, as nearly as possible, at some fixed depth best adapted to the needs of the crop to be grown.

In order to do this it is necessary to know how the ground-water table behaves under the influence of the tile drains. By means of rows of test wells (see Sec. 10.30) spaced at frequent intervals in a straight line at right angles to and crossing two or more parallel tile drains spaced at known distances, various investigators over a considerable period of time have measured the depth to the ground-water table and, by plotting these measurements on cross-section paper, have been able to determine, empirically, the general shape of the ground-water table in various soils and to show its variations under differing conditions of wetness in the soil following a rain. In general the ground-water table has thus been found to take about the conventional shape shown in Fig. 10.12.

10.30. Test Wells and Ground-water Piezometers. Originally, simply for determining the position of the water table, a test well was merely a hole in the ground extending well below the permanent water table. In stiff soils it might be merely an auger hole, but it is usually cased for permanence and convenience, with a wooden tube of ¾-inch lumber, a piece of 2-inch gas pipe or heavy galvanized sheet-metal pipe, or often pieces of small drain tile stacked one upon another. The sides of such wells, except those of drain tile, were perforated, for most of their length for the ready entrance of soil water. The water entered the drain tile wells through the joints. The top of the casing is always left projecting a few inches above the surface and is loosely capped to prevent the entrance of debris and vermin. But the test well now most widely used takes on also the character and function of a piezometer, enabling measurement of the pressure of the ground water at the point of installation.

The piezometer is a length of ¼- or ⅜-inch iron pipe driven into the soil so that there is no leakage into or out from the sides. All entrance of

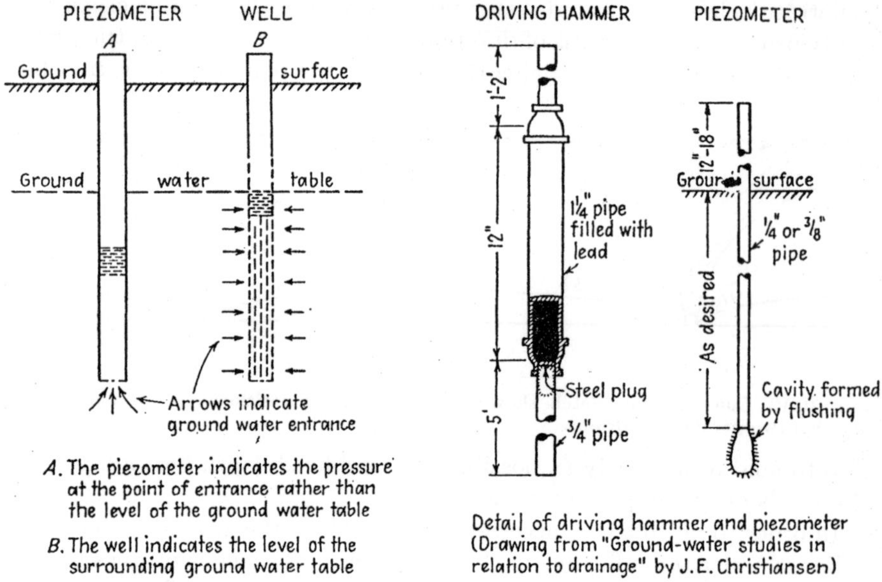

FIG. 10.13. Diagram of test well and field piezometer, and apparatus for installing the latter. (*Courtesy of Soil Conservation Service, Pacific Region.*)

water is from the bottom, an essential characteristic of the piezometer (see Fig. 10.13). A loose rivet in the bottom prevents soil entering as the pipe is driven, the rivet being punched out afterward. The pipe may be driven to a depth of 30 feet with proper driving equipment, a feature of great usefulness in irrigated country. (For a full discussion see contributions by Donnan and Bradshaw [36], Christianson [26 and 27], and Donnan and Christianson [37].)

10.31. Equation of the Ground-water Curve. Various investigators have sought to identify the mathematical form of the ground-water curve which is of growing interest in theoretical studies of tile spacing and depth of laterals on flat lands. The point at the tile line and that at the midpoint between two consecutive tile lines are the two points on the curve of greatest interest.

Carl D. Gross, working at the Engineering Experiment Station at Iowa

State College, found the shape of the water table in fine sand to be hyperbolic of the general form

$$y = \frac{x}{a + bx} \tag{16}$$

with the origin assumed at the bottom of the given tile drain; where x is the abscissa, y the ordinate, and a and b the constants for a specific subsoil condition.

Neal [94] of the staff of the Minnesota Agricultural Experiment Station, from a study of several actual tile drainage projects in specific soils of varied character, at widely separated locations, found the general shape of the ground-water table in mineral soils to be approximately that of a semiellipse of the form

$$y^2 = b^2 - \frac{b^2x^2}{a^2} \tag{17}$$

in which the origin is the halfway point on the straight line between the two drains; where a is one-half the major axis or one-half the distance between the two drains, b is one-half the minor axis or the height of the curve at mid-point above the origin, and y and x are the ordinate and the abscissa of any point P on the ground-water curve.

10.32. Theory of Ground-water Movement between Parallel Tile Drains. The shape of the ground-water curve in average mineral soils seems to be generally assumed to be a close approach to a semiellipse. On this assumption consider two such tile drains as A and B in Fig. 10.12 at a uniform depth d and a distance apart AB or S. Following a rainfall which has been just sufficient to saturate the soil to the ground surface, let curve 1 represent the ground-water surface as the rain ceases, and curve 2 this surface just as the gravity water from the crosshatched wedge directly over A has lowered to a point where it no longer obstructs lateral flow from points anywhere between A and C'.

The gravity water in the crosshatched area over drain A will have but a short distance to flow, almost all vertically, to reach the drain under the influence of gravity opposed only by the limited soil friction in the distance d. Its movement will be relatively rapid, partly dependent on the rate of entry into the tile drain through the joints.

On the other hand, gravity water at the mid-point C' in order to reach the drain at A or B must move under the influence of gravity through the soil and overcome the soil friction in obedience to Darcy's law, not only for the short vertical distance d but also for the many times greater horizontal distance AC or $S/2$. Near mid-point C' the flow will be relatively slow, opposed by the much greater soil friction in the longer path of travel. This total friction is a resultant of the vertical and horizontal components. The flow from C' is also obstructed by the faster-flowing water directly over A until the latter has been drawn off sufficiently to make room for some of the water coming in by lateral flow. When this occurs, the velocity of flow of both the vertical and horizontal components of the lateral flow will steadily tend to increase, the vertical component gaining on the horizontal, with the result that the integrated resultant heads acting on given particles of water between the two drains will lie on a curve (the semiellipse above mentioned) represented by curve 2. The actual path of travel varies with all the vagaries of the soil (worm holes, root cavities, embedded rocks, etc.).

If the rain stops entirely or continues only less than necessary to keep the surface soil supplied with water in excess of its capillary capacity, the peak C' of the ground-water curve will gradually drop, and the movement of water toward the drains will become steadily slower until the soil friction along the horizontal component balances the force of gravity. When this occurs, flow will cease and the ground-water curve will take a fixed position represented by curve 3. In sand or very sandy soils this final position of the ground-water curve, resulting from the given rain, will tend to approach closely the straight line between the two drains A and B.

Neal [94] makes the following comments on the actual ground-water curve as compared with the theoretical elliptical form:

> In nearly all cases, the measured y values are less than the calculated values, especially when the ground-water is near the surface. This may be attributed to the following pertinent facts about the surface soil: (*a*) A surface soil contains more non-capillary pore space than a subsoil with the same moisture equivalent. (*b*) The rate of moisture depletion by plants is much greater in the surface than in the subsoils, due to transpiration and evaporation. (*c*) Therefore the rate of drop of the ground-water through the first foot is much faster than through the subsoil. Consequently the surface of the ground-water table between the tile lines is flatter than a theoretical elliptical curve as given by Equation 1.[1]

10.33. Actual Ground-water Curves. Figure 10.14 shows several ground-water profiles as they actually occur in field soils and indicates the variations to be expected from any ideal curve.

These curves bring out clearly the effect of rainfall on underdrained soils. On Apr. 14, 1909, a rain of 0.31 inch occurred which apparently was entirely absorbed as capillary moisture, since the curve for Apr. 15 is practically identical with the one shown for Apr. 13. The rain of 1.34 inches on Apr. 18 fell on moist soil and was much more than enough to replenish the capillary supply. Consequently, the excess was stored as gravitational water, which raised the water table to the position shown on Apr. 19. By Apr. 20 the tile drains had drawn off enough water to lower the surface of saturation to the elevation indicated on the diagram for that day.

[1] Equation (17) of this book.

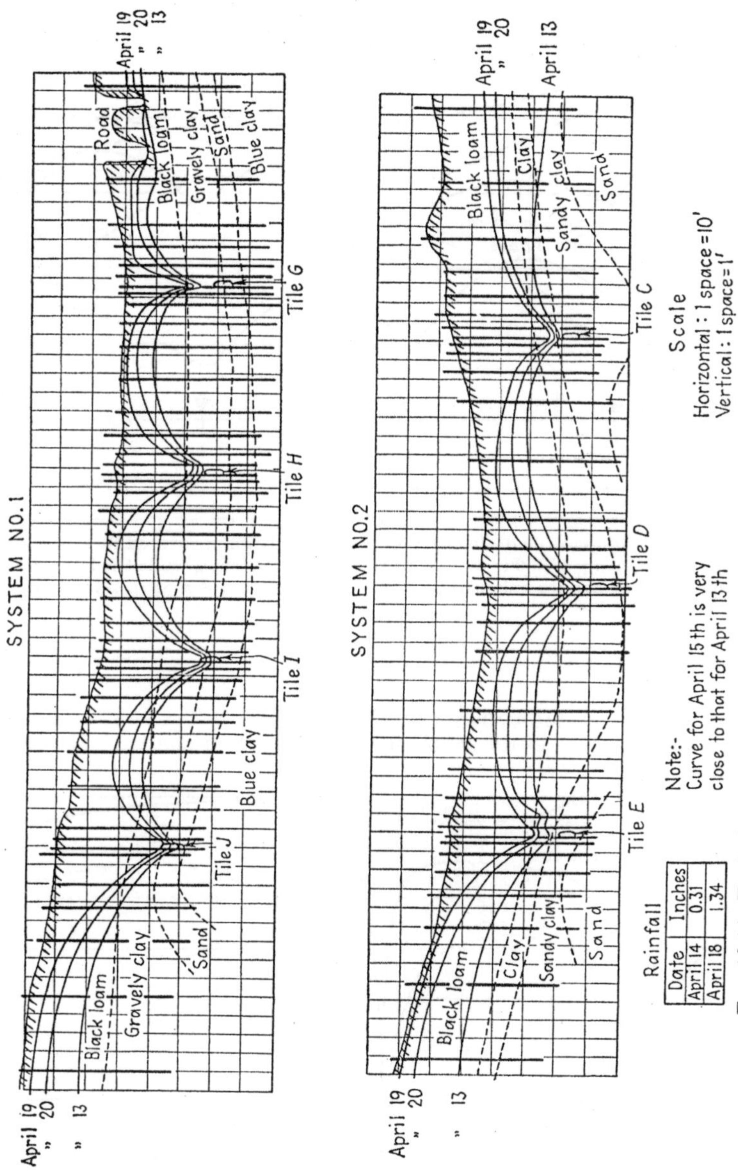

FIG. 10.14. Typical ground-water profiles. (*Eng. Expt. Sta., Iowa State Coll. Bul. 52.*)

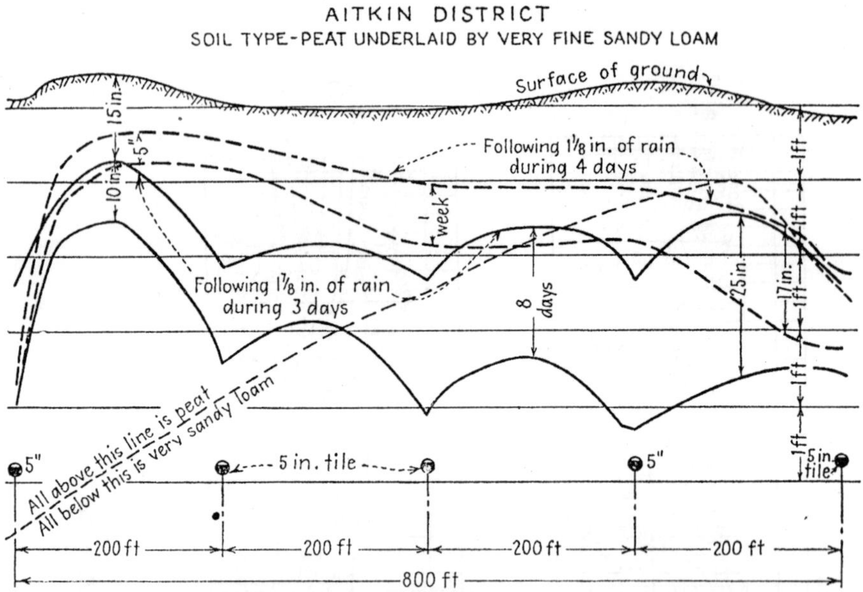

FIG. 10.15. Effect of different tile drain spacings in fine sandy loam covered by a blanket of peat. The original spacing of 800 ft was about right for grass crops, while the later spacing furnishes sufficient drainage for cultivated crops. (*Univ. Minn. Agr. Ext. Div. Spec. Bul.* 149.)

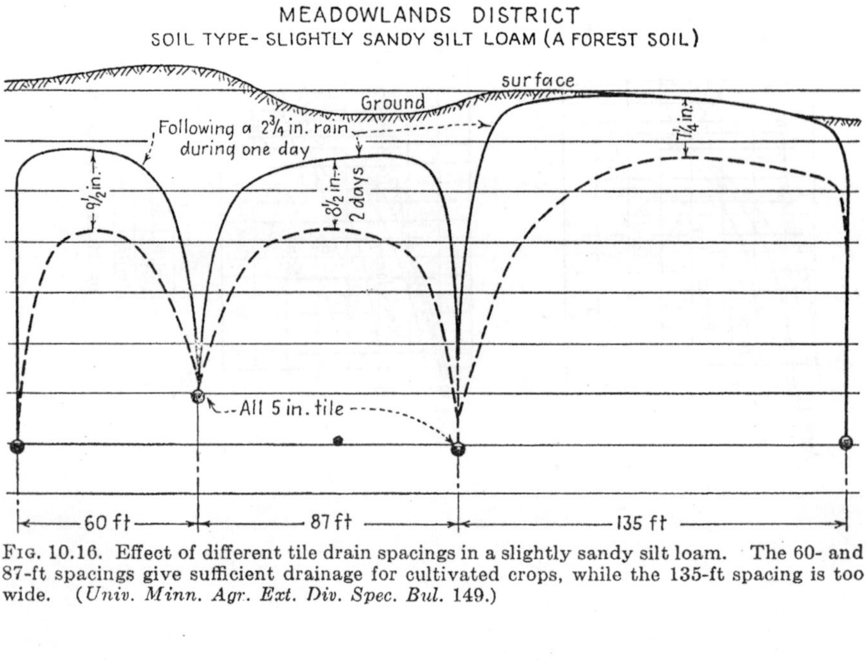

FIG. 10.16. Effect of different tile drain spacings in a slightly sandy silt loam. The 60- and 87-ft spacings give sufficient drainage for cultivated crops, while the 135-ft spacing is too wide. (*Univ. Minn. Agr. Ext. Div. Spec. Bul.* 149.)

It would appear from a casual study of these graphs that some of the drains of system 1 would have been even more effective had they been placed deeper so as to lie nearer the surface of the blue-clay subsoil. The heavy vertical lines represent the location of the test wells used in determining the position of the ground-water curves.

Figures 10.15 and 10.16 show actual ground-water curves and their fluctuations at different drain spacings in specific soils due to rainstorms of specific magnitude in central and northern Minnesota as carefully measured. Figure 10.15 is of especial interest as it shows clearly the effect of reducing the spacing of the tile lines in the given designated soil, the depth being maintained constant. The upper pair of curves, shown in broken lines, were the results of the trial spacing of 800 feet. The lower pair, in solid lines, show the great improvement in results following a comparable rainstorm 2 years later with spacings of drains reduced to 200 feet. The first trial produced only good grass land. The second trial resulted in an excellent crop of potatoes.

10.34. Theory of Depth and Spacing of Tile Drains on Flat Lands. The discussion thus far clearly indicates that proper tile drainage for flat lands is a double problem. The first part is concerned with the optimum drainage condition for each type of crop, that is, with the questions of just how long the given type of crop can withstand a fully saturated condition of the soil, how rapidly that condition must be overcome in order to obviate injury to the crop, and what depth of water table in the soil will give the best results with the given crop in both yield and quality. The second part of the problem is concerned with the proper depth and spacing at which tile drains must be placed in the given kind of soil in order to fulfill the above conditions. Both phases of this problem are difficult of solution.

Up to the early 1900's the only guides to tile depth and spacing were recommendations by engineers based on experience developed by years of observation. These recommendations are still widely followed by practicing engineers and agriculturists. The essential features of the recommendations by a few of the best known of these earlier authorities follow.

C. G. Elliott [41]

Depth: 2 to 2½ feet in dense heavy soils.
4 feet in light open sandy soils.

Spacings: 30 to 40 feet in close dense soils, largely clay; 60 feet in mixed clays with fine sand; 70 to 80 feet in alluvial gumbo or heavy soils with granular structure; 100 feet in alluvial, glacial drift and sandy loams with joint clay subsoils; 150 to 200 feet in sandy lands or those containing much vegetable matter having sandy or gravelly subsoil.

A. G. Smith [140]

Depth: General average of 3½ feet.

Spacing: In eastern general farming about 32 to 70 feet; Middle West in closer soils, 8 rods, on lands with pervious subsoils 10 to 12 rods, in specially favorable areas 20 rods; at Marion, S.C., on sandy loam soil with clay subsoil, 100 feet (depth 3½ feet); for truck crops 40 to 50 feet; general guide, uniformity of crop development.

W. J. Schlick [131]. (Iowa soils, only, considered.)

Depth: 3½ to 4 feet, usually the latter, using closer spacings rather than shallower depths to expedite early action of drains.

Spacing: 100 feet the maximum in areas used for crops; in close or average soils for truck crops 75, 50, or even 33 feet in some cases. In rare cases with sand or gravel subsoil a maximum of 200 feet (see further comment in Sec. 10.12).

G. W. Pickels [104]

Depth: 3½ to 4 feet in general to lower the ground water to 3 feet below the surface; 2 to 2½ feet in dense clay soils. Laying tile too shallow eliminates the advantages of wide spacing.

Spacing: 50 to 100 feet gives best results in the majority of cases; less than 50 feet is prohibitive in cost except for areas devoted to truck crops; in unusually open soils 150 to 200 feet spacing may be used; present tendency is closer spacing.

The period of guesswork, however, is passing—but all too slowly. The scientific character of the problem has been fully recognized only since the early part of the twentieth century and by only a few technical men. The part of the problem relative to the rate of drainage required to prevent injury to crops has been recognized only since about 1920.

It is evident that the leading element in the scientific consideration is proper depth of drains, the spacing being secondary but essential to the securing of the depth; hence both elements must be jointly considered.

NEAL'S FORMULAS

Several investigators have individually made a beginning at a study of the problem of drainage requirement according to soil character. Only one of these seems to have carried his studies far enough to provide a practical rule for depth and spacing of tile drains on flat lands in humid regions. In 1934, Neal [94] presented empirical formulas resulting from a statistical analysis of the relation between the moisture equivalent of a given soil and the depth and spacing of tile drains required in that soil to give a desired rate of drainage as measured by the rate of drop, in feet per day, of the ground-water curve at the mid-point. The formulas are

$$T_d = \frac{17.5}{(M_e)^{0.5}} \tag{18}$$

$$T_s = \frac{12{,}000}{(M_e)^{1.6}(R_d)^{1.43}} \tag{19}$$

where T_d = tile depth, ft
T_s = tile spacing, ft
M_e = average moisture equivalent of soil to depth of tile
R_d = rate of drop of ground-water curve at mid-point, feet per 24-hour day

Unfortunately circumstances limited Neal's studies on the problem to only four observation stations at widely separated points in Minnesota, only, representing but four different soil and climatic conditions. Still this method has been used with satisfactory results by the Soil Conservation Service in the northern Mississippi Valley and, with certain modifications later considered, in the northwestern United States in the Pacific Region. A full discussion of Neal's studies is available in the publication cited.

10.35. The Drop at Mid-point. Relative to the rate of drop at the mid-point Neal states:

> With the limited observations made, it was found that the crops were not seriously injured if the water table was held at least 6 inches below the surface and was lowered at the rate of one foot per day through the second 6-inch interval, and at the rate of 0.7 foot per day through the third 6-inch interval. Moreover, if the tile spacing required to give a rate of drop of one foot per day through the second 6-inch interval is plotted as ordinate and the moisture equivalent as abscissa, a graph is obtained giving the tile spacing necessary for the optimum rate of drop for any soil type.

The equation of this graph is

$$T_s = \frac{12{,}000}{(M_e)^{1.6}} \tag{20}$$

The original practical form in which Neal's results were given, based on a rate of drop of 1 foot per day, was in the form of the four following equilateral nomographs on a logarithmic scale (see Figs. 10.17 to 10.20). All four are offered because the field engineer often is not able to secure determinations of moisture equivalent, but he may be able, himself, to work out with a kitchen scale, the farm wife's oven, and a thermometer, fairly satisfactory determinations of the plastic limits (see Secs. 3.12 to 3.16).

ILLUSTRATIVE EXAMPLE

The following example illustrates the use of these nomographs.

1. Given that the moisture equivalent is 20 and the required drop at mid-point 1 foot per day; spot 20 on the base scale of Fig. 10.17. Follow

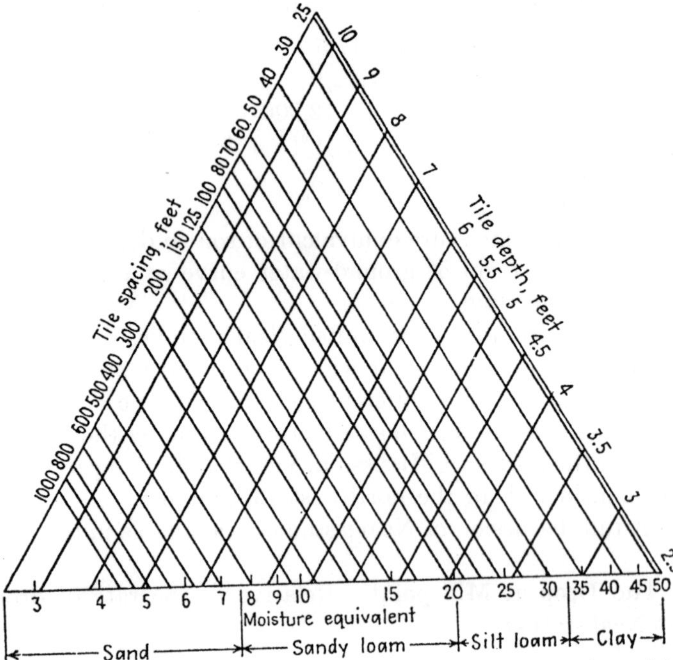

FIG. 10.17. Nomograph giving tile spacing and depth with respect to the moisture equivalent; for $R_d = 1$, only. (*Minn. Agr. Expt. Sta. Tech. Bul.* 101.)

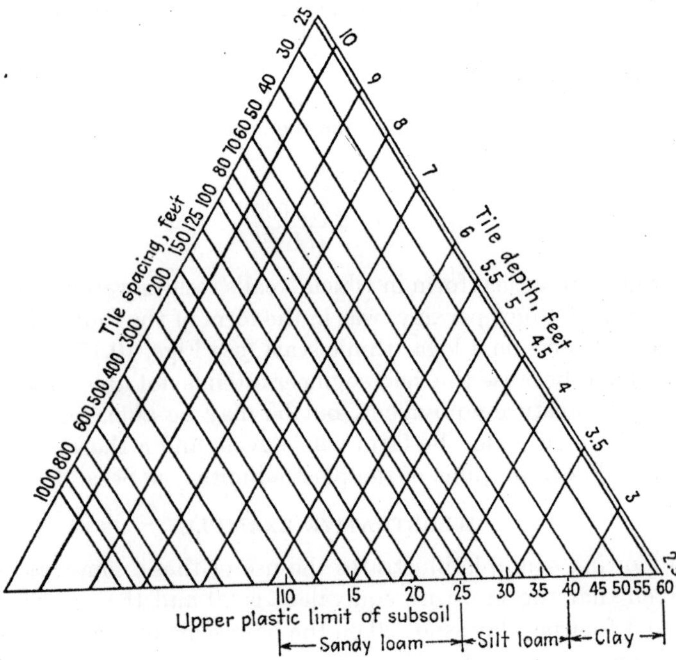

FIG. 10.18. Nomograph giving tile spacing and depth with respect to the upper plastic limit of the subsoil; for $R_d = 1$, only. (*Minn. Agr. Expt. Sta. Tech. Bul.* 101.)

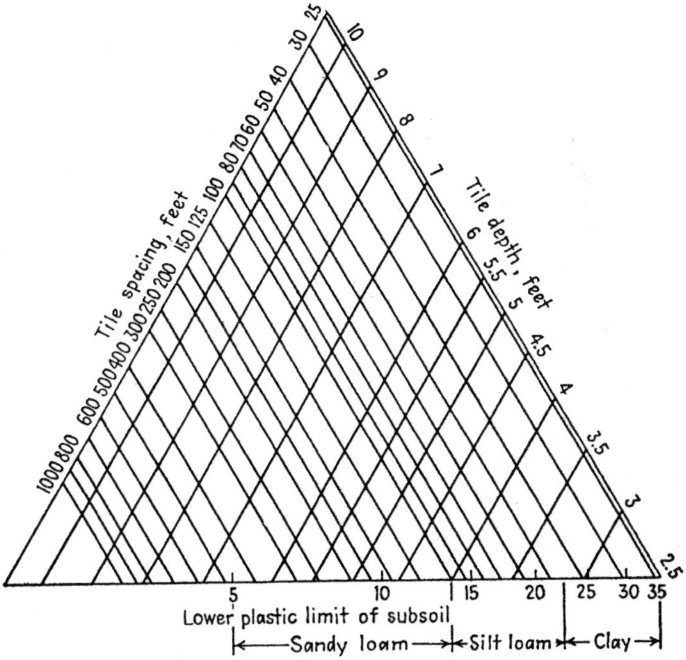

FIG. 10.19. Nomograph giving tile spacing and depth with respect to the lower plastic limit of the subsoil; for $R_d = 1$, only. (*Minn. Agr. Expt. Sta. Tech. Bul.* 101.)

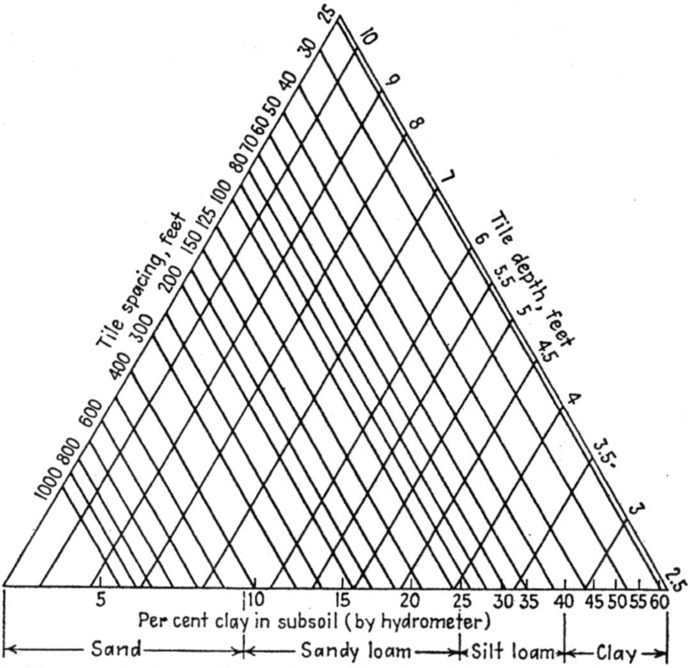

FIG. 10.20. Nomograph giving tile spacing and depth with respect to the per cent of clay in the subsoil; for $R_d = 1$, only. (*Minn. Agr. Expt. Sta. Tech. Bul.* 101.)

the diagonal line upward to the right to its intersection with the depth scale at 4.0. This is the required depth. Next, from 20 on the base scale, move along the diagonal line upward to the left to its intersection with the spacing scale at 100. This is the proper spacing for a depth of 4.0 in the given soil.

2. Similarly, using the lower plastic limit (70 per cent of the moisture equivalent, or 14) in Fig. 10.19, the diagonal on the right through 14 on the base scale again gives the required depth as 4.0, and the diagonal to the left gives the required spacing as 100.

3. Similarly with the upper plastic limit (120 per cent of the moisture equivalent) Fig. 10.18, or for the per cent of clay (125 per cent of the moisture equivalent) Fig. 10.20, the same values are obtained.

10.36. Other Rates of Drop R_d than 1 Foot. The triangular nomographs shown in Figs. 10.17 to 10.20 are applicable where the rate of drop at mid-point R_d is 1 foot. The required rate of drop may, however, be almost any other value, as, for example, 0.5 or 0.8 foot. In Neal's formula for T_s, already presented, this situation is provided for in the factor $R_d^{1.43}$, which becomes unity and is of no effect where $R_d = 1$, but in no other case. For solution of the problem where R_d has *any* given value, Neal presents the nomograph in Fig. 10.21, together with an example illustrative of its use.

In the use of this nomograph there are two possible cases.

1. Where the depth is determined by soil character, start with the given value on the vertical scale headed M_e. Then proceed as directed in the illustrative example accompanying the nomograph, Fig. 10.21.

2. Where the depth is determined by some other factor than the soil character, as, for example, the presence of a hardpan layer a short distance below the surface, Neal's nomograph alone will not give the solution. This case is fully discussed in the next paragraph.

10.37. T_d Fixed by Other Factors than Soil Character. Here the T_d that *must* be used will always be less than T_d fixed by soil character alone. The same will be true of spacing that must be used compared to T_s determined by soil character. The problem is to find the new values of T_s and T_d (designated as T'_s and T'_d) that will maintain a constant value for R_d. The average vertical cross section of flow toward the drain, or A, and the value of P, the coefficient of permeability, will both remain constant. Let v and v' be the velocities of flow through the soil corresponding to T_s and T'_s, respectively. Then by Darcy's law of flow through saturated soil

$$v = P\frac{h}{l} = P\frac{T_d}{\frac{1}{2}T_s} \qquad \text{and} \qquad v' = P\frac{T'_d}{\frac{1}{2}T'_s} \tag{21}$$

The time unit t is the same in both cases, say seconds in a 24-hour day.

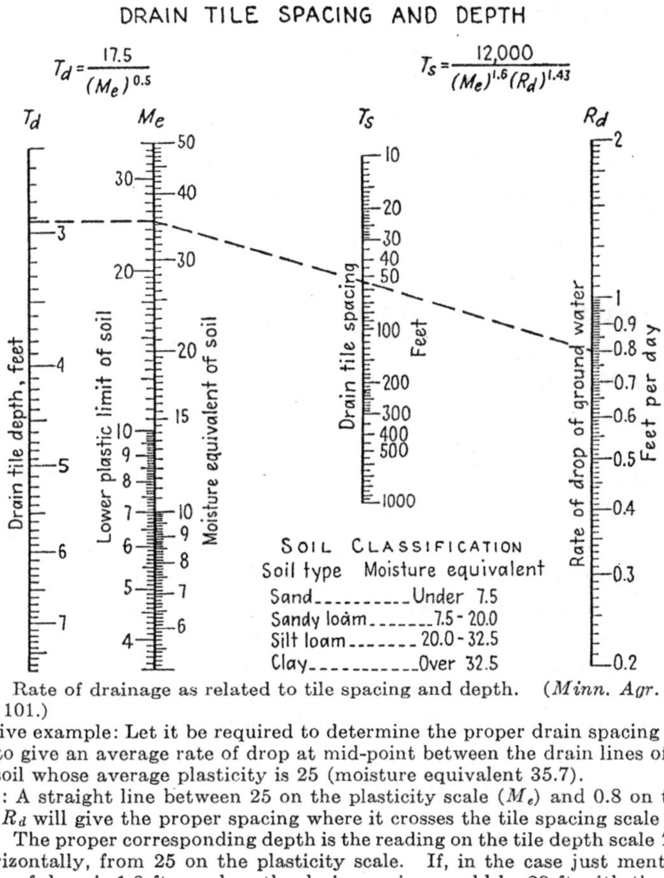

FIG. 10.21. Rate of drainage as related to tile spacing and depth. (*Minn. Agr. Expt. Sta. Tech. Bul.* 101.)

Illustrative example: Let it be required to determine the proper drain spacing and depth necessary to give an average rate of drop at mid-point between the drain lines of 0.8 ft per day for a soil whose average plasticity is 25 (moisture equivalent 35.7).

Solution: A straight line between 25 on the plasticity scale (M_e) and 0.8 on the rate of drop scale R_d will give the proper spacing where it crosses the tile spacing scale T_s, in this case 54 ft. The proper corresponding depth is the reading on the tile depth scale T_d directly across, horizontally, from 25 on the plasticity scale. If, in the case just mentioned, the desired rate of drop is 1.2 ft per day, the drain spacing would be 29 ft with the depth the same as before.

Let Q and Q' be the total flow in the two cases in time t, and let q and q' be the rate of flow in each case in cubic feet per second. Now

$$Q' = \frac{T'_s}{T_s} \times Q \tag{22}$$

and

$$q = Av = \frac{APT_d}{\frac{1}{2}T_s} \quad \text{and} \quad q' = Av' = \frac{APT'_d}{\frac{1}{2}T'_s} \tag{23}$$

Then

$$t = \frac{Q}{q} = \frac{Q'}{q'} = \frac{QT'_s/T_s}{q'} \tag{24}$$

which may be written

$$t = \frac{Q}{APT_d/\frac{1}{2}T_s} = \frac{QT'_s/T_s}{APT'_d/\frac{1}{2}T'_s} \tag{25}$$

Simplifying and solving for T'_s

$$T'_s = \left(\frac{T'_d}{T_d}\right)^{\frac{1}{2}} \times T_s \tag{26}$$

ILLUSTRATIVE EXAMPLE

When $R_d = 0.5$ and $M_e = 16$, by nomograph, $T_s = 385$ and $T_d = 4.4$. If a hardpan layer permits T'_d to be only 3.2, substituting in Eq. (26),

$$T'_s = \left(\frac{3.2}{4.4}\right)^{\frac{1}{2}} \times 385 = 328$$

10.38. Stratified Soils. Most of the soils in the northern Mississippi Valley and in other glaciated regions are roughly stratified, the M_e being different in each stratum. The following example shows how to handle such a case.

ILLUSTRATIVE EXAMPLE

Crops: sugar beets, corn, and alfalfa.
Heavy rooting depths 30 to 60 in., or average 3.75 ft.
Average required depth of drainage about 4.0 ft.
Required maximum rate of drop at mid-point, R_d, 0.5 ft in 24 hr.
Soil profile as shown below.

Depth of stratum		Physical characteristics			Corresponding calculated *LPL*
Inches	Feet	M_e	*LPL**	Per cent clay	
0– 8	⅔	21.5	..	....	15
8–24	1 + ⅓	....	9	....	9
24–30	½	....	..	14.3	8
30–48	1 + ½	....	16	....	16

* *LPL* = lower plastic limit.

Solution:

(1) Depth of stratum, ft	(2) Calculated *LPL*	(3) Col. (1) times col. (2)	(4) Standard spacing, ft	(5) Standard depth, ft	(6) Col. (3) times col. (4)	(7) Col. (3) times col. (5)
⅔	15	10	240	4.13	2,400	41.3
1 + ⅓	9	12	550	4.90	6,600	58.8
½	8	4	650	5.10	2,600	20.4
1 + ½	16	24	220	3.67	5,280	88.1
Totals.. 4.00		4)50			50)16,880:	50)208.6
	Av. *LPL*...	12.5		Av. T_s.	337.6	4.17 Av. T_d
Results by nomograph, using average *LPL**...					330.0	4.13

* This may be considered a very good check.

10.39. Regional Modification of Neal's Formulas. In view of the limited and local character of the projects on which they are based it has been found that regional modification of Neal's formulas, based on formal tests to fit soil conditions and characteristics of the given region, is apt to be necessary. In this connection, Fred H. Larson of the Soil Conservation Service staff of the Pacific Region, who has used Neal's method with success in the humid areas of the Pacific Northwest, says it has occasionally been necessary to apply the results for a given moisture equivalent or plastic limit in the Minnesota work to a value several per cent lower. For example, values of T_d and T_s shown in Neal's nomographs for an M_e, say of 35, may in some coastal areas be handled as though the M_e were 30. He feels that, if thus modified by a little study and experience, the Neal method will work well in tile drainage design in practically any humid area. As this book is being written there does not appear to have been found any other scientific index for tile drainage design generally applicable to humid-region conditions. Therefore, it appears that Neal's method at least points the way of progress and lifts the problem out of the realm of guesswork, based simply on the judgment of the engineer which, to be reliable, must rest on long years of experience that only a relatively few of the older engineers can be expected to possess.

THE DONNAN FORMULA[1]

A formula hereinafter called the Donnan formula has been developed and mathematically justified by Donnan [33]. It has been used with considerable success by Blaney, Donnan, and Bowen [14] for spacing of parallel tile laterals on the irrigated lands of the Imperial Valley, California. While this formula was expressly developed for use on irrigated lands it seems probable that, with some adjustments, it will also work satisfactorily, within the limits of the middle range of soils, on humid lands. (For further detailed discussion of this formula see Secs. 13.28 to 13.30.)

DRAINAGE OF LANDS WET FROM SEEPAGE

Drainage of lands wet from seepage has been earlier mentioned in Chap. 6 (see Secs. 6.21 and 6.22 and Figs. 6.9 and 6.10, where the proper treatment of the more simple forms of this problem are outlined). The following more serious and difficult cases occur.

If the surface blanket of pervious soil at the base of seeped slopes has a depth appreciably greater than good practice allows for tile drains in such soil, part of the water from the higher land will get by the marginal drains and gradually saturate the pervious layer on the adjacent low lands. In

[1] Derivation of this formula was first proposed by Aronovici and Donnan in 1945 and appears in Ref. 6.

this case a parallel system of drains will also be needed on the low lands, so placed as to prevent the escaped seepage from raising the water table into the root zone of crops.

Sometimes the pervious stratum, absorbing escaped seepage water from the higher lands, extends down *under* a less pervious blanket of soil. If it does not extend far enough under the less pervious surface layer on the low land to reach a *natural* outlet, pressure will be built up in the pervious layer which will force the soil water to the surface. If the less pervious surface soil is of a depth not greater than normal for tile drains, a parallel system of drains will relieve the pressure in the pervious substratum and prevent the saturation, due to pressure, of the surface stratum. However, in such a case if the depth of the less pervious surface layer is appreciably greater than normal for tile drains, in order to prevent the subsurface pressure from forcing some of the soil water upward past the parallel drains in the surface layer, it is frequently necessary to connect the surface system of tile drains to the deeper pervious layer where the pressure has built up by vertical *pressure-relief wells*, the spacing of which will be governed by the same factors of soil character which determine the design of the parallel system in the surface stratum. The relief wells and the parallel drains in the surface stratum may work together to allow a much wider spacing than normal for the latter alone.

SPECIAL STRUCTURES

All special structures should be included in the design of the system.

10.40. Surface Inlets and Catch Basins. On farm lands, almost everywhere many surface pockets occur ranging in area from a few square feet to several acres and in depth from a few inches to several feet. Following

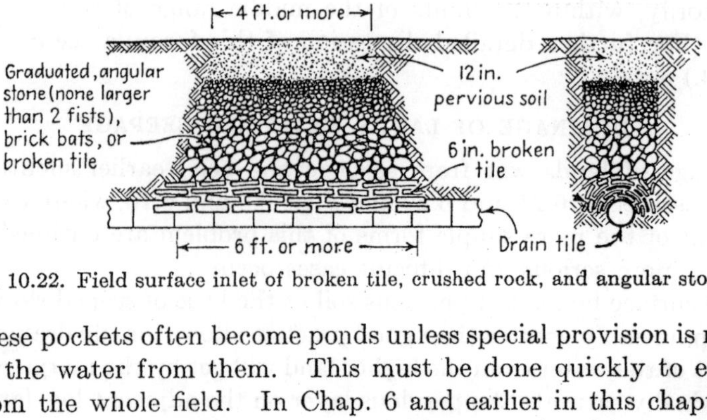

FIG. 10.22. Field surface inlet of broken tile, crushed rock, and angular stones.

rains these pockets often become ponds unless special provision is made to remove the water from them. This must be done quickly to ensure a crop from the whole field. In Chap. 6 and earlier in this chapter it is explained how the tile drains must be laid to care for this problem, but the tile drains alone will not remove the ponded water quickly enough to obviate injury to crops. Therefore, in cases not economically cared for by drainageways on the surface (see Secs. 6.4 and 6.5), a surface inlet into the tile directly over the tile line at the lowest point in each pocket should be provided. Figure 10.22 shows a type of field surface inlet widely and effectively used, easily installed, and inexpensive. It can be farmed over without lifting the tillage implement from the soil. For permanent pastures, barnyards, and residential grounds, Fig. 10.23 shows a catch basin that removes surface water quickly without danger of clogging the tile drains. In place of the corrugated metal standpipe, lengths of sewer tile or monolithic concrete may be used.

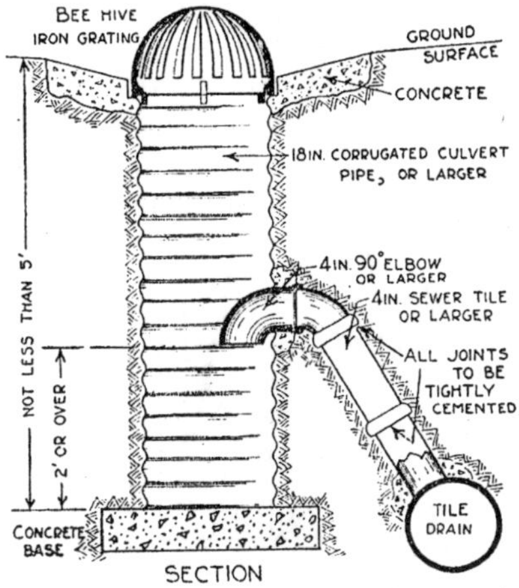

FIG. 10.23. Catch basin for use in yards and gutters.

10.41. Silt Basins. Where considerable silt is unavoidably washed into the tile, or where there is a change of grade sufficient to cause suspended silt to be deposited, silt basins are used to collect it (see Fig. 10.24). It is essential that the bottom of the basin extend 1 or 2 feet below the tile grade to provide a place for the silt to collect. The surface-inlet feature can be included if desired. Otherwise the basin should have a removable cover for inspection and cleaning which may be buried below plow depth if desired. It is sometimes advantageous to construct a concrete bowl around the top just below the iron grating to serve as a silt trap. Inlets and silt basins require inspection and cleaning at reasonable intervals to keep them in good working order.

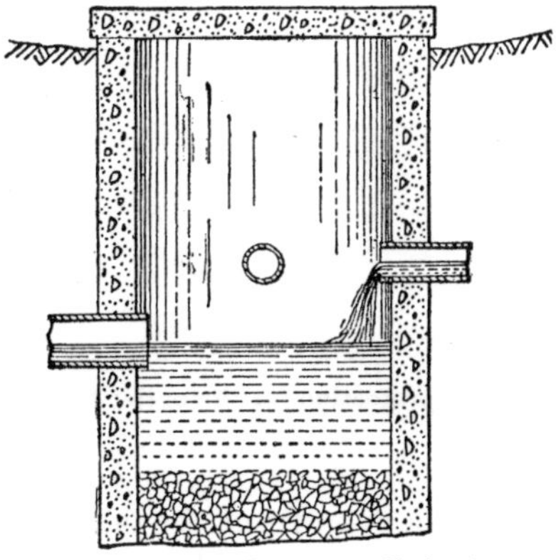

FIG. 10.24. Concrete silt basin.

10.42. Outlet Protection. The individual short tiles should never be extended to the outlet of a drain because the water running out of the end tile gradually washes away the soil under it and allows it to fall into the ditch. The same thing happens to each succeeding tile until the end of the drain recedes far into the field and starts what is apt to become a bad gully. The end of the drain should be protected against undermining for 15 or 20 feet back from the outlet. There are several ways to do this.

The essential thing is to make protections permanent. If the tile line ends at or near the bottom of the outlet channel with little or no drop from the tile to the stream or but little side current in the stream to cause scouring and undermining, a cheap and satisfactory protection is 15 or 20 feet of corrugated iron culvert of good grade at the tile outlet. The joint between the last tile and the culvert should be tightly cemented, and the culvert itself should be laid on a firm bed of natural soil. A ring of ½-inch round iron rolled into the outer end of the culvert protects against crush-

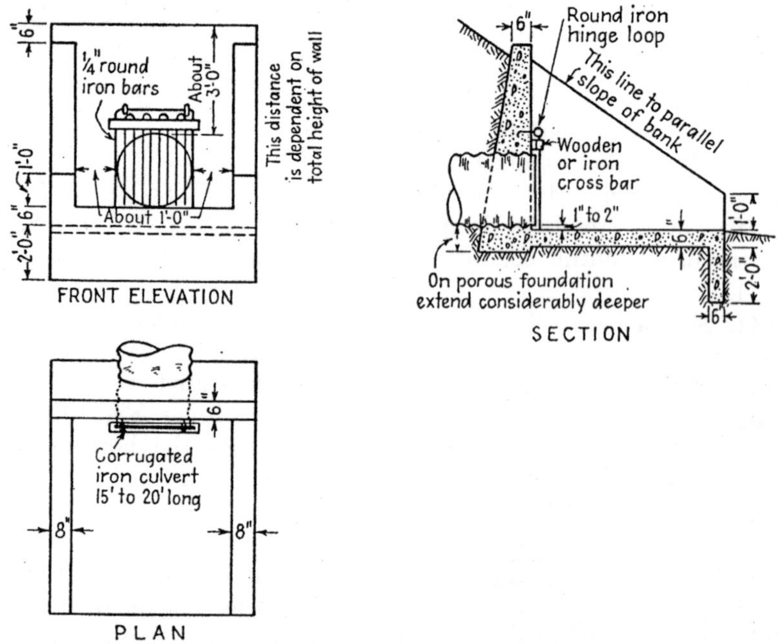

FIG. 10.25. A simple effective plan of a concrete bulkhead protection for the outlet to a tile drain.

ing by trampling of livestock. Where there is a drop from the outlet of the tile to the bed of the outlet channel, the tile outlet should be protected with a head wall or bulkhead and apron (see Fig. 10.25).

A screen is necessary to keep out living creatures that are, otherwise, a dangerous cause of clogging. The screen may be of fine-mesh poultry netting, or it may be a grating of slender iron bars spaced not over an inch apart. These bars may be set in the concrete of the bulkhead or in a frame so fastened that the whole grating may be removed or lifted to clean out the mouth of the drain and again replaced.

10.43. Protection against Backwater. This protection is necessary only where a submerged outlet is used. In placing tile drains, the question of the outlet should be carefully considered, and whenever possible, a submerged outlet should be avoided. But where it must be used, protection should be taken against the floodwater backing up into the tile. Floodwater carries considerable sediment, and if allowed to back up into the tile and become stationary, it will drop the sediment, thus filling up the tile.

The automatic backwater gate is the best protection against this. The gate is automatic in its action; when the water in the open ditch is of sufficient height to prevent the water in the tile from running, it closes. When not, it opens and allows the water in the tile to drain out.

Such a gate is manufactured commercially, is generally sold already attached to a long corrugated iron pipe, and has proved very satisfactory, especially where outlets are subjected to periodic inundation (see Fig. 6.8).

10.44. Relief Ditches. Even where surface inlets are provided, tile drains will not remove the surface water fast enough to prevent damage to crops from excessive rainstorms in rolling country nor even from ordinary rainstorms on flat lands infested with pockets that tend to become ponds. In such cases the main tile lines should be followed, directly over or near them, by the broad V-shaped field drainageways which are discussed in Sec. 6.4, to relieve the tile drains of an overload which reduces their efficiency. These drainageways will keep the water moving and will remove it quickly from the fields. Near their outlets the drainageways should be well to one side of the tile drains and have their own separate discharging points into the main outlet channel in order to avert danger of washing out the tile outlet.

10.45. Crossings of Railroads, Highways, and Open Ditches. Where a railroad, highway, or open ditch crosses over a tile main, the entire length of the main under such construction should be of either heavy corrugated iron pipe or extra-heavy culvert pipe of hard-burned shale or of concrete, and all joints on such tile should be strongly cemented. When such structures pass over a field served by or planned to be served by a complete tile drainage system, the latter should be so designed as to reduce the crossings to the fewest possible. By means of additional submains on either side of railroads, roads, and ditches, such crossings can often be reduced to a single one on any farm. The reaches of this one line under the other structure should be handled as just outlined for passage over a single main, or else they should be carried under the other structure within a conduit of corrugated metal pipe of considerably larger diameter than the tile to permit the latter to be sufficiently packed in earth to prevent the individual tiles from being crowded out of line by any such agency as vibration or frost. The ends of such conduit should be completely closed outside the tile with a plug of impermeable concrete. Such installations should be frequently inspected.

All crossings of this kind require approval, in advance, of the proposed design by the authorities responsible for maintenance of the right-of-way.

ILLUSTRATIVE EXAMPLE IN TILE MAIN DESIGN

Objective and Assumptions. The purpose of this example is to illustrate the use of the chart in Fig. 10.11 (solving the U.S. Department of Agriculture formula) dealing solely with mains and submains. As no scale is given on the map (see Fig. 10.26) measurements of the main seem to give a scale of 800 ft = 1 in., which will be assumed. As no contours are shown it will be necessary to assume that the area drained is uniform in each subbasin for each 100 ft of length of main and submains. Owing to the evident inaccuracy of this assumption the results worked out will not be strictly correct according to the actual shape of the various subbasins, but this fact will not affect the correctness of the method of using the chart. As the chart is based on an n value of 0.01077, whereas the example calls for an n value of 0.013, acreage values obtained direct from the chart must be multiplied by the reduction coefficient 0.829 (see Table 10.6). Determination of *depth* of main or submains does not enter into the problem as stated. Only the sizes of tile required and the length in feet of each size are required to be determined. The solution is carried out from the *head* of each line, downstream. The D value for the submains and for the main from D to C is $\frac{3}{8}$ in., and from C to A $\frac{1}{2}$ in., in 24 hr.

Solution:

The Main, AD. Section DC: gradient 0.12%, length 1,200 ft, n value 0.013, D – $\frac{3}{8}$ in. in 24 hr or 0.01575 cfs per acre.

Try 5-in. tile. Entering the chart (Fig. 10.11) on the slope axis (at bottom) at 0.12, following vertically upward to intersection with the 5-in. diagonal scale, thence on a line horizontally to the left to intersection with the discharge axis at 0.145 cfs, and to the right to the acre-axis in the $D = \frac{3}{8}$ in. column at 9.2 acres, check the discharge and acreage readings obtained from the chart, against each other as follows: for a 5-in. tile the maximum area it will drain or

$$A = \frac{0.145}{0.01575} \times 0.829 = 9.2 \times 0.829 = 7.6 \text{ acres}$$

This is virtually one-half the area of the subbasin served by the section CD of the main, which by inspection is estimated to extend 500 ft down from D toward C. That is, length of 5-in. tile required is 500 lin ft.

Entering the chart in a similar way for a 6-in. tile,

$$A = \frac{0.230}{0.01575} \times 0.829 = 14.6 \times 0.829 = 12.1 \text{ acres}$$

estimated to reach down from D a distance of 800 ft toward C, making the amount of 6-in. tile called for 800 − 500, or 300 lin ft.

Proceeding in the same way for a 7-in. tile,

$$A = \frac{0.35}{0.01575} \times 0.829 = 22.2 \times 0.829 = 18.4 \text{ acres}$$

which takes it to point C but not beyond as 40 acres enters in the lump discharge from submain 2 at the upper left; hence the length of 7-in. tile required is 1,000 − 800 = 200 lin ft.

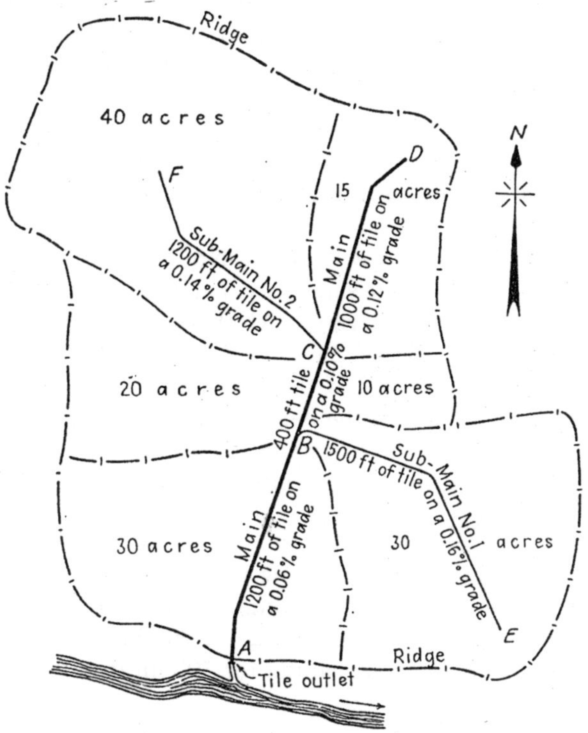

FIG. 10.26. Sketch to accompany illustrative problem in tile main design.

Proceeding in the same way from point C to B, just below C the total area to be cared for is 55 acres, the grade changes to 0.10% and the D value changes to ½ in. in 24 hr or 0.021 cfs per acre. Entering the chart again at the new values of grade and D, a 12-in. tile proves too small; hence, trying a 14-in. tile in the same way,

$$A = \frac{0.20}{0.021} \times 0.829 = 95.2 \times 0.829 = 79 \text{ acres}$$

which takes it from point C toward point B, (79 − 55)/(20 + 10) or four-fifths of the 400 ft reach, or the length of 14-in. tile called for is 320 lin ft.

Similarly for a 16-in. tile,

$$A = \frac{0.285}{0.021} \times 0.829 = 135.1 \times 0.829 = 112 \text{ acres}$$

which will evidently carry it to point B but not below it as the total area immediately increases at B to 115 acres; hence the amount of 16-in. tile required is 400 − 320, or 80 lin ft.

For the reach on the main from B to A (the outlet) the area starts with 115 acres at B and increases to 145 by the time it reaches the outlet. The gradient on this reach is 0.06%, and the D value remains at 0.021 cfs per acre. Successive trials with this new gradient in the manner shown above call successively with 18- and 20-in. tile for a total length of 18-in. tile of 120 lin ft capable of serving 118 acres; and for a total length of 20-in. tile of 1,080 lin ft capable of serving 162 acres but having to serve only 145 acres, the total of the project.

The Submains (No. 1 from E to B and No. 2 from F to C). These are each handled separately in the same way, beginning at the head, that is at point E and point F, respectively. At the head, especially on FC, the acreage served by each size of tile must be judged with discretion especially at the upper end, over half the area evidently being served by the portion of the submain above the bend. The process is the same as that repeated several times in the foregoing; hence need not be repeated here for the submains. Results worked out in each case as follows:

Submain 1 (E to B):

5-in. tile, serving 9.3 acres, calling for 500 lin ft
6-in. tile, serving 14.7 acres, calling for 100 lin ft
7-in. tile, serving 21.5 acres, calling for 200 lin ft
8-in. tile, serving 31 acres, calling for 700 lin ft

Submain 2 (F to C):

Anything less than 8-in. too small
8-in. tile, serving 28.2 acres, calling for 340 lin ft
9-in. tile, serving 39 acres, calling for 340 lin ft
10-in. tile, capable of serving 52.7 acres, but having to serve only 40 acres, calling for 520 lin ft

Five-inch tile is considered too small for any main, and the odd sizes of 7 in. and 9 in. are seldom obtainable as few manufacturers make them. Hence it is necessary to replace those called for by strict design as follows: 5 by 6 in., 7 by 8 in., and 9 by 10 in. On this basis the foregoing design can be summarized as follows:

Diameter of tile, in.	Main, lin ft	Submain *BE*, lin ft	Submain *CF*, lin ft	Grand totals, lin ft
6	800	600	.....	1,400
8	200	900	340	1,440
10	.....	.....	860	860
14	320	.....	.....	320
16	80	.....	.....	80
18	120	.....	.....	120
20	1,080	.....	.....	1,080
Grand total.......	2,600	1,500	1,200	5,300

10.46. Sample Drainage Report. The following report of a system of drainage worked out for a typical wet and fertile farm is given with the accompanying map to show what such a report should contain:

DRAINAGE REPORT FOR AUBURN PLACE (SEE FIG. 10.27)

Introduction

By arrangement with the owner, __________, drainage engineer of __________, visited a farm of __________, known as Auburn Place, for the purpose of making surveys and collecting such other data as might be necessary to plan a comprehensive system of tile drains. The farm in question is located in the south-west corner of Calhoun county, Iowa, and is described as a portion of sections 30 and 31, township 86 north; range 24 west. It contains approximately 387 acres.

Topography and Soil

Situated as it is in the heart of the Wisconsin drift area which stretches across 17 counties in north central Iowa, the topography of this tract is made up of a series of undulating ridges and flat depressions which have no natural drainage outlet.

The soil consists of a black sandy loam, tight and waxy in spots, underlain by a light-colored clay subsoil mixed with considerable sand, gravel, and small boulders. This soil is of a type that responds quite readily to the action of underdrains.

Existing Drainage Conditions

In making a study of existing drainage conditions, the matter of outlet facilities came up for first consideration. It was found that wherever an outlet was needed, it had been provided—either by a county drain or by the termination of a property line on a steep hill. In fact, a drainage assessment of nearly $2,000 stands against this property as its share of the construction expense of the outlet drains. Notwithstanding this fact, nothing has been done to make use of the outlets by constructing the tributary laterals so necessary for thorough field drainage. As a consequence, a large portion of the farm remains in its original state of open ponds and native slough grass.

Cultivated fields are cut up by numerous tracts too wet to sustain any growth of crops. These wet spots of irregular outline thus become a double liability in reducing the area of production and in yielding no return on the labor of planting and cultivation.

Catch Basins

Four intakes or catch basins for the ready accommodation of surface water have been constructed by the county and were inspected. Only one of these was found in good condition, the other three having been stopped up by débris or else washed out and undermined by floods. The repair of these inlets is a matter of serious concern and should be brought to the attention of the county supervisors without delay.

Stoppage by Roots

In some places, roots from willow trees have entered the tile and practically destroyed its capacity to carry water. Here the tile must be dug up, cleaned, and relaid, and the trees killed before installing any tributary drains. No tile closer than 125 feet to such a tree is safe from willow roots.

The Drainage Plan

A map has been prepared showing the usual topographic features with contours at intervals of 1 foot. The proposed lateral drains appear in bold relief, and each line has been inscribed with its name, length, and size of tile. The arrangement indicated has been worked out after a very careful study of the entire situation, and it is believed that they represent the most effective as well as the most economical system that can be devised. Nothing less than all the drains shown will yield maximum return for each dollar invested and provide adequate drainage for every square foot of the farm.

Piecemeal Construction

It is recommended that all the drains be installed under one contract for best results, but, if other considerations make this impracticable, piecemeal construction will afford some relief. It is important to keep the general plan always in mind and plan each season's work so that eventually the systems as shown will be complete. In selecting the drains for immediate installation, preference should be given those that skirt the foot of slopes and thus intercept the hill water before it has a chance to saturate the flat land below.

Special Problems

The outlets for systems *A* and *B*, on the north fence line, will be into steep draws leading to another county drain farther north. Substantial walls or equivalent should be built to protect the last few pieces of tile, and for 25 feet back from the outlets vitrified sewer pipe or equivalent should be used to guard against damage by frost.

The drains of systems *A*, *B*, and *C* could be made to operate most efficiently in conjunction with a series of terraces, as outlets for the water thus retarded. Some of the slopes in this region are as steep as 5 feet per 100, which is sure to cause trouble from washing sooner or later. It is at least extremely unwise to

CHAPTER 11

DRAIN TILES: KINDS, CLASSES, ESTABLISHED SIZES, AND STANDARDS OF QUALITY

By

Dalton G. Miller[1]

One of the principal materials used in drainage development is drain tile. There are two main kinds in common use, namely, clay tile and concrete tile. Their characteristics, both as to physical requirements and safeguards against deterioration, will be discussed in this chapter. It is important to observe the simple principles outlined in order to ensure a long-lasting tile-drainage system.

11.1. Origin of Specifications for Drain Tile. There were no standard specifications for drain tile prior to 1911. In March of that year, Committee C-6 of the American Society for Testing Materials, on Standard Tests and Specifications for Drain Tile, was established. This committee was active, from its inception, under the able leadership of Dean Anson Marston who remained as chairman until a short time before his death in 1949. It was through the work of this committee that the first standard specifications for drain tile were prepared and issued under the ASTM Designation C-4. These specifications, adopted in 1914 and revised in 1916, 1921, and 1924, were rewritten under the auspices of Subcommittee 10 of Committee C-15, Manufactured Masonry Units, and now bear ASTM Designation C4-50T [5]. Few of the essential provisions of C4-50T differ greatly from those of Designation C-4, as the principal change was deletion of the class, farm drain tile.

11.2. Classes of Drain Tile. The new specifications now cover two classes of drain tile as follows:

Standard drain tile, intended for ordinary land drainage where the tile are laid in trenches of moderate depths and widths.

Extra-quality drain tile, intended for land drainage where the tile are laid in trenches of considerable depths or widths, or both, and where extra quality is desired.

Physical requirements for these two classes are recorded in Tables 11.1 and 11.2.

[1] Materials Engineer (retired) Bureau of Public Roads, U.S. Department of Commerce.

TABLE 11.1.* PHYSICAL TEST REQUIREMENTS FOR DIFFERENT CLASSES OF DRAIN TILE

Internal diameter of tile, in.	Standard drain tile				Extra-quality drain tile			
	Minimum average crushing strength, lb per lin ft		Maximum average absorption by the 5-hr boiling method, per cent		Minimum average crushing strength, lb per lin ft		Maximum average absorption by the 5-hr boiling method, per cent	
	Test-bearing method		Kind of tile		Test-bearing method		Kind of tile	
	Three edge	Sand	Clay	Concrete	Three edge	Sand	Clay	Concrete
4	800	1,200	13	10	1,100	1,600	11	8
5	800	1,200	13	10	1,100	1,600	11	8
6	800	1,200	13	10	1,100	1,600	11	8
8	800	1,200	13	10	1,100	1,600	11	8
10	800	1,200	13	10	1,100	1,600	11	8
12	800	1,200	13	10	1,100	1,600	11	8
15	870	1,300	13	10	1,100	1,600	11	8
18	930	1,400	13	10	1,200	1,800	11	8
21	1,000	1,550	13	10	1,400	2,100	11	8
24	1,130	1,700	13	10	1,600	2,400	11	8
27	1,230	1,850	13	10	1,800	2,700	11	8
30	1,330	2,000	13	10	2,000	3,000	11	8
33	1,430	2,150	13	10	2,200	3,300	11	8
36	1,530	2,300	13	10	2,400	3,600	11	8
42	1,730	2,600	13	10	2,800	4,200	11	8

* Table 1 of ASTM Designation C4-50T.

Drain tile meeting these requirements are not necessarily safe against cracking in ditches wider or deeper than ordinary.

When the freezing and thawing test is made as provided in Section 25, ASTM Designation C4-50T, the number of reversals (freezing and thawing) shall be as follows: for Standard Drain Tile, 36; for Extra-quality Drain Tile, 48.

11.3. Established Sizes of Drain Tile. Established sizes of drain tile are also shown in Table 11.1.

As in the old specifications, C4-50T contains a clause stipulating that the purchaser may include special requirements for drain tile to resist damage where soils are markedly acid or where they contain unusual quantities of soil sulfates. In the new specifications, however, the expressions "markedly acid" and "unusual quantities of soil sulfates" are evaluated by the following notes:

"Soils or drainage waters with pH of 6.0 or lower may be considered to be markedly acid."

"Where the sulfates are chiefly sodium or magnesium, singly or in combination, unusual quantities may be assumed to be 3,000 ppm (0.30 per cent) for soil or soil water."

There may be differences of opinion as to the reliability of these exact values as indicators of corrosiveness, but they furnish specific bases from which conclusions may be drawn. It is not to be assumed that tile which meet the requirements of Table 11.1 for Extra-quality Drain Tile will satisfactorily meet all soil conditions and load requirements, as will be discussed.

TABLE 11.2. DISTINCTIVE GENERAL PHYSICAL PROPERTIES OF THE TWO CLASSES OF DRAIN TILE

Physical properties specified	Standard drain tile	Extra-quality drain tile
Permissible variation of average diameter below specified diameter, per cent	3	3
Permissible variation between maximum and minimum diameters of same tile, per cent of thickness of wall	75	65
Permissible variation of average length below specified length, per cent	3	3
Permissible variation from straightness, per cent of length	3	3
Permissible thickness of exterior blisters, lumps, and flakes which do not weaken tile and are few in number, per cent of thickness of wall	20	15
Permissible diameters of above blisters, lumps, and flakes per cent of inside diameter	15	10
General inspection	Rigid	Very rigid

11.4. Kinds of Drain Tile. On the assumption that the type of clayey raw materials is not of vital importance provided the finished product meets the physical requirements stipulated, the new specifications, C4-50T, list all drain tile as clay or concrete without attempting to define shale, fire clays, or surface clays.

SPECIAL REQUIREMENTS FOR DRAIN TILE

The following special conditions of exposure necessitate that drain tile have certain qualifications not definitely prescribed by C4-50T of the American Society for Testing Materials: (1) oversize trenches, (2) severe conditions of freezing and thawing, (3) soil acids, (4) soil sulfates.

Tile for Oversize Trenches

For a number of years the trend in the drainage field has been to install tile relatively deep in machine-dug trenches considerably wider than those formerly dug by hand. This is particularly true for deep cuts made with "back hoes" and other wide bucket-type excavating machines. These conditions all tend materially to increase the static loads on tile. Furthermore the use of much heavier farm machinery than formerly

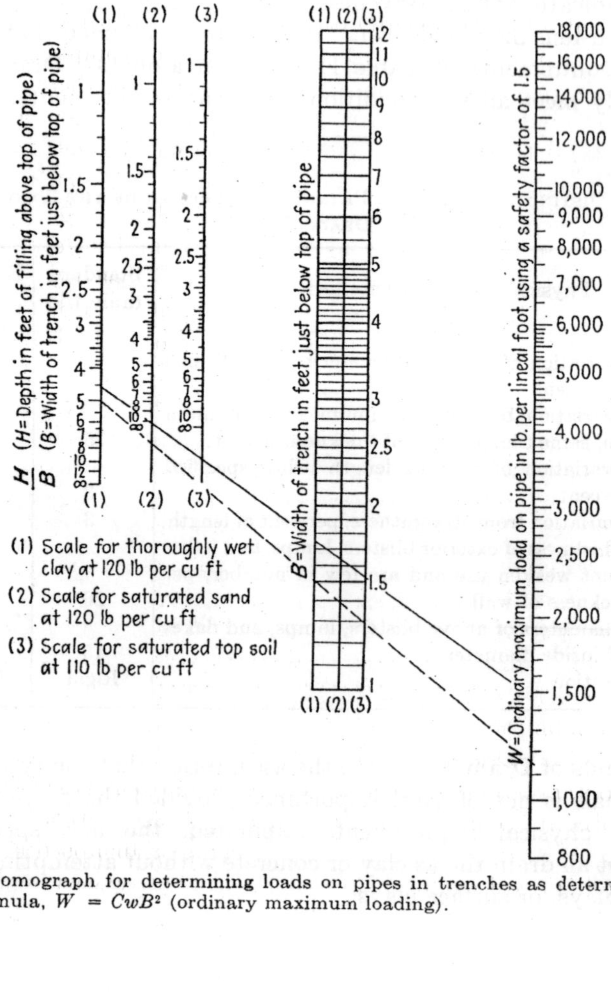

Fig. 11.1. Nomograph for determining loads on pipes in trenches as determined by the Marston formula, $W = CwB^2$ (ordinary maximum loading).

employed has increased the superimposed load factor. It is, therefore, of extreme importance that the probable loads be calculated and considered along with strength tests of the tile to be used whenever there arises any question of the sufficiency of tile strengths, in order that cracking of tile in the trenches be reduced to a minimum.

11.5. The Marston Formula. The basic work for the determination of loads on pipes in trenches was reported in 1913 by Marston and Anderson [80]. The conclusions of this bulletin are still accepted as fundamentally sound, and it is not known that any work has been done since which greatly modifies them. As a result of these early studies there was developed the following working equation, known as the Marston formula:

$$W = CwB^2 \tag{27}$$

where W = total weight on pipe, pounds per unit of length
w = weight of ditch filling, pounds per unit of volume
B = breadth of the trench in feet a little below the top of the pipe
C = coefficient of loads on pipes in ditches and is the average vertical pressure in pounds per unit area in a ditch of unit width under a ditch-filling material weighing unity per unit volume

Determination of the value of C is involved, mathematically; for actual calculations of loads, an approximate value of C may be taken from Table 11.3.

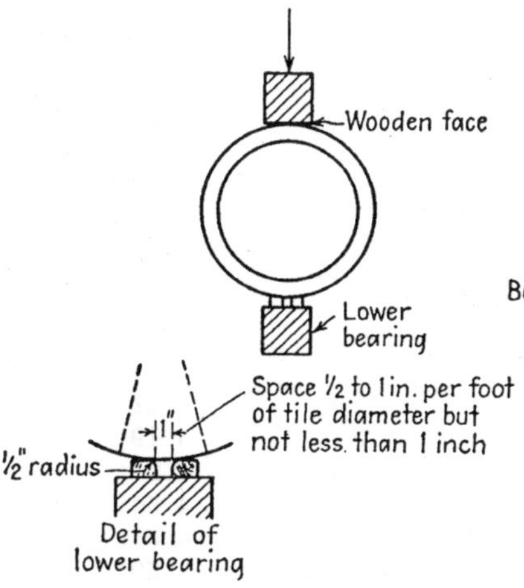

FIG. 11.2. The three-point bearing for testing tile.

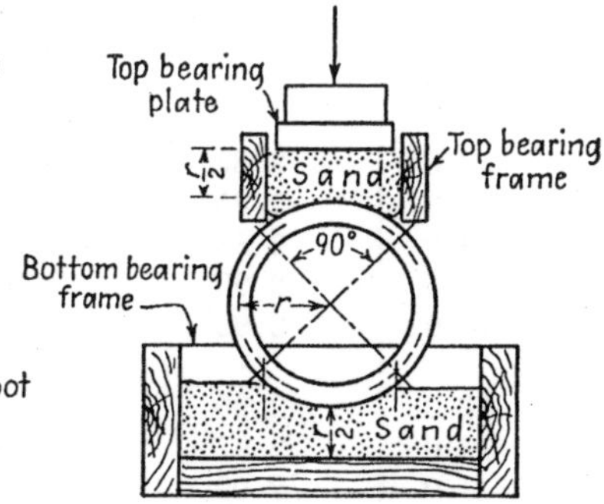

FIG. 11.3. The sand bearing for testing tile.

11.6. The Miller and Wise Load Nomograph. In order to make available a simple means for ready determination of the ordinary load requirements of pipes in trenches, Miller and Wise [91] in 1928 presented the nomograph shown in Fig. 11.1 applicable to thoroughly wet clay at 120 pounds per cubic foot, saturated sand at 120 pounds per cubic foot, and saturated topsoil at 110 pounds per cubic foot.

The values derived from this nomograph are the calculated loads multiplied by a safety factor of 1.5. In order to meet these load requirements,

TABLE 11.3. APPROXIMATE SAFE WORKING VALUES OF C, THE COEFFICIENT OF LOADS ON PIPES IN DITCHES

Ratio H/B	Approximate values of C			
	For damp topsoil and dry and wet sand	For saturated topsoil	For damp yellow clay	For saturated yellow clay
0.5	0.46	0.47	0.47	0.48
1.0	0.85	0.86	0.88	0.90
1.5	1.18	1.21	1.25	1.27
2.0	1.47	1.51	1.56	1.62
2.5	1.70	1.77	1.83	1.91
3.0	1.90	1.99	2.08	2.19
3.5	2.08	2.18	2.28	2.43
4.0	2.22	2.35	2.47	2.65
4.5	2.34	2.49	2.63	2.85
5.0	2.45	2.61	2.78	3.02
5.5	2.54	2.72	2.90	3.18
6.0	2.61	2.81	3.01	3.32
6.5	2.68	2.89	3.11	3.44
7.0	2.73	2.95	3.19	3.55
7.5	2.78	3.01	3.27	3.65
8.0	2.82	3.06	3.33	3.74
8.5	2.85	3.10	3.39	3.82
9.0	2.88	3.14	3.44	3.89
9.5	2.90	3.18	3.48	3.96
10.0	2.92	3.20	3.52	4.01
11.0	2.95	3.25	3.58	4.11
12.0	2.97	3.28	3.63	4.19
13.0	2.99	3.31	3.67	4.25
14.0	3.00	3.33	3.79	4.50
15.0	3.01	3.34	3.72	4.34
Infinity	3.03	3.38	3.79	4.50

tile tested by the three-edge bearing (Fig. 11.2) method should have crushing strengths at least two-thirds as great, while tile tested by the sand-bearing (Fig. 11.3) method should have equivalent or greater strengths.

TABLE 11.4. MAXIMUM LOADS ON TILE WHICH RESULT FROM A COVER OF THOROUGHLY WET CLAY

Depth of cover over tile, ft	Inside diameter of tile, in.																					
	4		5		6		8		10		12		15		18		24		30		36	
	Trench width, in.	Load, lb per lin ft	Trench width, in.	Load, lb per lin ft	Trench width, in.	Load, lb per lin ft	Trench width, in.	Load, lb per lin ft	Trench width, in.	Load, lb per lin ft	Trench width, in.	Load, lb per lin ft	Trench width, in.	Load, lb per lin ft	Trench width, in.	Load, lb per lin ft	Trench width, in.	Load, lb per lin ft	Trench width, in.	Load, lb per lin ft	Trench width, in.	Load, lb per lin ft
2	12	270	13	320	15	360	18	450	21	530	23	620	26	690	29	770	36	1,000	42	1,170	50	1,490
3	13	430	15	510	17	580	20	740	24	880	28	1,040	31	1,240	34	1,340	41	1,690	47	1,970	54	2,200
4	14	580	17	700	18	780	22	1,010	26	1,220	29	1,440	34	1,690	39	1,930	46	2,380	53	2,770	59	3,170
5	15	710	18	870	20	1,020	24	1,300	27	1,570	31	1,850	36	2,270	41	2,600	50	3,240	56	3,650	63	4,170
6	16	890	19	1,060	21	1,200	25	1,610	29	1,960	33	2,320	38	2,730	44	3,170	53	4,060	63	4,860	67	5,250
8	18	1,200	21	1,460	23	1,620	28	2,200	32	2,710	36	3,130	42	3,690	47	4,390	58	5,710	68	6,760	76	7,690
10	20	1,460	22	1,830	25	2,090	29	2,650	34	3,360	38	3,930	44	4,760	50	5,670	62	7,240	73	8,790	82	10,120
12	21	1,800	24	2,160	26	2,480	31	3,280	36	3,950	41	4,860	47	5,920	53	6,890	65	8,850	77	11,020	87	12,710
14	23	2,090	25	2,530	28	2,910	33	3,870	38	4,720	43	5,580	49	6,800	57	8,300	69	10,680	80	12,980	91	15,140
16	24	2,390	27	2,930	29	3,330	34	4,310	40	5,410	45	6,490	51	7,750	59	9,430	71	12,220	83	14,900	95	17,640
18	25	2,700	28	3,280	31	3,790	36	4,950	42	6,140	47	7,420	53	8,910	61	10,600	73	14,000	86	16,910	97	19,860
20	27	3,060	30	3,660	32	4,210	37	5,390	43	6,230	49	8,370	55	9,780	63	11,810	76	15,550	89	19,130	101	22,500

The trench widths shown are minimum widths for maximum loads. Greater widths are without significance.

Loads are calculated by the Marston formula, $W = CwB^2$. A factor of safety of 1.5 is included in values indicated. C = coefficient based on type of fill material and value of H/B. w = 120 lb per cu ft for thoroughly wet clay (130 lb per cu ft recommended for extreme conditions). B = width of trench at top of tile.

Ordinary pipe laying is that whereby the underside of the tile is well bedded on soil for 60 to 90 deg of the circumference.

Caution. There is a limiting width of trench beyond which the static load of backfill does not increase on tile of a given diameter under a given depth of cover. Therefore, when loads are determined by the Miller and Wise nomograph, they should be checked against the limiting values recorded in Table 11.4 for thoroughly wet clay at 120 pounds per cubic foot.

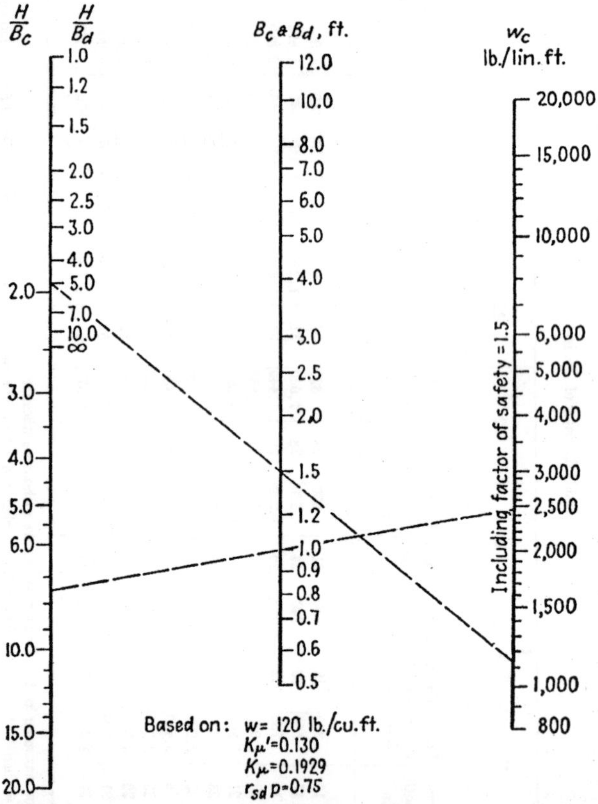

Fig. 11.4. Nomograph for loads on tile installed in thoroughly wet clay.

For other soil conditions the maximum loads may be calculated by the Marston formula, using the same maximum trench widths shown in Table 11.4 and the proper C indicated in Table 11.3. The C and w values are less for saturated sand and for saturated topsoil than for thoroughly wet clay; consequently, the loads will always be less for the same trench widths.

The limiting values in Table 11.4 were calculated on the assumption stated as follows on page 6 of Ref. 130:

As the width of the ditch increases, other conditions remaining constant, the load upon the conduit increases in accordance with the ditch conduit theory till it equals that by the projecting conduit theory, and then remains constant for all greater widths. . . . On this basis it is possible to calculate the transition width ratio for any given set of conditions by determining the width of ditch for which the ditch conduit load equals that by the projecting-conduit theory.

The formulas are the same general form for each condition, namely, $W = CwB^2$, but differ greatly in the value of C. When used together

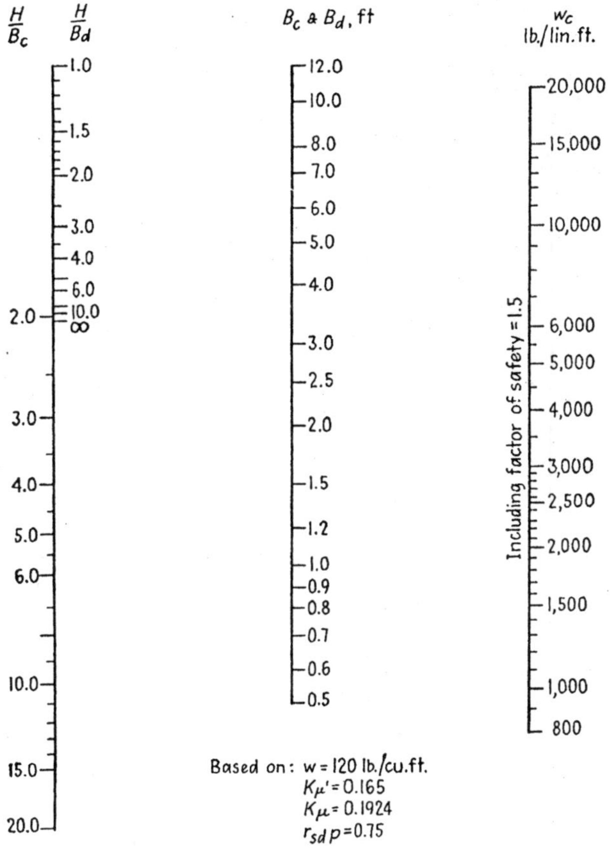

FIG. 11.5. Nomograph for loads on tile installed in saturated sand.

each factor in the formula is identified as to "ditch" or "projecting" condition by subscripts d and c, respectively.

11.7. The Schilfgaarde, Frevert, and Schlick Nomographs. Shilfgaarde, Frevert, and Schlick [129] deal with three aspects of the problem of tile loading. First, the Marston formulas for loads on underground conduits are presented so they can be solved by nomographs with a minimum of calculations. Second, some information is presented to show the probable loads to which tile may be subjected. The third aspect is the effect of bedding conditions on the supporting strength of the tile.

Three nomographs are presented in this paper (see Figs. 11.4 to 11.6). Figure 11.4 is for loads on tile installed in thoroughly wet clay, while Figs. 11.5 and 11.6 are, respectively, for tile installed in saturated sand and for tile installed in saturated topsoil. Each nomograph enables the user to determine both the loads on pipes in narrow trenches, using the "ditch-

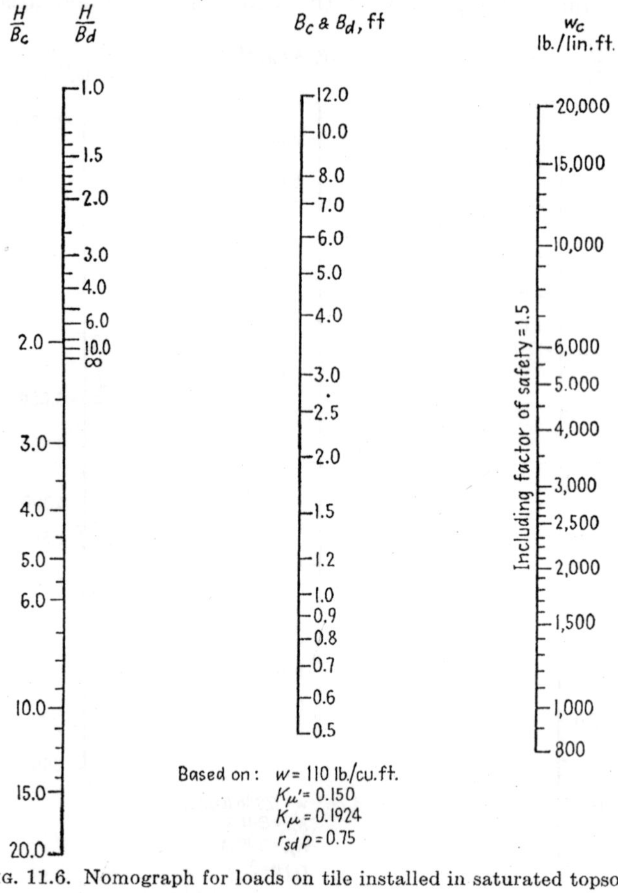

FIG. 11.6. Nomograph for loads on tile installed in saturated topsoil.

conduit load formula," which has been found, experimentally, by Schlick [130] also to apply to pipe in wide ditches. The lower value obtained represents the design load as it never will be possible to exceed the load calculated in accordance with the "projecting-conduit load formula." Note the following example.

EXAMPLE

A 10-inch tile is installed 8.5 feet deep in an 18-inch trench in thoroughly wet clay. The first nomograph (Fig. 11:4) applies in this case. Use $B_c = 1.0$ foot, $B_d = 1.5$ feet and, for the height of fill above the tile, $H = 8.5 - 1.0 = 7.5$ feet.

For $H/B_c = 7.5$ and $B_c = 1.0$, the nomograph gives $W_c = 2470$ lb. per lin. ft. For $H/B_d = 5$ and $B_c = 1.5 W_c = 1130$ lb per lin. ft. The lower value is the one to be used. Therefore the design load on the pipe is 1130 lb. per lin. ft.

When Fig. 11.1 is used, it will ordinarily expedite the procedure to refer to Table 11.4 first and find the load on a tile of the size and depth of cover under consideration, remembering that the trench widths shown are always those for maximum loads and that greater widths are without significance. Therefore, when the design trench width is less than shown in the table, the load on the tile will be less and the nomograph (Fig. 11.1) will give the correct load. Note the following examples.

EXAMPLES

1. Suppose an 8-in. tile is to be installed with a cover of 8 ft in a trench 21 in. wide at the top of the tile in thoroughly wet clay. What will be the load on the tile? Inspection of Table 11.4 shows that for an 8-ft cover the maximum load will be 2,200 lb per lin ft for a minimum trench width of 28 in. Since the width of the proposed trench is 21 in., or 7 in. less than that shown in the table, the nomograph (Fig. 11.1) should be used. When this is done, the calculated load is found to be between 1,400 and 1,500 lb per lin ft, indicating extra-quality drain tile or equivalent improvement in the bedding conditions.
2. Suppose a 6-in. tile is to be installed with a cover of 6 ft in a trench 24 in. wide at the top of the tile. What will be the load on the tile? Inspection of Table 11.4 shows that for a 6-ft cover the maximum load would be 1,200 lb per lin ft for a minimum trench width of 21 in. and the proposed trench width is greater. In this case the value given in the table should be used for the load.

It matters little whether the loads are determined by Fig. 11.1 with Table 11.4 or by Figs. 11.4 to 11.6 as the calculated loads will be essentially identical because, as shown by Chen [25], the bases are the same for these nomographs. It should be noted that all the calculations are based on "ordinary pipe laying" in which the underside of the pipe is well bedded on soil for 60 to 90 deg of the circumference. The permissible depth of cover will be different for other bedding conditions.

For all practical purposes Table 11.5 will be most helpful to field engineers and contractors.

Severe Conditions of Freezing and Thawing

11.8. Clay Tile. Clay tile must be of satisfactory quality to withstand freezing and thawing. The frost resistance of clay tile is largely dependent on the quality and handling of the raw clay or shale before and after burning. To resist frost action, tile must be well burned because doing so increases the density of the raw materials, resulting in a finished product

TABLE 11.5. ALLOWABLE DEPTHS FOR DRAIN TILE AND IRRIGATION PIPES

Tile size, in.	ASTM class	Crushing strengths, lb per lin ft	Width of trench, in.							
			18	20	22	24	26	28	30	36
5	Standard	1,200	8.7	7.5	7.5	7.5	7.5	7.5	7.5	7.5
	Extra Quality	1,600	Inf.	12.2	9.5	9.5	9.5	9.5	9.5	9.5
6	Standard	1,200	8.8	6.8	6.6	6.6	6.6	6.6	6.6	6.6
	Extra Quality	1,600	Inf.	12.3	8.7	8.4	8.4	8.4	8.4	8.4
	Irrigation Pipe	1,500	18.6	10.3	8.0	8.0	8.0	8.0	8.0	8.0
8	Standard	1,200	9.0	7.0	5.8	5.4	5.4	5.4	5.4	5.4
	Extra Quality	1,600	Inf.	12.5	8.8	7.3	6.9	6.9	6.9	6.9
	Irrigation Pipe	1,500	18.8	10.5	8.1	6.6	6.6	6.6	6.6	6.6
10	Standard	1,200	9.2	7.2	6.0	5.3	4.9	4.9	4.9	4.9
	Extra Quality	1,600	Inf.	12.7	9.0	7.5	6.6	6.0	6.0	6.0
	Irrigation Pipe	1,880	Inf.	24.4	12.4	9.6	8.2	7.0	6.8	6.8
12	Standard	1,200	9.4	7.3	6.2	5.5	5.0	4.7	4.7	4.7
	Extra Quality	1,600	Inf.	12.9	9.2	7.7	6.8	6.1	5.6	5.6
	Irrigation Pipe	2,250	Inf.	Inf.	26.8	14.0	11.1	9.2	8.1	6.3
15	Standard	1,300	....	8.6	7.1	6.3	5.8	5.2	4.9	4.7
	Extra Quality	1,600	....	13.1	9.5	8.0	7.1	6.3	5.8	5.2
	Irrigation Pipe	2,550	....	Inf.	Inf.	21.3	14.3	11.7	10.0	7.5
18	Standard	1,400	....	....	....	7.0	6.5	5.9	5.5	4.7
	Extra Quality	1,800	....	....	....	9.6	8.5	7.5	6.9	5.7
	Irrigation Pipe	2,850	....	....	....	Inf.	20.0	14.1	12.0	8.6
21	Standard	1,550	....	....	....	....	....	6.7	6.2	5.0
	Extra Quality	2,100	....	....	....	....	....	9.3	8.3	6.8
	Irrigation Pipe	3,150	....	....	....	....	....	18.3	14.3	10.0
24	Standard	1,700	....	....	....	....	....	....	7.1	5.9
	Extra Quality	2,400	....	....	....	....	....	....	10.0	7.9
	Irrigation Pipe	3,300	....	....	....	....	....	....	16.1	11.0

NOTES: 1. Values given for the various trench widths are depths of trench in feet.

2. Crushing strengths given are averages in pounds per linear foot based on sand-bearing method, Specifications for Drain Tile and Concrete Irrigation Pipe, ASTM Designations C4-50T and C118-39, respectively.

3. These values allow a safety factor of 1.5.

4. Width of trench is measured at top of tile.

5. Loadings were computed for wet clay soil at 120 lb per cu ft. Somewhat greater depths are permissible for lighter soils.

6. "Ordinary pipe laying" whereby the underside of the tile is well bedded on soil for 60 to 90 deg of the circumference.

7. Inf. indicates infinity.

of high strength, low absorption, and excellent frost resistance. Many clay-tile plants have raw materials of so high an order that the entire output is almost immune to frost action. Other plants can produce frost-resistant tile only by burning at the highest temperatures the clay will stand without too much deformation. Plants having the poorer clays cannot make frost-resistant tile at all. It is evident, therefore, that there are definite limitations to the degree of frost resistance which may be expected of clay tile from many tile plants (see Fig. 11.7).

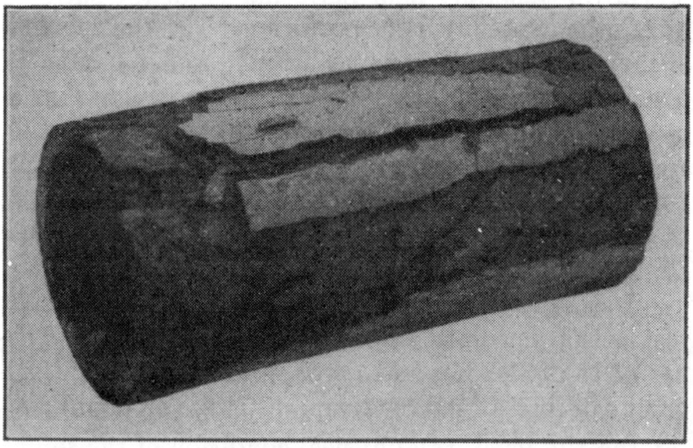

FIG. 11.7. A poor-quality clay drain tile with open laminations in its walls, disintegrated by freezing and thawing.

11.9. Frost-resistance Tests. The only reliable tests for frost resistance of drain tile are actual freezings and thawings made along the lines prescribed in the ASTM Designation C4-50T [5]. Tests of this type are tedious to make and require several days for completion. As a consequence, numerous efforts have been made by different laboratories to correlate the water absorption of clay tile with frost resistance. The general opinion seems to be that absorption-test results of clay tile from one plant compared with similar results on tile from other plants have little practical value. On the other hand, comparisons of absorptions of individual tile from any given plant are reasonably reliable as indicators of the relative frost resistance of the tile from that particular plant. Therefore, once this relationship has been established, tests for frost resistance of tile from a particular plant can be reduced to the absorption test, utilizing standard freezing and thawing tests only when necessary.

11.10. Absorption in Relation to Frost Resistance. The permissible maximum absorptions shown for clay tile in Table 11.1 are 13 per cent for standard drain tile and 11 per cent for extra-quality drain tile. These are based on tests of clay and shale tile from many plants, and it has been found that they are reasonably safe upper limits for frost resistance.

Effort was made in 1948 and 1949 to secure information as to the frost-resisting properties of clay drain tile installed in farm drainage systems in the upper Mississippi Valley. Tile were dug up from 19 systems which had been in service for 18 to 40 years, averaging slightly under 33 years. Tile from 10 of the 19 systems were presumably made of shale and 9 from surface clays. This work was reported in detail in February, 1951, by Miller and Manson [89]. Based on tests given in this report on 244 tile, three points stand out in particular: (1) the high resistance to frost action of the shale tile examined even though the depth of cover averaged but 1.7 feet; (2) the generally poor performance of the surface clay tile; and (3) the more satisfactory service from drain tile with fairly high absorptions, except at outlets, than would be predicted from tests made in the laboratory. These findings lead to the conclusion that exact data are needed as to depth and frequency of frost penetration in farm fields where most drain tile is installed.

Nearly all available frost-penetration data are extremes compiled for use in the design of foundations for buildings and other types of above-ground structures. Extremes, while of some use in connection with probable frost action on drain tile, do not by any means tell the whole story because of the following and perhaps other factors: (1) frequency and duration of extremes of winter temperatures; (2) depth and duration of snow cover; (3) depth and type of soil cover; (4) moisture content of the soil at the tile level; (5) physical condition of the soil cover, fall-plowed, etc.; (6) vegetative cover; (7) quantity and source of water carried by the tile. Much research is necessary before it will be possible to establish a fully satisfactory relationship between the frost resistance of clay drain tile as determined by laboratory tests and durability under field conditions.

It is always good insurance to install all clay tile without needless and undue abuse. They should not be allowed to lie on the surface of wet ground for long periods of time during cold weather. Such exposures will not appreciably affect high-quality tile but may be disastrous to borderline tile which otherwise might give satisfactory service for many years (see Fig. 11.7).

11.11. Concrete Drain Tile and Frost Resistance. Concrete drain tile which meet American Society for Testing Materials requirements [5] for standard drain tile may ordinarily be expected to have satisfactory frost resistance.

SOIL ACIDS

Alway [3] states that

It has long been known that the peat of many bogs shows an intensely acid reaction. Until recently it was generally accepted that this was due to the presence of very slightly soluble organic compounds of definite but undetermined composition, and these were referred to as humic acids, while the products obtained by neutralizing them with bases were known as humates. The water from such bogs shows various characteristics of acids, liberating hydrogen with metals and carbon dioxide from carbonates, and also decomposing insoluble phosphates and silicates. In recent years, however, it has become a subject of much controversy as to whether the acid characteristics both of such bog waters and of the peat itself are not rather due to expression of the colloidal properties of compounds in the peat, some of which are held in suspension in the water.

In this same paper Alway also cites (1) the occurrence of iron sulfide in peat mentioned by Rennie[1] in discussing "pyritous or vitriolic peat," and (2) the suggestion by Tacke [149] that hydrogen sulfide gas may form in the deeper nonaerated layers of every kind of peat as well as in mineral soils rich in organic matter.

Lea and Desch [72] reported that, "the main causes of acidity in natural waters are carbon dioxide and humic acids" and that, "the amount of free carbon dioxide which exceeds that necessary to maintain the chemical equilibrium is called aggressive carbon dioxide acid and this is what does the damage."

The following is a suggested procedure for determination of aggressive carbon dioxide acid.

TABLE 11.6. THE CORRESPONDING VALUES OF CALCIUM BICARBONATE AND FREE CARBON DIOXIDE WHEN IN EQUILIBRIUM

CO_2 present as $Ca(HCO_3)_2$, mg per liter	Free CO_2 required to stabilize bicarbonate, mg per liter	pH value of solution	CO_2 present as $Ca(HCO_3)_2$, mg per liter	Free CO_2 required to stabilize bicarbonate, mg per liter	pH value of solution
44	0.32	8.64	264	58.80	7.13
88	2.45	8.05	308	92.00	7.00
132	8.00	7.70	352	133.50	6.89
176	18.30	7.46	396	187.00	6.79
220	35.00	7.28	440	253.00	6.70

If the total content of CO_2 and calcium bicarbonate [$Ca(HCO_3)_2$] in a water is estimated, then with the aid of the following data, the amount of aggressive CO_2 present can be calculated. The $Ca(HCO_3)_2$ can be determined by direct titration with one-tenth normal[2] hydrochloric acid

[1] "Essays on the Natural History and Origin of Peat Soils," Edinburgh, 1810.

[2] A normal solution contains one gram-molecular weight of the dissolved substance divided by the hydrogen equivalent of the substance (that is, one gram equivalent) per liter of solution.

(N/10 HCl) with methyl orange as indicator: 1 cubic centimeter N/10 HCl = 0.0044 gram CO_2. The free CO_2 is determined by adding excess $BaCl_2$ and then excess $Ba(OH)_2$ and titrating back the excess $Ba(OH)_2$ with N/10 HCl, with phenolphthalein as indicator. The difference between the liter of the $Ba(OH)_2$ solution added and that of the final solution gives the free CO_2: 1 cubic centimeter N/10 HCl = 0.0022 gram CO_2. Table 11.6 gives the CO_2 requirement for stabilization for a considerable range of values of $Ca(HCO_3)_2$.

Certain soils which are not acidic in reaction themselves are capable of exchanging a hydrogen ion for a base ion by reaction with a salt solution and so forming an acid. The exchange acidity is determined by shaking 100 grams soil with 200 cubic centimeters normal sodium acetate solution for 1 hour and then titrating the solution with N/10 NaOH with phenolphthalein as indicator. The exchange acidity is defined as the number of cubic centimeters N/10 NaOH required.

11.12. Acid Action on Concrete Drain Tile. Acid solutions attack cement mortars or concrete by dissolving some of the cementing constituents. Lea and Desch [72] report that results of experiments by Werner showed that the leaching of lime from crushed set cement was about 14 per cent of the original weight of the cement. This, no doubt, is the reason for the etched appearance of the concrete drain tile which have been subjected to the action of soil acids (see Fig. 11.8).

For many years not much effort was made to answer the fundamental question of why concrete tile of some systems in acid soils disintegrated badly while harmful action on the tile of other systems was relatively slight. However, in 1923 a project was initiated to deal with this problem under a cooperative agreement between the University of Minnesota, the Minnesota State Department of Conservation, the U. S. Department of Agriculture and, informally, the University of Wisconsin. This project was active for 26 years, as originally set up except that Federal support, for the last 4 of the 26 years, was through the Federal Works Agency and the Department of Commerce. The final report by Miller and Manson [86] was issued in June, 1948. The conclusions of this report are based on tests of about 1,100 drain tile and 10,000 experimental concrete and mortar cylinders after field exposures in Minnesota, Wisconsin, and North Carolina for periods up to 26 years for some of the tile and for 17 to 20 years for the cylinders.

The following are some of the conclusions of this work abstracted from a later paper by Miller and Manson [87]:

1. Peat acidity as measured by pH determinations is a fair indicator of the degree of corrosive action to be expected on concrete.

2. Acidities of three mineral soils, as measured by pH determinations, did not furnish so reliable an index of corrosive action on concrete as for the peats. No exact explanation is offered for this apparent difference.

3. None of six admixtures of different types increased resistance of the cylinders in which used.

4. Variations in the chemistry of 20 different cements had little effect on resistance to the soil acids with the exception of alumina cements, which were somewhat more resistant than were any of the portlands.

5. Specimens cured in steam at temperatures of 155, 212, or 285 degrees Fahrenheit did not resist peat action appreciably better than did water-cured ones, but those cured at 345 degrees Fahrenheit held up in strength throughout the 17 years they were under observation.

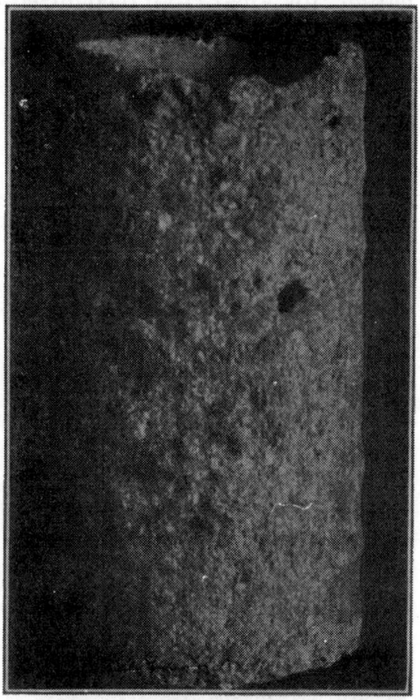

FIG. 11.8. A concrete tile destroyed by the acid in a high-acid peat soil.

6. The only surface treatments used were linseed oil and a cutback bituminous paint. Both of these somewhat increased the resistance of the concrete cylinders to which they were applied.

It was further concluded that the degree of corrosion varied with the acidity of the peat and the unit strength of the test piece, whether drain tile or cylinder. It is, therefore, not good practice to lay concrete drain tile of the smaller sizes, as ordinarily made, in the more acid peats. But concrete drain tile of extra quality or better should give satisfactory service for many years when laid in the low-acid grass and sedge peats that favorably respond to cultivation without much liming.

11.13. How to Make Acid-resistant Concrete Tile of Small Diameters. There are five basic requirements to be met in the manufacture of concrete drain tile of the smaller diameters, made on the "packer-head" machines in common use, if the output is to be high in quality as indicated by tests for supporting strength. These are (1) well-graded, sound and clean aggregates; (2) rich mixes; (3) proper amount of mixing water; (4) adequate compaction of the materials in the tile machine; and (5) adequate curing. These are manufacturing requirements; therefore, detailed discussion of them is outside the scope of this book. Those interested should consult Refs. 88, 90, and 84.

11.14. Acid Action on Clay Drain Tile. So far as known there is no deleterious action of ordinary soil acids on clay tile which meet the requirements of ASTM Specifications C4-50T [5].

SOIL SULFATES

11.15. Action of Soil Sulfates on Concrete. Since the invention of portland cement, about 1824, many studies have been directed to solving the problem of deterioration of concrete in contact with sulfate waters. Some of these studies by both the French and English were made even earlier; for Vicat [155] recognized the problem as far back as 1812 in connection with work, reported in 1818, on limes and natural cements, and many years earlier Smeaton [139] experimented to secure a mortar that would best resist the action of sea water when the Eddystone Lighthouse was built in 1756–1759. Since early in the twentieth century the combined efforts of many workers and the continual improvement in the techniques of cement chemistry have markedly extended the earlier knowledge of the complex nature of portland cement. As a result we can now speak with assurance concerning some of the factors which improve the resistance of cement to soil sulfates where previously it was possible only to theorize.

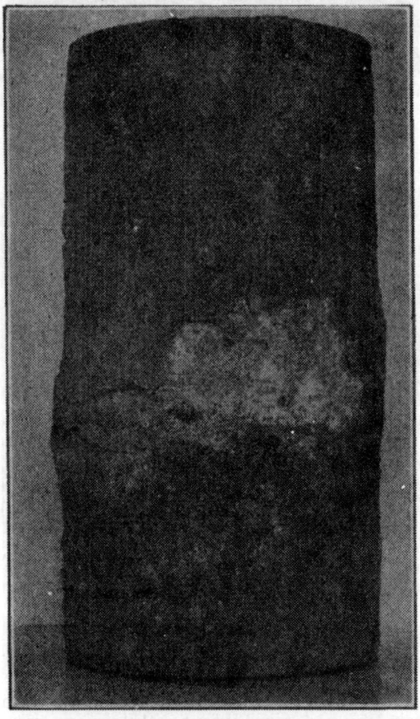

FIG. 11.9. A low-grade concrete tile disintegrated by alkali action.

Shelton [135] found that sodium and magnesium sulfate solutions, added to tricalcium aluminate ($3CaO \cdot Al_2O_3$), commonly abbreviated (C_3A), resulted in the formation of certain sulfur-aluminum compounds which, when formed in hydrated cement, cause marked increase in volume and are the chief reason for physical rupture of concrete and mortars made of sound aggregates exposed to sulfates (see Fig. 11.9).

As will be shown later, this whole problem is a leading consideration in the manufacture and use of concrete pipe—especially concrete drain tile and irrigation pipe.

11.16. Sulfate Action in Relation to Solution Strength and Types. The most deteriorating soil sulfates are of magnesium ($MgSO_4$) and sodium (Na_2SO_4). Miller and Manson [88,90] found that the destructive action of these salts increased rapidly with the strength of the solution up to 0.5 per cent, then at a diminishing rate. See Fig. 11.10 for strength decrease of cylinders stored in $MgSO_4$ solutions from 0.25 to 4.0 per cent.

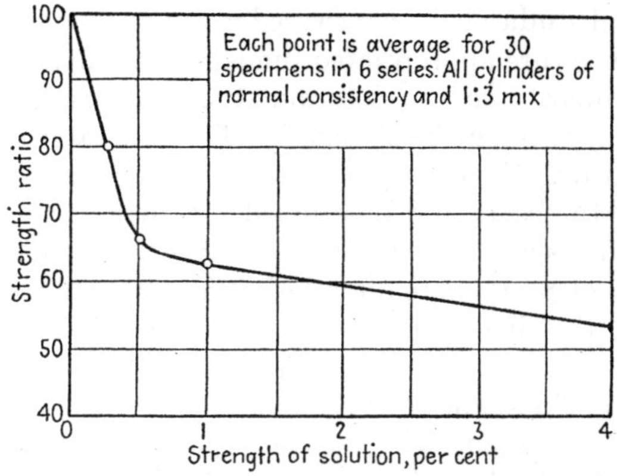

FIG. 11.10. Relative effects of magnesium sulfate solutions of different strengths on the compressive strengths of cylinders.

It was also observed that in reaction with the C_3A in the cement, the disruptive action of the Na_2SO_4 solutions was more severe than that of the $MgSO_4$ solutions for calculated values of the C_3A above 9 per cent, while below 9 per cent the situation was reversed and aggravated as the per cent of C_3A drops still lower (see Fig. 11.11). Thus, in general, the susceptibility of concretes and mortars to the action of $MgSO_4$ and Na_2SO_4 solutions is markedly lowered by reducing the C_3A content of the cement. The more severe action of the $MgSO_4$ solution on cements relatively low in C_3A may be attributed to the fact that the $MgSO_4$ reacts not only with the C_3A but also with the hydrated calcium silicates. However, for cements high in C_3A the disintegration is so rapid that the secondary action of $MgSO_4$ on the silicates becomes a minor matter.

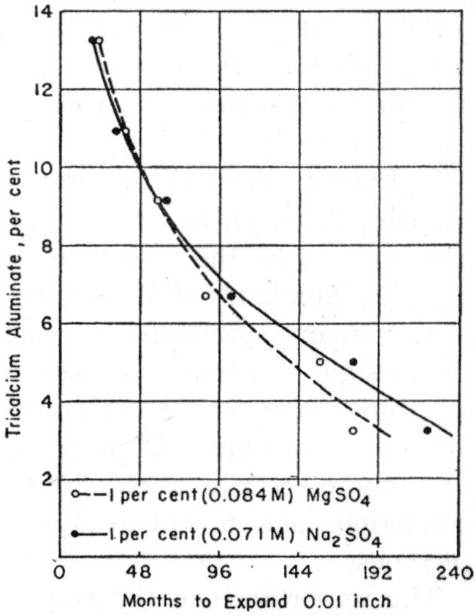

FIG. 11.11. Effect of variation in the C_3A content of portland cements on the expansion of cylinders stored in magnesium or sodium sulfate solutions.

11.17. Investigations and Research in Minnesota. Comprehensive studies were conducted in 1919 and 1920 of the failure of some concrete drain tile which occurred in a number of tile systems in southwestern Minnesota. The tile which failed ranged in size from 6 to 30 inches inside diameters. In July, 1921, a 74-page bulletin by Miller [84] was issued based on these studies which included chemical analyses of 1,062 water samples and 150 soil samples.

These original studies left unanswered many questions, and in 1921 the Minnesota Legislature appropriated funds for creation of a laboratory to be directed primarily to the determination, if possible, of how future failures of this type might be obviated. This laboratory was established in the Agricultural Engineering Building on the farm campus of the University of Minnesota and functioned for three decades through support of the agencies mentioned in Ref. 86, page 3, except that the state of Wisconsin did not participate directly in the sulfate phase of the studies. In May, 1951, the final report of this work was issued by Miller and Manson [90].

Based on the field studies of 1919 and 1920 it was concluded that

1. The trouble was caused by excessive quantities of sulfates in the soil in direct contact with the outer surface of the tile wall.

2. The danger limit of total salts in direct contact with the outer surface of concrete drain tile of the quality of the ordinary southwest Minnesota product at that time was about 3,000 parts per million (0.30 per cent) where the salts consisted almost wholly of $MgSO_4$, Na_2SO_4, and $CaSO_4$ (calcium sulfate), with the first two of these running near two-thirds or three-fourths of the total. The most pronounced failures occurred where the total salt content of the soil water exceeded 4,500 parts per million (0.45 per cent).

3. Except where disintegration had resulted from the action of soil sulfates, the concrete tile were giving satisfactory service throughout the area covered by the investigations.

11.18. Summary of Minnesota Laboratory Studies. Observations and tests of many specimens (see Ref. 90) have shown convincingly that, in the design of concrete mixes for maximum durability under sulfate exposure conditions, the primary consideration must be low permeability, usually a corollary of high unit strength. The concrete must be properly placed to avoid honeycombing, irrespective of the cause, as deleterious action will always be most pronounced at such points of weakness.

The importance of the chemistry of the cement has been shown so many times that it is now accepted as a fact, and standard specifications have been prepared for sulfate-resisting portland cements in both the United States and Canada, the essential provisions being maximum limitation on the percentage of the calculated compound C_3A. Based on the many tests reported in Ref. 90, the permissible limit for C_3A should not exceed 5.5 per cent.

The special type of cement known as alumina, or aluminous, proved to be extremely resistant to soil sulfates under conditions of exposure where the temperatures were mostly 70 degrees Fahrenheit and below.

High-temperature steam curing proved to be an extremely effective means of increasing the sulfate resistance of portland-cement concrete and mortar specimens. The best results followed curing at 350 degrees Fahrenheit for 6 hours and more.

Results with admixtures were largely disappointing. About the only exceptions were certain organic oils, such as linseed, soybean, and tung oil, emulsified as they were added to the batch.

None of the accelerated or short-time tests proved to be entirely satisfactory for the determination of the sulfate resistance of portland cements, independent of the chemical analyses. Seventeen such tests were explored. The lean-mortar bar made of a 1 to 5 mix of Ottawa standard sand, exposed to laboratory solutions, using measured length changes as the criteria for sulfate action shows the most promise. All the details for a completely satisfactory test of this type have not yet been determined.

The actual testing of experimental drain tile by direct exposure to soil sulfates has been limited in these studies. The specimens used, in the experiments, however, were mostly 2- by 4-inch cylinders of aggregate grading and mixes which roughly approximate those used in the commercial manufacture of high-quality concrete tile of 4- to 12-inch diameters. These cylinders had 28-day compressive strengths which mostly averaged between 5,000 and 6,000 pounds per square inch. It has been shown in Ref. 86 that the equivalent crushing strength of drain tile of the smaller diameters and normal wall thickness is about 40 per cent of these values, at least 2,000 pounds per linear foot.

11.19. Sulfate-resistant Drain Tile. The primary requirement for sulfate-resistant concrete is a sulfate-resistant cement, and the first requisite of a sulfate-resistant cement is low C_3A. This conclusion has been drawn from the tests of 122 portland cements reported by Miller and Manson [90]. These cements were rated for sulfate resistance by giving equal weight to the following three exposure conditions: (1) compression tests of cylinders exposed to the waters of Medicine Lake, South Dakota, which, during the exposure period of the cylinders, had an average salt content near 5 per cent, consisting of about two-third $MgSO_4$ and one-fourth Na_2SO_4; (2) length changes of cylinders in the laboratory in 1 per cent (0.084 molar[1]) solutions of $MgSO_4$; and (3) length changes of cylinders in the laboratory in 1 per cent (0.071 molar) solutions of Na_2SO_4. The cements were rated from best to poorest, from 1 to 122 under each test, and the final rating for cement is the result of the three tests totaled; consequently the cement with the lowest total was presumed to be highest

[1] A molar solution contains one mole or gram-molecular weight of the solute in one liter of the solution.

in resistance. In Table 4 of Ref. 90 the cements were finally arranged in descending order of sulfate resistance in 12 groups, each group containing 10 cements except group 12 which has 12 cements.

In Fig. 11.12 is shown the average, maximum, and minimum C_3A percentage for each of the 12 groups. Study of this figure reveals that the average C_3A content of the cements of a group increased with the group numbering with the minor exceptions for groups 3 and 8. Individual cements displayed some deviation from the group pattern, although, on the whole, the position ratings in Table 4 of Ref. 90 and the C_3A trends are similar.

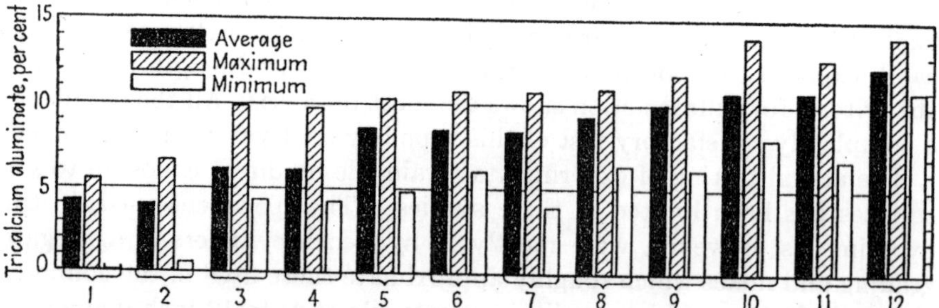

FIG. 11.12. Range in tricalcium aluminate (C_3A) in each of the 12 groups as the 122 cements are assembled in Tables 4 and 5 of *Minn. Agr. Expt. Sta. Tech. Bul.* 194.

Any possible effect which variations in the manufacturing practices at the different mills may have had on the sulfate resistance of these cements is obscured by the greater effect of variation in the C_3A content. A few of the cements showed an appreciable variation from the general trend as regards the influence of the C_3A content. These exceptions indicate the reasonable possibility of corresponding variations in the same directions as a result of improper processing in manufacture somewhere along the line; but these relatively few variations furnish no good reason to believe that the concept of low C_3A for sulfate resistance would be altered appreciably. Further research may clarify this part of the picture.

11.20. Modified Portland Cements. Among the 122 cements on which were made the long-time tests for sulfate resistance, 19 had been modified in composition so as to raise the iron-alumina ratio from an average of 0.45 to 0.96. This resulted in a reduction of the calculated compound C_3A from an average of 10.0 to 4.7 per cent. Chemical analyses of the 19 modified cements and of the 19 comparison cements from the same mills, as they were before being modified, show that practically the only consistent and significant differences between the modified and regular cements were in the oxides Al_2O_3 and Fe_2O_3 and the calculated compounds C_3A and C_4AF, which latter two have already been

TABLE 11.7. NINETEEN MODIFIED AND NINETEEN REGULAR PORTLAND CEMENTS FROM THE SAME MILLS, COMPARED FOR SULPHATE RESISTANCE

Cement		C_3A, %	Fineness, surface area, sq cm per g	Position rating*	Time, months, required to increase in length 0.01 in. after storage in 1 per cent solutions	
Lab. Nos.	Type				$MgSO_4$	Na_2SO_4
1599-631	Modified I	3.3	1,450	3	220	300
1600-632	Regular I	11.4	1,370	115	23	14
1551-531	Modified II	3.5	1,450	4	200	260
1550-528	Regular II	6.7	1,750	32	88	140
1568-591	Modified I	3.9	1,470	6	260	250
1567-590	Regular I	9.7	1,430	52	66	103
1576-582	Modified I	5.2	1,160	7	220	210
1575-579	Regular I	11.6	1,540	97	20	21
1571-576	Modified V	3.7	1,380	8	190	280
1570-573	Regular I	10.7	1,470	61	52	82
1583-605	Modified II	2.9	1,660	11	160	280
1582-602	Regular I	9.9	1,400	98	23	24
1543-507	Modified II	2.2	1,750	14	160	180
1542-504	Regular I	10.0	1,470	100	29	18
1554-543	Modified I	4.0	1,690	15	160	190
1553-540	Regular I	7.2	1,660	34	122	170
1546-513	Modified I	0.7	1,910	17	160	127
1545-510	Regular I	8.6	1,750	58	66	60
1593-621	Modified IV	2.5	1,690	18	180	220
1592-620	Regular I	11.4	1,490	92	27	25
1537-489	Modified II	4.9	1,610	21	130	150
1536-486	Regular I	9.0	1,540	40	76	97
1628-665	Modified I	5.8	1,720	22	150	160
1626-663	Regular I	10.0	1,730	83	40	29
1579-608	Modified I	5.6	1,560	23	115	180
1538-492	Regular I	9.8	1,650	73	51	45
1503-381	Modified V	4.2	1,650	24	140	108
1502-378	Regular I	8.7	1,500	36	96	130
1508-396	Modified II	5.4	1,880	28	119	150
1509-399	Regular I	9.0	1,500	66	55	52
1531-471	Modified I	4.8	1,700	42	76	77
1530-468	Regular I	12.0	1,550	120	16	10
1564-564	Modified I	6.0	1,640	49	68	91
1513-411	Regular I	11.0	1,450	108	27	17
1618-645	Modified I	8.5	2,060	103	26	26
1617-644	Regular I	9.5	1,710	111	24	22
1573-594	Modified I	12.9	1,710	107	29	28
1572-593	Regular I	13.8	1,690	118	17	12
Averages:						
19 modified cements		4.7	1,640	27	145	169
19 regular cements		10.0	1,560	79	48	56

* Table 4 of Ref. 90.

pointed out as being in the class of compounds most strongly influencing the sulfate resistance of the cements—and especially C_3A.

The essentials of the tests of the 19 regular and 19 modified cements have been assembled in Table 11.7. Reference to this table supports the statement that, without exception, resistance to sulfate attack of each of the 19 modified cements was greater than that of its companion cement. This is illustrated by Fig. 11.13. In both Table 11.7 and Fig. 11.13 the

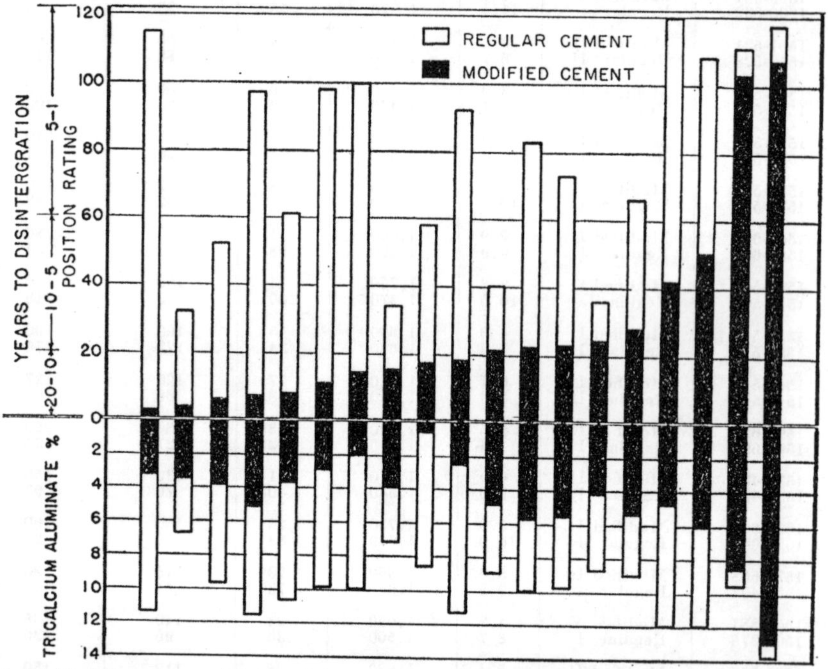

FIG. 11.13. Nineteen modified and 19 regular portland cements from the same mills compared for sulfate resistance.

order of arrangement of the cements is that of sulfate resistance from best to poorest of the modified cements. The importance of low C_3A cements for sulfate resistance is clearly illustrated by Fig. 11.13 based on the 19 modified cements compared with 19 companion regular cements processed identically in manufacture.

11.21. How to Make Sulfate-resistant Drain Tile. A sulfate-resistant cement is essential for durable drain tile that is to be installed in soils high in sulfates, but from here on, every possible consideration must be given to the tile manufacture in order that the tile be low in permeability and high in strength. In order to do this the same general procedure should be followed as outlined in Sec. 11.13, How to Make Acid-resistant Drain Tile of Small Diameters. Because certain cements are more resistant to soil chemicals than others, it is necessary to know the characteristics of any type of cement to be exposed in soils containing appreciable percentages of the harmful chemical agencies considered in this chapter.

SPECIAL CEMENTS

11.22. Pozzolan (High-silica) Cements. Long-time tests of seven cements of this type, with variations in SiO_2 from 28 to 34 per cent and in C_3A from the still-too-high 6.2 to 25.6 per cent, show practically as wide variations in sulfate resistance as the 122 portland cements. This type stood up best in the 1 per cent Na_2SO_4 laboratory solution and poorest in the water of Medicine Lake. The specimens showed no disruptive swelling but only wasted by attrition, as much, in some cases, as 30 to 50 per cent by weight. No close correlation appeared between their behavior and their chemical composition. Their characteristics, as revealed by these studies, do not indicate that they have any special value for use in the manufacture of sulfate-resistant drain tile.

11.23. Alumina Cements. Cements of this type are distinctly different from portland cements as they carry but 25 to 40 per cent lime and about 40 per cent Al_2O_3 compared with about 62 per cent lime and 5 to 10 per cent alumina for ordinary portland cements. The most resistant cylinders exposed to the sulfate waters of Medicine Lake were made with alumina cements. Many of them gave highly satisfactory strength tests after 20 years in the lake and appeared to be in almost perfect condition after this long, severe exposure.

Some specimens made with alumina cements, however, displayed definite tendencies to expand and lose strength in the laboratory solutions held at ordinary room temperatures (70 to 100 degrees Fahrenheit). It appears, therefore, that concrete drain tile made with alumina cement would give highly satisfactory service below the ground line almost irrespective of the sulfate concentration of magnesium and sodium, whereas the same tile, if left lying on the ground, might show signs of distress due to increase in volume brought about by ordinary summer temperatures.

11.24. High-temperature Steam Curing. Any discussion of sulfate-resistant concrete is not complete without mention of curing in water vapor at elevated temperatures, because marked resistance follows such curing when it is properly done. Below are some of the conclusions which were drawn from the work at the University of Minnesota (see Ref. 90):

1. The data show that 350 degrees Fahrenheit was consistently the most effective temperature used for increasing sulfate resistance (see Fig. 11.14). Based on the 17-year tests of cylinders exposed in Medicine Lake, it does not appear that prolonging the curing period beyond 6 hours

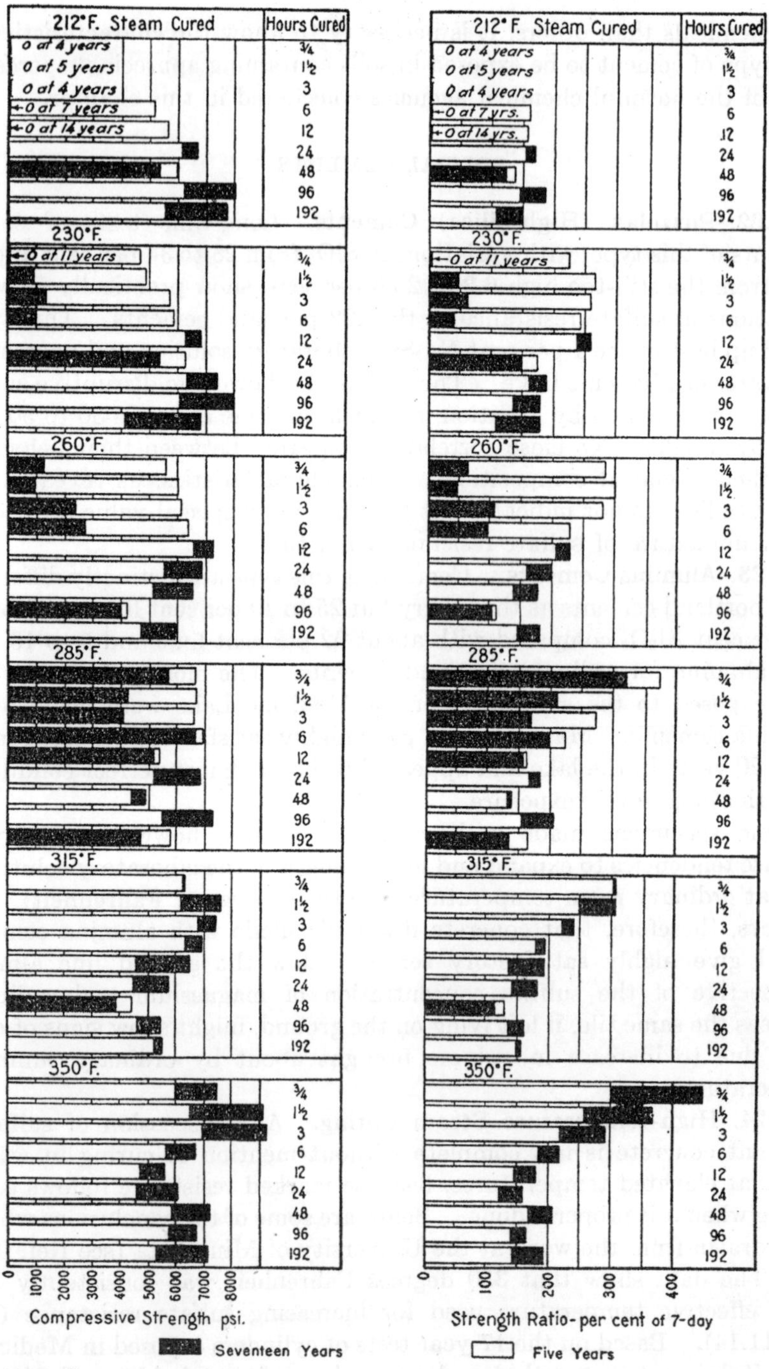

FIG. 11.14. Seventeen- and five-year tests of Medicine Lake cylinders cured in water vapor as indicated.

increased the resistance of the cylinders cured at 350 degrees Fahrenheit. A shorter period appears to be preferable at this temperature.

2. Not until curing temperatures of 212 degrees Fahrenheit and over were reached did steam-cured cylinders display resistance to sulfates essentially different from that of cylinders water-cured at room temperatures.

3. It was exceptional that the strengths of any steam-cured cylinders were as high at 28 days as comparable ones water-cured at room temperatures.

4. The percentage improvement in sulfate resistance brought about by steam curing at elevated temperatures is ordinarily much greater for cements high in C_3A than for cements relatively low in this compound. In other words, steam curing at elevated temperatures tends to equalize cements with respect to sulfate resistance caused by difference in chemical composition.

11.25. Action of Soil Sulfates on Clay Drain Tile. Clay drain tile which meet the American Society for Testing Materials requirements for standard drain tile or better are almost immune to the action of soil sulfates. However, clay drain tile with absorption well above the limits of 13 per cent fixed for standard drain tile may spall to the point of disintegration if left lying on the surface of the ground with a high salt content, regardless of the nature of the salts. Also, high-absorption clay tile will spall badly if subjected to salt or frost action, singly or in combination, at exposed outlets. It is, therefore, unwise to use clay drain tile extremely low in quality in soil relatively high in alkali salts on the erroneous assumption that there are no possibilities of deterioration, because it may develop that such is not the case.

Requisites for Drain Tile Installation

1. Anticipate as nearly as possible the exposure conditions to which the drain tile of a proposed system will be subjected.

2. Specify tile to meet the anticipated requirements.

3. Have the tile tested by a reputable laboratory to be certain that they conform to the requirements specified.

4. Each tile should be inspected before laying and rejected if it does not meet the general physical requirements shown in Table 11.2. In brief, these are as follows: (*a*) reasonably uniform in cross section; (*b*) reasonably straight side walls; (*c*) reasonably uniform wall thickness; (*d*) reasonably smooth inside surface; (*e*) reasonably true ends; (*f*) freedom from cracks, checks, chips, etc., which will appreciably reduce the normal strength of the tile or leave openings which will readily admit undue quantities of soil into the tile line; (*g*) when dry, should give a clear metallic ring when held in the hand, or stood on end, and lightly tapped with a steel tool.

5. Handle with care at all times to guard against breakage.

Where Drain Tile Are Tested. There are commercial testing laboratories in most of the large cities. Some of the state agricultural experiment stations and extension services test drain tile for farmers and drainage boards without charge, or for a nominal charge, when samples are submitted at the laboratory all delivery charges prepaid. Five tile of each size, selected at random from a lot, should be submitted. Contact your county agricultural agent or your state college extension director in regard to the details of such services.

PROBLEMS

1. A shipment of 6-in. concrete tile showed an average absorption of 12%, but an average crushing strength by the three-point bearing method of 1,100 lb per lin ft; the average depth in clay loam must be 6 ft in a trench 15 in. wide. Would you accept the tile? What classification would you give it?

2. Twelve-inch clay tile delivered on a job withstood a maximum crushing load of 2,500 lb per lin ft. How would you classify it? What other extreme test would you insist upon being applied?

3. What are the visual field test requirements of a good tile?

4. What will be the load on a 10-in. tile laid with an 8-ft cover in a trench 24 in. wide at the top of the tile in wet clay loam? What class of tile should be used?

5. What is the allowable depth for

(*a*) Standard 18-in. tile where the width of the trench at the top of the tile is 34 in. in ordinary loam damp soil?

(*b*) An extra-quality 15-in. tile where the width of trench is 27 in., same soil conditions as under (*a*)?

6. In a peat bog subject to deep freezing, which kind if tile, clay or concrete, in a 6-in. size would you use? Explain your decision.

7. A 6- to 10-in. tile is to be used in a clay-loam soil in which the soil solution tested 0.5% sodium sulfate. What special manufacturer's specifications would you insist on being fulfilled?

8. Using a fence post, two pieces of 2- by 4-in. lumber, and two spikes, devise and draw a sketch of a simple design whereby a load of 600 lb can be applied to a drain tile by a man weighing 170 lb.

CHAPTER 12

CONSTRUCTION OF UNDERDRAINS

In this chapter it is assumed that the tile drain lines have been staked out according to the method illustrated in Fig. 10.6 with stakes set at least at every 100-foot station on tangents and every 10 to 50 feet on curves according to their sharpness. Changes of direction in a tile line should never be by an abrupt angle, but always by a smooth curve which may be staked out by the simple method explained in Sec. 10.8 and illustrated in Fig. 10.7.

TRENCHING BY HAND

Up to about 1915, practically all construction of tile drains was by manual labor. Tile-trenching machines, discussed in a later section, were first introduced about the beginning of the twentieth century. About 1940 they came into general use, since which time very little tiling has been done by hand, and since about 1945, skilled tilers have become very scarce. However, tile trenching by hand is a highly skilled type of labor and should be classed as a trade. In any case trench junctions must be dug by hand, the trenches dug by some types of trenching machines must be finished to exact grade by hand, and as limited reaches frequently occur on proposed tile drains where trenching machines cannot operate, the hand-trenching process should be well understood.

12.1. Hand Tools. These tools are shown and listed in Fig. 12.1. Most of the spading is done with a tile spade. In very loose or very wet soils the solid-blade spade is needed, but in most soils the skeleton blade is handier, as it is lighter and the soil drops from it more readily. The muck shovel is very serviceable for cleaning out crumbs in sticky wet soils.

12.2. Digging to Guide Line. Digging must begin at the outlet and proceed upstream. The trench must be dug to straight lines and smooth curves. Top spading should be done to a guide line stretched along the stakes. The workman always faces the outlet and casts his material into a heap fairly well back from the edge of the trench. He may, if most convenient, cast it all on one side if there is no objection to mixing soil and subsoil when the trenches are refilled, or he may cast to both sides if more convenient. This last method is imperative when it is desirable to finish the refilling with the original topsoil. If the top spading lines up poorly, it is practically impossible to smooth up the line in later spadings (see Fig. 12.2). On curves, the guide line, stretched reasonably tight past the regular stakes, should be drawn outward to a smooth curve between stakes and fastened by pegs (see Fig. 12.3).

12.3. Minimum Width of Trench and Method of Spading. Experience has shown that, for ease of digging and a minimum of earth to be moved,

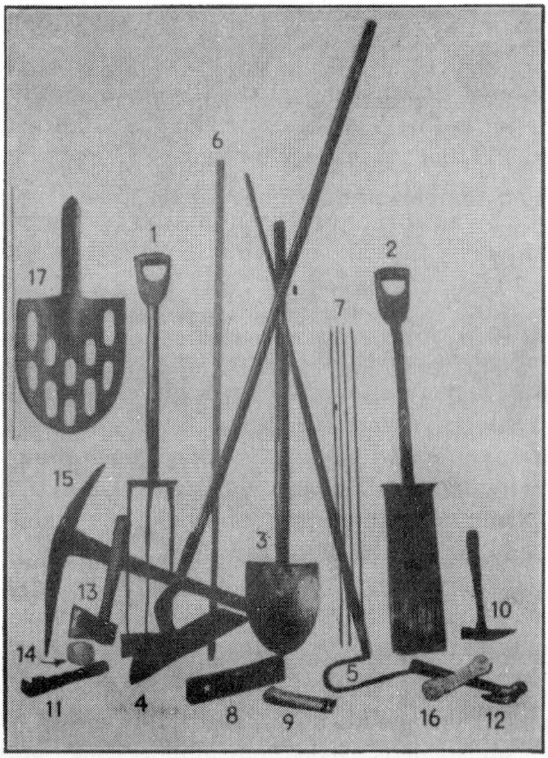

FIG. 12.1. **Tools used in tiling by hand.** (1) skeleton-blade tile spade; (2) solid-blade tile spade; (3) long-handled round-nosed shovel; (4) drain cleaner; (5) tile hook; (6) 5-ft gage rod; (7) target rods; (8) carpenter's level; (9) tiler's 6-ft zigzag rule; (10) prospector's pick; (11) monkey wrench; (12) claw hammer; (13) hand ax; (14) ball of white hard-twisted cotton twine; (15) pickax; (16) heavy chalk line for guide line (at least 100 ft long); (17) muck shovel for use in wet, sticky soils.

for ordinary depths of trench, the top width should be about equal to the outside diameter of the tile plus one-eighth of the depth of the trench to the center of the laid tile. This is about 12 inches for a trench that is to receive 4-, 5-, or 6-inch tile. In spading, one should not set the tile spade square across the trench or attempt to take either a full width or a very thick bite. When the blade is settled to its full length, the handle should be thrust forward slightly to break the side bond before the spade is pried backward to lift out the slice.

Cleaning Out the Crumbs. The crumbs left in the trench by the tile spade are removed with a long-handled round-nosed shovel, the man handling it standing in the trench and facing upgrade.

The Second Spading. The second spading should not reach lower than 2 or 3 inches above the grade line of the trench. These last few inches are to be removed from the trench by means of the *drain cleaner* or *scoop* (No. 4, Fig. 12.1) which is used to plane the bottom of the trench to exact grade by the method next discussed (see Fig. 12.9).

FIG. 12.2. Spading to guide line. Only the unskilled workman will try to dig the ditch straight without a guide line. Note that the workman faces the outlet.

FIG. 12.3. Guide line on curve. Note the short pegs holding the line to a smooth curve.

12.4. Trenching to Exact Grade by Hand. Finishing the bottom of the trench to exact grade involves, first of all, the establishing of an auxiliary grade line in the air, parallel to and a fixed distance above the established grade of the trench bottom. This fixed distance is usually set at 5 feet because the workman standing in the bottom of the trench must be able to see along the air grade line as he tests the bottom of the trench for grade. This air grade line will be delineated by a continuous cord held in it by high stakes at the grade hubs or by a series of horizontal targets extending out over the trench with their upper edge *in* the air grade line.

METHODS OF USING THE AIR GRADE LINE

There are three well-tried methods of using the air grade line to check the trench grade, namely, the sighting method, the line-and-gage method, and the line-and-rod method. Each is described in the following paragraphs. Another method of checking grade, known as the water method, was used extensively in the early tiling work, but few workmen were skillful in its use. It has since been discarded as unreliable.

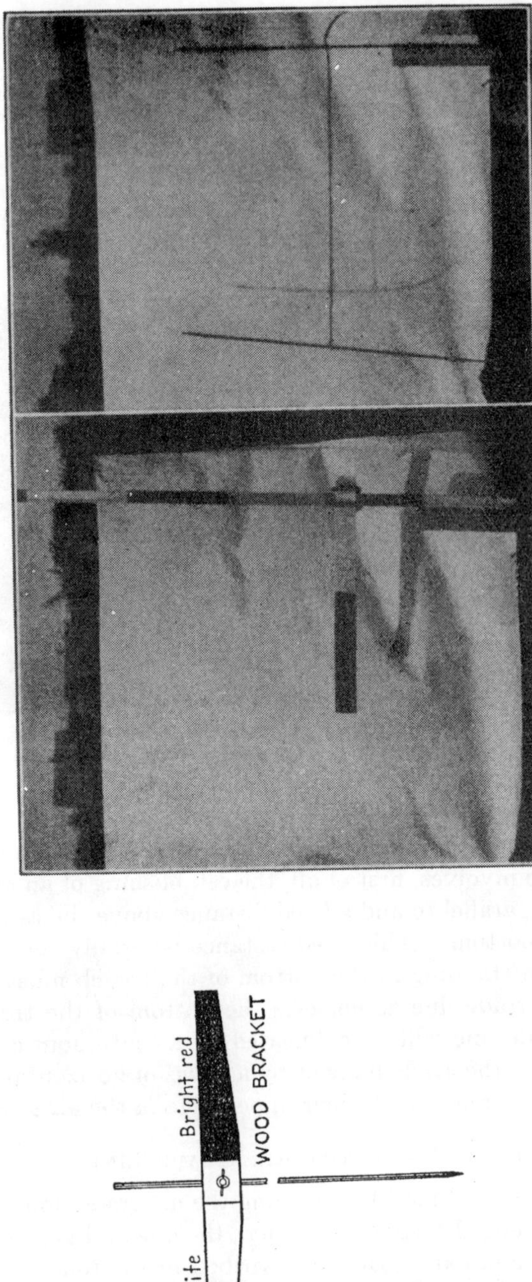

FIG. 12.4. Types of targets used in tiling by hand: *a*, swivel, wood or metal; *b*, one way, wood or metal; *c*, white string target.

NOTE: The white string target is the most satisfactory for use in tiling by hand. A dark object should be set up behind to make the string readily visible.

12.5. The Sighting Method. This method involves the use of targets of which the best-known types are illustrated in Fig. 12.4. On types *a* and *b* one-half the length is painted bright red and the other half white. In setting these the white and red faces of successive targets should be set alternately against each other as red will not show a sharp line against red nor white against white, but either will show plainly against the other. Skilled tilers have generally found type *c*, the hard-twisted white string target, the most reliable for tiling by hand, as a series of such targets, carefully set in a given grade, will always appear against a dark back-

FIG. 12.5. Setting a white string target 5 ft above bottom of trench.

ground as a single fine white line in the horizontal plane of the air grade, unaffected by refraction which always tends to dim exact vision over the edges of types *a* and *b*. A dark coat or a piece of tar paper set up as a background for a line of string targets makes the line stand out sharply.

Setting Targets for Hand Work (Cut Less than 5 Feet). Setting the targets for handwork is illustrated by Figs. 12.5 and 12.6 for white string targets. The steps are as follows:

1. Drive a small rod ($\frac{5}{16}$ or $\frac{3}{8}$ inch) or a stiff tall stake just back of or beside the hub on the outside and a similar rod square across the proposed trench from the first rod. [Where either type *a* or *b* targets (Fig. 12.4) are used, for obvious reasons only the first rod by the hub stake is needed.]

2. Subtract established cut at given station from 5 feet.

3. Measure up from the top of the hub the 1.58 feet obtained in step 2. Tie the hard-twisted white twine at this point. Then stretch it across to the second rod as nearly level as the eye can judge and tie it. A tiler's zigzag rule is the handiest measure to use.

4. Level the string with a carpenter's level. Check the measured height (1.58 ft) on the first side. After adjusting the height, check the leveling.

Number of Targets Needed. Targets should be set at not less than three successive stakes along any given strip of continuous grade at any one time—five or six is better. Each target must be set from the cut required at its own particular hub stake (see Fig. 12.6). If the engineer's work and the target setting are both perfect, all the targets along any given grade will line up and be projected against the background as one line, and the line of sight along such a series of targets, as line AB (Fig. 12.6), at all points will be just 5 feet above the established bottom of the trench.

FIG. 12.6. Targets at two consecutive stakes establish the line of sight 5 ft above grade.

Correction of Apparent Errors in Line of Targets. For several reasons a series of targets along a given stretch of grade may not line up perfectly. In such a case, assuming that the greater number of targets that do line are on the correct grade, move the others up or down until they come into line, being sure to keep all targets level. Before such adjustment, be sure that none of the hub stakes has been disturbed. If any great apparent error occurs at any stake, it is probable either that the recorded cut is wrong or that there is a mistake in the measured height of the target.

When Cut Exceeds 5 Feet. If the cut exceeds 5 feet, targets for use with the same gage rod cannot be set until the trench has been excavated below a cut of that amount. Then the rods holding the target must be driven slanting into the sides of the trench, 5 feet must be subtracted from the cut, and the remainder measured *down* from the hub or from a level line set across the top of the hub.

Targets at Change of Grade. The hand tiler works upgrade from and facing the outlet; hence the target line guiding him is always downgrade

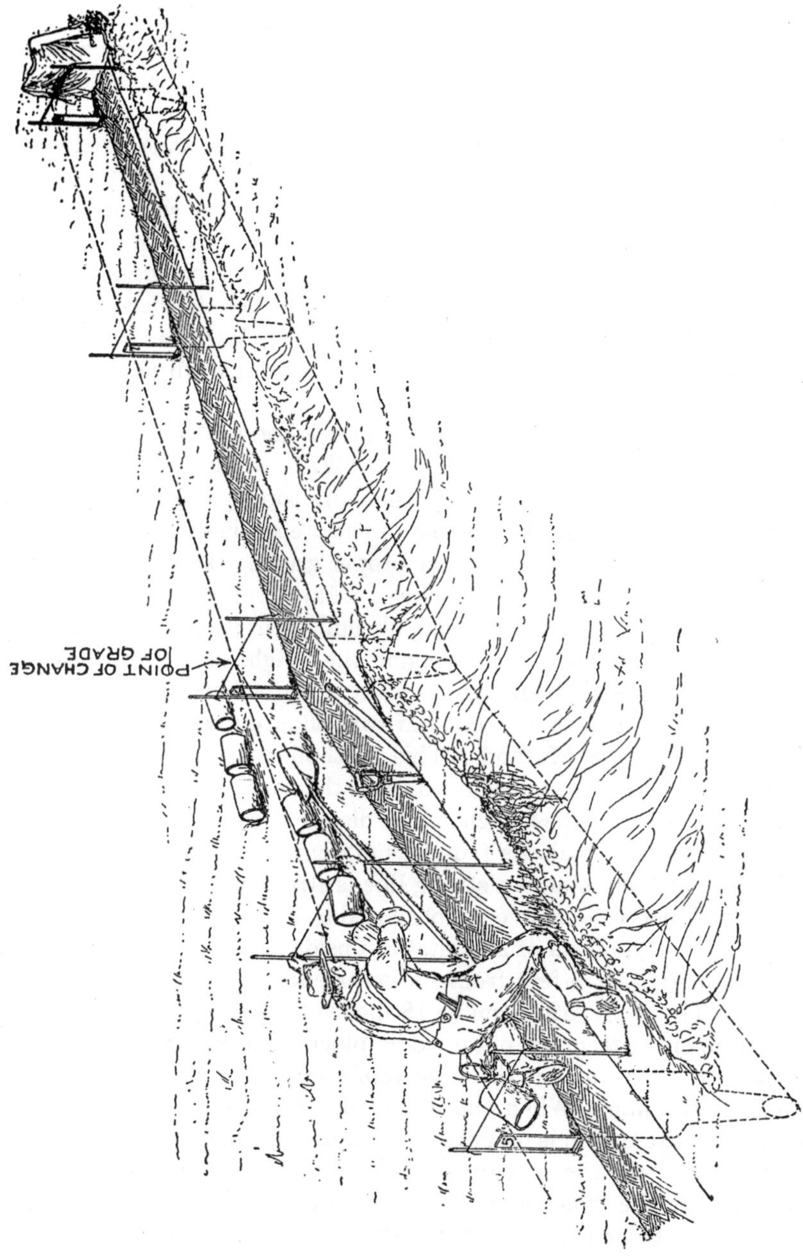

FIG. 12.7. Lining in additional target below outlet or point of change of grade.

from him. Therefore, below the outlet or below a point of change of grade a target must be set, by eye, in line with those already set upgrade from the point, so that the workman will always have below him a target line from which to check until he passes the second regular stake on a given grade (see Fig. 12.7).

Setting Targets on Sharp Curves. On sharp curves on steep grades any two targets being used on any stretch of trench should straddle the trench parallel to each other as the flare of radial lines from one side of a trench to the other often makes enough difference in distance between stakes seriously to affect the rate of grade and the correct bottoming of the trench. If the hubs are on the inside of the curve, the flare will be a divergence and the trench will be graded too deep. If the hubs are on the outside, the flare will be a convergence and the trench will be graded too shallow.

How to Use the Targets (Gage Rod). Line AB, Fig. 12.6, is a sighting line through all the targets on a given grade. With targets set 5 feet above the proposed bottom of the trench, when the top of a 5-foot stick held vertically with one end on the bottom of the trench lies in the line of the targets, the trench at that point is at proper grade. The tiler may thus test the grade of every foot of the trench. Many workmen use a notch in the handle of the drain cleaner at gage height from its point instead of a separate staff, but in loose or soft ground the weight of the drain cleaner will settle its cutting edge enough to make a serious error in testing the grade. Therefore, a slender wood staff with square-sawed ends is better.

12.6. Line-and-gage Method. This method consists of driving tall stakes alongside the hub stakes down until the tops are just 5 feet above the proposed bottom of the trench. The distance from the top of the hub to the top of the tall stake is found as is done in the sighting method. Over the tops of these stakes a strong line is stretched. This line will be parallel to and just 5 feet above the proposed bottom of the trench. Next a gage is made in the form of a capital L, upside down. The long arm of this gage is made just 5 feet from its lower end to the underside of the short arm. The short arm is made 18 inches to 2 feet long. The grade of the bottom of the trench is obtained by placing the end of the long arm, held exactly vertical, on the ditch bottom and noting whether the short arm touches the line. When the underside of the short arm does just touch the line, the trench is on grade. Because this method is simple it is frequently used. However, care should be used to see that the gage is held vertical at the time of making the test. If the long arm is not vertical, serious error is apt to result, the amount of error depending on the extent of the divergence from the vertical. If the vertical leans away from the line, the trench bottom will be finished too low; if it leans toward the line, the trench bottom will be finished too high.

12.7. Line-and-rod Method. This method (see Fig. 12.8) consists in driving a tall stake on each side of the proposed trench at each hub stake and nailing or clamping a crossarm to both of these stakes with its top edge just 6 feet above the proposed bottom of the trench. Care must be taken that this crossarm is exactly level. On the center of these crossarms a strong cord is stretched which will be parallel to and just 6 feet above the proposed bottom of the trench. A 6-foot gage rod like that

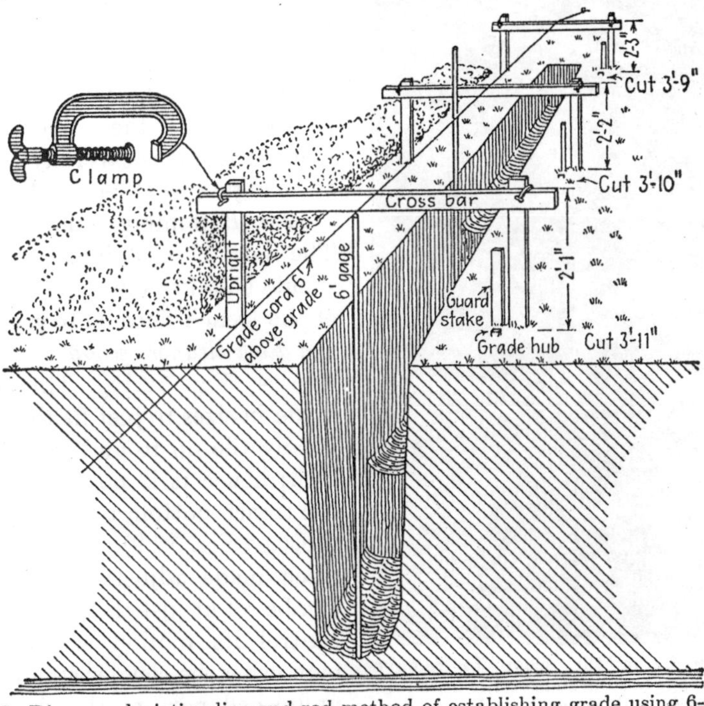

FIG. 12.8. Diagram depicting line and rod method of establishing grade using 6-ft rod or gage. (*USDA Farmers' Bul.* 2046, *L. A. Jones.*)

used in the sighting method is used for checking grade. A drawback to this method is that the cord is in the way. The most convenient length of gage is determined by the depth of cut—5 or 6 feet.

12.8. Drawbacks in Methods Described. Both the line-and-gage and the line-and-rod method have one serious drawback in common. A cord line, no matter how tightly stretched, will sag appreciably on a span of 50 feet, which is the longest that should ever be used with these methods. With a marked increase in atmospheric humidity or during a rain, a cord will become wet and sag badly. Most workmen will not give the proper attention to restretching the cord under such conditions, and resulting errors will occur at times that will wipe out the fall for short reaches or even result in slight backfall. A fine steel wire, used in place of a cord, is better because it is not so seriously affected by changing weather conditions; but even a wire will sag to a certain extent by its own weight, and it is much more difficult to place and adjust. Any of these methods, which are all in common use, will serve well if great care is used to offset natural drawbacks, but when all is said and done, the sighting method, especially with white string targets, will give the best insurance against faulty installation in handwork. However, it is not adaptable to machine installation, later discussed, but the line-and-gage method is if it is carefully adjusted to changing weather conditions.

FIG. 12.9. Using tile scoop to remove crumbs, shape bottom to keep tile from rolling sidewise, and shave trench to exact grade. (*Courtesy of Mason City Brick and Tile Company.*)

12.9. Bottoming the Trench. The last 2 or 3 inches of earth should be taken out and the bottom of the trench finished to grade with the *ditch cleaner*, *scoop*, or *crumber* (No. 4, Fig. 12.1). This tool is made in sizes to just fit the outside of any tile up to and including 12 inch; hence it is possible to finish the trench bottom to exactly fit the tile (see Fig. 12.9). The handle is often made adjustable so that it can be set at just the right angle to control its motion exactly. The man using the cleaner stands in the unfinished part of the ditch and, placing the cleaner snugly against the trench bottom in the last section where the grade is correct, with this section serving as a guide, draws the cleaner toward him (see Fig. 12.9). In this way the cleaner trims the next section of unfinished trench to true grade much as a jointing plane trims the edge of a board. Care must be used to cut just to grade. If any spot is cut too low, that spot must be filled back with well-pulverized soil, firmly tamped with the crumber and again planed to grade.

TILE TRENCHING BY MACHINE

As earlier stated, the bulk of tile trenching is now done by machine, handwork being largely confined to short lengths of trench in cramped quarters where it is practically impossible, for one reason or another, to operate a machine, or in excavating trench junctions or sharp curves often used on drains with small tile.

12.10. Types of Trenching Machines. Several machines have been developed especially for tile trenching and have been in use for years past in drainage work. Only two types have proved thoroughly effective for a finished job. These two, both of which are widely used, are heavy traction machines moved and operated by their own power which is usually that of one of the standard forms of internal-combustion engine. They are built in several sizes to cut from about 4 to a little over 8 feet deep or even deeper in some of the larger sizes. Minimum width of cut is

FIG. 12.10. Bucket-ladder type of tile trencher set for travel between jobs. (*Courtesy of the Parsons Company.*)

12 inches or just about enough to receive a 10-inch tile without forcing, normal maximum width being about 24 inches. Some makes may have that width increased by removable lugs. Both of these machines cost several thousand dollars and are subject to heavy maintenance cost. Such an investment practically bars them from ownership by individual farmers. However, they are excellent equipment for tile-drainage contractors whom it is usually practical and wise to employ, especially on large single jobs or for several jobs in the same community. One of these machines is the *bucket-ladder type* (see Figs. 12.10 and 12.11). With this machine it is difficult, especially on uneven ground, to finish the trench to sufficiently exact grade for tile-drainage work; hence the bottoming must often be checked by hand. However, this machine is a faster worker than the other type and will follow a much sharper curve. The other machine is the *bucket-wheel type* (see Figs. 12.12 and 12.13), which is of more rigid construction, and a skillful operator can carry within ¼ inch of the established grade line if the soil is comparatively free from large rock

particles; but it is not possible, with this machine, to follow a very sharp curve such as is often used in tile-drainage systems. The shovels leave a smooth groove in the bottom of the trench in which the smaller sizes of tile may readily be laid by a workman standing inside the shoe shown in the rear of the machine. This shoe prevents caving until after the tile is laid. It thus protects the tile layer from being buried in deep trenches by

FIG. 12.11. Bucket-ladder type of tile trencher in action. (*Courtesy of the Parsons Company.*)

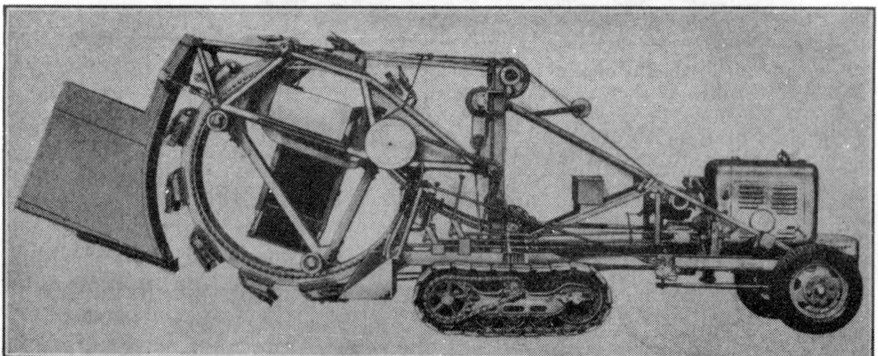

FIG. 12.12. Bucket-wheel type of tile trencher set for travel between jobs. (*Courtesy of Gar Wood Industries, Inc. Findlay Division.*)

cave-ins and thus obviates the necessity for stays and trench braces always needed in deep handwork and later discussed. For best progress under average conditions with either of these machines there should be a minimum crew of at least three—the operator, one general handy man, and one skilled hand tiler, this latter to excavate the junctions between any two lines as it is impossible to do this with any trenching machine.

A number of other tile trenchers of the plow type have been developed from time to time. None of these will finish the trench to grade but will excavate only to within 6 inches to 1 foot of the bottom, leaving the work to be finished by hand. None of them has proved satisfactory, and they tend to become obsolete.

Another type of machine extensively used for tile trenching is the *trench hoe*, a machine of the old steam-shovel type, but powered with an internal-combustion engine and with the bucket reversed to draw *toward* the power unit much like a hoe in the hands of a gardener (see Figs. 12.14 to 12.16).

FIG. 12.13. Rear view of bucket-wheel tile trenching machine in action.

This type of machine will not finish a trench to grade. Care must be taken to keep several inches to a foot above grade with the machine excavation, the finishing to be done by hand. However, this machine is a fast worker even under rather difficult conditions and will cut a clean true trench all the way. It, also, is made in different sizes, its range of width and depth being comparable with those of the regular tile trenchers. Some contractors prefer it to either the bucket-wheel or the bucket-ladder type because of its greater adaptability to wet or stony conditions. One contractor of many years experience says this machine should have a minimum crew of four men (he uses five), but at that he considers it the best all-around machine for general tile-drainage work. Any of the three

FIG. 12.14. Trench hoe—general side view—in action. (*Courtesy of Insley Manufacturing Company.*)

FIG. 12.15. Trench hoe—rear view—in action. (*Courtesy of Insley Manufacturing Company.*)

machines thus far described will cover, in a given time, several times the work of an experienced hand crew under similar conditions. This feature is more fully discussed in Chap. 15 on Cost of Drainage.

Conditions sometimes occur where a regular small dragline of the type discussed under open ditching and illustrated in Fig. 8.21 is the most satisfactory machine for tiling work. This is apt to be especially true under wet conditions, particularly where there is serious surface ponding and where large stones, extra stiff hardpan, or areas of rock requiring frequent blasting obstruct the path of the machine. The dragline is thus

Fig. 12.16. Trench hoe equipped with backfill blade, filling trench. (*Courtesy of the Schield Bantam Company.*)

used extensively in some areas, although it cannot rightly be classed as a tile trenching machine as it will not cut a clean-walled trench of specified width, and the bottom must always be finished to grade by hand.

Carrying Correct Grade in Trenching by Machine

The surveying and construction procedure, the specifications, precautions, and other details of trenching to grade by machinery correspond in general outline to the procedure earlier described in this chapter under Trenching by Hand, with the following variations therefrom necessary in machine work.

Any one of the three machines above described is heavy, each weighing several thousand pounds. They must straddle the trench, and their width of base in each case is several feet. Therefore, in case of the bucket-ladder and bucket-wheel types, as stated in Chap. 10, page 244, under

Laying Out Tile Drains, and illustrated in Fig. 10.6, the following requirements must be taken into account:

1. The line of grade hubs from which the targets are set must be placed 2.5 to 8 feet, according to the particular machine used, to one side, usually the right, looking upgrade, of the actual center line of trench to allow the machine to pass by the stakes and hubs without disturbing them. Therefore, the engineer must know the type and size of machine to be used on the given job before he surveys out the lines, in order that he may place the hub line at the proper distance from the trench center line.

2. The target stems must be tall, usually of ¾-inch gas pipe, in order that the air grade established by the targets is where it can readily be held by the eye of the operator, standing or seated on a fixed platform on the machine frame opposite the digging unit and on the same side of the machine, looking ahead, as are the hub stakes.

3. Type *c* of Fig. 12.4 (the white-string target) cannot be used in trenching by machine for obvious reasons. Hence, either type *a* or type *b* must be used, resulting in what is commonly known as the "type-and-bar" method (corresponding closely to what has earlier been described as one form of the sighting method) only that the gage rod cannot be used but is usually replaced by the device described in the following paragraph.

4. The machine is so equipped with a grade-control device under the vertical control of the operator that a sighting-bar gage, rigidly connected to the vertical control of the digging unit, is kept always in the approximately horizontal plane of what has earlier been described as the air grade line, on any given gradient, extending either forward or backward from the machine. Thus the lowest point of the digging unit is held at the established grade.

In the hands of a skillful operator this *target-and-bar* method of grade control results in fair accuracy, but it has certain limitations and tendencies to error which are difficult or even impossible of adjustment. As they should be recognized and understood, they are listed and discussed in the following paragraphs.

12.11. Operator Fatigue. The operator's platform follows the contour of the ground, coordinate with the traction base of the machine, while the sighting bar is kept, as exactly as is humanly possible, at a constant distance above grade. This forces the operator, at the same time that he is adjusting both the vertical control levers and the steering mechanism, to take continually changing positions from squatting to tiptoe in order to keep the sighting bar in the exact line of the targets. These complicated activities, carried on continuously for many hours at a time, cause a great strain on the eyes and are generally very fatiguing. The result is that, with the customary spacing of targets of 100 feet on tangents and 50 feet on curves, appreciable inaccuracies in grade occur of from 0.10 to 0.40

foot or more, according to the sensitiveness of the individual operator. On some machines this trouble is partially eliminated by provision of an operator's seat that moves vertically in synchronism with the digging unit. A still more effective device especially adapted to the bucket-wheel type of machine is discussed in the following paragraph.

12.12. The "Pipe-and-Cord" Grade-control Device. This device, which overcomes the disadvantages of the target-and-bar method, was

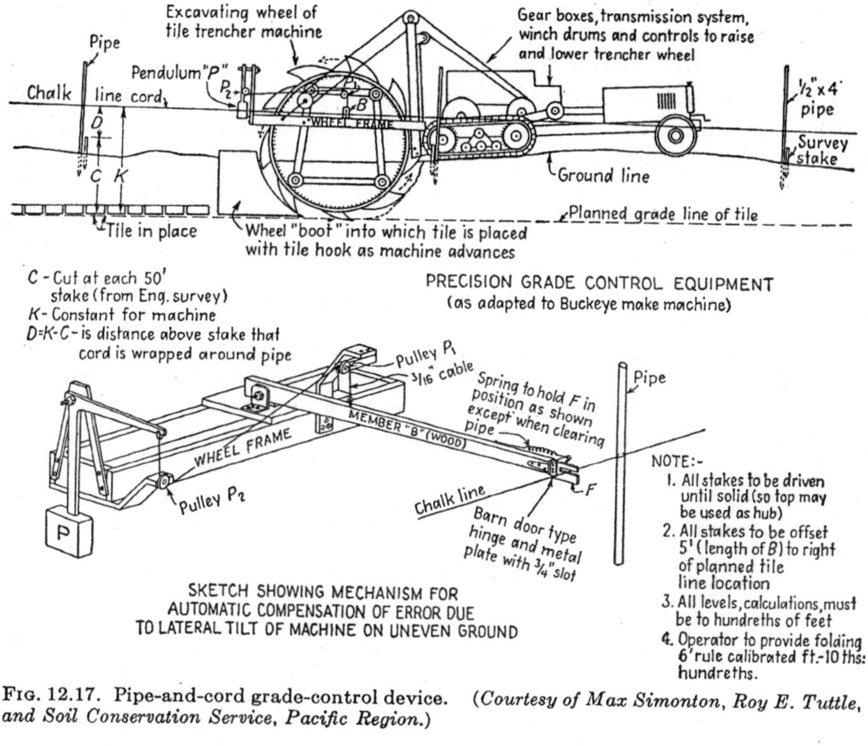

FIG. 12.17. Pipe-and-cord grade-control device. (*Courtesy of Max Simonton, Roy E. Tuttle, and Soil Conservation Service, Pacific Region.*)

recently developed by Max Simonton, tile drainage contractor of La Conner, Wash., for use on the bucket-wheel type of trencher, and it has been successfully used for several seasons in western Washington. It is illustrated and described substantially as follows by Roy E. Tuttle, Engineering Specialist of the Soil Conservation Service, Pacific Region (see Fig. 12.17).

This device is attached to the wheel frame of the machine in the vertical line of the center of the bucket wheel. To avoid sag in the guide cord, it is essential that the spacing of the hub stakes in the line shall never exceed 50 feet, this distance being appreciably less on curves according to their sharpness. The hub line in this case must be set 5 feet to the right of and parallel to the proposed center line of the trench. It is possible, with this device, to maintain the finished grade within 0.05 foot of the planned grade elevation at all points along the tile line. Such a degree of accuracy is essential for good results, especially with the low gradients that must be used on flat lands. The following accessories to the device are necessary:

1. Six to ten ½-inch galvanized pipes 4 feet long with a plug welded into one end and sharpened. These pipes are to support the lining cord.

2. Three hundred feet or more of good-quality chalk line. This is the lining cord to be drawn taut on the air-grade line. Some operators prefer piano wire as being more rigid and having less sag.

3. A 6-foot folding tiler's rule, graduated on at least one side in feet and tenths and hundredths of feet, for measuring up the line pipes from tops of hubs to cord position.

4. A blacksmith's hammer for driving the line pipes, and a piece of lumber crayon.

The following instructional comments are offered by Tuttle:

> A vertical hanging bar (not shown on sketch) supported from the right hand corner is used to assist the operator in aligning machine and making a truly straight trench. This is a ½-inch diameter by 40-inch long piece of steel supported from one end at a distance of 4¾ feet from the center of machine. The machine is so driven that this bar is 2 to 4 inches from the taut string.
>
> **12.13. Lateral Tilt Error Compensation Device.** The arrangement of pendulum, pulleys, and lever as shown on the print effectively compensates for this error. It maintains the member "B" in a truly horizontal position at all times. "F" is 16 gauge sheet metal with a slot ¾-inch by 10-inch. By steering control and vertical digging control the slot is kept straddle of but not touching the taut cord. "F" is hinged so that the pipes are allowed to flip it back until pipe is passed.

The great advantage of this device is that it does away with sighting on the part of the operator, thus eliminating operator fatigue which always accompanies the target-and-bar method, as all the operator has to do is to steer properly and so handle the digging control that the slot in the end of member *B* does not touch the guide cord. The device also takes care of the other features found undesirable in the target-and-bar method, and the performance at change of grade shown in Fig. 12.7 is eliminated. Figure 12.18 shows a Buckeye trencher equipped with the Tuttle device operating in western Washington. It will be noted that, in this case, the operator stand and guide line are on the left-hand side of the machine.

While the Tuttle device was originally developed to use with the bucket-wheel type of trencher, there does not seem to be any valid reason why it cannot be adjusted for use on the bucket-ladder type also.

This type of outfit requires a three-man crew for most efficient work—operator, tile layer, and general helper.

12.14. Tiling with Trench Hoe. Different makes of trench hoe dig on different sides of the stake line. The following discussion is based on the practice of a contractor who has operated for over 40 years in the upper Mississippi Valley, and his latest machine work has been with the trench hoe shown in Figs. 12.14 and 12.15. Only the top digging is done by machine, finishing to grade being done by hand. The aim is to keep the machine digging about 21 inches or one full spading above finished grade.

FIG. 12.18. Buckeye tile trencher operating with Simonton's and Tuttle's pipe-and-cord grade-control device in western Washington. (*Courtesy of Soil Conservation Service, Portland Office.*)

Some operators prefer an overhead grade cord, but owing to errors from sag and to the fact that the cord is somewhat in the way of the operation of the machine, this operator prefers the original target line, set 4 feet to the left of the center line of the trench. For bottoming by hand behind the machine a second set of targets is set over near the edge of the trench by leveling from the first set with a spirit level. These targets must be longer than those used in purely handwork because the trench made by this machine has a minimum width of 25 inches and the targets must reach a point over the center line of the trench. This minimum trench will receive tile up to 20 inches inside diameter.

This operator uses a five-man crew for the entire job with the trench hoe—the operator, the target setter who also hands the tile into the trench, the bottom man who finishes the trench to grade, the tile layer, and one handy man who blinds the tile. The contractor insists, however, that four really good men can handle the job.

The trench hoe has several advantages over the regular trenching machine; for example, it can back up and pick up cave-ins, and it is usually equipped with various units which are interchangeable with the hoe bucket. An illustration of this is shown in Fig. 12.16 which is a small-sized machine equipped, among other things, with a backfill blade that is widely used effectively for this phase of the work.

12.15. Stays and Trench Braces. In trenches over 8 feet deep it is necessary to use staging and two or more lifts, the bottom man casting to

FIG. 12.19. Trench braces and plank stays. *Left:* Trench brace alone. *Right:* Trench braces and plank stays in position in the trench.

the first staging, from which another passes the soil to a second staging or to the top. In loose material or in clay liable to slip along cleavage lines or cracks, it is usually necessary in trenches over 6 feet deep to use plank stays and trench braces to prevent caving (see Fig. 12.19).

In deep trenches where there is a tendency to slide or cave over a considerable length of the trench, the only safety either for the work or the workmen lies in using continuous curbing held in place by bracing that will not give way, such as customarily used in deep sewer work. Such curbing should be removed after the tile is blinded. If necessary to leave any of it in the trench, there should be at least 4 inches space for backfilling on both sides between curbing and tile.

12.16. Curbing Shoe on Machines. Some trenching machines, especially those of the bucket-wheel type, are equipped with a metal shoe the width of the trench, located just behind the digging unit. This shoe generally obviates the need for plank stays and trench braces as it gives a space in which the tile layer can work with the same protection to both himself and the tile afforded by plank stays and trench braces (see Figs. 12.12 and 12.13).

CONSTRUCTION ITEMS THAT FOLLOW THE TRENCHING

12.17. Laying the Tile. Beginning at the outlet the tiles are laid upgrade to permit drainage of any water that may accumulate in the trench.

FIG. 12.20. Use of tile hook in placing tile.

On almost all jobs where the trenches are dug by machines that are equipped with the shoe above discussed, practically all sizes of tile are laid by hand in the shoe, the tile layer having a helper who passes or lowers the separate tiles to him. In nearly all other cases small tile up to and including 8 inch may most readily be laid with the tile hook (No. 5, Fig. 12.1). With the tile distributed along the trench within easy reach, the tile layer stands astride the trench, picks up a tile from the side on the tile hook, tries it for close joint against the last tile laid, turns it, if necessary, by means of the tile hook to get the best fit, then swings it into place, finally tapping it with the heel of the tile hook to a snug contact (see Fig. 12.20).

Some tilers, standing in the unfinished trench, prefer to finish the bottom about two tile lengths at a time and then lay the tile by hand, snugging them with a tap of the boot heel. This method is really the most satisfactory when the sand shield, later shown, must be used.

The tile should be firmly bedded in the trench bottom and laid with close joints to an even line that looks, before covering, like a continuous pipe (see Fig. 12.21). If a tile does not fit close against that previously laid it should be turned on its long axis, as above noted, until the position is found in which it fits most snugly. The small sizes of concrete tile should not be laid with such tight joints because they have such perfect ends that, laid very close, they tend, in clay or clay loam soils, to seal the joints thus shutting out the water. In numerous instances such sealing has occurred for lengths of several feet, giving the effect of a continuous solid pipe.

FIG. 12.21. A line of tile in place ready for inspection and blinding. (*Minn. Agr. Ext. Bul.* 149.)

Broken ends of tile should be placed on the bottom. Where the joints lie open more than ⅛ inch in ordinary loam soils or more than ¼ inch in clay soils, they should be covered with broken pieces of tile (tile bats), grass clods, hay, or untarred building paper. (Tar paper tends to stick tight and seal the joints.) In fine sands all joints should be thus protected.[1]

Tile larger than 8 inch must be laid by hand. Sizes too large to be lifted and lowered readily by hand must be handled by means of a rope, a derrick or other mechanical hoist.

The tiler should reject all tile that are cracked, too soft, or so ill-shaped that a smooth line cannot be obtained.

12.18. Making Junctions and Trimming Tile for Curves. For sizes of tile up to and including 12 inch, factory-made junctions are usually obtainable, but the tiler should know how to make his own junctions by hand (see Fig. 12.22). To start the hole for a junction in a small tile, one may either chip it through with the point of a prospector's pick (No. 10, Fig. 12.1), or he may use the face of the pick or any light hammer as follows:

[1] For a thorough theoretical analysis and discussion of crack spacing, see Don Kirkham, "Potential Flow into Circumferential Openings in Drain Tubes," *Jour. App. Phys.*, Vol. 21, No. 7, pp. 655–660, July, 1950.

Completely fill a sound tile with dry sand to prevent vibration, holding the tile between the knees with two small boards to keep the sand in. Then give a prolonged series of light taps in the same spot with the face of the hammer. A small hole will soon result without cracking the tile if

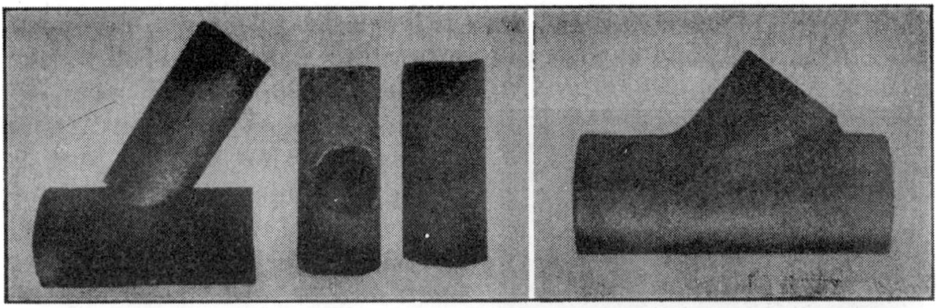

FIG. 12.22. Tile junctions. *Left:* Handmade. *Right:* Factory made.

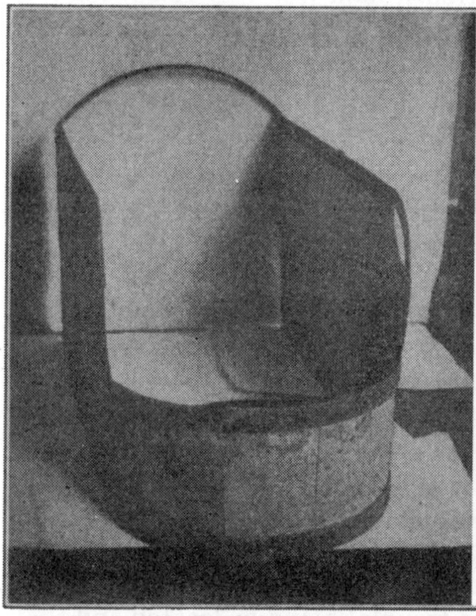

FIG. 12.23. Sand shield, standing.

FIG. 12.24. Sand shield in position.

care is used and the job is not hurried. To enlarge the hole, use a monkey wrench just as one does a glass cutter in trimming the edge of a pane of glass. The same tool is best, also, for trimming the ends of a tile to secure close joints on sharp curves.

12.19. The Sand Shield. In dry sand or in quicksand in a wet trench, a sand shield of galvanized iron (see Figs. 12.23 and 12.24) is used to prevent the sand from running in and spoiling the bed of the trench. The bed of the trench is finished inside this shield and the length of tile laid, the sand shield is then pulled forward its own length, and the process repeated. This process ensures laying the tile to grade.

12.20. Keep the Tile Line Clean. Keep tile clean while laying. Earth in a tile line attracts roots and may cause a stoppage. Earth is easily dumped out of a small tile. With large tile, tie a large burlap swab to a short length of small rope and draw it forward through each newly laid tile. The tile should be kept laid up to within 2 or 3 feet of each day's finished trench, and the upper end should be closed to prevent fouling of the tile whenever the work of laying is temporarily stopped. A piece of board or a close-fitting cobblestone held in place by a little loose earth may be used for this purpose. Fully equipped tilers often use the trench dam (see Fig. 12.25) for this purpose, especially in wet work.

FIG. 12.25. Trench dam in position.

After a line is completed the upper end should be permanently plugged with broken pieces of tile, a flat slab of burnt clay or of concrete, or a flat stone to keep out earth and small animals.

12.21. Checking Grade. After the tile is laid on any line and before it is blinded, the grade of the tile should be checked. This is done by taking level readings on the top of the tile every 25 feet of the line, preferably starting at the highest end. If each successive reading shows the proper drop from the preceding reading, it may be taken for granted that the tile between the two readings is properly laid. The entire line should be covered in this way. If any successive set of readings shows that the tile does not have the proper fall, the trouble should be located and that portion of the line taken up and the bottom refinished to grade and the tile replaced. Reasonable allowance should be made for small variations from theoretical grade and for slight unevenness in the cross section of the individual tiles. This checking is a necessary precaution and should always be done to make certain that the work has been properly done. Running water in a tile line is a valuable aid in checking its grade.

12.22. Blinding the Tile. A newly laid line of tile should be covered 4 to 6 inches deep with topsoil or other open soil, close up to the end of any day's work, to hold the tile in place against any disturbance before the trenches are filled. This is called *blinding the tile*. In tight soils a layer of several inches of coarse hay or dry grass, spread over the tile before blinding, prevents joints from sealing and facilitates drainage. *In the narrow hand-dug trenches,* blinding is done by walking astride the trench and cutting down topsoil from both sides below the surface (see Fig. 12.26). One must not disturb the batted joints or the alignment of the tile. As *machine-made trenches* are usually much wider than the outer diameter of the tile, blinding is usually much more of a job than it is in hand-dug trenches. Owing to the excess width of trench much more material must be moved and much of it at the sides of the tile must be packed in more or less firmly, with considerable care not to crowd the individual tile out of line and surely to prevent the same thing happening from outside disturbances such as stock falling into the trench, cave-ins, and violent rainstorms before the final filling of the trenches is done. With excessively wide trenches in order to get sufficient material for packing and blinding it may be necessary to use some of the methods that apply in filling the trenches, in which case, extreme care should be used not to disturb the tile with the falling material. This precaution may necessitate a partial packing or blinding first with a small amount of earth that can be cut in from the lower part of the trench wall before any is plowed in from the waste banks. Tile once pushed out of line are hard to reset at grade without much and rather aggravating hand labor.

FIG. 12.26. Blinding the tile by carefully shaving off topsoil from the sides of the trench. (*Courtesy of Mason City Brick and Tile Company.*)

12.23. Refilling Trenches. Almost any method that will move the material may be used in refilling the trenches. It may be done by hand with shovels, by plow, by team and scraper, by road or terracer grader drawn by tractor power, by bulldozer, by trench hoe, or by numerous other special methods and devices. Short sections near buildings, fences, and other places inaccessible by teams or machinery must be filled by hand, as must also short lengths of trench at the outlet to provide turning ground for teams and machinery. Usually the quickest and cheapest way to refill small or shallow trenches on limited farm jobs where horses or mules are available is with a common walking plow drawn by two teams, one on each side of the trench, with the apparatus arranged as shown in Fig. 12.27. The long evener should have a series of holes 6 inches apart

FIG. 12.27. Filling trenches with teams and plow.

FIG. 12.28. Filling trenches with buck scraper.

near the center for the kingpin, to permit quick readjustment to fit changing surface conditions, as the plowing is done up and down either side of the trench, regardless of the position of the spoil bank.

If a scraper is used, the team must be attached to it by means of a long light chain, as the work is done crosswise of the trench, the team being on one side and the scraper on the other. The ordinary slip scraper is clumsy and rather slow. Figure 12.28 shows a small buck scraper especially built for trench filling.

On large jobs any one of the following is satisfactory equipment for trench filling: the regular road grader or terracer grader with 10- or 12-foot blade and drawn by tractor power, bulldozer, or trench hoe with back-filler blade (see Figs. 12.29, 8.11, 8.28 and 12.16). In soils reasonably free from big roots and stones, the drag grader shown in Figs. 8.6 and 8.7 is also a good trench filler. Any of these types of equipment is also well suited for leveling the surface after trenches are filled.

FIG. 12.29. Filling trenches with road grader drawn by tractor power. Note the long hitch.

However moved, the filling material should be heaped up over the trench to allow for settling.

12.24. Protecting Tile Outlets against Undermining and Crushing. The essential thing is to make outlet protections permanent. If the tile line ends at or near the bottom of the outlet ditch or stream gorge, so that there is no appreciable drop from the tile to the stream or but little side current in the stream to cause scouring and undermining, a cheap satisfactory method of protection is to finish the tile line with 15 or 20 feet of continuous corrugated iron culvert of good grade. In this case it is necessary to place the culvert itself on a firm natural bed of clay or other soil and to cement the joint tightly between the last tile and the culvert. A ring of ¾-inch round iron rolled into the outer end of the culvert gives additional stiffening against crushing. Figures 12.30 and 12.31 show effective outlet culverts of this type. Where the tile is carried to the outlet without a culvert, the last 15 or 20 feet of the line should be constructed of bell-and-spigot sewer pipe with the bell ends turned upstream and the joints tightly cemented. In such a case, a bulkhead of the general type of Fig. 10.25 to protect the outer end of the tile is indispensable, and one may be necessary even where a culvert is used where the traffic is heavy, or where there is a decided drop from the end of the tile to the bed of the outlet channel, or where the current in that channel

FIG. 12.30. This corrugated metal outlet for a tile system in Iowa is said to have given perfect service for 25 years with no repairs whatever. (*Courtesy of Mason City Brick and Tile Company.*)

(*a*) (*b*) (*c*)

FIG. 12.31. Corrugated iron tile-outlet culverts: *a*, bell-ended corrugated culvert at outlet of a 16-in. tile in Stearns County, Minnesota; *b*, corrugated iron culvert extending through a highway embankment at the outlet of a 15-in. tile in Scott County, Minnesota; *c*, corrugated iron culvert, outlet for a 12-in. tile main in the bed and direction of the outlet stream; St. Louis County, Minnesota.

causes scouring at the tile outlet. Figure 12.32 shows an effective bulkhead of this type in perfect condition after 45 years of service. If the bulkhead is not provided, eventual destruction of the outlet is almost certain to occur. This is well illustrated in Fig. 12.33.

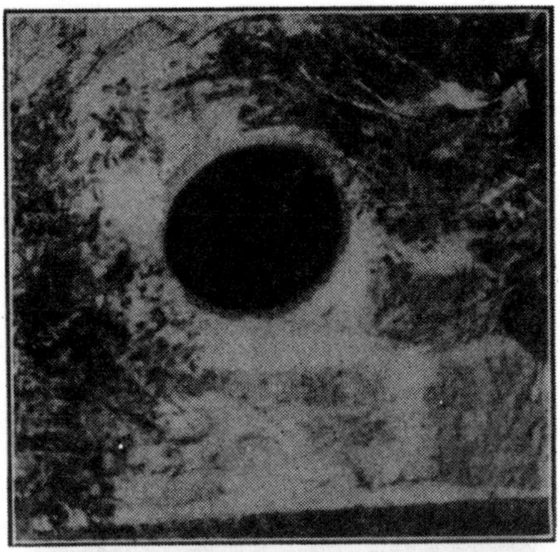

Fig. 12.32. A well-built masonry bulkhead around an 18-in. tile; constructed on a firm foundation in 1909 and as good as ever in 1944. Traill County, North Dakota.

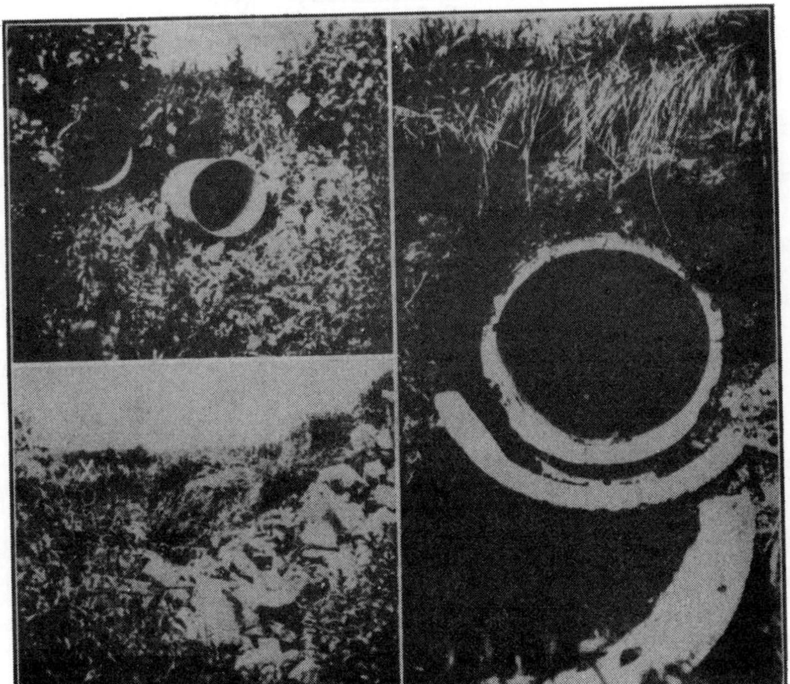

Fig. 12.33. An unprotected corrugated iron outlet culvert to a 22-in. tile drain, set in Clay County, Minnesota, in 1916. *Upper left:* Original culvert extending through a road embankment and surrounded and supported only by loose rocks, as it appeared in 1920. *Lower left:* Where the culvert had been—after the flood in May, 1922. *Right:* A close-up view of the end of the 22-in. tile after the flood, 30 ft back of the original tile outlet; point *X* in lower left view.

The footing of the bulkhead should be set deep enough to obtain a firm, natural foundation of gravel, rock, or hard clay, and it should be fully protected against undermining by seepage. When the soil on which the foundation of the bulkhead rests is of a character that is liable to be

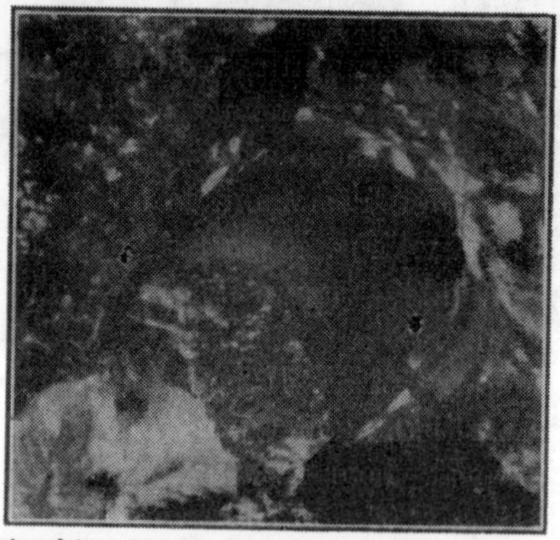

FIG. 12.34. All that remained in 1922 of a masonry bulkhead similar to that shown in Fig. 12.32, built around a 20-in. tile but on a yielding foundation of silt in Itasca County, Minnesota, in 1910.

FIG. 12.35. Improperly designed and partially failed head wall at tile outlet in Webster County, Iowa. Typical of many in the Middle West. Outlet buried under silt due to neglect in maintaining outlet ditch.

gradually cut away by seepage or eroded by the flowing water in the outlet ditch, undermining of the structure is almost sure to occur eventually, with consequent destruction of the entire tile outlet. Such failure is well illustrated by Fig. 12.34. The only sure way to prevent such failure is to provide a complete envelope of sheet piling around the entire foundation, driven deep enough if possible, to reach below the permanent water table. This precaution cannot be too strongly emphasized.

Often a straight head wall is built across the tile outlet as a protection thereto. In such a case the head wall should be properly designed to withstand the maximum pressure that may develop in the fill behind it; otherwise failure will eventually occur (see Fig. 12.35).

A screen is necessary to keep out living creatures that are, otherwise, a dangerous cause of clogging. The screen may be of fine mesh poultry netting, or it may be a grating of slender iron bars spaced not over an inch apart. The bars may be set in the concrete of the bulkhead or in a frame so fastened that the whole grating may be removed to clean out the mouth of the drain and be again replaced. Screens of these types are shown in Figs. 10.25 and 12.36.

FIG. 12.36. Removable screen of small iron rods in wooden frame and held in place by thrusting bottom of rods into ground and hooking the top over the bulkhead. The bulkhead is simply a well-built end wall, still good after 29 years of service. North Dakota Agricultural Experiment Station at Fargo.

12.25. Miscellaneous Items. On almost every tile-drainage project in the course of installation, conditions exist or develop that have no direct connection with any of the items thus far listed. Such items can, therefore, only be classed under the general head of "miscellaneous," and each such item usually has to be handled by itself in a manner best suited to the individual problem. Some items of common occurrence are temporary ditching for removal of flood water from areas whose flooded condition, until removed, prevents the regular progress of construction, the removal of stumps or boulders from the path of a drain line, the removal of excess earth that has clogged the trenches as a result of cave-ins or slides, and the placing and removal of cribbing or plank stays to prevent such cave-ins. Any well-managed drainage construction job will have made available in advance an allowance of time, labor, lumber, and tools to care for such special needs as they may occur.

12.26. Maintenance of Tile Drains. Maintenance of tile drains properly installed with high-grade tile is a minor but important requirement. (1) Open catch basins should be inspected and cleaned out after every heavy rainstorm and at least twice a year under ordinary rainfall conditions. (2) Tile outlets should be inspected in both fall and spring and obstructions of any kind removed. (3) Surface evidences of stoppage or failure of a tile drain should be investigated promptly upon their appear-

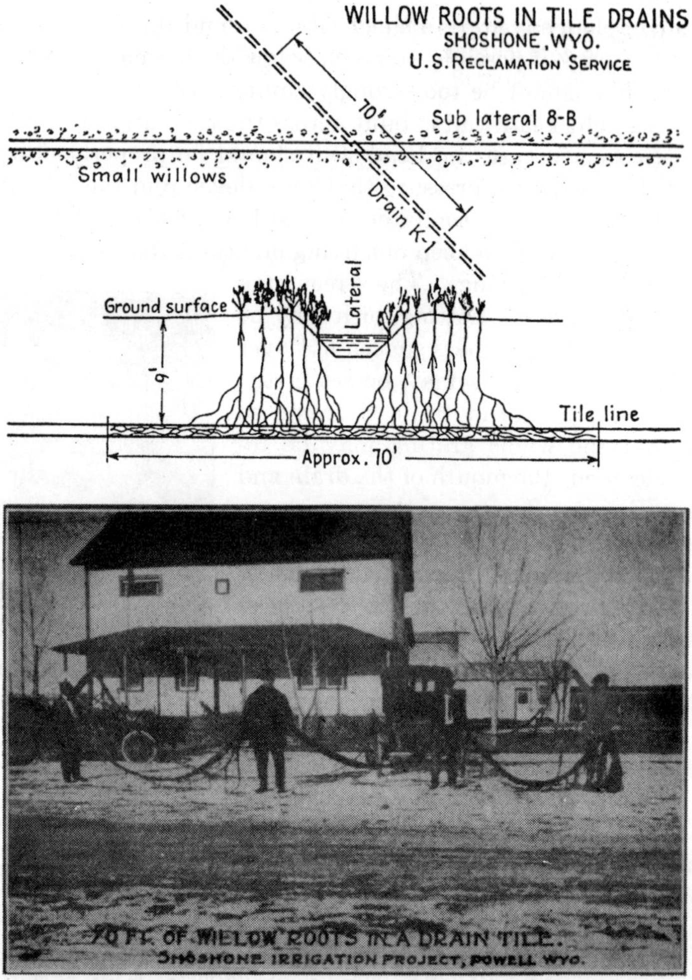

FIG. 12.37. Willow roots clog drain tile. *Above:* Location of willows and the tile lines. *Below:* Roots from the tile line. Shoshone irrigation project, Wyoming.

ance and necessary removal of obstructions or replacement of broken tile at once attended to. (4) Willows or other soft, fast-growing, water-loving types of brush or trees should be prevented from growing near or over tile lines. Otherwise choking of the tile by roots is almost sure to occur sooner or later (see Fig. 12.37).

CHAPTER 13

DRAINAGE OF IRRIGATED LANDS

Strange as it may seem to the uninitiated, thorough drainage is practically a universal need on irrigated lands if they are to continue to be productive. This is due largely to the facts that soils in arid regions and the conditions and waters attending their use are, as a rule, quite different from those in humid regions. These differences need to be understood.

13.1. Arid and Humid Soils Contrasted. In arid regions the agencies of soil formation are almost wholly physical (largely wind and temperature effects) rather than chemical (such as water and the salts dissolved in it). Hence arid soils are usually coarser textured, permeable to greater depths, and better aerated than are humid soils. Virginally they are usually deficient in nitrogen owing to the sparseness of native vegetation, but the nitrogen content of the virgin humus present in them is usually higher than in humid soils. In the latter a much larger part of it is leached out by rainfall and drainage. In arid soils the nitrogen content must be built up. This is easy to accomplish because of the open character and good aeration favorable to bacterial action in such soils.

In arid soils both the desirable salts of calcium and phosphorus and the undesirable ones of sodium and magnesium are apt to be present in excess owing to their presence in the soil-forming rocks, light rainfall, high evaporation, and upward leaching action from strongly alkaline substrata. In humid regions the reverse of the situation in arid regions obtains, and the chief soil-forming agencies are chemical or partly chemical. The principal agency is water which, as the universal solvent, furnishes chemical solutions that, aided by often repeated wetting, drying, and frost action, convert the rock materials into soil by disintegration, solution, and sedimentation. This results in a finer texture, greater compactness, and often poorer drainage conditions. The alkaline salts such as those of sodium and potassium tend to leach out. On the other hand because native vegetation is usually much more abundant, nitrogen coming from organic decomposition is carried into the soil by the moving water in great quantities, although much of it is leached out again as the water moves out into drains or into the deeper soil strata. The net result, however, is greater virgin fertility in the arid soils as evidenced by the much higher yields secured from them under irrigation than are secured from the same crops in humid regions.

The impervious clay soils that are of frequent occurrence in humid regions are usually lacking in arid regions, but they do occur, and at such points they interfere with deep percolation and the drainage opportunity so necessary in western irrigation and tend to render farming operations unsuccessful unless they are broken up by the expensive operation of blasting or by deep tillage. Either of these operations should be followed by addition of abundant organic matter.

In arid soils secondary impervious layers of calcareous material and clay sometimes form within the rainfall penetration zone of virgin soils. According to Thorne and Peterson [150] these result first by the leaching of the lime out of the surface horizon into the deeper strata where a new calcareous layer results: An excess of lime in the soil hinders the movement of clay, but the lime having been leached into the lower strata, the finer clay follows and forms a separate zone *above* the zone of lime accumulation. In tight soils such secondary relatively impervious layers tend to poor drainage and a high water table; but this condition sometimes *improves* the farming quality of a region, where originally a very open virgin soil has existed to great depths, by plugging against serious water loss by deep percolation and seepage. An illustration of this kind of development (but artificially activated, not the result of natural geological development) is the region tributary to the Qualley Ditch taking out of the Deschutes River at Bend, Oreg.

13.2. The Function of Drainage on Irrigated Lands. There are many notable cases on record where recognition of the need of drainage on irrigated lands, and its prompt installation thereafter, have produced excellent results in elimination of the most serious problems of irrigation. A few cases are here cited.

Brown [19,20] cites several large projects in central and north-central Utah where, around the turn of the century, seepage, high water table, and accumulation of alkali resulting from several decades of irrigation without contemporary drainage had forced abandonment of thousands of acres originally highly productive; but through drainage installed on these abandoned lands early in the twentieth century, they were restored in 3 or 4 years, at very low cost, to their original state of high productivity.

In 1925, Hancock [52] reported on a 22,000-acre project in the Gila River Basin in Arizona, which, with a 289-day growing season, an average temperature of 69.4 degrees Fahrenheit, and ample water to irrigate 17,000 acres, was highly productive, originally. Thirty years of irrigation without drainage had waterlogged and ruined 3,000 of the most productive acres. Ample outlet ditches provided in 1923 quickly lowered the water table well below the root zone and, in a very short period, restored the waterlogged area to 100 per cent productiveness.

A similar problem originally existed in the large agricultural area in the Imperial Valley, California. The 10-year report of the Imperial Valley Drainage Investigations at the end of 1950 states that 126,000 acres of this waterlogged area has already been reclaimed by tile drainage and that this reclaimed acreage is being increased each year by several thousand acres.

A committee of the Irrigation Division of the American Society of Civil Engineers [4] in 1930 summarized the problem of drainage of irrigated lands substantially as follows: Many of the ancient irrigated lands of the Middle and Far East, including Chaldea, Goshen, Israel, and Egypt, and even the Salt River Valley in what is now known as the state of Arizona in the United States, all once very fertile and finally abandoned for centuries because of the accumulation of alkali from overirrigation and lack of drainage, are now coming back under modern treatment, including drainage, into a prosperous agriculture. In spite of this fact it has been estimated that in the United States during the early third of the twentieth century 20 per cent of the irrigated land has again been rendered unfit for producing good crops because of waterlogging and accumulation of alkali resulting from excessive irrigation and from seepage from canals and higher-lying overirrigated lands, *largely because of lack of drainage.* The *Reclamation Record* of April, 1921, reported that on 22 irrigation projects scattered over the arid West, through 7 years, seepage varied from 6.1 to 53.2 per cent of all the water diverted from basic sources, with a general average of 33 per cent. It was further reported that the mean seepage losses from U.S. Bureau of Reclamation canals in 24 soils and combinations of soils showed varying rates per day per square foot of wetted surface of 0.29 to 2.53 feet of depth, with a mean of 0.87 foot. In discussing the report M. R. Lewis emphasizes the necessity for recognition of the influence of alkali on the need for drainage and warns that, without leaching and deep drainage, high water table will occur and alkali will accumulate. Brown and Hart [21] emphasize the fact that drainage of irrigated lands, as protection against high water table and alkali deposit, is more certain and reliable than reclamation and far less costly.

13.3. Drainage Functions Listed. In the light of such experiences and conditions as those just listed, the report cited in the *Reclamation Record* summarizes the function of drainage of irrigated lands substantially as follows:

1. Prevention and elimination of *high water table* and *waterlogging* of the soil.
2. Prevention of *excess salt and alkali accumulation* and removal of existing accumulations.
3. Prevention and control of *soil erosion* from excess and waste irrigation water.

13.4. Adaptable Types of Drains. These three major functions of drainage in irrigated lands embrace numerous types of drainage problems as will be seen from the subsequent discussion. They also involve many classes of soils; but, in many cases, somewhere in the area affected and within the depth zone of the drains, will be found some deposits of sticky clay soils. These clays usually have a type of structure that impedes vertical movement of water and thus hinders drainage but which permits reasonably free horizontal movement. When wet they become a pasty gluey mass that will clog and eventually wholly seal the joints of pipe drains unless such drains are laid in a gravel envelope of a thickness sufficient to prevent sealing the drains. Therefore, practically all accepted types of drains are called into play, each case having to be treated as a special problem.

Usually an underdrain of some kind is the most satisfactory, but not always. For example, if at its first appearance in a given area the excess water is concentrated within narrow limits and not far from a good outlet, a short open ditch heading right into the wet spot, where it can quickly pick up and remove the excess water before it can spread much laterally, usually gives the most satisfactory results. Gradients should be nonerosive and side slopes suited to the natural stability of the undisturbed soil.

If, on the other hand, the breakthrough is spread for some distance along the slope, an intercepting *underdrain*, approximately along the water contour, will usually be best. The trench should be left open for a short time to enable measurement of the flow from which to determine the size of drain needed. If the substratum is nonerodible material at a depth suitable as a base for the drain, a triangular wooden box drain, like the slat-bottom type which is shown in Fig. 6.1, can be used effectively and thus avoid the clogging that tends to occur with pipe drains in clay soils once they become wet; but this type of drain must not be used on an erodible base. It is fairly economical as it has a normal life of about 20 years and is inexpensive in first cost; but it is subject to destruction by moles and larger rodents who bore large holes from the ground surface down into the drains, thus ending their usefulness.

Any high-grade drain tile or No. 2 concrete irrigation pipe with tongue and groove ends makes good underdrains. The last-named is being used increasingly and is gradually crowding out the box drains because it is practically permanent and usually readily available. It is laid with joints open about ¼ to ½ inch at the outer circumference, and often each 3-foot length is perforated with a 1-inch hole, located in laying, slightly below the lateral mid-point. Each joint should be protected by a small piece of burlap. In the stiff pasty clays just mentioned concrete pipe must be laid in a gravel envelope to avoid sealing the joints. Ordinary drain tile may be laid in the same manner with open joints and in a gravel envelope, and a small opening may be made in one end of each length of tile, the opening being placed on its bottom when laid.

Good judgment must be exercised by the engineer in selecting the type of drain best suited to any given case.

HIGH WATER TABLE

High water table results from prolonged presence of excess water from whatever cause and persists where adequate drainage is lacking. The generally coarse texture of irrigated soils usually induces a high permeability rate and a consequent tendency to a rapid rise of water table with three damaging influences on plant life:

1. *Stagnant saturation and waterlogging*, generally known to be strongly harmful and eventually fatal to upland plant life through diluting the plant food solutions and tending to reverse the osmotic process.
2. *Increased salt accumulation in the root zone*, resulting from the higher rate of evaporation induced by the nearness of the water table to the ground surface. Most of the irrigation water thus evaporated has an appreciable salt content which is left behind on evaporation.
3. *Development of chlorosis*, a weakening of the plant tissues by bleaching the chlorophyll through diluting the plant-food solution and thereby weakening the plant-growth processes.

This last result is extremely serious even in the annuals which complete the life cycle each season, but it is especially detrimental in the woody perennials (shrubs and trees) where the weakness is carried forward from one year to the next and becomes a continuous and expanding disease. Many an originally valuable orchard has been thus ruined within a year or two. A notable example of this kind of damage from high water table is discussed by Larson and Bowman [71] relative to the Sunnyside Valley Irrigation District within the West Benton Soil Conservation District in Benton County, Washington. The trouble occurred as the result of the development of the Roza Irrigation District on the higher slopes of the Yakima Valley above the districts just noted. These districts must either take care of the ultimate outflow from the Roza or permit its passage through them and receive the soil eroded therefrom along with the alkali burden deposited from this outflow. The result has been the dying out of orchards and vineyards and the serious damaging of originally fine meadowlands and pastures. The valleys in the intermountain country on the east slope of the Cascade Mountains present many cases of this kind.

By its very nature irrigation tends periodically to produce a saturated condition of the soil in the upper layers which constitute the root zone of plants. Hence, over long periods of time, where overirrigation or other sources of excess water occur, even deep soils tend to become saturated and waterlogged where adequate relief by drainage is lacking.

13.5. Specific Causes of High Water Table. Some of the most commonly occurring sources of high water table and waterlogging are the following:

Normal Overirrigation. Usually about 10 to 20 per cent.

Surface Waste. From allowing water to run too long. Records show total seasonal applications as high as 12 to 15 acre-feet per acre, or three to five times actual need. Or from applying too large heads, thus frequently causing water to accumulate at the lower end of runs without adequate facilities for prompt removal of this waste water.

Lengths of Runs from ⅛ to ¼ Mile. Resulting in waterlogging the upper end to great depths while the lower end may not receive enough water. Runs of 200 on very open sandy soils to 600 feet on tight clay loams or clays, in most cases for land slopes of about 2 per cent, are now known to be about the upper limit of good practice in most cases.

Surface Washing. Removal of surface crusts of alkali requires heavy applications of surface water which must be removed quickly over the surface, carrying the salt with it, before there has been time for any great quantity of the water to soak into the soil. Soil erosion is apt to result. A common lack is surface ditches of adequate grade and capacity.

Leaching. Another source of excess water in the soil is the large amount applied on the surface to leach out the excess salts in the soil as it percolates through the root zone to drains in the deeper subsoils. Such treatment is useless and worse if both natural underdrainage through permeable subsoils and artificial underdrains of sufficient depth and capacity to receive and *quickly* remove the leaching water are lacking. Otherwise this water will simply saturate the soil and not remove the alkali, and leaching water must often amount to several acre-feet per acre per season. In a later section of this chapter it will be shown how to determine the amount of leaching water and the depth and capacity of underdrains required, to be effective in any given case.

13.6. Leaching and Loss of Plant Nutrients. Thorne and Peterson [150] list a number of well-known investigators who make the claim, said by some authorities to be not yet strongly authenticated, that leaching also removes a large quantity of plant nutrients, thus tending to offset the benefit of diminishing the alkali content. Imperial Valley experience strongly supports this claim, results there probably being due to the high permeability of their subsoils. In a recent unpublished report G. R. McDole gives a brief abstract of the studies of several scientists on loss of plant nutrients from overirrigation and leaching. This report shows that there is as yet no general agreement in this matter beyond the fact that damage is caused by overirrigation but that the character and amount of the damage is not yet fully understood and that standards of measurement are lacking. It seems, however, to be generally agreed that, if heavy leaching does carry away serious amounts of plant nutrients, normal irrigation does also. Furthermore, there seems to be fair agreement that the most serious losses from this cause are in calcium and nitrogen, beyond the losses of the soil itself, but that the *greatest* apparent damage is in changes in character of sodium salts and in soil erosion. The final majority conclusion appears to be that damage from overirrigation and leaching, which is a deliberate form of overirrigation calculated to reduce excess salinity, can be corrected, the cost being the deciding factor as to its feasibility. In the end the unchanged assumption is that leaching is essential for the purpose for which it is used and that prevention of alkali accumulation is far better than attempted reclamation of lands already thus badly affected by its presence and that the practice of leaching—with drainage accompaniment—appears to be entirely justified in the experience of Western land operators. However, it should be pointed out that the more permeable the soil, the more effective leaching will be in the removal of alkali and excess salts. To this end a greater use of manure and other organic matter is strongly urged. Some scientists of wide observation and experience feel confident too that if an infiltration and vertical permeability rate of 1 inch in 24 hours can be obtained, by this or any other legitimate method, even strongly alkaline lands can be reclaimed.

13.7. Upward Flow from Underground Sources under Artesian Pressure. Artesian water under pressure is a serious cause of waterlogging if relief-well outlets into underdrains are not provided.

13.8. Seepage. Seepage occurs to lower-lying lands from natural streams, around irrigation canal structures, the anchorage of which is not tightly sealed, and from all sorts of irrigation canals and ditches (much greater from unlined than from lined ones), and from leakage of artesian wells, experienced in various parts of the West.

Seriousness of the Seepage Problem

In a report covering the results of some 400 measurements of seepage losses from irrigation canals, both lined and unlined, in several of the states of the arid West and in a wide variety of soils from open sandy loams to tight clay loams and clays, Rohwer and Stout [124] state that, "the data show that the losses range from 17 to 67 per cent of the total diversions." Table 13.1, a rough analysis of the data shown in the report cited, on the basis of simple arithmetic averages, gives an approximate summary in more specific irrigation units, according to the sizes of the ditches or canals and their classification as lined or unlined. The

TABLE 13.1. APPROXIMATE AVERAGE SEEPAGE LOSSES PER MILE PER 24-HOUR DAY FROM PROJECT OR DISTRICT DITCHES OR CANALS

Size class of ditch	Length of wetted perimeter, ft.	Unlined channels, losses in			Lined channels, losses in			All channels, losses in		
		Cubic feet	Acre-inches	Acre-feet	Cubic feet	Acre-inches	Acre-feet	Cubic feet	Acre-inches	Acre-feet
Small:	Under 8									
Maximum	....	795,000	...	18.2	71,500	.....	1.6	582,000	.....	13.8
Minimum	....	200	1/16	......	+700*	+3/16	...	+700	+3/16	
Average	....	123,000	...	2.8	13,600	4	...	65,500	.....	1.5
Medium:	8–15									
Maximum	....	1,630,000	...	37.25	229,500	.....	5.25	1,630,000	.....	37.25
Minimum	....	420	1/8	......	1,020	1/2	...	420	1/8	
Average	....	183,800	...	5.8	34,000	.....	1.1	252,000	.....	4.2
Large:	Above 15									
Maximum	....	5,380,000	...	123.0	420,000	.....	9.7	4,500,000	.....	106.1
Minimum	....	2,860	3/8	......	2,400	2/3	...	1,200	1/4	
Average	....	833,000	...	19.0	86,600	.....	2.0	506,000	.....	11.5
All sizes:	2.7–91.9									
Maximum	....	5,380,000	...	123.0	420,000	.....	9.7	1,163,000	.....	37.25
Minimum	....	200	1/16	......	+700	3/16	...	200	1/16	
Average	....	380,000	...	9.0	44,700	.....	1.2	274,800	.....	5.7

* + means inflow.

evidence shown in the table is far from complete and, consequently, far from exact because the ditches covered in the study include only a small fraction of all the canals in the irrigated states. However it is sufficiently accurate to show that the seepage problem is so vast as to call for effective drainage systems to collect and carry off the seepage water and reduce to a minimum the waterlogging therefrom of the lower-lying lands. The general average of nearly 6 acre-feet of water thus lost per 24-hour day per mile of canal is a lot of water cut off from the lands it was originally diverted to irrigate. It is sufficient per day to furnish an average total season's irrigation requirement for 2 acres of land per mile of irrigation canal. Therefore, its value is sufficiently ample to justify the cost of a drainage system capable of collecting and carrying it to other lower-lying lands needing it and lacking a sufficient water supply from other sources.

Another fact of great economic interest is that, on the average, seepage loss from lined canals is only about one-seventh of that from unlined canals of otherwise similar character, which fact alone, in view of the general scarcity and cost of water in the arid states, justifies a considerable expenditure for lining canals and ditches. A similar problem, on which little exact knowledge is available, arises out of seepage losses from natural streams that flow through permeable soils and openly laminated rock channels.

The question may arise why there should be any seepage loss from lined canals. The answer is that the foundation on which such lining is laid is the rather unstable natural soil of the bed of the channel, which instability often results in time in the development of cracks and fissures in the lining. This weakness is further augmented by the fact that many of these linings are very thin and the lining material in many cases is not 100 per cent effective.

SOIL SALTS

13.9. Alkali. Serious as is the high-water-table problem, the alkali problem, especially aggravated by the high-water-table problem, is much more serious. However, it must not be assumed that all irrigated lands contain alkali, for they do not. Probably there is about as much alkali-free land as there is alkali-infected land in the irrigated West. In agriculture the harmful chemicals called alkali, as well as other salts, must be dealt with as they occur in soils and in irrigation water. To deal intelligently with this problem it is necessary to distinguish between salty soils that are *not* alkaline (*saline soils*) and those which *are* alkaline (*alkali soils*). Hayward and Magistad [56] make this distinction substantially as follows: Saline (white alkali) soils usually have good structure, take water well, and drain well. Alkali (black alkali) soils run together or "puddle," have poor structure, take water very slowly, and are difficult to leach. A saline soil may be made productive by leaching out the salt, but the alkali soil must be changed chemically by replacing the excessive sodium held by the soil with calcium and then leaching out the replaced sodium salts.

Generally included in the category of *saline soils* are those in which such neutral and not readily soluble salts as the carbonates, phosphates, and sulfates of calcium predominate. The alkali soils are those in which the hydroxides and chlorides of calcium, sodium, potassium, and magnesium and the carbonates and often the sulfates of the last three predominate.

13.10. Sources of Soil Salts. The sources of soil salts are various. An original source is the parent rock as the fundamental elements are in much of that rock. In many areas quantities of the salt were brought in with the soils themselves, transported by wind, water, or ice from their points of origin to many other points. In many areas, as explained by Scofield [128], much of the salt has been brought there in solution and deposited by evaporation or is filtered out as the water moves into and through the soil, as most of the irrigation waters have absorbed an appreciable salt content by the time they reach the land they are to irrigate. Some of this salt is already in the water in its place of storage; some of it is picked up from the rocks or soils over or through which the water flows in reaching the reservoir or the irrigated lands. Hydrostatic pressure often forces salt water from great depths up into the root zone. The part of this water from these various sources which evaporates leaves its salt content behind in the soil. These are all natural sources the total of which may be very little or a great deal—even more than most plant life will tolerate. The irrigated soils may retain most of it, or they may retain but little, according to the character of the soil, the effectiveness of the underdrainage, and the strength of the evaporation opportunity. In addition, man-made sources occur, such as mixing already impregnated water with the original supply, salts added in soil treatment to modify the character of existing salts in the soil, and cultural activities that may stimulate evaporation or interfere with drainage. These man-made sources may add to the soil 2 to 4 tons of salt per acre per year.

13.11. Plant Tolerance for Alkali. Soil alkali and often a great excess of so-called neutral salts is inimical to plant vigor and life, but the tolerance for salts and especially for alkali varies widely as between different crops. For effective handling of the salt problem it is, therefore, necessary to know the tolerance of the crops to be grown in a given region for salts, expressed in measurable terms.

13.12. Salt Content: How Expressed. There are four accepted ways of indicating salt content in the soil:

1. As a percentage of the dry weight of the soil.

2. As parts per million (ppm), by weight, in the soil water, or as a percentage of the total weight of the soil water (that is, 10,000 parts per

million is the same as 1 per cent; 1,000 parts per million the same as 0.1 per cent; etc.).

3. In terms of electrical resistance, in ohms, of the soil sample or soil-water sample compared with the corresponding parts per million noted in item 2.

TABLE 13.2.* THE RELATIVE TOLERANCE OF SOME WESTERN CROP PLANTS TO SALT CONSTITUENTS IN THE SOIL SOLUTION

Strong tolerance, approx. 1.00–0.50%‡	Good tolerance,† approx. 0.50–0.30%‡	Medium tolerance,† approx. 0.30–0.10%‡	Weak tolerance, approx. 0.10–0.05%‡
Sugar beets	Barley (hay crop)	Onions	Beans, wax
Garden beets	Tomatoes	Squash	Beans, navy
Milo	Alfalfa	Carrots	Field peas
Bermuda grass	Corn	Asparagus	Horsebean
Rhodes grass	Sorgo	Cow peas	Red clover
Alkali sacaton grass	Kale	Rice	Vetch
Fescue grass	Rape	Rye (grain crop)	Proso
	Perennial ryegrass	Barley (grain crop)	Oats (grain crop)
	Wild ryegrass	Wheat	Emmer (grain crop)
	Italian ryegrass	Oats (hay crop)	Peaches, Elberta on Lovell rootstock
	Crested wheatgrass	Grain sorghums	
	Slender wheatgrass	Foxtail millet	
	Mountain bromegrass	Flax	
	Tall fescue	Guayule	
	Reed canary grass	Sunflower	
	Sudan grass	Ladino clover	
	Blue grama grass	Strawberry clover	
		Sweet clover	
		Smooth bromegrass	
		Meadow fescue	
		Orchard grass	
		Dallis grass	
		Tall oatgrass	

* Table 1 of "The Salt Problem in Agriculture," *USDA Misc. Pub.* 607.

† A few crop plants listed in the medium and good groups do not die in strong salt concentrations, but they fail to make appreciable growth.

‡ Figures inserted by the authors.

Purely from close observation but not from scientific test, Magistad and Christiansen [74] comment substantially as follows on tolerance of alkali by fruit crops:

Date palms are the most salt resistant of all cultivated plants.

Grapes are highly tolerant.

Apples and pears are sensitive to salinity; will not grow well in over 2,000 ppm (0.20%).

Peaches are very sensitive.

Of the citrus fruits:

Grapefruit will tolerate most salt.

Lemons are the most sensitive to salt.

4. In terms of the specific conductance of the solution extracted from the soil—in millimhos the conductance of a body through which one ampere of current flows when the potential difference is one volt.

13.13. Recorded Plant Tolerance. The U.S. Salinity Laboratory at Riverside, Calif., through Hayward and Magistad [56], list the tolerance of a large number of crops for soil salts as shown in Table 13.2.

13.14. Plant Injury by Saline Soils. It should be understood that even salts not classed as alkali may injuriously affect plant life if present in appreciable excess over plant needs, just as too much at one time even of the best food may not agree with a person. In plant life this is especially true of exchangeable forms of sodium salts such as, for example, sodium sulfate. This may appear from the following discussion.

The fact that most Western waters available for irrigation have come to contain appreciable salt content poses a serious problem. According to Magistad and Christiansen [74], water with a salt content up to 700 parts per million or with sodium per cent less than 60 may be classed as excellent to good; with 700 to 2,000 parts per million salt content or with sodium per cent 60 to 75 as good to injurious, probably harmful to the more sensitive crops; with over 2,000 parts per million or with sodium per cent over 75 as generally injurious and unsatisfactory, probably harmful to most crops and unsatisfactory to all but the most tolerant. The second class just listed, where the annual water requirement is 1½ to 3 acre-feet per acre, will in some areas, according to some authorities, change a salt-free soil to a saline soil throughout the root zone in 3 to 5 years.

Control of High Water Table and Alkali

13.15. Prevention. Prevention of the occurrence of high water table and of accumulation of excess salts in the soil always results in the reduction of the amount of drainage required on irrigated lands. Either or both of these undesirable conditions may occur on almost any type of area from causes earlier discussed. The probabilities of such occurrence should appear from a thorough survey of the area, preceding its opening up to agriculture.

Water from naturally wet higher-lying lands should be intercepted at the upper edge of the given area by an intercepting drain normal to the line of flow and leading to an adequate outlet, in the manner discussed in Sec. 6.22 and illustrated in Fig. 6.10. As a rule an underdrain is preferable to an open ditch in such a case, but the choice must rest upon the relative cost, probable effectiveness, soil and topographic conditions, and location of the outlet relative to the affected area. Often a single drain of ample capacity will suffice, but a series of parallel drains may be required, spaced at intervals down the slope and of a size and gradient dependent on the character of the soil and subsoil and the amount of water to be cared for as shown by the survey.

If the source of the water is seepage from canals, the first consideration should be to stop or reduce the seepage by lining the canals. This subject is treated in detail in a later section. If lining the canals is not sufficient, it should be supplemented in the manner just discussed, by an intercepting drain on the lower side of the canal, near and practically parallel to the general direction of the canal, the depth to be fixed by determining the depth of the subsurface water flow.

If the source of the water is an artesian basin from which the water is forced to the surface by artesian pressure, the proper procedure is to provide relief wells at points where the upward flow is greatest, outletting into either existing surface drains or underdrains, already serving the area or built especially to receive the artesian water and carry it to an adequate and convenient outlet. An underdrain is preferable and should be designed with ample capacity for the purpose.

Where the water is from adjacent higher-lying lands, whether as subsurface flow and seepage or as surface flow either from overirrigation or as normal waste water, the first proper step is better controlled use of his irrigation water by the operator of the higher-lying land, including orderly disposal of unavoidable waste water. A problem of this kind can usually be thus met if the disposition of the neighboring farmer is cooperative. Otherwise the steps just outlined for disposal of excess water from other sources must be resorted to.

Prevention of accumulation of excess or undesirable salts will usually be automatic in the cases just discussed.

13.16. Elimination. When the effects of excess water on an irrigated area already exist in the form of high water table and/or alkali, according to the topography, soil, availability of outlets, and kind of crops to be raised, the methods of control and elimination are the same, often with slight variations to fit some purely local condition, as those outlined for prevention of these evils. The various methods may be summarized as follows, accompanied, in individual cases, by a brief statement of the steps called for and the function of each method:

1. Elimination of overirrigation.

2. Provision of adequate outlet drains for the prompt removal of unavoidable excess water from whatever cause.

3. Open ditches for quick removal of surface flooding calculated to eliminate surface accumulations of excess salts and alkali.

4. Cutoff drains, preferably underdrains, to collect and remove seepage from whatever cause; such drains, especially on flat or relatively flat lands, often needing to be a complete system of parallel drains, designed to fit the type of soil, kinds of crops raised, and amount of water to be removed.

5. Deep underdrainage systems to eliminate high water table and to carry off water from heavy surface flooding required to leach out alkali and other excess salts from the root zone. These drains should never be less than 6 feet deep, and their capacity should be ample for quick removal of the excess water.

The subject of leaching and of underdrainage for the removal of high water table and salts is discussed in much fuller detail in a section immediately following this summary.

6. Pressure-relief wells as above discussed, for prevention of waterlogging through pressure from artesian basins.

7. Adequate outlet ditches to receive and carry all drainage water from the area to a normal outlet.

8. Pumping outlets from sumps and shallow wells, for underdrainage systems on flat lands, as for example, in the San Joaquin Valley and other areas in central California and pumping from deep wells for direct elimination of high water table when local conditions are especially suited for such a system, as in the Salt River Valley, Arizona.

Pumping for drainage is discussed in fuller detail in a later section.

SOIL EROSION

Soil erosion has a tendency to occur in irrigated fields from any of the following causes:

1. Use of excessively high heads in applying irrigation water.
2. Use of surface gradients which result in erosive velocities.
3. Removal of surface crusts of alkali by surface flooding, which must be heavy with grades steep enough to remove it quickly over the surface if it is to be successful. It is not always effective (see Sec. 13.5).

Prevention and control of erosion from these causes is accomplished by reduction of irrigation heads and field gradients in irrigation runs and by use of smaller amounts of floodwater at more frequent intervals before heavy crusts of alkali develop on the surface. Nearby surface outlet ditches of ample capacity are also essential.

CANAL LININGS, LEACHING AND TILE DRAINAGE, AND PUMPING FOR DRAINAGE

The following section is devoted to a detailed discussion of these three related topics.

13.17. Canal Linings. It has been shown in an earlier section that seepage from canals and reservoirs and around irrigation structures is one of the leading causes of high water table and the accumulation of excess salts and alkali in the root zone of most low-lying areas. It has also been shown that the seepage from such sources has been reduced, on the average about 85 per cent, by lining the irrigation canals. The U.S. Bureau of Reclamation [115] state the requirements of a good canal lining, after Etcheverry [44], as follows:

> In his text "Irrigation Practice and Engineering," B. A. Etcheverry established the following requirements for a good canal lining: "It should be practically water-tight to prevent seepage loss and the resulting water-logging and rise of alkali; its cost should not be excessive; it should prevent the growth of weeds; resist burrowing of animals; be strong and durable, preferably not affected by trampling of cattle; adapted to the construction of a smooth canal of the proper shape to increase its carrying capacity and permit the use of high velocities." These, together with a reasonable amount of flexibility and extensibility, are the desirable characteristics of an ideal lining.

In Ref. 34 is also given a fully illustrated and complete discussion, beyond the scope of this book, of the methods of preparing the canal surfaces to receive the lining and the comparative value and method of handling the various canal-lining materials mentioned in the following paragraphs. Those charged with the responsibility of placing canal linings are advised to have this pamphlet at hand for continued reference and guidance along with the Utah bulletin by Israelsen and Reeve [63]. A number of materials have been used for this purpose with varying success. Prominent among these may be listed portland cement concrete, bentonite, asphalt, various clays, and oil. The effectiveness and permanence of any of these materials is largely dependent on the density and firmness of the earth side slopes and bottom of the channel on which the lining is placed and on the care exercised in placing the lining. One additional precaution that is necessary for safeguarding the life of a lining is sterilizing the soils of the ditch bed before placing the lining, because organic materials in soils, coming in contact with the lining materials, have a strongly disintegrating action on some of them. This is especially true with concrete and may be with some asphalt membranes. The final choice of materials to be used in a given case is, as a rule, dependent on the local availability and relative cost, both materials and labor considered.

Concrete. Portland cement is generally available, and linings of portland cement concrete are undoubtedly the most reliable although their cost is relatively high; but their relative economy can be judged only in comparison with the value of water based upon the urgency of the demand for it. Concrete linings of thicknesses varying from less than 1 inch to 4 inches or more, according to the importance and size of the canal and firmness of its earth bed, have long been used with considerable satisfaction. Methods and implements of placing are not greatly different from those used in sidewalk, highway, and other slab construction, slightly modified to fit the ditch work. Machines are also available expressly for laying concrete linings in canals. Gunite linings, often reinforced with poultry or other wire netting, have long been in use (see Fig. 13.1).

Asphalt. Various plastic asphalt mixes have been used with fair satisfaction, with no other support than the earth bed of the ditch, for lining small ditches through uniformly fine firm soils. But they have not given entire satisfaction in large canals through coarse materials because the asphalt softens in warm weather and yields under pressure of water, livestock, etc., and eventually develops holes. Leaks, once started, are hard to stop. Experimentation is gradually improving this condition. On the other hand, asphalt membranes, built on a base sheet of coarse felt, plant fiber, or thin noncorrosive metal such as aluminum have overcome these objections to asphalt alone and give good satisfaction. For canals a type of membrane should be selected which will be resistant to the disintegrating influences of the organic material in the soil of the ditch bed.

FIG. 13.1. Demonstration of lining an irrigation canal with gunite. Montecca, Calif., June, 1926.

Bentonite. Bentonite also seems to give good satisfaction as a canal lining, but it is not so generally available. While small deposits are found in many parts of the North American continent, according to Powell [106], about the only extensive deposits are found in eastern Wyoming, South Dakota, New Mexico, California, Utah, and South Carolina. Bentonite is a fine clay classified as a hydrous silicate of alumina with the following characteristics: absorbs water freely, becoming a gelatinous, impervious mass; expansion when wet is ten to thirty times its original dry volume but can be restrained by a pressure of 0.04 to 0.06 ton per square foot of surface and is reversible below the fusing point. It is commercially processed and packaged in powdered form in bags of 100 pounds equivalent to 1 cubic foot. In 1940 the price was \$7 to \$10 per ton at the plant plus transportation cost to point of use.

In massive slab form bentonite is usually applied as a powder at a varying rate of 1/4 to 1 pound per square foot and rolled. A final consolidation by rolling has been known to cut seepage 85 per cent. The publication cited gives more exact location of deposits and rather full directions for placing the material. It makes a very satisfactory canal lining and is also an excellent material for waterproofing hydraulic structures and repairing leaks around their anchorages.

Various Clays. Israelsen and Reeve [63] report that, "in the Delta Area, natural clays having low permeabilities suitable for lining canals to reduce seepage are abundant." They list 10 in their studies with coefficients of permeability varying from 0.013 to 0.800 feet per year, which seem practically negligible. It seems probable that exploration would reveal many other deposits suitable for lining canals or ditches, within the North American continent, for example, in the shale hills of the Red River of the North and the Hudson Bay Drainage, as well as on the Colorado Plateau. Puddling the insides of canals with selected clays has long been practiced. Its effectiveness largely depends upon the colloidal content of the puddling clay and the fineness of texture of the main body of the clay. If sufficient care is taken in its selection and its application to the wetted portion of the canal surface, it can be made remarkably effective, as shown by the tests of Israelsen and Reeve. A clay lining should be made at least 4 to 6 inches thick, according to the character of the bed on which it is laid, and firm conpaction by a heavy roller is essential. Thoroughly effective machinery for compacting is still lacking and is urgently needed.

Oil. In small canals through uniformly fine, very compact, and tight soils, heavy crude oils, such as are used on temporary road surfaces, have served as temporary or very light service canal-lining materials, but they possess all the disadvantages of asphalt linings and some additional ones as, for example, deterioration of the water for irrigation. Their use in this way is not recommended.

Although the initial cost of canal lining is heavy, the long-time saving is great. Israelsen and Reeve find justification for canal lining from the standpoint of the irrigation company only on the basis that, "the annual cost of lining must not exceed the value of the water saved each year." On that basis, however, they also report that, with water values existent in 1944 and interest rates and canal lining maintenance not over 9 per cent annually, Delta irrigation companies can justify initial lining costs of 9 cents per square foot on some canals.

LEACHING AND TILE DRAINAGE

Such well-known investigators as Magistad and Christiansen [74], Hayward and Magistad [56], Marr [78], Israelsen [61], and many others agree that the real solution of the alkali problem is leaching with abundance of water, accompanied by thorough and deep tile drainage strongly reinforced by abundant use of certain chemicals, whose function is to convert the black alkali and other not readily soluble forms to forms readily soluble in free soil water. The most important chemicals are gypsum, lime, and other exchangeable calcium compounds. Marr [78] notes that the use of chemicals is relatively expensive and is apt to make reclamation prohibitive, especially in severe cases.

The following reports of investigations, among many others, show clearly that deep underdrainage, accompanied by heavy flooding, is the best means of combating alkali.

In various parts of Utah, Dorsey [38], Harris [55], and Israelsen [61] report marked decreases in total salts through 4 feet of depth of soils in 2 years time, through deep, carefully designed and installed tile drainage followed by heavy flooding.

Fortier and Cone [49] report similar satisfactory results from tile drainage and leaching in the San Joaquin Valley in California with excellent crops following in raisins, sorghum, and alfalfa on previously unproductive areas.

Powers [109] reports removal of maximum accumulations of alkali from the third foot of the heavy loams of eastern Oregon by drains laid 6 to 8 feet deep to tap the permeable strata.

Miller [85] reports remarkable reductions in total salts to depths of $5\frac{1}{2}$ to 6 feet in the silt loams and adobe clays of the Colorado River Valley in Colorado from a single 90-day continuous flooding in 1913 over tile drains laid at various spacings and depths up to 6 feet.

Bloodgood and La Ferrierre [16] report similar results with excellent following crop yields in the Mesilla Valley in New Mexico.

Reeve, Allison, and Peterson [116] report good results in leaching out of excess salts through permeable subsoils without artificial drains in the Delta area in Utah, resulting yields varying closely with the amount of flood water applied. Good results were clearly dependent on an ample supply of leaching water and adequate drainage, whereas without adequate drainage, attempted leaching may be detrimental rather than beneficial.

Burgess [22] also testifies to the marked reduction of alkali by successive leaching at the Yuma Valley Experiment Station.

Along with leaching, Knight [70] also suggests a blanket of 3 or 4 inches of sand over dense soils to stimulate water penetration.

In conference in late 1946, W. W. McLaughlin emphasized drainage as the real and only effective cure for alkali.

13.18. Influence of Per Cent of Sodium Salt in the Water on Soil Permeability. As a possible aid in determining this influence more exactly, the studies by Fireman and Magistad [48] on five western soils shown in Table 13.3 have a definite significance. The report of these studies resulted, in substance, in the following findings:

1. The permeabilities of the given soils vary with the percentage of the sodium salt content of both the irrigation and the leaching water applied.

2. There was no *critical* percentage of sodium salt, either above or below which deleterious effects were apparent.

3. All soils do not respond alike to increasingly great percentages of sodium in the irrigation water.

4. The permeabilities of all the soils decreased when the water was changed from that with a high salt content, where the sodium was 50 per

TABLE 13.3.* CHARACTERISTICS OF SOILS USED

Lab. No.	Soil series and type	Location	pH value	Remarks
60	Aiken clay loam	Paradise, Calif.	5.9	Kaolinitic
85	Reagan clay loam	Roswell, N. Mex.	7.8	Calcareous and gypsiferous
376	Hanford sandy clay loam†	Riverside, Calif.	7.6	Calcareous and high in organic matter
79	Cajon silty clay loam‡	Glendale, Ariz.	7.7	Calcareous
247	Altamont clay loam	Los Angeles, Calif.	6.2	High in organic matter

* Table 1 of Ref. 48.

† Mapped as Hanford silt loam; but Hanford soils are, by definition, calcareous. Correct designation not known.

‡ A recent official communication says this soil is now called Glendale silty clay loam.

cent and up, to ordinary tap water with a much lower total salt content, where the sodium was about 45 per cent.

5. For appreciably favorable physical change to take place at any stage the soil must be allowed to dry.

6. Decrease in permeability is of more vital significance than increase in percentage of sodium.

7. With all the foregoing points noted, no hard and fast results could be anticipated with any given water applied, and sometimes action intended to bring about reclamation resulted in just the reverse situation. Still, when structure is good and organic content high, surprisingly high permeability may result.

8. Various soil minerals, such as lime and gypsum, tend to maintain soil permeability as, by contributing appreciable amounts of calcium in the leaching water, they lower the percentage of sodium and increase the general salt content which tends to increase flocculation.

At this point reference is made to the statement in Sec. 2.11 under

Colloids that, *in a sodium clay, colloids tend to plug the soil completely against the passage of water*. In the light of this fact the studies just discussed are of particular interest because the inference is clear that the findings may and probably do quite generally apply to most western soils and particularly to the various silt and clay loams.

13.19. Importance of Increasing Permeability of Irrigated Soils. All scientists concerned with irrigated soils agree that it is impossible to overestimate the importance of increasing the permeability of irrigated soils by every legitimate means, one of the best of which is abundant use of barnyard and green manures. Some scientists of long experience believe that, if an infiltration and permeability rate of 1 inch in 24 hours can be obtained and maintained, reclamation is economically feasible.

13.20. Thorough Underdrainage Also the Solution of the High-water-table Problem. It should be borne in mind that deep underdrainage and improvement of soil permeability, as above discussed, is also one of the best means of solution of the high-water-table problem whether or not it is needed for the removal of excess salt. The two problems are related in their development. They are quite as closely related in their solution.

13.21. Surface Flooding Requirements for Leaching. In order to know how much additional water above the purely irrigation requirement is needed to leach out the excess salts in a given case to a point where the local crops will tolerate the residue and to provide a drainage system of sufficient depth, capacity, and frequency of drains, it is necessary to determine the following facts:

1. The tolerance of the predominant crops for alkali in terms of the alkali content of the soil expressed as a per cent of the dry weight thereof.

2. The alkali or total salt content of the soil expressed as a per cent of the dry weight to a depth slightly more than that of the proposed tile drains.

3. The alkali or total salt content of the water applied and its capacity to take on more salt expressed in terms of parts per million or as a per cent of the water content.

4. The average permeability rate of the subsoil to a depth at least equal to the required depth of the underdrains (not less than 6 feet and possibly as much as 10 feet below the ground surface).

ILLUSTRATIVE EXAMPLE

Computation of the Amount of Leaching Water Required. If the facts as to the water-holding capacity, permeability, and salt content of the soil and that of the available water are known, the amount of excess water needing to be applied to leach out excess salt can be computed. The method is shown in the following simple example.

Statement of the Example. On a given farm the average alkali content of the soil for the first 4 ft of depth is 0.62% of the dry weight of the soil, whose A_s* is 1.20. The available irrigation water has a salt content of 210 ppm (0.021%). The drainage water contains 1,200 ppm (0.12%). If an annual total of 6 acre-ft per acre is applied, of which 2 acre-ft per acre is required for the production of the crop, how many years will it take to reduce the alkali content of the soil to the average maximum salt concentration (0.25%) tolerated by wheat (grain crop)?

Solution. Considering 1 sq ft of surface area to a depth of 4 ft below the ground surface

Weight of dry soil = $62.5 \times 1.20 \times 4 = 300$ lb.

Original weight of salt in the soil = 0.62% of 300 = 1.86 lb. To be reduced to 0.25% of 300 lb or to 0.75 lb by taking out by leaching $1.86 - 0.75 = 1.11$ lb.

Six feet of water applied yearly = $6 \times 62.5 = 375$ lb.

Total salt in the water as originally applied = 0.021% of 375 = 0.078 lb. All of which must be removed in the drainage water each year in addition to a portion of that to be taken from the soil itself.

Total salt in the drainage water each year = 0.12% of the total yearly drainage water, or $0.0012 \times 250 = 0.300$ lb.

Hence the drainage water removes *from the soil itself* each year 0.300 − 0.078 or 0.222 lb.

Therefore the total number of years required to reduce the alkali content of the soil to 0.25% of the 300 lb of soil or to 0.75 lb would be 1.11/0.222 or 5 years.

Actual design of tile drains for irrigated lands is treated in a subsequent section.

PUMPING FOR DRAINAGE OF IRRIGATED LANDS

Broad low-lying relatively flat lands extending out from the base of slopes wet from seepage and having no good natural outlet for drainage below their own level are to be found on the floors of many broad river valleys in the West. A high water table is a common development in such cases. One widely used solution of the outlet problem is pumping from wells into canals, reservoirs, or natural streams of higher level than that of the high water table. Two distinct methods are used: pumping from sumps or *shallow wells* or pumping from *deep wells.*

13.22. Shallow Wells. In the shallow-well method, tile drains are made to discharge into a shallow well or a reservoir from which, with a relatively low lift, the drainage water is pumped into a nearby irrigation canal, reservoir, or natural stream for reuse in irrigation. This method

* A_s is the apparent specific gravity, the ratio of the weight of a given volume of the oven-dry soil to that of an equal volume of water.

Fig. 13.2. Drainage pumping station returning drainage water for reuse to the main irrigation canal of the Sutter Basin Company, north of Davis in Sutter County, California. *Left:* Interior view of pump house. *Right:* Discharge into receiving bay of irrigation canal.

Fig. 13.3. Nunes pumping station in south San Joaquin Valley, California, showing pump house, drainage sump, and adjacent receiving bay of the main irrigation canal. The drainage water is reused for irrigation.

is well illustrated in a number of areas in California, for example, the Sutter Basin lying north of Davis and the floor of the San Joaquin River Valley (see Figs. 13.2 and 13.3). The practice of pumping from *shallow* wells not supplemented by tile drains, for the purpose of preventing or eliminating high water table, has also been tried. Investigation of this method by Weir [159] showed definitely that it is not economically justifiable to place pumps in shallow wells close enough together to obtain results with pumps alone, comparable with those obtained with tile drains.

13.23. Deep Wells. Pumping from deep wells is done both to provide additional irrigation water and to make room for drainage water from the irrigated areas where the geology is such as to permit steady flow to the wells, as in the Salt River Valley, Arizona, where this type of drainage, developed in 1918, was widely used by the Salt River Valley Water User's Association until it became so effective that pumping is no longer needed for drainage. The water table is now about 50 to 200 feet below the surface, and the water is now pumped purely for irrigation. This system is fully described as to plan, performance, and results in the able discussion by Marr [77] to which the reader is referred for details.

Marr closes his discussion with the following statement concerning the use of similar systems in other districts.

> The success of drainage by pumping in the Salt River Valley has induced other irrigation enterprises to undertake similar developments. Several such enterprises, although their records are not so long as those in the Salt River Valley, never the less serve to prove the practicability of this method of drainage. A few, in fact, have disclosed advantages not found in the Salt River Valley, such as water having very low alkalinity, and a possible greater flexibility in choice of location of pumps with reference to irrigated lands, permitting practically complete utilization of the pumped water for irrigation. Hence . . . it is possible that the success achieved there may be duplicated or even exceeded in many other irrigated sections.

TECHNICAL DESIGN OF DRAINAGE OF IRRIGATED LANDS

13.24. Essential Preliminaries. Marr [76] presents a discussion of the investigations that should be made in preparation for design of drainage of irrigated lands and of the elements that should be covered in such design. This discussion includes the leading features covered by this chapter up to this point. It is, therefore, made a basis for the following summary of essentials that must precede a final determination of such design. The items are as follows:

1. Examination of the soil and subsoil to a depth of 6 to 10 feet, as to its texture, structure, water-holding capacity, permeability coefficient

at different depths through the root zone, alkalinity, and other salt content.

2. Alkali and salt content of the water that will be used to irrigate and to leach the soil of the area.

3. Probable kinds of crops to be raised on the area and their respective tolerances for alkali and excess of neutral salts and the normal rooting depths of the crops. (This latter largely settles the required depth to water table. "Land that is shaded by crop and regularly watered will withstand the highest water table.")

4. Normal maximum height of water table before installation of drainage.

5. Method or methods of application of irrigation water apt to prevail and quality of irrigation control.

6. Probable extent of encroachment on the area, of seepage and waste water from other areas and sources, including artesian pressure.

7. Probable maximum application of leaching water required.

8. Probable character and amount of chemicals required to so modify the character of existing soil alkalies as to secure maximum removal by leaching.

9. Deep geology of the region as related to disposal of excess water, through gravity action or by pumping from deep wells or through relief wells from artesian basins.

10. Drainage outlets available or capable of development, both surface and subsurface.

11. Physical, legal, and administrative relationships to adjacent areas and water rights.

12. Trend of depth of impervious barrier stratum, if such exists, below the ground surface.

In the same paper, Marr offers in substance the following suggestions pertinent to drainage design and installation on irrigated lands:

1. Deep drainage is essential from a minimum of 6 feet to a possible maximum of 12 feet or even more in rare cases.

2. Some sort of district organization is essential for best promotion and administration of the community drainage problem.

3. According to a mimeographed report by L. T. Jessup in 1929, average annual drainage-water yield per acre in 13 of the Western states is, according to soil character,

Predominantly gravel	2.86 acre-ft
Sand	2.39 acre-ft
Dense formations, as clay or hardpan	1.28 acre-ft

4. Uniformly widespread test-well and piezometer investigations of soil water conditions is essential, covering monthly readings at time of year of peak water level.

5. The physical and economic feasibility of drainage by pumping from deep wells may be a paying matter. It is so under favorable conditions.

6. Use of relief wells, following piezometer investigations of hydrostatic pressure, is a must.

7. It has been shown in various areas over the West, "that soils containing large percentages of sodium salts become deflocculated and impervious when only drained and leached . . . that they are restored to production when, *in addition*, they are treated with large quantities of certain chemicals."

These chemicals include gypsum, sulfur, iron sulfate, and gypsum and iron sulfate together. As authorities are not all in close agreement on the amounts of each to use, no suggestions on this practice are offered here.

The presence of a large per cent of calcium and magnesium salts, as was found from the pump drainage in the Salt River Valley, Arizona, resulted in rapid reclamation. However, these types of reclamation with neutral salts have not generally proved economically justified.

Carbon dioxide has been found helpful against black alkali in Idaho, and it is believed to have been effective against sodium chloride in the rice fields of central California.

13.25. Surveys. All the features listed in Marr's summary should be covered in the survey on which the design is to be based, in addition to all the features discussed under drainage surveys for humid regions in Chaps. 7 and 10. These discussions should be reviewed before drainage design for irrigated lands is undertaken.

The survey will require the extensive and intelligent use of the soil-sampling apparatus which is discussed and illustrated in Sec. 3.2 and Figure 3.3, as well as the field piezometer described in Sec. 10.30 and Fig. 10.13, for determining ground-water levels, hydrostatic pressure, and rates and direction of ground-water movement as discussed by Christiansen [26], Donnan and Christiansen [37], Donnan and Bradshaw (36), Pillsbury and Christiansen [105], and Reeve and Jensen [117].

From the data secured by piezometer, both water-table and hydrostatic pressure contours should be plotted on the area map wherever variation in water table and movement of ground water is found to exist. The use of the field permeameter, discussed in Chap. 3, Secs. 3.2 to 3.5, will also be involved unless the surveyor is able to make reliable use of O'Neal's method of determining permeability by study of the soil structure, which is discussed in Chap. 3, Sec. 3.1. Topographic surveys may be necessary much more often than in humid regions, owing to subsurface water supplies and movements involved, but the most influential governing feature in them will still be the degree of unevenness of the area surface.

The designer having given full attention to the essential preliminaries given in Sec. 13.24, all that then remains is to determine proper depth, location (spacing in case of flat lands or lands of very low, uniform slope), gradient and size, according to the methods laid down in Chaps. 7 and 10 on Design of Open Ditches and Design of Underdrains.

13.26. Depth. Lateral drains in irrigated lands will be, with only rare exceptions, some type of underdrain. The depth at which the lateral drains should be laid is a first and most important consideration. Even after a static water table between irrigations has been obtained, at the mid-point between parallel drain lines it will always be higher than at the drain lines; just how much higher will depend largely on the permeability of the soil and somewhat on the rate of consumption of the water by the crop. But the rise will seldom be less than 1 foot and may often, even under best conditions, be 2 to 2½ feet where soil conditions limit the spacing of drains to 500 feet or less.

It has already been emphasized that the minimum effective depth of underdrains in irrigated lands is 6 feet. This might leave the depth of the water table at the mid-point only 3½ to 4 feet which is too shallow for most irrigated crops and for control of excess salts in the soil of the root zone. Cost not being considered, best practice is thought to be 7 to 8 feet, preferably 8 in average soils. This will leave the depth at mid-point a probable minimum of 6 feet in soils of average permeability.

Financial agencies who lend money on irrigated lands will not usually look with great favor on those where the depth of drains is less than 6 feet. On the other hand they will generally be liberal where the drains are 7 to 8 feet deep. In the matter of depth, however, it must be remembered that, with each additional foot of depth, cost of installation increases greatly and is apt to prove economically infeasible below a depth of 6 feet, especially under a waterlogged condition of the subsoil. Good practice calls for a careful balancing of probable returns against initial cost, in preliminary planning of the project. An extra lateral not over 6 feet deep may be the solution.

13.27. Location. Lateral underdrain lines should be run as nearly as is reasonably possible normal to the direction of flow of the subsurface water. Single lines will be placed where surface topography and water-table contours indicate a position of maximum concentration of subsurface flow.

13.28. Spacing. In parallel systems, width of spacing will usually be uniform for a given soil and will depend largely on the total amount of water to be moved in a given unit of time as, for example, in a 24-hour day, on the average permeability P of the soil for the predetermined depth, and on the rate at which the drain will receive the water. These facts may all be determined in the field by methods discussed in Chap. 3, Secs.

3.1 to 3.6, which the designer will do well to review at this point. Having made the necessary measurements in the field, the experienced engineer will determine the proper spacing largely in line with that experience supported by his own judgment. Several spacing formulas especially adapted to use in irrigated soils have been developed. Only two of them will be mentioned here. One of these two is the Donnan formula already briefly referred to in Chap. 10, Sec. 10.39. The other was developed by Israelsen [62]. Both are based on the same fundamentals. However, space limitations in this book do not permit a complete comparison of the two formulas; hence further discussion of this problem is illustrated by the Donnan formula, which the Soil Conservation Service engineers in the Imperial Valley, California, have used extensively with satisfactory results and which furnishes all that is needed as a basis for the general discussion of design of tile drains in irrigated lands.

THE DONNAN FORMULA[1]

This formula is based on the assumption that a barrier or practically impervious stratum exists in the soil blanket a considerable distance below

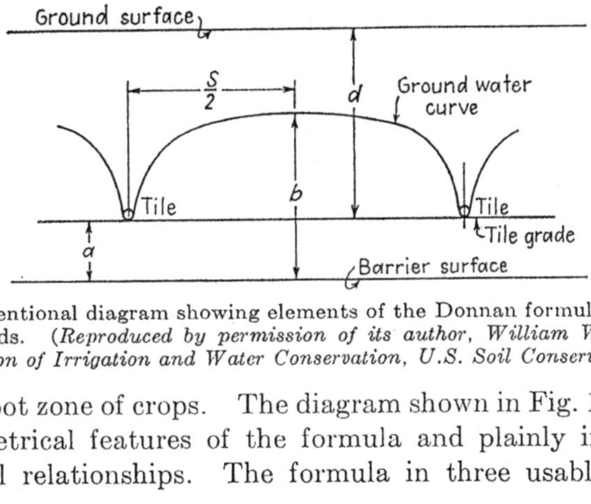

FIG. 13.4. Conventional diagram showing elements of the Donnan formula for tile spacing on irrigated lands. (*Reproduced by permission of its author, William W. Donnan, Civil Engineer, Division of Irrigation and Water Conservation, U.S. Soil Conservation Service.*)

the normal root zone of crops. The diagram shown in Fig. 13.4 shows the purely geometrical features of the formula and plainly indicates their mathematical relationships. The formula in three usable forms is as follows:

$$S^2 = \frac{4P(b^2 - a^2)}{q} \tag{28}$$

or it may be written

$$S = \frac{4P(b^2 - a^2)}{Q} \tag{29}$$

or, solved for Q,

$$Q = \frac{4P(b^2 - a^2)}{S} \tag{30}$$

[1] Derivation of this formula was first proposed by Aronovici and Donnan (see Ref. 6).

where S = spacing of tile drains, ft
P = average coefficient of permeability, or cu ft per sq ft per day
b = distance from average ultimate water table at mid-point between the tile drains to the barrier stratum, ft
a = distance from average tile depth to barrier stratum, ft
q = quantity of water to be drained by the tile, gals or cu ft per sq ft of land surface per day (it is, therefore, a function of D)
Q = total quantity of water to be drained by the tile line, gal or cu ft per lin ft of tile line per day, or $Q = Sq$

Before accepting the formula as applicable in practice in all cases, it should be noted that, as S increases, both q and Q become smaller. It is, therefore, evident that this formula, especially as transformed above, gives a value of Q that may, at the limit, be misleading because, when S reaches infinity, Q = zero, which is an impossible situation, as will be shown. Theoretically, Q is the total amount of water received per day per foot of tile line from the strip of land 1 foot wide, extending laterally, both right and left a distance $S/2$ from the given foot of tile line. But the maximum inflow into that foot of tile line may, under some conditions, be greater than the theoretical Q because of the inflow from an area directly above the tile, as illustrated by the hatched portion in Fig. 10.12. Because of inflow from this source Q could never become zero when there was any free water in the soil above the tile grade, although it may be said, in the possible extreme case, to approach it. In practice, therefore, the impossible ultimate will not hinder the use of the formula because, in drainage practice, S never takes on any but very limited *finite* values such as, say, 50 to 2,000 feet in the extreme case. The error, then, in resulting finite values of Q, considering that Q may approach zero as S *approaches* infinity as a limit, as compared with the corresponding value of Q at the given finite value of S, would be so small as to be wholly inappreciable in practice.

With this reasoning accepted it seems clear that the value of the Donnan formula and its safe use in design of drainage of irrigated lands rests primarily upon the accurate predetermination of the following features, followed by extreme care in construction:

1. The average value of P for the given area to a depth slightly greater than the proposed depth of the tile laterals.
2. The value of the drainage coefficient, D or q, which is considered suited to the given region, physical character of soil and subsoil, and the crops to be raised.
3. The depth of the barrier stratum and the values of d, a, and b.

Blaney, Donnan, and Bowen [14] suggest that, "If no substantial barrier is located the designer should assume a depth to barrier of twice the depth to the tile lines, that is about 12 or 14 feet." Slater [137][1] suggests in such a case that a in the formula be assumed at 2.0 feet which would generally make the depth to barrier about 8 or 10 feet. A fair adjustment between the two methods by cut and fit is possible. (However, more research on this point is desirable.) In Ref. 14 it is further advised that the value of q be increased about 10 per cent over the measured value to provide against unforeseen elements in required drain inflow.

At this point review of the discussion of permeability in Chap. 3 is suggested. It should also be remembered that, in irrigated lands, depth of lateral drains should never be less than 6 feet.

The factors P, q, and Q must all be expressed in the same kind of units, for example, either in gallons or in cubic feet per day. As a rule cubic feet will be more convenient than gallons. When the essential factors have been determined, use of the formula is simple (see illustrative example below).

ILLUSTRATIVE EXAMPLE

Statement of the Example. It is desired to design a parallel system of tile drains to fit the following conditions:

Area relatively flat; soil sandy silt loam; measured average permeability P to a depth of 7 ft, 2.5 in. per hr, being equivalent to 5.00 cu ft per sq ft of land area per day; $d = 7.0$ ft; depth to barrier $= 12.0$ ft; and $a = 5.0$ ft.

Therefore, depth, water table to barrier at mid-point $= 6.0$ $(= b)$
$D = \frac{1}{16}$ in. per 24-hr day over and above the crop requirement and normal evaporation, making $q = 0.0052$ cu ft per sq ft per day.
Tile gradient $= s = 0.002$
Longest tile lateral 2,000 lin ft

Solution:

$$q + 10\% = 1.10 \times 0.0052 = 0.00572$$

$$S^2 = \frac{4P(b^2 - a^2)}{1.1q} = \frac{4 \times 5.00(36 - 25)}{0.00572} = 38{,}462$$

Then $S = (38{,}462)^{1/2} = 196$ or virtually 200 ft

$Q = 1.1Sq = 1.1 \times 200 \times 0.0052 = 1.144$ cu ft per day per ft of tile line

For a 2,000-ft line this $= 2{,}288$ cu ft per day
$= 0.0265$ cfs

[1] To those interested in further study of this type of spacing formula this reference will be found interesting and helpful.

Reference to Fig. 10.11, page 259, Chap. 10, shows that this is only about one-fourth the capacity of a 4-in. tile on a gradient of 0.2%. Hence a 4-in. tile, used as good practice, allows nothing smaller than 4 in.

13.29. Donnan Formula and Humid Lands. The Donnan formula was originally developed with especial reference to irrigated lands and for a special region. Neal's formulas (see Chap. 10, Secs. 10.34 to 10.40) were developed exclusively in and for humid soils. Very limited studies, not thus far carried to a conclusion, were made with the former on humid soils in the middle range in comparison with results obtained on the same soils by use of the latter. These studies seemed to give promise that, with some modifications, not yet clarified, the Donnan formula may also be used to a limited extent in design of drainage for flat lands in humid regions, at least as a fair check on results by Neal's formulas.

The barrier stratum is apt to occur as often in humid soils as in irrigated soils, but it is usually only a small fraction of a foot below the established grade of the lateral drain in humid soils.

Permeability is as much a feature of the humid soils as it is of the irrigated soils, although for the same general class of soils it may vary widely between the two groups.

The approximate maximum effective depth of lateral drains in humid soils is uniformly two-thirds of that for irrigated soils.

Partially based on the apparent facts just stated, results close to those obtained through use of Neal's formulas have been obtained in the middle range of humid region soils through the use of the Donnan formula, with a D value of ½ inch and a and b values of 0.25 and 3.75, respectively. However, both light and heavy soils showed wide and rapid deviations between the results obtained by the two formulas.

Computations of S for a limited number of Imperial Valley soils by the Donnan formula seemed to give values uniformly about three times as great as T_s values by Neal's formulas for comparable humid region soils.

Though far from conclusive, these experiences may point to a possible means of correlating the two methods as a check on each other for soils of both humid and irrigated regions. Hence, further studies to this end by interested investigators are suggested as justifiable.

13.30. Parallel Systems Not the Rule. Although a standardized system of treatment is a practical essential in design of parallel systems of drainage on flat or uniformly and gently sloping irrigated lands, such lands and such systems are not the general rule. On many areas the drainage problem is a special one, differing in various respects from any other, a random system, so to speak, often composed of a series of minor problems of limited scope that may or may not be loosely related merely because all are located on the same agricultural unit and often have a common outlet. Each of these minor features often requires its own separate and independent solution. Hence no general rule of procedure can be laid down, although all the requirements in any case are embraced in the preliminaries to design listed in Sec. 13.25. Under the circumstances no typical example can be presented, but the following case may serve to indicate the wide range of possibilities in such projects, and the general method of attack.

ILLUSTRATIVE RANDOM PROJECT

Outline of the Project. The project is an irrigated 20-acre pear orchard on the Perham Ranch near Selah in south-central Washington, on the second terrace on the east side of the Naches River. The description of the project will be somewhat clarified by a close scrutiny of the map (see Fig. 13.5).

Soil, a fine sandy loam varying in thickness from about 2½ to 7 ft.

Substratum, an open pervious water-bearing gravel 3 to 9 ft below the surface and of indefinite thickness. Limited study indicated a sharp drop in the depth to this gravel substratum approximately along the broken line shown on the map near the originally proposed tile line marked *T*. Above this line, depth to the gravel stratum averaged only about 3 ft.

A subsurface dike, 50 to 60 ft wide, of caliche, clay, and other practically impervious material, diagonally across the south end of the area, cuts the orchard off from the low-lying swamp land to the south which eventually receives the drainage from the orchard area. The dike backs up the water in the soil uphill from it to a dangerously high level throughout the orchard. Some seepage occurs at the toe of the steep slope at the edge of the dike.

Topography of the area is that characteristic of the region—a fairly smooth steep slope to the southwest as indicated by the surface contours taken in August, 1950.

Both the orchard and the land all around it are irrigated. The main canal is located some distance from the ranch, the irrigation water for the orchard coming from a head ditch along the north property line. The method of irrigation is by furrows. The ditch shown running south along the west boundary is intended to care for the waste irrigation water. It is about 5 ft deep and could be deepened to serve also as a drainage outlet on the west, but some right-of-way adjustments would be necessary. This ditch is designated Main *B* and is so marked on the map.

Tree pattern. The trees are 20 ft apart in the rows which run east and west.

Condition of the orchard in 1950. The ground water is so high that the trees are badly affected with chlorosis and are unfruitful.

Ground-water contours, as shown on the map, were run in August, 1950, before the east boundary ditch was dug. These contours show that

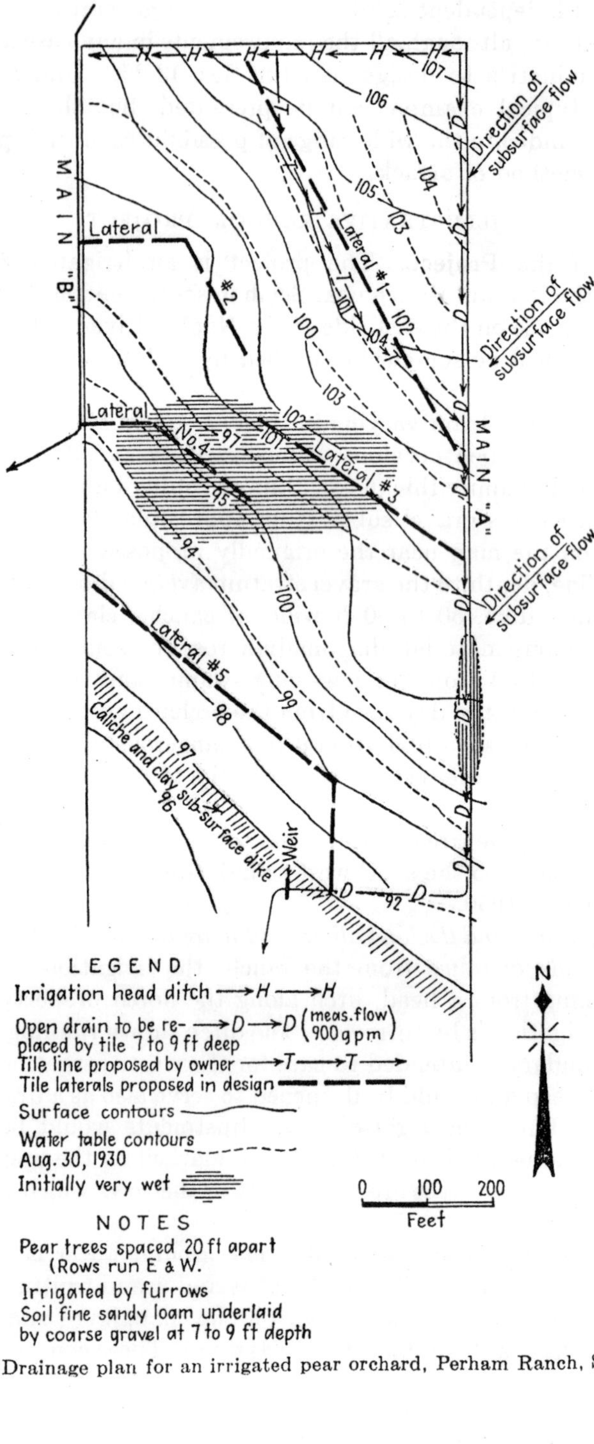

FIG. 13.5. Drainage plan for an irrigated pear orchard, Perham Ranch, Selah, Wash.

the ground water comes from the northeast. The oval shaded area in the center of the map was the wettest area before the east boundary ditch was dug, but this ditch has lowered the ground-water table to 5 ft below the ground surface and holds it there. The long narrow oval astride the surface 101 contour on the east boundary is a bad spot. About the only reasonable outlet for its drainage is the boundary ditch which picks up a lot of water from this spot. This ditch is designated Main *A* and is so marked on the map.

Study and Action Taken in 1950. A main intercepting drain was evidently needed about one-third of the way down from the north boundary but could not be put in along the water contours because the digging of it would seriously disturb the root systems of the trees. It appeared that such a drain would not be much more effective than one running straight south along the east boundary of the orchard. This drain is the one which was put in in August, 1950, and is shown on the map as Main *A*. It was run on a continuous gradient from 7 ft deep at the lower end to 9 ft deep at the upper end which gradient carried it a short distance into the gravel substratum. The surface of the gravel stratum, as tapped, was very uneven, showing more or less uniform undulations. The heaviest flow into the ditch was from the upward swing of the gravel undulations where the cut through the gravel was the deepest. The initial plan called for a large tile in place of an open ditch, but the trench was left open through 1951 to permit study of the flow from which to determine exactly the size of tile that would be required. A measurement of the flow in August, 1950, at the weir shown on the map at the lower end of the ditch showed a discharge of 900 gpm or 2 cfs. Approximately the same flow was obtained on Dec. 28, 1951 (909 gpm).

The result of this flow was a marked lowering of the ground water during 1951 but not enough completely to solve the problem. Two other advantages of this east boundary ditch are noted: the evident one is that it also offered an outlet for drainage of the land to the east and, owing to this fact, an easier adjustment of the right of way problem. On this latter account the owner of the orchard preferred this location of the ditch.

Even at the surface near the center of the north boundary, excessive wetness was still indicated. To improve this situation a single 6-in. tile line was contemplated along the line marked T on the map, but this line had not been installed by late 1951.

13.31. Final Design for Complete Drainage. At the time this account is written the project is incomplete and no final design can properly be made without additional information which is not available to the authors at this time. It would be difficult to make a reliable final design without basing it on progressive steps over a considerable period of time, by trial and error. Even for such a development, the following additional information seems essential: average rate of permeability of the soil blanket and of the gravel stratum, average porosity of the same, more extended measurements of discharge through Main *A* surely covering the time of year when the heaviest flow occurs, additional check of water-contour determinations, reliable determination of the sources of the excess water and the maximum amount from each source (what part of it is seepage from the head ditch along the north boundary), the probable normal water requirement of the trees (annual and maximum), and the time of year that the maximum occurs.

13.32. Tentative Design on Basis of Present Information. On the basis of the incomplete information available at this writing supplemented by certain assumptions of needed additional data, the following tentative design is offered, purely to illustrate an effective method of attack. It should not be used as final unless a later check, based on exact field determination of the data herein assumed shows it to be what is needed.

Assumptions:

Average free-water porosity: soil blanket (by section)	20%
Gravel stratum (by section)	40%
Average permeability rate: soil blanket, cu ft per sq ft per hr	0.30
Gravel substratum, cu ft per sq ft per hr	0.60

Practically vertical side walls on east boundary ditch

Width of ditch bottom 2½ ft

Initial average distance of ground-water table above average surface of gravel substratum 2.6 ft

Average depth of ditch (Main *A*) into gravel substratum 2.0 ft

All free soil water above a 6-ft depth is excess drainage water. As the east boundary ditch is known to have pulled the ground-water table down to 5 ft below the surface, that leaves only 1 ft of additional excess water coming in steadily to be removed by additional drains which should probably be tile drains, and this would all be in the gravel substratum

Inflow equal from both sides of the outlet ditch (Main *A*)

Total length of outlet ditch, (Main *A*) head to weir, 1,600 ft

Ratio of velocity of subsurface flow through soil to that through gravel substratum same as ratio of permeabilities, or 1 to 2

Average depth of outlet ditch 8.0 ft; hence, as the initial average depth of drained soil above the water table was 3.4 ft this left 4.6 ft above the ditch bottom from which drainage water initially flowed of which 2.6 was in soil and 2.0 in gravel

With all these assumptions allowed it is clear that the initial flow areas through the soil and gravel into the ditch were as follows:

Soil on both sides of the ditch 1,600 × 2.6 × 0.2 × 2 = 1,664 sq ft
Gravel, both sides and ditch bottom 1,600 × 6.5 × 0.4 = 4,160 sq ft
From this point it is a matter of simple algebra to compute the velocities of subsurface flow which are

Through the soil............	0.0002 fps
Through the gravel.........	0.0004 fps

The tile lines of the future, whether flowing to an outlet ditch on the east side or on the west side will probably run about at 45 deg to the meridian if they extend clear across the area and will be about 900 ft long. The final excess water which they are calculated to carry off will be that lying between the 5- and 6-ft depth or a 1-ft depth of water; therefore all are in the gravel stratum, with a velocity of flow (always on a certain pressure because on a fairly steep slope) of 0.0004 fps. Hence the steady discharge will be 0.0004 × 1.0 × 900, or 0.36 cfs into any tile line thus placed.

If *D* were planned as ½ inch in 24 hr, the equivalent of this per second over the whole area will be (600 × 1,200)/(24 × 60 × 60 × 24), or 0.35 cfs, which is in practical agreement with the flow through the gravel *to* the tile.

On a 0.2% gradient that amount would require a single 7-in. tile line 6 ft deep at the base of the area or more practically an 8-in. tile as 7-in. is seldom obtainable. It would be far better design to have about three smaller tile lines equally spaced, probably at about 400 ft and giving a safety factor of at least 2. Three 6-in. lines are suggested with a capacity of 0.750 cfs. This area being on quite a steep natural slope a spacing formula will not apply with exactness but, assuming the Donnan formula as a rough check works out about as follows, on the assumption that the excess draw from the uphill side is balanced by the diminished draw from the downhill side:

Depth of tile line, 6.0 ft
Depth of barrier, 20.5 ft
b = 16.0 ft
a = 14.5 ft
P = 1.60 cu ft per sq ft per hr
D = ½ in. in 24 hr, whence
q = 0.0018 per sq ft per hr

Then

$$S^2 = \frac{4 \times 1.60(16^2 - 14^2)}{0.0018} = \frac{6.4(256 - 210.25)}{0.0018} = 163,244$$

whence S = 404 ft or practically 400 ft

PRACTICAL VARIATIONS FROM THE FOREGOING TENTATIVE DESIGN

Details apparent from a study of the map strongly indicate that it would undoubtedly be unwise to stick closely to the more or less rigid design just worked out. Approximate locations for what seem to be desirable tile laterals, which do not follow the parallel system above suggested, are shown on the map in broad heavy lines. Any of these would no doubt need to be shifted materially to fit in more closely with the topography, shape of the water table, and tree pattern. It seems desirable to deepen the lower half of the old waste ditch on the west boundary (Main *B*) to serve as outlet for laterals 2 and 3 or, still better, replace the ditch by a large tile. It probably would not be wise to put in the entire system here indicated all at one time but to spread the installation over several seasons to study the effect of each newly installed lateral for a time before taking the next step. Maybe some of them will not be needed. The situation as shown by the map seems strongly to indicate that the best order to follow in construction would be to start with lateral 1 and proceed in the following order: 4, 3, 2, and 5. None of the laterals need be larger than 6-inch tile, but neither would it be wise to use smaller sizes on any of them. In planning to replace the trench on the east boundary with large tile, it should be made large enough to serve as outlet for the drainage from the east in addition to the known drainage from the Perham orchard. A tile system of laterals from the east, which would undoubtedly add fully as much as that figured herein, will be added by the tile system in the orchard, making a total of at least 950 gallons per minute or 2.2 cubic feet per second at least, at the outlet. A uniform gradient of 0.7 per cent is quite possible, and a 12-inch tile would be ample to carry the total discharge at the outlet above named. But, in view of the uncertainty as to future needs it would probably be wise to make this main tile a 15-inch, at least at the outlet. It probably could be considerably smaller at the head, but more information as to probable needs is required to make a safe determination on this.

As long as the trench on the main is left open, the weir at the outlet should be retained and systematic measurements of discharge made.

It is probable that seepage from the head ditch will bear investigating and that lining of this ditch will be justified.

A warning is again sounded that no final design for the drainage of this area should be accepted until a much more complete survey is made than is at present available.

PROBLEMS

1. What are the principal functions of drainage in irrigated lands?

2. What are the leading sources of soil salts?

3. (*a*) Why is underdrainage the surest cure and controlling agent of an alkali condition in the soil?

(*b*) The average alkali content of the first 4 ft of soil in a given field is 0.7% of the dry weight of the soil whose A_s is 1.25. The irrigation water has an alkali content of 200 ppm. The drainage water tests 1,000 ppm. What annual depth of water above the crop requirement must be applied to reduce the alkali content of the soil to the average maximum salt concentration (0.20%) tolerated by the crops raised, in 4 years time?

4. Using the Donnan formula, design a parallel system of tile drains to fit the following conditions: area practically flat; soil moderately sandy silt loam; measured average P to a depth of 7 ft, 2 in. per hr; depth of tile, 6.5 ft; depth to barrier, 10.5 ft.

5. Assuming the M_e in Prob. 4 to be 22 and the required R_d to be 0.5 ft, check the results in Prob. 4 by Neal's method. Is the comparison close enough to serve as a practical check on the design? If not, how well do the correlating factors between the two methods, suggested in the text, serve?

6. On a given 1,000-acre tract, the alkali content of the soil, for the first 4 ft of depth, was 1% of the dry weight of the soil whose A_s was 1.4. Irrigation water containing 800 ppm of salts was applied, which in 5 years time reduced the alkali content of the soil to the maximum allowable concentration for plant growth, the drainage water testing 900 ppm. What depth of water was applied to the surface, yearly?

7. In Prob. 6, if the original alkali content of the soil was 2½%, water applied and time for producing results being the same, (*a*) what became of the rest of the alkali? (*b*) If one-half of the original salt was removed in the drainage water in 5 years, what would be the salt test of that water?

CHAPTER 14

THE SOIL-MOISTURE CONTROL PROBLEM ON PEAT AND MUCK LANDS

EXTENT AND IMPORTANCE OF PEAT AND MUCK LANDS

Swamp lands, largely covered with peat and muck, constitute a considerable part of the farming areas in many parts of the world. In the Soviet Union there are reported to be over 300 million acres of peat and muck, approximately 60 million of which are in European Russia alone. In each of the other countries of Central and Northern Europe there are many thousands of acres; in Great Britain there are over 3 million and about the same in Ireland. In Canada there are many millions of acres; the exact extent is not reported. The total in the United States is approximately 300 million acres, of which some 12 million are in vast continuous areas in the Lake States; but the main bulk lies in more or less scattered areas throughout northern Iowa, Illinois, Indiana, Ohio, and New York and along the Atlantic and Gulf coasts as well as over a relatively narrow strip up the lower two-thirds of the Mississippi Valley. There is also a gross total of appreciable extent in scattered areas throughout the entire Pacific slope. The extent and importance of agricultural peat and muck lands is sufficient to justify giving careful consideration to the soil-moisture control problem on such lands, as it differs widely, in various respects, from the same problem in mineral soils.

14.1. Physical Types of Peat. There are several types of peat readily distinguishable from each other by their evident origin and structure. Some of these types make good agricultural soils; some do not, at least without extensive modification or treatment; and still others lie between these two extremes. As ordinarily considered and named in what is generally accepted as the ascending order of their agricultural value, peat lands are commonly classified according to their physical character as follows: Dachnowski [30,31] moss peats (Fig. 14.1), sedimentary peats (Fig. 14.2), woody peats (Fig. 14.3), and sedge and reed peats (Fig. 14.4).

The moss peats, and especially the sphagnum peats, tend to a strongly acid reaction with a pH value as low as 4 in some cases. They do not give satisfactory results in agriculture without marked addition of lime and fertilizers or barnyard manure. The essentially sphagnum peats of this class are characteristic of the more northern latitudes and of cold, exces-

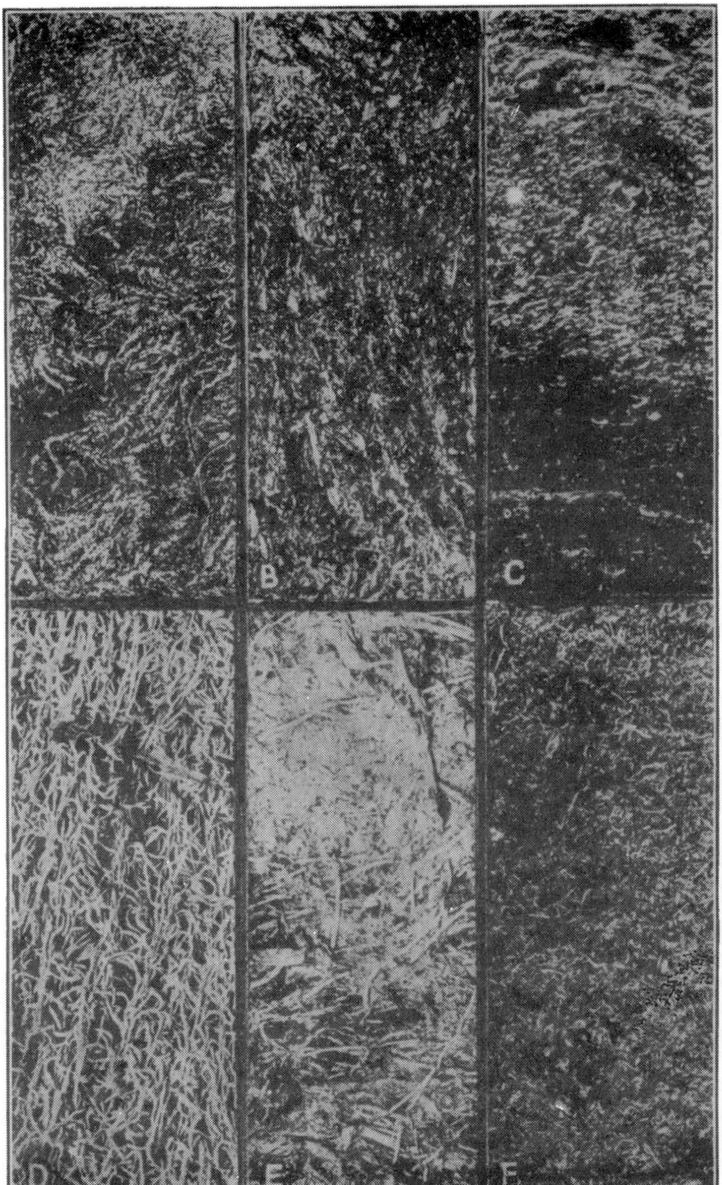

FIG. 14.1. Varieties of moss peats. *A*, hypnum peat; *B*, partly disintegrated; *C*, over sedimentary material; *D*, sphagnum peat; *E*, partly fibrous; *F*, partly disintegrated. All natural size from air-dry samples. (*Courtesy of Dr. A. P. Dachnowski.*)

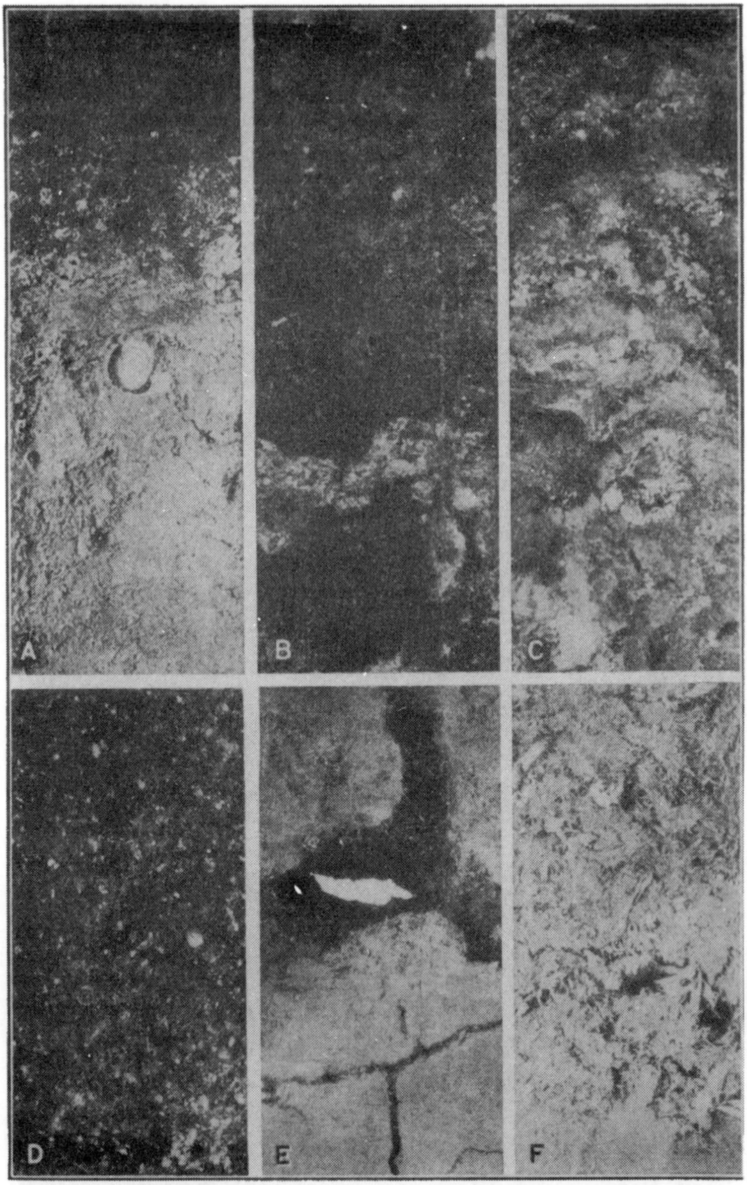

FIG. 14.2. Varieties of sedimentary peats. *A*, over shell marl; *B*, with iron salts; *C*, with calcium sulfate crystals; *D*, algal; *E*, colloidal; *F*, mixed phase. All natural size. (*Courtesy of Dr. A. P. Dachnowski.*)

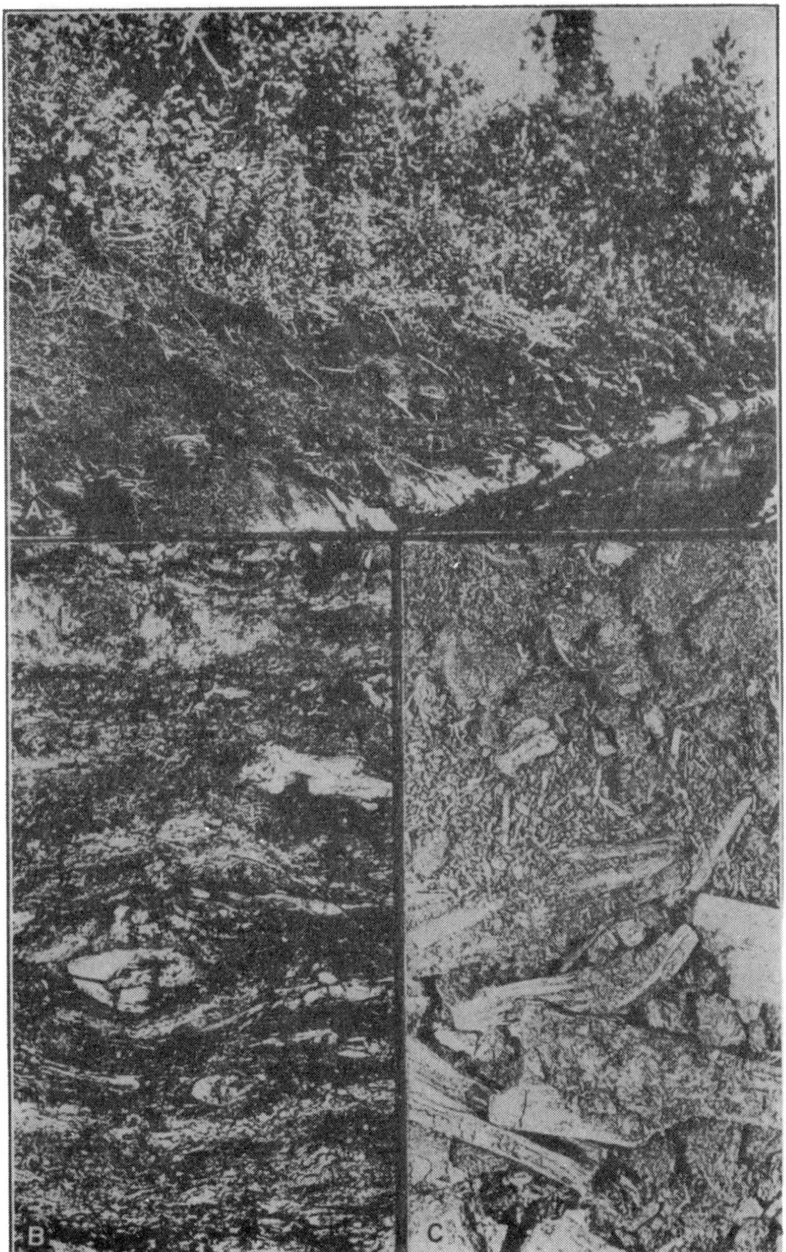

FIG. 14.3. Varieties of woody peat. *A*, profile section of an area of woody peat; *B*, laminated woody peat from shrubs; *C*, partly decomposed woody peat from a swamp forest; *B* and *C*, natural size from air-dry samples. (*Courtesy of Dr. A. P. Dachnowski.*)

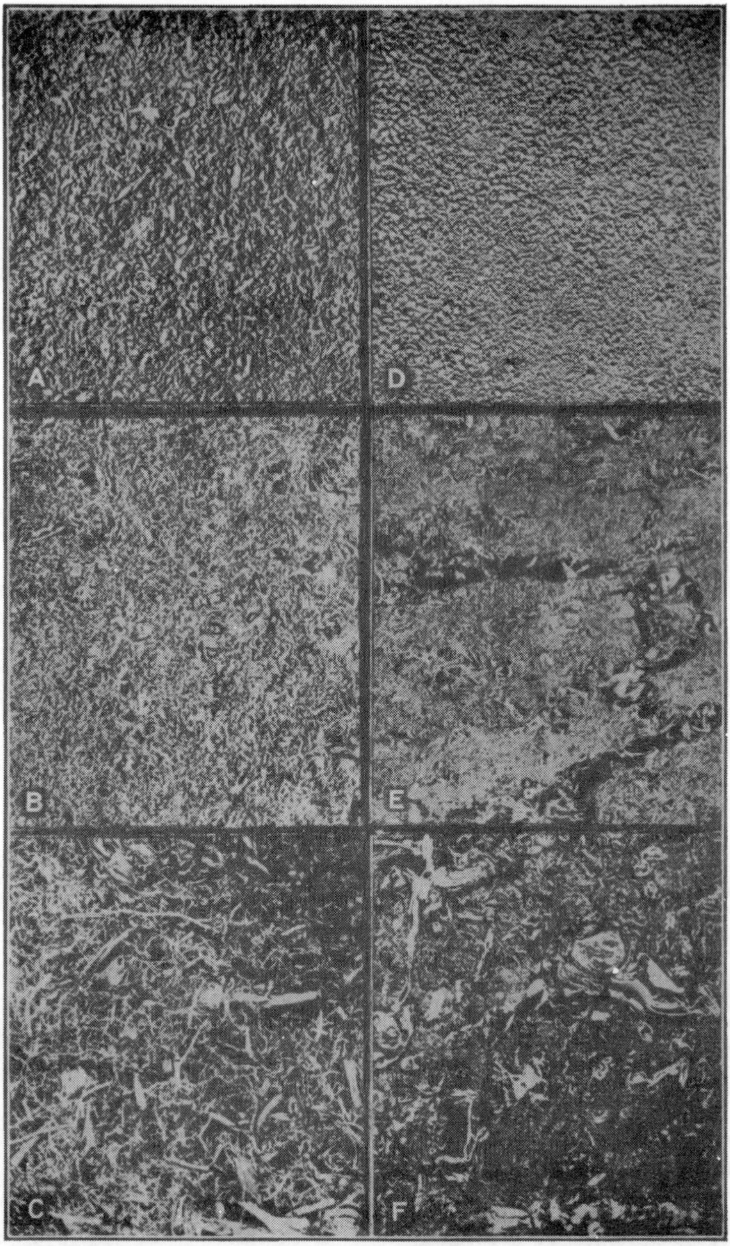

FIG. 14.4. Varieties of sedge and reed peat. *A*, sedge muck; *B*, radical late sedge peat; *C*, coarsely fibrous sedge peat; *D*, reed muck; *E*, partly fibrous reed peat; *F*, coarsely fibrous reed peat. All natural size from air-dry samples. (*Courtesy of Dr. A. P. Dachnowski.*)

sively wet areas. They have a strong affinity for water which makes them hard to drain. Furthermore, once drained, they are seriously subject to damage from drought, as they do not compact readily.

The sedimentary peats are generally waterlogged and not readily drained. When they are overdrained, a condition which may readily occur on the surface, they tend to harden and cake. If deep and not well mixed with fairly coarse mineral soil or more desirable coarser peats, this type is ill suited to agriculture.

The woody peats are built up largely from forest litter and seem given to marked shrinkage, probably because of less complete oxidation of the original materials before being brought into agricultural production. However, they often seem to possess a larger percentage of directly available fertility than the sedge and reed peats. They are drained with fair readiness.

The sedge and reed peats are more compact and exhibit less tendency to shrinkage and subsidence than most of the other forms, and their natural fertility is high. They tend to a fibrous structure which makes them readily drained and well aerated. They are, therefore, the most important for agricultural use. They also stand up well in excavation which is an aid in drainage operations.

Both the woody peats and the sedge and reed peats usually give fairly satisfactory results in agriculture.

14.2. Muck Soils. Muck soils are variously defined as impure peat containing 25 per cent or more of inorganic mineral matter, as soils containing 15 to 50 per cent of organic matter, and as soils containing more than 50 per cent of ash. Muck is related closely to physical disintegration, humus to chemical decomposition.

Muck soils are much more desirable in agriculture than true peats as they are not so subject to settling, are more easily drained, are more readily handled in field operations, and their plant food constituents are more fully available with less demand for commercial fertilizers.

14.3. High-lime and Low-lime Peats. The classification of peats into *high-lime* and *low-lime* is so important in their agricultural relations that they cannot be passed by without mention; but no extensive discussion is justified here, as this classification has no especial relation to drainage and soil-moisture regulation. A *high-lime* (low-acid) bog is one in which the reaction is only slightly acid, neutral, or even slightly alkaline so that, for agricultural use, the artificial application of lime is unnecessary. A *low-lime* (high-acid) peat, on the other hand, is one in which the acid reaction is strong and, for agricultural uses, requires, primarily, a strong application of lime to neutralize the acid. Moss peats (and more particularly sphagnum peat) belong in and largely compose the latter class. Woody peats and reed and sedge peats are mostly high-lime.

14.4. Surface Settling of Peat Bogs under Drainage and Agricultural Use. When raw peat bogs are first broken and put under cultivation, the surface 2 to 5 feet of depth, and especially the first 2 feet, is very loose and capable of much compacting. After drainage such compacting or subsidence goes on naturally for a long period of time, measurably for the first 5 or 6 years, after that more and more slowly for an indefinite period, under the influence of various agencies. As agricultural use of the peat land to the best advantage requires a compact surface, artificial means are found to stimulate subsidence. According to Alway [3] and others, agricultural use of peat soil for satisfactory results requires that, following drainage, it be compacted with a heavy roller. This has become common practice, which is, without doubt, a leading cause of subsidence.

Most peats will absorb and hold tremendous amounts of water, even four or five times their dry weight, greatly increasing their volume while wet; conversely, on drying out, they shrink rapidly to minimum bulk. Hence drainage is a main cause of their subsidence. Slow oxidation from bacterial action is also seen as a cause of shrinking, and drainage stimulates bacterial action. Fire is not properly considered a cause of subsidence as it simply destroys a portion of the peat solid matter.

Cultivation of the average raw bog is impossible until drainage has been accomplished to a considerable degree, followed by rolling to firm the surface. However, below plow depth, peat never dries out completely except under the influence of fire. No definite limiting figures can be stated for the total subsidence of peat bogs because it varies widely according to the depth of the peat and climatic and other local conditions. It is generally agreed that surface settlement of a bog tends to be proportional to the depth of the water table. Powers [108] states that, in his experience, peat, after drainage, may shrink one-fourth of its volume, which will, of course, appear mostly in decreasing the depth over the tile. Powers [107] further reports the widely varied experience of a number of observers in different parts of this country and of the world on the subsidence of peat (see Table 14.1).[1]

In 1915, under the Minnesota Agricultural Experiment Station, Alway [2], instituted studies in cropping of peat on a 20-acre experimental tract in the heart of a deep peat bog many square miles in extent, at Fens in southwestern St. Louis County, Minnesota, about 20 miles west and 40 miles north of Duluth. During normally wet years this was a quaking bog of great depth. According to the annual rainfall and temperature, the depth to the water table varied between 2 and 6 feet. From this the impression was gained that the great mass of the peat was a continuous floating body in a subsurface lake. The high water table on the main bog never permitted very deep drainage on the diminutive experimental tract

[1] The last two items in the table were supplied by the authors.

nor, because it kept the moisture content relatively high, did it permit the customary and expected total subsidence.

Open trench drainage to a depth of 3 to 3½ feet was installed in 1915, levels being first taken over the surface of the raw bog. In the summer of 1918 additional but no deeper drains were installed, and the surface levels then taken showed an average subsidence, in the three years, of 1.3 feet. Subsequent levels, taken in 1918, 1919, 1920, 1921, and 1927, showed slight fluctuations but no appreciable further general settling. This is an unusual circumstance caused by the attendant conditions above stated.

TABLE 14.1. SUBSIDENCE OF PEAT

Location	Reported by	Subsidence, in.	Length of time	Initial depth of peat
English fens near Ely. .	. . . .	100	100 years	
English fens near Whittlesay*.	. . . .	39	1848–1875	
		3	1 year later	
		120	By 1913	18 ft
Louisiana and Florida. .	Okey	18	8 years	A few inches to 16 ft
Florida Everglades. . . .	Dachnowski	11–13	. . . .	8.8 ft (saw grass peat)
Near Batavia, N.Y. . . .	J. E. Knott, Cornell Univ.	14	20 years	
Near Clatskanie, Oreg.	Powers	18	10 years	7 ft (muck soil, following drainage and cultivation)
Near Salem, Oreg.	Powers	24	2 years	A few inches to 21 ft (willow sedge peat)
Various parts of the United States east of the Mississippi	Clayton	20–56	17–18 years	Variable but mostly deep
University of Wisconsin Marsh	Zeasman	9, all above tile	5 years	4 ft and over

* General range of settlement, 93 in. in 27 years to 24 in. in 60 years. *Average initial* depth, 6 ft.

Another extended Minnesota experience was the following on a small experimental peat bog located about 5 miles east of Excelsior, Hennepin County, Minnesota, officially designated as "The Water Controlled Peat Plots" and hereinafter referred to as the "Peat Plots." This tract of about 2½ acres was subdivided into six pairs of plots on each of which the ground-water table was controlled, theoretically, at successive depths of 1, 2, 3, 4, and 5 feet, by a series of loops of subirrigating tile around the experimental plots, combined with control weirs on transition plots

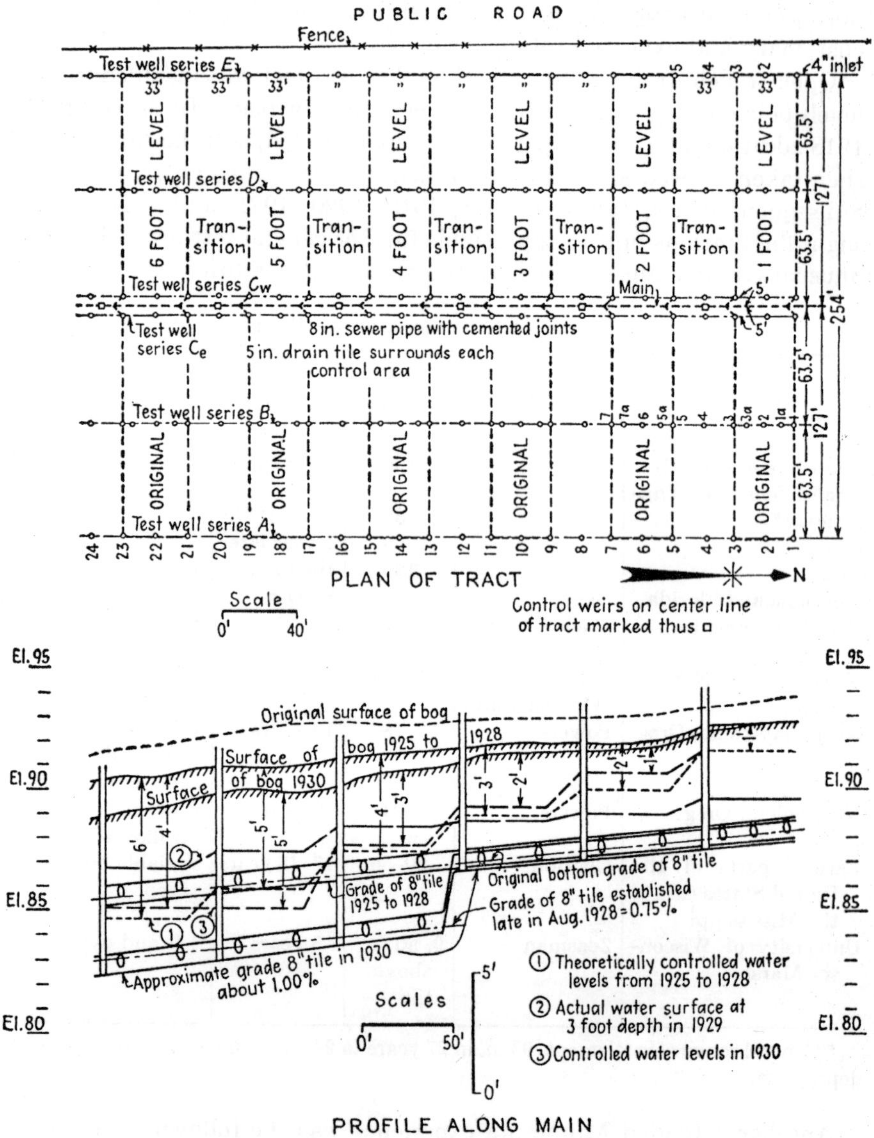

FIG. 14.5. Plan and profile of Minnesota Peat Plots, showing layout of control zones and theoretical control levels.

separating the experimental plots (see Figs. 14.5 and 14.6). These plots are hereinafter referred to as the 1-foot control, 2-foot control, 3-foot control, etc. Levels were taken on the surface before installation of the control tile in 1921 and again at least once each summer, 1925 to 1930.

The original surface of the bog after clearing was moderately smooth, with a fairly uniform fall toward the south. The surface of the bog settled measurably after the installation of the tiling system in 1921. The levels taken each summer season from 1925 to 1928 indicated that the settling of the surface due to the influence of the tile lines *in their*

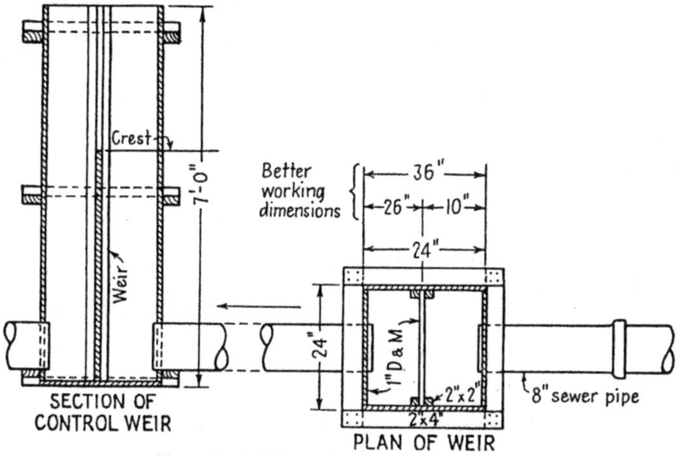

FIG. 14.6. Plan of control-weir boxes.

original position was completed before 1925, but was sufficient seriously to upset the previously controlled depth of water table. While deepening of the tile lines in 1928 was partially successful in overcoming this difficulty, it also disturbed the apparently completed condition of surface settling, as an additional settling occurred thereafter. The amount of settling is given in Fig. 14.7. The figures in this illustration indicate that the total settling above the tile up to August, 1928, is evidently the most significant. It varies from an average of about 0.5 foot at the north end to about 1.1 feet at the south end of the tract, the variation being approximately proportional to the depth of the water table, as might have been expected.

14.5. Influence of Settling of Peat and Muck on Drainage Installations and Performance. While, as already noted, as a general rule the greater part of the subsidence occurring in peat bogs takes place above the level of the drains, there are known marked exceptions to the rule as noted by Powers [108] and Roe [122]. Even in ordinary cases with only a mild amount of settling below the drains, serious upsetting of the gradients of tile drains does occur, especially in extra-soft spots in the bottom of the

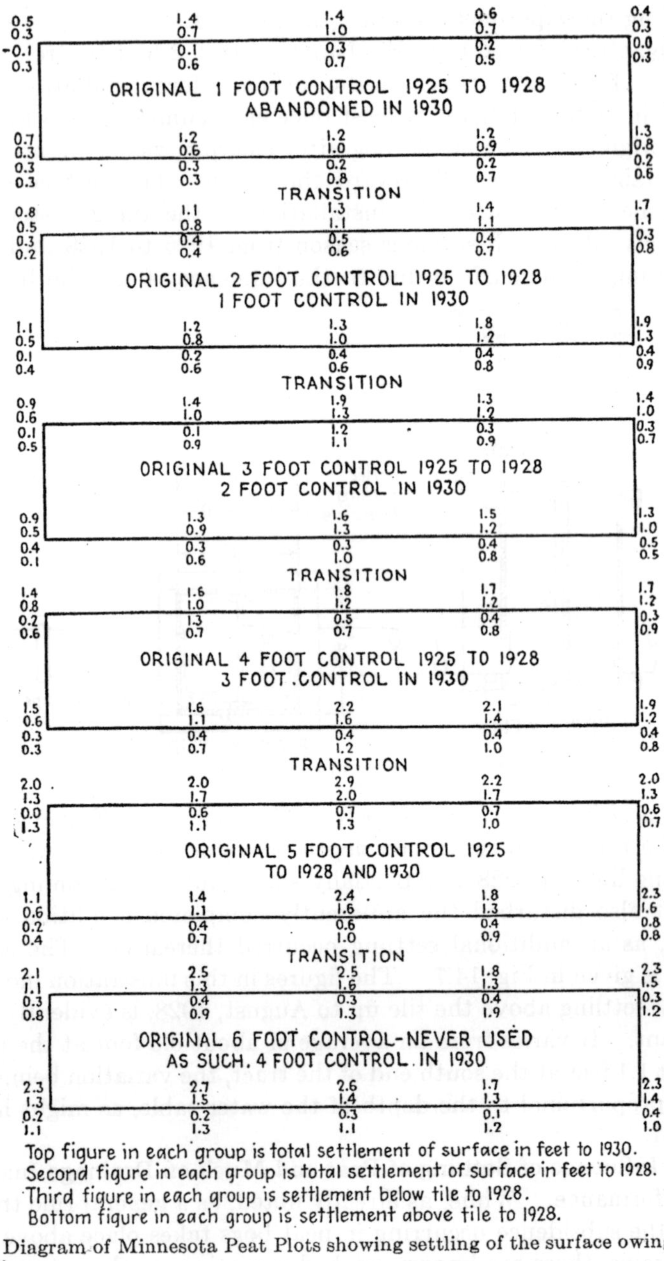

FIG. 14.7. Diagram of Minnesota Peat Plots showing settling of the surface owing to drainage and tillage.

trenches. In the exceptional cases where excessive settling occurs below the drains, the complete system is upset by heavy sags occurring, in time, in the tile lines. The more usual influence of settling on drainage practice is further discussed in several later sections and especially in Secs. 14.14, 14.16, and 14.29.

14.6. Drainage a Prime Requisite. The agricultural development of a peat or muck area having been decided upon, if the tract is clear of timber and brush, the first thing needed is that thorough drainage should be provided. The wide range in depth of bogs and in the character of the underlying formations makes proper drainage of peat and muck soils an intricate and often a difficult problem. Harmer [54] points out that drainage of muck soils must be rapid but not excessive in amount as where they become too dry they resist rewetting.

TYPES OF DRAINAGE PROBLEMS IN PEAT

Several definite cases occur in drainage of peat in each of which the drainage development must be different.

14.7. Shallow Peat over Nonwater-bearing Sand. Where the peat is shallow, not deeper than the average root zone of crop plants, or a maximum of 4 feet, if directly underlaid by nonwater-bearing sand, experience has shown that no subdrainage should be required. However, in such cases, provision should be made for prompt removal of excessive surface water resulting from heavy showers or protracted steady rainfall. This may be accomplished by provision of broad shallow surface runs, *grown to sod to prevent erosion*, with a slight rate of fall in the direction of natural flow. However, if the water to be removed quickly at any one time should be considerable in amount and if the surface is muck or peat that is already well decomposed and finely divided, excessive and harmful erosion may occur in spite of the sod blanket. Hence, a better way is to use open trenches like shallow tile trenches, 8 or 10 inches wide and spaced 60 to 200 feet apart according to the texture and structure of the peat and as determined by trial and experience with the given field. Usually the agricultural peat is of a sufficiently fibrous and tenacious structure that the sides of these trenches can be left practically vertical. With the passing of time there will be some filling in, but trenches are readily cleaned by hand or by spud ditchers of the bucket-wheel type (see Fig. 14.8). The necessary depth for such trenches will depend upon the depth of the peat. As a rule, no such drains should be needed unless the peat is over a foot deep or unless the substratum is such, in itself, as to call for drainage, in which case the general rules for tile drainage of mineral soils as discussed in Chap. 10 govern. In peat 2 feet deep the trenches may safely be about 18 inches deep or even reach nearly to the sand subsoil. In peat 4 feet deep, trenches 24 to 30 inches deep probably will be required

FIG. 14.8. Spud ditcher on the Sacramento–San Joaquin delta in California. *Above:* Rear view, showing spud ditch. *Below:* Side view of machine in action.

and will be deep enough. Extensive observations by the authors support this method.

14.8. Shallow Peat over Silt or Clay Subsoil. There are two classes of conditions under this case. The first is that in which there is no constant or even reliable intermittent source of soil water at the surface of the subsoil. The required method is similar to that of the immediately preceding case, but care should be taken that none of the trenches should be deep enough to reach the subsoil layer. The layer of undisturbed peat between the bottoms of the trenches and the surface of the clay subsoil serves as storage space for reserve moisture provided by melting snows or by rainfall during the spring and early summer, and also prevents the seeping away of moisture from summer rains along the plane of cleavage between the peat and the mineral subsoil.

The second class in this case is the one in which there is a constant or reliable intermittent source of incoming water along the plane of cleavage between the peat and the clay subsoil, such as seepage from higher-lying adjacent land or from nearby shallow streams. In this class the open narrow trenches are still the best for depths of peat up to about 3 feet. If, however, the peat still retains a depth of 3½ to 4 feet after subsidence has practically ceased, lines of tile lying, for half their outer diameter, in the surface of the subsoil, 4 to 8 rods apart and leading on a minimum gradient of at least 0.10 per cent to a sufficient and reliable outlet, are desirable. Stevenson, Brown, and Boatman [144] practically support this plan. They also warn against using heavy mineral soil in refilling the trenches over the tile, as it tends to check movement of water into the drains, and they emphasize the need for a *good* outlet. Such drains have the additional advantage characteristic of all tile drains in that they eliminate the inconvenient cutting up of the surface of fields by open drains.

Some Especial Advantages of the Open Trenches. In the case of a shallow peat over nonwater-bearing substrata or over clay, there is always danger of overdrainage, especially during periods of drought and extreme heat with the resultant injury to crops and the development of a serious fire hazard. If, in such cases, a source of surface water can be made available, the primary function of open trenches may be reversed, and they may become subsurface irrigation channels, the water being run into them through hose, surface pipes, or flumes, and held there any desired length of time simply by damming up the trenches at both ends. This may be done either with peat clods or mineral soil hauled in, or by construction of conventional plank, plank and concrete, or concrete dams that may be opened or closed at any time. This makes of the lower layers of the peat a vast reservoir of stored moisture as most peats act like a sponge when in contact with water and absorb large quantities which cannot be removed by gravity. McCool and Harmer [81] and Powers [108] suggest this scheme even in large or outlet ditches, to supply water for subirrigation. Powers suggests the use of flash boards in such cases. Beebe [13] holds that water dammed back in trenches and ditches affects the soil as far as 8 rods away.

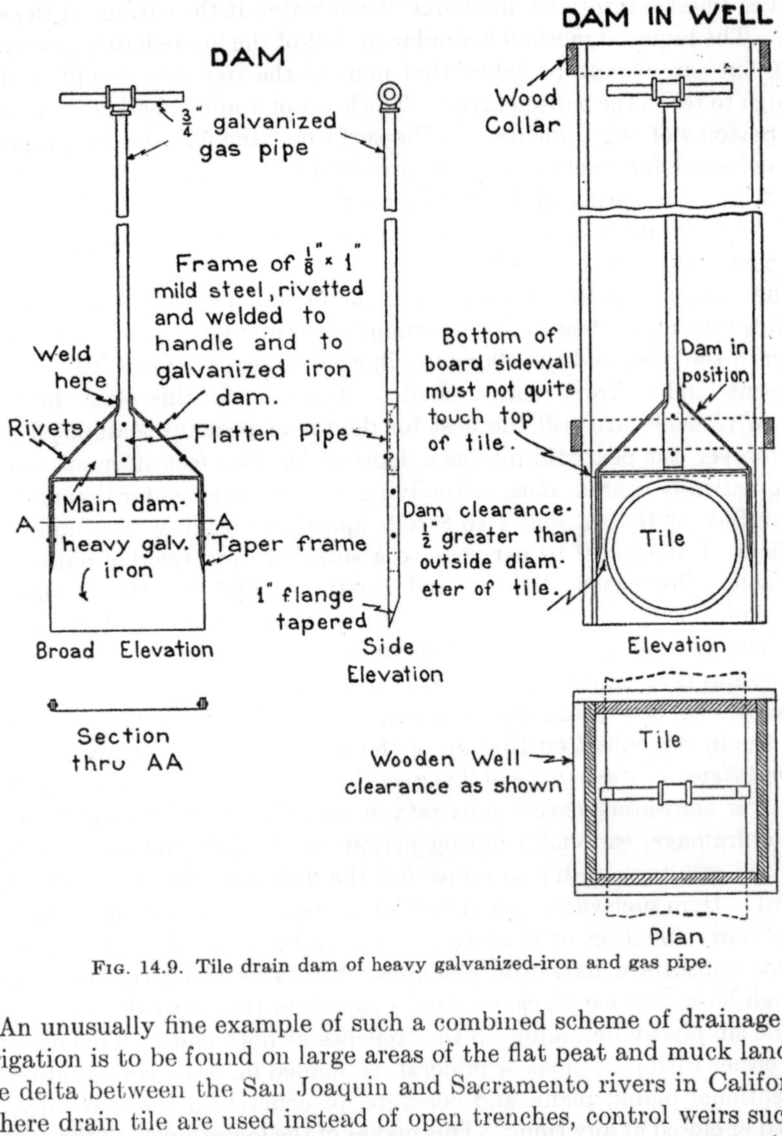

FIG. 14.9. Tile drain dam of heavy galvanized-iron and gas pipe.

An unusually fine example of such a combined scheme of drainage and irrigation is to be found on large areas of the flat peat and muck lands of the delta between the San Joaquin and Sacramento rivers in California. Where drain tile are used instead of open trenches, control weirs such as shown in Fig. 14.6 are recommended. However, another simpler scheme than the use of control weirs, to check the flow in the tile when some subirrigation is desired, is to introduce into each tile line, near its lower end, a galvanized iron dam such as is shown in Fig. 14.9. It is inexpensive, easily made and manipulated, and will usually give the result desired.

Another advantage of the open trenches is that they can readily be filled in and relocated as changing situations may demand. Their installation is inexpensive and easily accomplished.

14.9. Deep Peat over Silt or Clay Subsoil. Where peat or muck occurs over mineral subsoil other than sand, at depths of 4 or 5 feet up to many feet, outlet drainage channels if not naturally present must be artificially provided to care for spring floods and excess water in summer seasons that are abnormally wet. For the latter condition, especially, supplementary tile drains are also essential. However, this situation, in dry seasons, may readily result in overdrainage. Where such a liability seems probable, therefore, the main outlet ditch should run either across the tract to be served or along one side of it. Then that portion of the main ditch within or adjacent to the tract may readily be converted into a storage reservoir by building in it at the downstream edge of the tract a dam that will retain the water in that portion of the ditch lying upstream from it. The height of the water level upstream from the dam is readily controlled by a pump installed at the dam to lift the excess water over the dam and discharge it on the downstream side. The pump, best run by electric power, should be automatically controlled by means of a float valve that starts the motor when the water in the reservoir ditch rises above a certain height and that shuts off the motor as soon as the water level drops again below the desired elevation. Control of the soil-moisture conditions over the adjacent tract is had by means of parallel lines of supplementary tile drains, or by open trenches such as those earlier discussed, outletting in the ditch reservoir and extending laterally away from it across the tract. These lateral tile lines or trenches thus automatically serve to apply to the entire affected tract the control of the soil-water level at the height set in the main ditch by the automatic pump. Capacity of the pumping plant must, of course, be designed separately for each individual case. The general plan is recommended for such cases, where water-table control is deemed essential.

Various commercial projects of this character, covering 20 to 80 acres each, have long proved effective and profitable in the celery and truck-raising areas in western Michigan. On some of these, open trenches 4 feet deep with 10- or 12-inch pipe outlets into the main ditch effectively replaced the lines of small lateral tile, although the trenches and the necessary berms along each side of the trenches occupy an appreciable area of land (see Fig. 14.10).

Somewhat similar systems but of much simpler design are being extensively used with profit on small areas of 2 to 6 or 8 acres. On any of these latter a 12-inch propeller-type pump of local manufacture belt-driven by a motor of small horsepower is wholly effective for maximum lifts of about 4 feet and is profitably used. This outfit is far less expensive in first cost and operation than are most of the larger standard installations which might prove economically prohibitive for small truck projects.

FIG. 14.10. Narrow lateral trench for control of water level on a celery farm near Muskegon, Mich.

14.10. Shallow Peat over Deep Water-bearing Sand. Peat soil over water-bearing sand presents several variations, each calling for its own special treatment. The case of shallow peat over deep water-bearing sand in which the natural depth of water table varies greatly with the type of season is well represented by the Coddington Marsh in central Wisconsin about 100 miles north of Madison and forming an appreciable portion of the east side of the upper Wisconsin River Valley. This area, which is under the supervision of the Wisconsin State Water Regulatory Board with headquarters at Portage, is a shallow bog many miles in extent. Its eastern border is a beach moraine following the general course of U.S. Wisconsin highway 51. For a narrow strip ½ mile to a mile wide bordering the moraine, the depth of the peat is 3 or 4 feet, but beyond that, westward for many miles, it thins out to a depth of 8 to 18 inches, with an average probably not greatly over a foot. The subsoil, to great depths, is a medium to medium-coarse water-bearing sand which crops out spottily in small areas at the surface. The general land slope, southwesterly at a rate of about 4 feet to the mile, is also the average grade of the public open ditches running from east to west and serving the entire tract. These ditches, 6 to 8 feet deep, were installed by the Portage County Drainage District in 1904 and 1905. They are a mile apart, north and south, and cover the whole district which is about 35 miles wide and 50 miles long. As the source of the continuous water flow in the ditches, enhanced at times by general heavy rainfall in the region, is evidently seepage from the eastern beach moraine, a cutoff ditch of similar size along its west base was also dug to intercept the seepage.

To endeavor to control soil-moisture conditions and, if possible, to stabilize agricultural possibilities on this tract, the Water Regulatory Board, in 1934 to 1936, installed in the ditches above mentioned 54 standard masonry and concrete control dams averaging about a mile

FIG. 14.11. One of a large number of stop-log water-level control dams on one of the deep ditches of the Coddington Water Control Project near Coddington, Portage County, Wisconsin.

apart (see Fig. 14.11). The dams are equipped with stop logs which enable the control of the water level in the ditches at any desired elevation from the ditch bottoms to the land surface. This further makes possible the shifting, within a short time, of the water table between the ditches approximately to any desired level, the rate of rise or fall of the water table having been found to be uniform laterally from the ditches for distances up to a half mile. The farmers' requests and recommendations are taken into consideration in the regulation of the water levels, but the Water Regulatory Board is responsible for such regulation. The Board, through its chief engineer, consults with the drainage officials in regard to the regulation which is required or is most effective and action is taken accordingly. As a result, farmers cooperating with the Board in securing such control of the water levels as is necessary to prevent injury from either floods or drought are able, by careful attention to preparation of seedbed and proper fertilization, to raise at a good profit, even on this shallow peat, excellent crops of potatoes, corn fodder, oats, and hay as well as truck crops, especially the small root crops, where the market justifies them. Some excellent dairy farming is being done in this region. The great drawback to the whole project is the expensive character of the control works.

Under conditions somewhat similar to those on the Coddington Marsh, without the aid of tile drains, excellent results are also obtained with potatoes and onions near Dousman in southeastern Wisconsin, where peat averaging 4 feet deep is underlaid by sand. A much simpler, less expensive type of dam is used than on the Coddington Marsh. At distances of a half mile therefrom, the 6-foot open ditch furnishes sufficient drainage. No poor results either from excessive water or from drought have been experienced on this tract during several years of operation.

14.11. Moderately Deep Peat over Water-bearing Sand Which Occasionally Crops up As Sand Islands above the Peat Surface. Another special type is that in which peat occurs moderately and more or less uniformly deep over deep water-bearing sand which rises at intervals in limited areas as islands above the general surface of the bog. Where this condition exists, the water in the sand, below the surface of the peat, is usually under more or less pressure and adequate drainage is often difficult to achieve. At any point upstream from the portion of the bog to be drained, where the sand substratum rises nearly to or above the surface of the bog with considerable abruptness, a seepage line of tile or, in many cases, two such lines parallel to each other, spaced 5 to 10 feet apart, should be located near the border of the sandy shore line. They should be laid just at or only slightly within the surface of the sand at a depth of peat not less than the depth to which it is desired to drain the bog and should be led to the nearest adequate outlet on a gradient of at least 0.3 per cent. No sand should be filled over the tile. Such seepage lines, as a rule, sufficiently reduce the upward pressure of the water against the bottom of the peat blanket so that a regular tile drainage system for the entire tract, designed for suitable depth of drainage and types of crop to be raised, will complete the rest of the job of soil-water control. It is impossible to be more specific on this point owing to the wide variation in the texture and structure of the peat on different bogs. A striking illustration of this type of case is that of the experimental peat bog of the Minnesota Agricultural Experiment Station at Coon Creek, Anoka County, Minnesota. Figure 14.12 shows a mensuration plat of the tile drainage finally installed thereon. The portion of lateral 1 extending northwesterly along the edge of the bog on the west side was not installed until some time after the rest of the drainage system. Previous to the installation of this extension of lateral 1, the subsurface flow into the bog through the sand knoll along its west edge was so heavy that the system of parallel tile drains across the line of natural subsurface flow did not keep the water table far enough below the surface of the peat to permit the crops to thrive. When lateral 1 was extended at a depth almost wholly within the sand substratum, it intercepted enough of the subsurface flow to enable the system of parallel laterals to lower the general water table in the peat sufficiently to permit normal growth of crops. This type of experience points to the fact that, under the case discussed in this section, extremely thorough investigation of subsurface-flow conditions is essential to effective design of the drainage system.

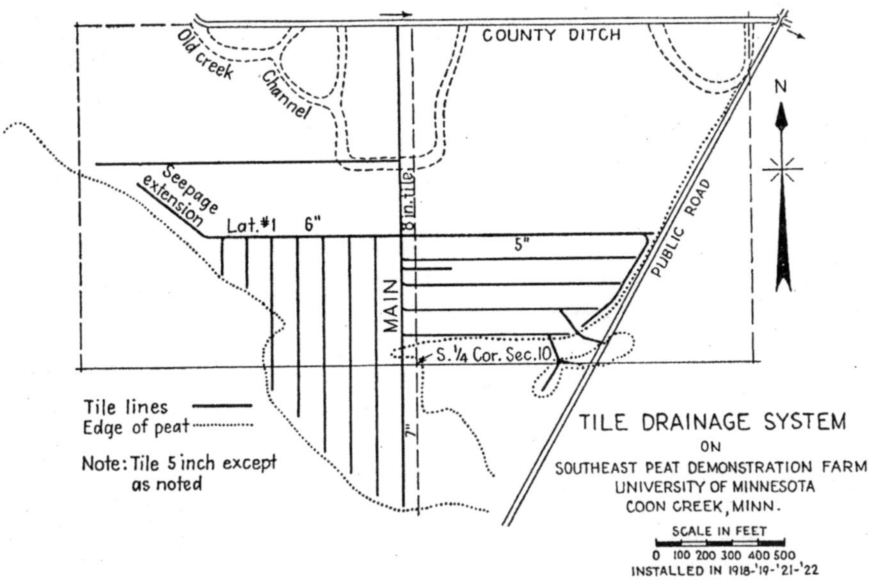

FIG. 14.12. Coon Creek drainage map showing seepage extension of lateral No. 1. Coon Creek, Anoka County, Minnesota.

Where the peat, after settling is practically completed, is 5 feet or more deep over the water-bearing sand substratum, there will be an undisturbed peat layer of considerable thickness between the base of the tile lines and the surface of the sand. If this undisturbed peat is not soft and pervious (which, however, it frequently is), it will tend to retard the upward flow of the water enough that a tiling system carefully designed to meet the drainage needs of the given peat will usually prevent waterlogging of the soil above the tile lines. However, even in this case, if the shore line of the bog is near the drained portion and is itself an extension of the sand upward to elevations above the bog surface, the same precautions in the investigation of the amount of subsurface flow and in the installation of intercepting drains should be taken as those outlined in the preceding paragraph.

14.12. Treatment of Free Water under Pressure beneath the Peat. Often on bogs in the vicinity of higher water-bearing lands deep pockets of water or springs under pressure are encountered, naturally restrained only by their own overburden of partly cemented material or by deep tough layers of dense colloidal peat. These permit sufficient distributed seepage into the peat area to prevent thorough drainage, within the root zone, by the usual tile drainage system. To overcome this condition the water under pressure must be relieved and a free channel for the excess water provided without its having to be forced through the soil. This can usually be done in one of two ways.

If the final outlet is sufficiently below the given pocket and the tile gradients steep enough to allow such action, one of the tile lines past the pocket can be laid deep enough to tap the source of the water, or a new line can be laid deep between two existing lines for this express purpose. This will relieve the pressure and allow perpetual free flow of the entrapped water without accumulation of pressure, and the problem will be solved. This method was employed by E. R. Jones years ago in the southeast corner of the University of Wisconsin Marsh at Madison, and recent investigation shows that the cure has been permanent. Such pressure areas may also be relieved by properly located open ditches, but they present the usual objection of interference with field operations.

Where outlet depths and drain gradients do not permit the method of relief just discussed, relief wells of tile, the same size as the adjacent tile drains, may be sunk directly into the pressure pocket and led upward directly into the tile drains. This method is often used effectively in western irrigation projects to relieve heavy seepage from higher lands or artesian flow tapped by subsurface drainage or irrigation conduits. A thoroughly successful case of this kind is that in a deep peat bog on the Turtle Valley Farms of J. B. Swan, near Delavan, Wis. (see Fig. 14.13). The relief wells were originally sunk into the pocket about 2 feet below tile drains passing over them at a depth approximately 4 feet below the surface. These wells empty vertically upward into the tile drains. Although the pocket was recorded during construction as a gusher, there was no evidence 14 years later—an exceptionally wet season—of upward seepage or excessive surface moisture at this point, and inspection at that time showed only a small stream of water not over 1 inch deep flowing quietly away from the spot through the tile drain tapped by the relief wells.

14.13. Peat over Marl. It is the usual experience that a deposit of marl in its original undisturbed bed almost completely obstructs the passage of soil water; hence, drainage of a marl subsoil is frequently impossible. This fact results in two opposing phenomena in dealing with peat underlaid by marl. As pointed out by Roe [122,123] this condition in the case of very shallow peat usually limits the use of such land to production of grasses and other shallow-rooting crops because the presence of the marl subsoil prevents artificial drainage deep enough for deeper rooting crops. On the other hand, marl does not give up moisture within itself by upward capillary action at a sufficient rate to act as a reserve

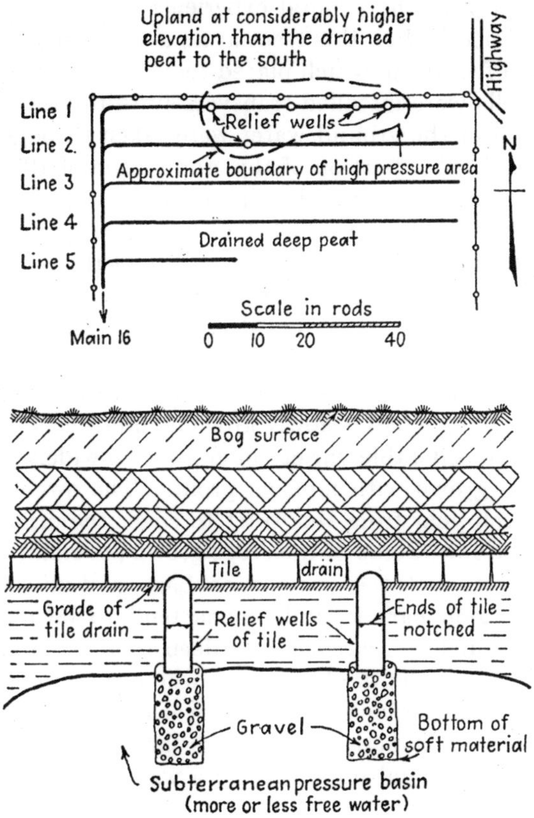

FIG. 14.13. Portion of Turtle Valley Farms near Delavan, Wis. *Above:* Relative location of relief wells under a portion of the tile-drainage system. *Below:* Method of setting the relief wells.

moisture supply for crops. Therefore, in dry seasons, overdrainage must be avoided, as pointed out in the case of shallow peat over clay subsoil, unless a permanent source of water at the surface of the marl is present.

The Peat Plots already referred to furnish a good illustration of the influence on drainage of marl under peat. After surface settling on this bog was practically complete, marl subsoil was at a depth below the surface not over 4 feet, over most of the experimental tract. This permanently upset the experimental drainage control and prevented securing a fixed water table deeper than the surface of the marl.

Another excellent illustration of this type of problem that eventually ran counter to general experience is that near Eden Prairie, Hennepin County, Minnesota, elsewhere reported on by Roe [123] in part substantially as follows. The surface is 12 to 18 inches of peat underlaid, to an indefinite depth, by almost pure marl. Results reported in this case seem contrary to the general experience. In the summer of 1925 tile drains were installed on this tract at an average spacing of about 125 feet and 2½ to 4 feet deep, which placed them 1 to 2½ feet into the marl. Series of test wells were installed transversely to the tile lines, and readings of the depth of the ground water were taken throughout the fall of 1925 and a considerable portion of the crop season of 1926.

These measurements showed that in no case, except immediately over the tile lines, the trenches of which had been largely refilled with peat, was the water table lowered below the surface of the marl. This condition seemed definitely to limit the crops that could be grown on the area to those which do not normally root deeper than 12 to 18 inches. However, later reports from the owner and from a tile drainage engineer familiar with this region for over 20 years state that through 13 years the drains seem to have had a slowly growing influence, even on the marl, so that finally a considerable variety of crops with widely varying rooting depth have been grown with unusual profit. The owner states that he clears enough each year from this tract to cover the original cost of the drainage installation. Some of the crops grown are small root crops such as beets and carrots, potatoes, onions, cabbage, and unusually good field corn. In this region carrots and onions tend to be moderately deep rooters (30 to 36 inches), cabbage is a deep rooter and so is field corn, which tends to reach a depth of 48 inches where conditions are favorable.

Another unusual experience with peat over marl is the following: Near Little Falls, Minn., a 90-acre tract of high-lime peat, running quite uniformly 4 feet in depth and underlaid by a water-bearing marl, has been most successfully drained.

This bog forms the center of a large watershed, surface runoff flowing onto it from three sides. The bog was drained in the year 1937 on the following plan. Open ditches were dug along three sides of the tract to intercept the floodwaters from the high lands. Six-inch tile drains were installed 200 feet apart and 3½ to 4 feet deep over the whole bog. During the excessively wet years, this spacing was too wide to ensure good drainage midway between the tile lines. This was rectified by placing additional laterals midway between the original lines.

The open ditches discharge directly into a shallow creek, and the tile empties into a sump, from which the water is pumped into the creek. The final result has been good drainage, but never overdrainage.

The complete drainage system cost about $6,000. According to the owner, the additional profits each year more than repay this, as this bog produces 400 to 500 bushels of potatoes and 700 to 800 bushels of onions per acre per season. (This claim is based on cash returns from *normal* years.)

Ordinarily the surfaces of marl deposits under potential agricultural peat do not seem to occur at greater depths than 4 to 5 feet from the bog surface. If they were deeper, governing drainage conditions and practices would fall under the general case of drainage of deep peat hereinafter discussed.

14.14. Deep, Landlocked Peat Bogs of Limited Area. It is largely unknown that drainage for agricultural uses of landlocked peat bogs of limited area, whose depth in the middle is great, is often a wholly uncertain, if not economically prohibitive, undertaking. This is because, without natural drainage outlet, frequently the central part of such a bog is a soft pulpy mass, extremely wet, with a consistency to a great depth scarcely greater than that of the confined water that causes its condition. It has no firmness and, even if most of the water can eventually be drained out, the settling following such action often measurably exceeds the depth to which tile drains may have been or may be placed. Frequently the tile itself settles greatly and may even sink beyond recovery. In such a case no support either of artificially placed firmer soil or any wooden cradle will be able to hold the tile permanently in place. The only solution is to let the water drain out all it will and deepen the expensive cut through the surrounding mineral ridge to permit deeper drainage of the center of the bog. It is this procedure which, as a rule, is utterly prohibitive, economically considered.

There are many such bogs in the northern Mississippi basin. On some of these bogs ill-advised drainage has at some time been attempted, usually resulting in total loss of investments that vary from a few hundred to several thousand dollars. Three widely separated cases carried out under entirely unrelated auspices are cited to give a clearer picture of the character of the problem.

The first is on an isolated bog of about 6 acres, 2 miles west of Mound, Hennepin County, Minnesota. It is entirely surrounded by a high mineral soil ridge. It is composed of very finely divided peat of a sedimentary character running from a depth of about 6 feet near the shore line to a center depth of 16 to 20 feet, very soft and wet. This bog was tiled some years ago, the main being cut through the ridge at a depth that would allow about 3 feet of settling in the bog without wiping out the tile gradient. This involved a 16-foot cut for 50 feet through the deepest part of the ridge. Across the bog the main was laid well to one side, not far from the shore line, and 6 feet deep. Laterals were then laid across the bog 4 feet deep, supported by boards in the center to prevent their sinking out of sight and reach. The main itself did not sink, but the peat covering it settled over half its depth. The peat over the laterals subsided until the tile showed on the surface, and then the laterals themselves settled for a considerable portion of their length well below their original gradient. The bog never did drain, and the project was a total loss but proved a lasting lesson both to the owner and to the engineer.

The second case is that of a deep 40-acre landlocked bog crossed by the located line of county ditch No. 40 in Wright County, near Monticello, Minn. It was composed of fibrous peat but very soft and slushy with no sustaining consistency to bear the weight of the 18-inch tile main. No sufficient sounding of the bog was made before construction started. The width of the bog along the main was 1,000 to 1,200 feet. The tile were laid on cradles of 2- by 8-inch planks cleated together with a space between to keep the tile from rolling off. Tile tried in the trench without plank support sank entirely out of sight and beyond recovery. The surrounding ridge was cut to a depth of 10 to 12 feet for a distance of several hundred feet through clay on a gradient of 0.10 per cent with no allowance for settling of the bog after the drainage should have been effected. Surface settling left no appreciable cover over the tile. A prod rod 20 feet long did not find bottom after the initial settling of the surface. The plank cradles held the tile on grade for only a short time, the drain was ineffective, the bog was unusable, and the job was never completed after the contractor refused to continue so hopeless a task. The state legislature finally reimbursed the farmers for what they had spent on the project which was adjudged by its committee to be wholly ill-advised.

The third case is that of an 8- to 10-acre bog of deep, soft, fibrous peat near Oak Knoll, Hennepin County, Minnesota. The ridge was cut only 4 feet deep without any thought of settling of the bog. The cut should have been 10 feet. Settling of the bog surface began as soon as the top water started to drain off, and the main tile ran out of cover 200 to 300 feet short of the center and deepest part of the bog. The contractor quit the job which was a complete loss to the owner.

Every bog of this type is an individual problem in itself. Where such uncertainties exist that cannot be clarified beforehand, there is only one economically safe procedure. An open ditch of sufficient capacity and depth to carry off from the bog all the apparent free water should be opened up and the project allowed to stand for 2 or 3 years to see how the bog material will respond. If by then the surface becomes dry and sufficiently firm to hold the weight of an average man, there is time enough to consider the advisability of tile drainage. The first essential in any case is a thorough examination of conditions in the bog at least to a depth as great as that to which it is planned to lay tile. This rule, moreover, may well apply to all bogs of whatever character.

Preliminary work of this kind costs something, but in the end, the bog may be worth it many times over. If the final examination reveals further development as ill-advised, the relatively light preliminary expense has at least prevented a heavy investment that will never yield any return.

14.15. Deep Peat. Bogs where the depth of the peat layer exceeds 5 feet are usually designated, from the standpoint of drainage, as deep peat. Where the nature of the project is such as to make effective drainage possible, proper drainage for deep peat somewhat resembles that for flat lands in mineral soil. However, the *moisture equivalent* of peat soils is not a definite figure, hence has no appreciable significance. Therefore the design of tile drainage for peat soil is not so relatively simple as for mineral soils because the direct relationship which Dr. Neal has established between the drainage requirements of the latter and their moisture equivalents or other physical properties has no known parallel in peat soils.

14.16. Spacing and Depth of Tile Drains in Deep Peat. This is a problem that is met with the world over, and considering the wide variation in conditions on different bogs and the wide difference in climatic conditions, the close agreement among authorities on practices that give the best results is interesting. This agreement is best shown in Table 14.2.

Farlam [46] found that, in German practice where the maximum depth of the peat is 12 yards (36 feet), 7-foot ditches 530 yards (approximately 1,600 feet) apart are furnished as outlet ditches into which the parallel tile laterals discharge.

Fippin [47] holds that many shallow drains rather than large deep ditches are best, although he agrees that the deeper peat is the best for crops. The authors do not wholly agree with the idea of many shallow drains. The rooting habit of crops raised must be the determining factor (see Chap. 10, Table 10.7). Ranson [114] calls for deep drainage below the root zone in Florida peats, and this practice seems to be agreed on by all with experience in deep peat, especially where there is a reliable ultimate source of needed moisture.

Although no spacing figures are shown for Ref. 122 in Table 14.2 for Minnesota, the following experiences are offered. In dense peat containing much colloidal matter where the water moves very slowly and where good underdrainage is needed, it has sometimes been found necessary to lay tile as close as 2 rods apart. Under more favorable conditions where peat drains easily and the area is used for general farming and grows almost any field crop, it has been the experience in Minnesota that spacings of 130 feet, accompanied by depths of 3½ to 4 feet, after settling has practically ceased, have given excellent results. This is borne out in experience in the Scandinavian countries as, related by a drainage engineer of several years experience in those countries. It is economy to lay the parallel laterals at the widest allowable spacing in the beginning, for, if subsequent experience on the given bog shows need of closer spacing in one or two places, it is a simple and relatively inexpensive matter to drop

TABLE 14.2. RECOMMENDED SPACING AND DEPTH OF LATERAL TILE DRAINS IN DEEP PEAT

Ref.	Author	Country or state	Suggested spacing			Suggested depth, ft	Remarks
			Rods	Yards	Feet		
46	Farlam	Germany	15–21	.....	....	5	Substratum sand, gravel, and muck soils. Spacing according to subsoil and depth of peat. Boards under the tile will support them in very marshy places. Gradients used were not less than 0.5 per cent
81	McCool and Harmer	Michigan	.....	.....	50, 150–300, 2,640	3½	Depth given after subsidence is practically complete. Final depth of 3.0 ft right for most farm crops; 2 to 2½ ft for pasture. Should be deeper in raw and fibrous material
96	Ogg	Northern Europe	.....	13–32, av. 21	......	3½–4½	
97	Ogg and others	Scotland	.....	22	......	......	No depth mentioned
100	Orchiston	New Zealand	.....	.....	......	3½–4	No spacing mentioned
108	Powers	Oregon	3–4	.....	......	4 min	Peat may shrink one-fourth volume
121	Roberts	Rhodesia	.....	.....	50–200	3½–4	
122	Roe	Minnesota	.....	.....	......	1½–2½ 3–3½	Continuous grass General cropping All recommendations after subsidence is complete
138	Smalley	N. Indiana and S. Michigan	5–12	.....	......	3½–4	In muck soils to allow for subsidence. At depths of 2½ ft tiles uncover in 8 to 10 years and must be relaid. At this depth they also settle out of line and fill up

in an additional line of tile at any time between any two of the lines as originally located.

It should always be borne in mind that, when drainage is installed on raw bogs which have neither been pastured nor otherwise used agriculturally, the final depths of laterals desired, after settling is practically complete, should be increased by the probable amount of the total settling.

Harmer [54] calls attention to the fact that "high water level crops" tend to die out on muck after drainage and that this should be taken into consideration along with costs of drainage before carrying out the actual drainage. He also maintains that when the muck has settled a few years before being tile-drained, tile laid 3½ feet deep will be all right for general crops. The authors, however, cannot entirely agree with this view as will be noted below.

Powers [108], in general, estimates that the total settling will amount to approximately one-fourth of the depth of the peat above the tile. This seems a fairly reliable rule to follow. However, Minnesota experience is that the subsidence *above the tile* will be approximately proportional to the original depth of the peat layer above the tile on raw bogs beginning with about 1 foot at an original depth of tile of 3.0 feet. This means that, in raw deep peat drained for general farming, the lateral tile should never be placed less than 4½ feet deep, and in rare cases where particularly loose surface conditions occur, the original depth may need to be as much as 6 or 7 feet where the final depth desired is 3½ to 5 feet. No hard and fast rule can be laid down. Good judgment and experience, coupled with a knowledge of the use to which the bog is to be put, must enter into the calculations.

14.17. Treatment of Soft Spots in Bottoms of Tile Trenches. In deep peat, spots are often encountered, in trenching for tile, where the natural material in the bottom of the trench is so soft that it will not support the tile at grade. There are several ways of handling such cases. If the soft material is not very deep and there is firm mineral soil at hand, the soft material may be removed and replaced by mineral soil well tamped in and the tile grade finished on this filled-in material. If the soft spot is deep, a considerable quantity of dry and tough fibrous material from the upper layers of the trench may be forced down into it and a bed firm enough to bear the tile thus secured. In addition, number 3 common boards the width of the outer diameter of the tile may be laid continuously in the bottom of the trench with butted ends supported by a shorter length of board (3 feet long) placed underneath to support the butted ends. This will give good support to the tile even on a very flat gradient. Boards alone may often be used in this way in soft bottoms without first replacing any of the soft material with firmer soil. This use of boards has been found effective by the authors as well as by Farlam [46], Harmer [54], Work [164], and others.

14.18. The Outlet to a Drainage System in Peat Soil. In the drainage of peat and muck soils the importance of a good outlet cannot be overemphasized. In this connection, reference is again made to Chap. 10. All the essentials discussed therein under the general topic of Outlet, Sec. 10.1, apply with especial force in drainage of peat and muck soils.

14.19. Drainage of Peat Swamp Forests. The matter of the drainage of peat swamp forests is virtually covered in Sec. 6.26, Forest Drainage. At this point the reader will do well to review that section, to consult the report by Averill and McGrew [8] therein cited, and again to inspect Figs. 6.11 to 6.13.

INVESTIGATION OF A PEAT DRAINAGE PROJECT

All suggestions offered in the preceding chapters concerning preliminary investigation of drainage projects in general apply with equal force to peat drainage projects, to which the following additional suggestions also apply.

14.20. A Convenient Peat Sounding Rod. One piece of equipment especially convenient and useful in investigation of peat areas is a sound-

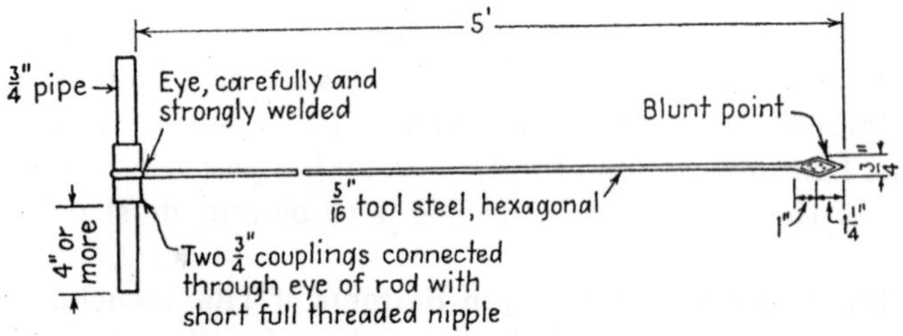

FIG. 14.14. Prod rod of tool steel with expanded blunt point—a very convenient tool for sounding peat bogs underlaid by sand.

ing rod such as is illustrated in Fig. 14.14. This rod should be at least 5 feet long under the handle. A longer one in a single piece is awkward to handle. The rod can be made of ½-inch mild steel, but it is better made of 5⁄16-inch tool steel. If made of tool steel, it can also be provided with extensions connected by wrought iron pump-rod couplings, provided the blunt-ended conical tip on its base is given a maximum diameter appreciably larger than the outer diameter of the couplings. In this case, however, two small crescent wrenches should be a part of the equipment. The main purpose of the enlarged conical tip is to make so large a hole in the soil that the wet soil will not "set" around and grab the rod making it difficult to pull out. Its blunt end will not penetrate firm sand subsoil, and this fact enables ready determination of the depth of the sand surface.

14.21. Time of Examination of a Peat Bog. Apart from observation of the location and general character of high water marks, high-lying seepage areas or lines, and extra soft areas, the best and most suitable time to make detailed examination of conditions on and around the bog is the driest time in the driest season that can be selected. The same may be said, also, as to the best time to install any drainage decided upon. At such a time, if ever, before the bog is drained, it will be possible to walk over the entire surface and examine it, to sound its lowest and wettest spots for depth and character, to determine the contour and character of the substrata and the character and condition of the peat itself, as well as to secure firmer bedding for tile in trenches during construction, with a minimum of temporary ditching and of artificial support for the tile.

THE FROST PROBLEM ON PEAT AND MUCK SOILS

It is generally known that peat and muck soils neither absorb nor conduct heat as rapidly as do mineral soils. Mineral soil, on the other hand, readily absorbs considerable heat from the adjacent atmosphere during the day and gives up much of it with equal readiness on cool nights. The peat and muck areas do not do this to any appreciable extent but act as a sort of low receiving basin for the cooler air. The warmer air in contact with the mineral soil, being lighter, is displaced by the heavier and cooler air settling in from the upper reaches of the atmosphere. This interchange above the mineral soil sets up air currents that increase the rate of flow of cold air down over the already cooler surface of the peat bog. The eventual result often is that the atmosphere continues to cool through the night and, reaching a temperature at or below 32 degrees Fahrenheit, causes frost on the surface of the bog when the higher mineral land surrounding the bog is entirely free from frost. On peat and muck soils, therefore, frosts are apt to occur almost any month of the year and, owing to the fact that moist peat both absorbs and conducts heat more rapidly than does dry peat, the occurrence of these frosts has a definite relationship to the moisture conditions in the surface and upper few inches of the peat soil.

14.22. Minnesota Studies Relative to Drainage, Temperatures, and Unseasonable Frosts on Peat Soil. With a view to determining a more reliable basis for predicting these unseasonable frosts and more effective methods of protection against their harmful action, studies were carried out by the Minnesota Station each season from 1925 to 1930 on the Peat Plots. Results of these tests are reported elsewhere in considerable detail by Roe [122,123] and by Howe [57]. The most important and directly applicable findings are summarized in the following paragraphs.

14.23. Field Studies and Equipment. Complete precipitation, wind velocity and direction, and temperature-recording equipment was used, sufficient to cover the limited area of the Peat Plots with tests on all depths of drainage control. This equipment included continuous-recording devices as far as obtainable as well as mercury soil thermometers up to 2 feet of depth as well as maximum-minimum surface thermometers, and all temperature-recording equipment was calibrated against a government-tested standard thermometer.

Records were kept along the *B* and *D* lines of test wells (see Fig. 14.5), the mercury thermometers being read daily throughout each entire growing season. Weekly charts were used on all continuous-recording instruments. Records covered each control plot on both halves of the tract.

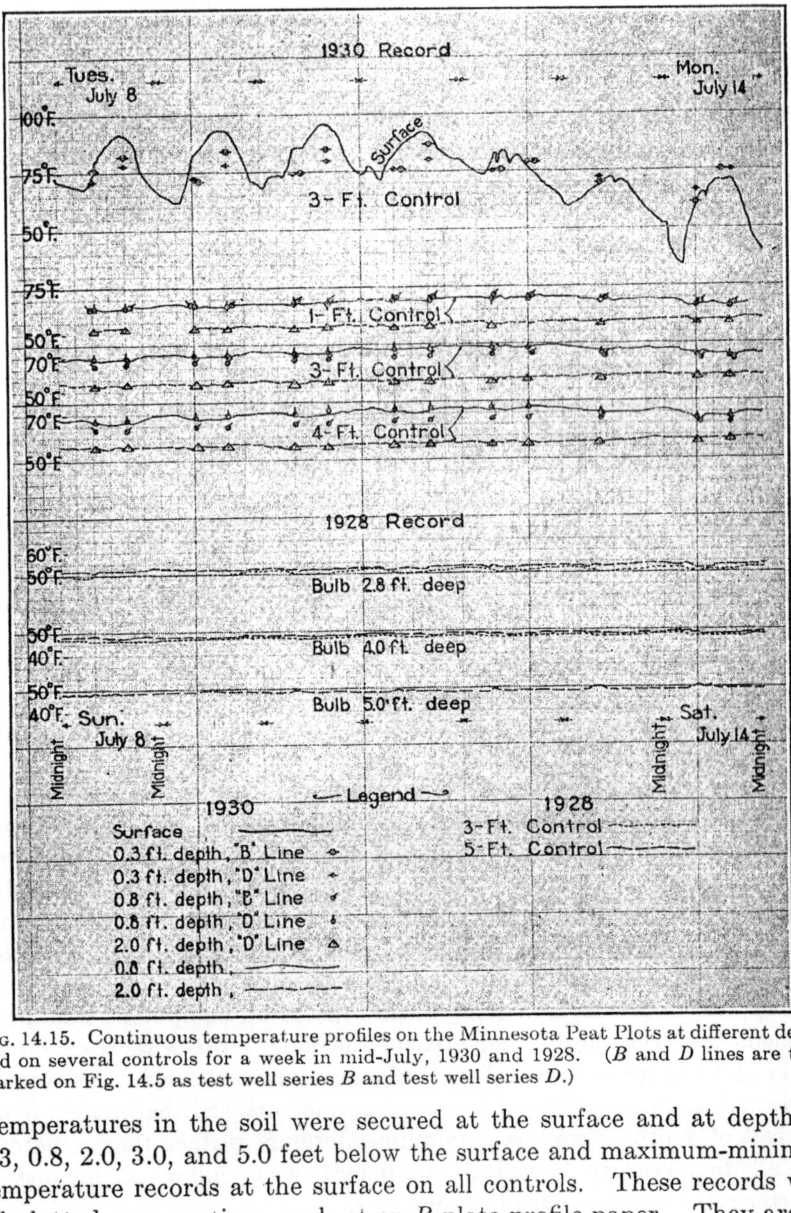

FIG. 14.15. Continuous temperature profiles on the Minnesota Peat Plots at different depths and on several controls for a week in mid-July, 1930 and 1928. (*B* and *D* lines are those marked on Fig. 14.5 as test well series *B* and test well series *D*.)

Temperatures in the soil were secured at the surface and at depths of 0.3, 0.8, 2.0, 3.0, and 5.0 feet below the surface and maximum-minimum temperature records at the surface on all controls. These records were all plotted on a continuous chart on *B*-plate profile paper. They are far too voluminous to all be shown here, but part of the chart showing a typical week in mid-July for 1930 and 1928 is shown as a sample in Fig. 14.15.

A careful analytical study of the 6-year record enables the drawing of many useful conclusions among which the following are here offered because they are particularly applicable to the behavior of temperatures, as influenced by depth of drainage, and to the problem of frost occurrence and control.

Before stating these conclusions it should be noted that all the peat on all the controlled depths of drainage was at least moist except a thin layer at the surface on the lower controls, when sufficient time had elapsed between rains to permit drying of the surface by evaporation.

14.24. Conclusions. Daily rise and fall in temperatures resulting in daily maximum and minimum temperatures occurred throughout each season *at the surface.* Similar maxima and minima occurred *in the peat soil* to a depth of 0.3 foot (practically 4 inches) below the surface but not appreciably at any greater depths. These latter corresponded closely in time of day with the surface maxima and minima.

These facts also indicate that 4 inches marks the approximate depth at which the peat-soil temperatures would have any appreciable influence on the liability to untimely frosts.

Progressively lower temperatures, both at the surface of the bog and at the 0.3-foot depth below the surface, tended to occur as one followed from the higher to the lower drainage controls.

The lowest minimum temperatures and, therefore, the freezing temperatures tended to occur 1 to 3 days *after* a rain, followed by clearing skies and a swing of the wind to the north or the northwest.

The fact that both of these last two conditions tend to produce a dry film on the surface practically confirms the earlier statement that moist peat tends to absorb heat from and to give up heat to the adjacent atmosphere much more readily than does dry peat. Thus, too, the claim made by the authors, Harmer [54] and other investigators, is well supported that the higher one can keep the moisture content of peat and muck soils *at the surface,* consistent with good crop development, the less the untimely frost hazard.

14.25. Controlled Indoor Laboratory Studies. As a check on the reliability of the foregoing conclusions, indoor laboratory tests were carried out on five separate 1-foot cubes of field peat soil taken from adjacent positions in the same field and undisturbed except for the cultivated 3 or 4 inches. In these tests it was endeavored to keep the average heat and barometric pressure conditions as near as possible to those of the bog at latitude 45 deg (St. Paul, Minn.) on a clear midsummer day with little or no wind.

The check on the field tests was surprisingly close, and the following final comprehensive conclusion seemed entirely justified:

As long as there is sufficient moisture in the peat to be appreciable to sight and touch, regardless of its percentage value on a dry weight of soil basis, the influence of variation in moisture content on changes in temperature in the peat, other physical surroundings being normal, is slight and the danger of untimely frost is minimized.

14.26. Occurrence of Frosts on Peat during the Growing Season. It is the general experience where crops are grown on peat and muck lands that damaging frosts are apt to occur sometime in every growing season month from May to September and that practically all the crops grown on peat are either destroyed or seriously damaged by such frosts. Owing to poor heat conductance of the dry surface film, frost damage was greater on the plots with the lowest water table where, apart from the frost menace, the largest and best yields were to be expected.

14.27. Summer Frost Prevention and Control. Various artificial methods of frost prevention and control have been used with some success in different parts of the United States.

One of the best preventives is provision of good air drainage or its improvement where it is already present. On the Peat Plots, after a dense area of timber along the east and south was removed, thus making possible free air movement across the entire region close to the ground surface, summer frosts were extremely rare.

Keeping the soil well compacted improves heat conductivity and is, therefore, of some definite value in preventing frost injury because it induces warming of the lower layers of the atmosphere.

Stirring the air by mechanical means has been claimed by some to be an effective frost preventive, but this claim has not yet been well substantiated.

In the fruit-growing sections of the West smudge pots or heaters are used effectively for frost prevention in orchards. This method was tried out by the Minnesota Station for vegetables on the Peat Plots in 1932. It was found, however, that the heat from the smudge pots rose rapidly away from the ground, leaving a surface mantle of cold air still around the plants and giving little or no protection to low-growing truck crops, strawberries, etc. Plants not over 6 feet away from the burners were often frozen. The method is expensive and requires much hard and disagreeable labor. Sparks and spatters of burning oil sometimes start serious fires in the bog. Homes in the immediate neighborhood are fouled with soot from the smudge pots. For these reasons this method was judged by the Minnesota Station to be unsatisfactory for protection of vegetable crops on peat lands.

Use of irrigation other than by sprinkling has occasionally been found to aid to some extent in warding off frost, but it is not nearly so effective as where the plants themselves are kept wet during the danger period. This can be accomplished only by overhead sprinkling which has proved especially satisfactory for this purpose both in experimental tests by several state experiment stations and also on the farms of commercial growers in various parts of the northern United States.

Avoidance of overdrainage of peat soils, in any form, is manifestly one of the surest means of protection against untimely frost.

CROPPING OF PEAT AND MUCK LANDS ESPECIALLY IN RELATION TO DEPTH OF WATER TABLE

As shown in Table 10.7, page 264, different types of crops have widely varying natural depths of rooting, and the depth to which the roots of many crops will penetrate depends largely on the depth of the water table below the land surface. Furthermore, the depth of water table often fluctuates on any given agricultural area, owing to lack of sufficient rainfall during the growing season, far more than is suited to the growth and maturing of the crops grown thereon. This undesirable degree of fluctuation often is even more serious on peat and muck lands than on mineral soils. Particularly, therefore, on peat and muck areas capable of producing high returns, especially in truck crops, it is frequently desirable to invoke the aid of controlled drainage already briefly mentioned in earlier sections of this chapter discussing open trench drainage. Controlled drainage is accomplished by so designing the drainage system as to be able to check the drainage action and maintain the depth of water table at the most desirable level throughout the entire season. There must be available from some source a sufficient supply of water to accomplish this. Outlet channels must be equipped with control dams of the stop-log or crest-board type. Lateral tile drains or submains must usually be equipped with control weirs (stop wells) similar to that shown in Fig. 14.6 or at least with tile drain dams similar to that shown in Fig. 14.9. The smaller control structures such as the control weirs are inexpensive and easy to construct and install. The control dams in open ditches, especially if the ditches are large, may be quite expensive in first cost. While these dams may be constructed of various materials, they must be well anchored in the sides and bottoms of the ditches to prevent failure from undercutting or lateral seepage. For permanence, a properly designed concrete structure will usually be preferable even at greater first cost. This type of structure is that used on the Coddington Marsh, Portage County, Wisconsin, and shown in Fig. 14.11.

Experience in Illinois, Indiana, Iowa, Michigan, Minnesota, Ohio, Oregon, and Wisconsin indicates that, with proper regard to drainage, tillage, and fertilizer requirements, peat and muck bogs will produce good yields of many crops commonly grown on uplands of the same region. Harker [53] tells in general terms of several extremely interesting and highly satisfactory results in production of potatoes, mint, and general truck crops under controlled drainage in Indiana and Michigan which

Fig. 14.16. Two typical crops commonly grown in peat and muck soils. *Above:* **Typical peppermint field on moderately deep muck with shallow water table, Livingston Farms near St. Johns, Mich.** *Below:* Typical early celery field on deep peat over water-bearing sand under pressure, with high water table, near Hamilton, Mich.

amply justified the expense of installing the necessary works. The following projects of this group viewed by Roe may be of interest here.

On large farms near St. Johns, in south-central Michigan, the muck is largely 4 feet or less deep over variable substrata of silts, silty clays, sands, fine gravel, and some marl. Lateral tile drains, as a rule 7 to 8 rods apart, discharge into large outlet ditches in which the desired high water level is maintained by means of dams equipped with adjustable flood gates. The mints, spinach, and Swiss chard, all very shallow rooters, are leading and profitable crops. Plowing is deep, 9 to 12 inches. For mint the water table is maintained at about 12 inches below the surface, or at 6 inches for peppermint which is an even shallower rooter with the bulk of its roots in the first 4 inches of surface soil. Figure 14.16 shows a mint field typical of this area. Early spinach also calls for a high water table and is usually followed by more spinach, red beets, Swiss chard, or other shallow-rooting crops. In western Michigan celery and miscellaneous truck crops are grown continuously year after year on deep peat with lateral tile drains generally spaced about 4 rods apart, outletting into large open channels in which the water level is controlled by bulkhead and pump. In some cases the underlying sand carries water under pressure that brings it to the surface, in which case the water level is controlled by tile drains leading to a sump emptied by an automatically controlled pump. The water level is usually maintained at 12 to 18 inches below the surface; heavy fertilization, mostly with phosphate-potash combination and little or no nitrogen, is practiced. Sprinkler irrigation is sometimes used only to start out the young plants immediately after transplanting. Figure 14.16 is typical of the celery fields of this region.

Clayton and Jones [28] also give an interesting account of controlled drainage, its methods, and the results accomplished in the Florida Everglades.

The projects on the Coddington Marsh and at Dousman, Wis., mentioned in an earlier section, are really under controlled drainage of a rather simple character where, owing to especially favorable subsoil conditions, control dams in the ditches suffice without the aid of lateral drainage systems.

14.28. Minnesota Experiments on Controlled Drainage. During the growing season each year from 1925 to 1930 at the Minnesota Station, the Divisions of Agricultural Engineering, Agronomy and Plant Genetics, and Horticulture carried out studies on the Minnesota Peat Plots, earlier discussed in the section on surface settling of peat bogs under drainage and agricultural use, to determine the influence of depth of water table on yields of crops grown on peat lands.

These studies included the following crops: *field crops*—field corn, flax, oats, soybean hay, sugar beets, and timothy and clover hay; *truck crops*—beans, cabbage, carrots, cauliflower, celery, head lettuce, onions, potatoes, spinach, sweet corn, summer radish, tomatoes, and winter radish.

The summary of results for the entire 6-year period is reported in considerable detail by Roe [122]. The depths of water table shown in these reports as giving the most satisfactory results with the various crops agree with the general statements in Table 10.7.

14.29. Recommendations for Depth of Drainage in Peat. Experience in cropping peat and muck lands does not show any marked difference between the *average optimum* depth of drainage for field crops and that for truck crops. Frequently either or both will be raised from time to time on the same field. In many cases it will not be feasible or profitable to attempt an elaborate ground-water control system that permits maintaining the ground-water level at different depths in the same season on different fields or portions thereof on the same bog. Hence, for general farming, it is recommended that an approximately uniform average depth of ground-water level be decided upon for the entire field, such as will most nearly approach the requirements of the majority of crops to be raised, exclusive of the true grass crops. This approximate average is evidently 3 to 3½ feet after settling of the bog is practically complete. Hence, the lateral control tile should be placed at this depth plus the allowance for settling of the bog—if it is a raw bog—and for a small upward arching of

TABLE 14.3. CROPS WHICH HAVE BEEN RAISED ON PEAT AND MUCK SOILS

Type of Crop	*General Drainage Requirement*
Crops with which frost is not a problem:	
The grasses	Poor to fair; 0.5 to 2 ft of depth
The clovers	Good; 1.5 to 3 ft of depth
Most root crops	Good; 3 to 5 ft of depth
Crops with which frost is not a serious problem:	
Celery (frost hard on young plants)	Good but shallow; plenty of moisture must be available under good control
Onions	Good; fairly deep, 3 to 4 ft
The mints	Good but very shallow; 4 in. to 1 ft and well under control
Crops with which frost is apt to be a considerable problem at certain stages of growth:	
Small grains except wheat	Good; 3 to 5 ft deep
Small fruits	Good but shallow; 1 to 2 ft
Tomatoes	Good; 2 to 3 ft deep
Crops with which frost is always a serious hazard:	
Corn, field	Good; 3 to 5 ft deep
Corn, sweet	Good but only moderately deep; about 3 ft
Legumes, except alfalfa	Good; moderate to deep, 2 to 4 ft
Potatoes	Good, but shallow; 2 to 2.5 ft
Salad crops	Cabbage and cauliflower good; 2.5 to 4 ft
	Chard, spinach, etc., good but shallow; 0.5 to 1.5 ft deep under good control
Silage crops	Good; 3 to 5 ft deep
Vinous crops	Good but rather shallow; 1.5 to 3 ft under good control

the ground-water surface during protracted rainy periods. The total original depth in a raw bog at the time of installation in most cases, therefore, will amount to 4 to 5½ feet according to the looseness of the original surface of the raw bog.

On bogs permanently devoted purely to grass crops, a depth of drainage of 1.5 to 2.0 feet, after settling of the bog is practically complete, is ample and is recommended.

A greater refinement of control will be justified in case of bogs given over wholly to the intensive cultivation of special crops of unusually high return per acre. In such cases normal rooting habits of the particular crops desired to be grown should be the leading factor in determining proper depth of drainage. The discussions in this chapter together with a study of the rooting habits of crops as shown in Table 10.7 should, therefore, serve as a fairly reliable guide to depth of water table to be provided for a given crop.

Table 14.3 also gives a fair idea of the general possibilities, but owing to the wide variation in moisture supply and severity of the frost menace in different regions, it should not be used as the *final* guide to crop selection and treatment on any specific bog.

The danger of frost in the more northern areas and the fact that the higher water table tends to lessen damage from frost indicates that in many cases, other things being equal, the shallow-rooted crops would be more adaptable than the deeper-rooted ones, and under such conditions their selection is recommended. If the deeper-rooted crops, requiring a lower water level for root development, are grown, the greatest possible precautions should be taken against frost injury.

14.30. Rolling of Peat after Breaking or Plowing. Rolling of peat after breaking or plowing and disking is essential for optimum results in cropping of peat and muck lands. The naturally loose, fluffy soil is compacted by the rolling thus helping to ensure good germination and to resist wind erosion. Rolling is also an indirect factor in protection against occurrence of summer frosts owing to better heat conductivity of the compacted soil. The rolling should be done when the peat is moist but not wet. The roller should have a weight of about 700 to 800 pounds per linear foot. Roller shells of iron that can be filled with concrete to give weight have been offered commercially. Rollers of concrete are also often made by the local user. Figure 14.17 shows plans for a concrete roller designed and built by the late John T. Stewart, who, however, warns that it is a little heavy for haul by team which, ordinarily is limited to about 1¼ tons over peat soil. Modification in weight should be made by changing length, not diameter, as smaller diameter greatly increases the difficulty of hauling. While making the roller in two sections eases the heavy drag on turns, experience resulted in a final recommendation that the concrete roller be made all in one piece because of the serious difficulty of lubricating and repairing the central bearing on the double roller.

14.31. Growers Advised to Seek Soils and Agronomic Counsel. The seedbed preparation, fertilization, and cropping of peat and muck lands

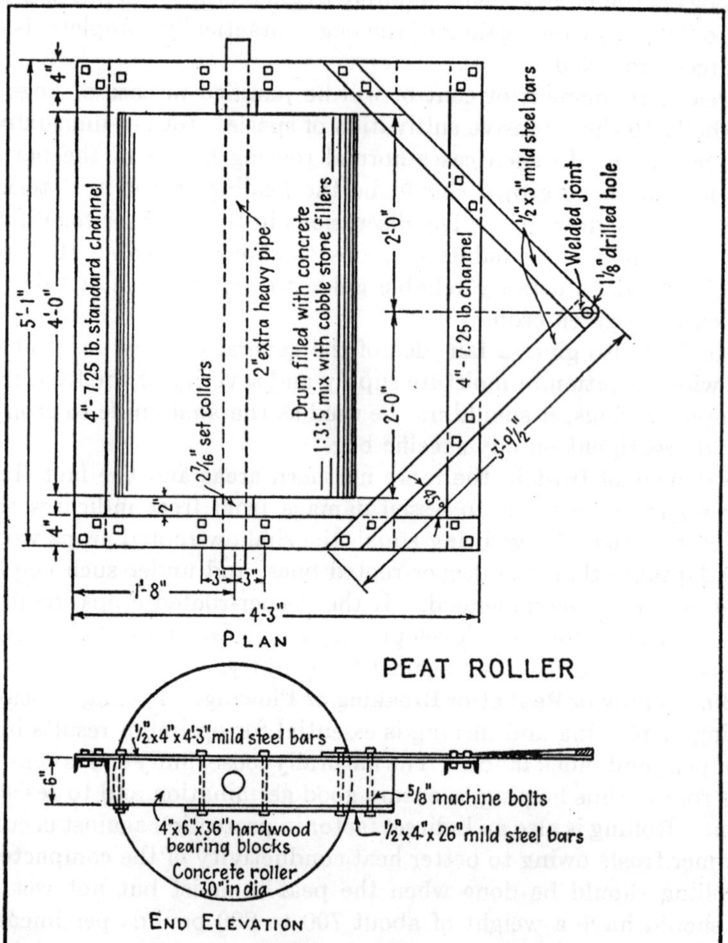

FIG. 14.17. Plan for peat roller designed by the late John T. Stewart.

is a special and extremely intricate problem in scientific soils and agronomy and is outside the scope of this book except in regard to its soil-moisture relations. Hence the reader interested in cropping such lands and needing counsel thereon should contact the soils and agronomy specialists of his own state agricultural experiment station and consult their specific literature on the subject.

PROBLEMS

1. What is the practical distinction in agriculture and in engineering between high-lime and low-lime peat?

2. What is the distinction between a peat and a muck soil?

3. In what essential ways does the determination of proper depth of tile in peat differ from that in mineral soils?

4. How does the practical treatment in drainage of the deep land-locked peat bogs of limited area differ from deep bogs of wide extent, and why?

5. What precautions should be taken in designing tile drainage of peat bogs of moderate depth over a substratum of clay?

6. How do the requirements of drainage of swamp forests differ from drainage of agricultural lands which bear annual crops?

7. (*a*) What is the general effect, apart from other considerations, of drainage on the summer frost menace on peat soils?

(*b*) What is apparently the best agency for control of such frost tendency?

8. (*a*) Is the proper spacing of tile laterals in peat as yet worked out as in mineral soils?

(*b*) What are the leading agencies that determine spacing of laterals in peat?

9. What is the essential relation of drainage in peat soils to their agricultural use?

CHAPTER 15

COST OF DRAINAGE

With economic conditions constantly shifting, mere arbitrary units of monetary cost have little significance for a long enough period to serve as a permanent guide to estimates of the cost of any endeavor. On the other hand, the physical units of time, weight, volume, and distance, which are the basis of money charges, are the same always. The local and temporary market value of any one of these units being known, it is a simple matter to estimate total costs of a project if the total of the physical units of each class involved therein is known. Furthermore, because of the constant character of these physical units on which the money-cost units are based, relative contemporaneous values of commodities, labor, etc., tend also to be stable regardless of economic shifts. These facts apply to drainage costs along with those of other types of projects. Therefore, the guiding principle in this chapter will be to analyze drainage costs on the physical bases of man-hours, tons, miles, etc., so that the results may readily be applied in any economic period provided going money rates for these units are known.

15.1. Factors in Cost of Drainage. The usual elements in the total cost of a drainage project are, in general, the following:

1. Engineering and supervision; usually centered in the same individual or small group of individuals; as a rule paid for at a professional rate per diem but sometimes on a lump-sum contract basis or percentage of total cost
2. Right of way; seldom encountered in private farm drainage, but always a feature of public outlet systems; usually purchased at a definite price per acre across a specified tract, but sometimes provided under a privilege agreement for a fixed sum, across the given tract
3. Materials; including such items as drain or sewer tile, culvert pipe, lumber, cement, sand, gravel, etc., purchased usually under lowest reliable bid at specified rates per accepted unit
4. Delivery of materials; including railroad freight charges, truck or team haul costs including the man-labor of loading and unloading, etc.
5. Execution of the undertaking, including
   a. Cost of financing the contract job
      (1) Those costs paid by the contractor and embraced as hidden elements in his unit contract price, such as financing fees, inter-

est on revolving funds, compensation insurance, and surety bond

(2) Those costs which must be met by the district on public work or by the individual owner on private work, embracing discount on ditch bonds sold for cash, interest on carrying funds, etc.

b. Miscellaneous items not specifically covered by contract prices; including labor on minor temporary jobs, special equipment, materials, and devices required to make possible performance of the job

c. Provision of special structures not contemplated in the original project but later found essential to its success

d. Wear and tear and depreciation on plant; always properly a hidden factor in the contractor's unit or lump price

e. Labor and management; the major basis of any contractor's bid

6. Damage claims; usually avoidable on private jobs concerned with only one owner or group of organized owners, but always encountered on public drainage work and usually having to be adjudicated in court or by a special official or board
7. Legal and administrative expense; including lawyer's fees, court and administrative fees; often largely avoidable on private jobs but a regular part of public drainage undertakings

15.2. Cost of Open Ditches. The cost of small ditches excavated by hand or of broad, shallow field drainageways constructed by road or terracer grader, either self-propelled or drawn by tractor power, is usually reckoned on the basis of length in rods or hundred-foot stations or, more exactly, on a man-hour and/or horsepower-hour cost basis.

The only records available at this writing indicate the possible ranges of cost of digging field drainageways according to type of equipment to be approximately those shown in Table 15.1. In determining the proper pay rate per hour it should be understood that the man-hours are properly those of skilled labor. If only the time rate per mile is known, that per 100 linear feet is the mile rate divided by 52.8 and that per rod is the mile rate divided by 320. The time rates given in the table are only general approximations largely based on terracing work, therefore applicable only for use in preliminary estimates. Final estimates should be based on side slopes and depths of the drainageways established as the result of field surveys, type of power unit used, and the local experience of distance covered per hour, if knowledge of the latter is obtainable from any reliable source.

On such ditches as will ordinarily be dug by hand a good experienced ditcher will take about 7 to 20 hours to dig 100 feet of ditch under average digging conditions.

On large ditches dug with teams and scrapers, or more often with mechanical ditchers of one type or another, the cubic yard is the accepted unit on which cost is based. Only in the case of relatively small ditches are such costs, as a rule, closely concerned with farm drainage, and they are not conveniently reckoned on an acre basis. Large ditches are usually a part of a public outlet drainage system. In the early 1940's the cost per cubic yard on large ditches ranged from about 5 to about 50 cents per cubic yard, depending on the size of the job, the seasonal conditions, the character of the digging, the type and efficiency of the excavating equipment, the prevailing scale of wages and machine charges, and the

TABLE 15.1. APPROXIMATE MAN- AND TRACTOR-HOURS REQUIRED PER UNIT OF LENGTH TO DIG FIELD DRAINAGEWAYS

	Man-hours			Tractor-hours		
	Per mile	Per 100 ft	Per rod	Per mile	Per 100 ft	Per rod
Light drag graders drawn by tractor power						
Maximum	84.48	1.60	0.26	42.24	0.80	0.13
Average	65.43	1.24	0.21	31.21	0.59	0.10
Minimum	45.00	0.85	0.14	18.20	0.35	0.06
Heavy graders, self-propelled or drawn by tractor power						
Maximum	43.90	0.83	0.14	21.65	0.41	0.07
Average	22.80	0.43	0.07	13.11	0.25	0.04
Minimum	6.77	0.13	0.02	6.77	0.13	0.02

experience of the contractor. As a rule, the larger the job, the lower the unit price quoted for the work.

From about 1920 to 1951, quite different from other cost trends, the range of cost per cubic yard for heavy ditching has not greatly changed. The reason commonly given for this fact is the great and rapid improvement in excavating machinery, resulting in far greater efficiency. The engineers of the Soil Conservation Service in the northwestern part of the Pacific Region have stated that the price for open ditch work during 1950 and 1951 has ranged from about 8 to 40 cents per cubic yard, the greater part of the work having been done in those years for about 18 cents per cubic yard on ditches with bottom widths from 4 to 20 feet and depths from 4 to 12 or 14 feet, often including spreading of the spoil banks down to depths of 2 to 3 feet and leaving a fairly uniform surface. This latter practice has been common throughout the country since about 1910. As a rule it has added only $\frac{3}{4}$ to $1\frac{1}{2}$ cents per cubic yard to the price charged for excavation alone. The higher rates in the ranges given apply only in the rare cases of ditches too small and in too cramped conditions to admit of clear use of trenching machines, hence involving a large proportion of hand labor in the removal of roots, stumps, large rocks, and other types of obstacles that require much use of crowbar, mattock, and dynamite.

On relatively small jobs necessitating frequent moves of dredges between jobs or considerable team and scraper work, a common range of price will be about 20 to 50 cents per cubic yard. A common range of price on large jobs done with heavy mechanical equipment under average economic conditions anywhere in the country is about 8 to 15 cents and has been thus for a long time. Powers and Teeter [110] list one job of 225,000 cubic yards in South Dakota, presumably in the early 1920's, which was done for 5.2 cents per cubic yard; no mention was made of spreading the spoil banks.

In a letter dated Oct. 19, 1943, Lewis A. Jones states that a contract was recently let in Arkansas for 12.4 cents per cubic yard on ditches ranging from 4 to 40 foot bottom width and from 7 to 12 feet deep, the bulk of the excavation being on ditches with 20-foot bottom width or more. This price also included spreading the spoil banks at a maximum depth of 3 feet or less and making their slopes, with a steepness of not more than 1 on 4 (or 4 to 1), merge into the adjacent fields. However, the final smoothing up of the spoil banks was left to the farmer owners to be done at their own discretion and expense.

15.3. Cost of Tile Drainage. From 1908 to 1922, the Minnesota Agricultural Experiment Station conducted, from time to time, private tile-drainage demonstrations on individual farms in various parts of the state and in eastern North Dakota on which exact cost data were secured. In 1927 the Minnesota Station published an analysis by Roe [121] of the cost of installation of these systems on a time and materials basis. The results should, therefore, be applicable in any period. Although mostly handwork, they form the basis of this discussion. Table 15.2 shows the number, range, size, and character of the projects. Table 15.3 shows the average per cent of total cost represented by each element, for the entire 18 projects included in the analysis.

The following pertinent facts should be kept in mind concerning Table 15.3.

1. As the projects covered in this analysis were designed and their construction supervised by members of the engineering staff of the Agricultural Experiment Station for purposes of investigation and demonstration, none of the items of cost include the elements of *insurance* or *profit,* but only actual costs of necessary travel, services performed, and materials purchased.

2. The charges for engineering and supervision covered only the regular salaries of the station engineers for the part of their time given to the

TABLE 15.2. SIGNIFICANT AREAS IN MINNESOTA TILE-DRAINAGE PROJECTS INCLUDED IN ANALYSIS OF COST OF INSTALLATION (1908 TO 1922)

Comparative items	Areas concerned, acres			Character of digging
	Watershed within farm limits	Land drained	Land reclaimed	
Rolling land, 8 projects: installation season, May to November. Soil: dark loam surface over stony and gravelly ground moraines and boulder clay				
Total	1,308½	352	257	Very hard on 2 projects totaling 176 acres drained, 21 acres reclaimed; other 6 projects, average digging
Maximum	500	160	120	
Average	163½	62	40	
Minimum	3½	2	1	
Flat land, 8 projects: installation season, May to December. Soil: widely varied from lake-bed clay approaching gumbo, wet and sticky; overridden ground moraines, marsh and peat pockets, some deep peat, and sandy till				
Total	1,162	791	465½	All average digging
Maximum	300	160	100	
Average	145	99	58	
Minimum	7	5	2½	
Peat land, 3 projects: installation season, April to November. Soil: deep peat, one project very deep, filled with tamarack roots; one underlaid at 4- to 6-foot depth with wind-blown sand				
Total	60	43	43	Easy digging except as interfered with by roots and caving sand
Maximum	40	28	28	
Average	30	21½	21½	
Minimum	20	15	15	
All classes of land, 18 projects: installation season, April to December, 1908 to 1922				
Grand total	2,530½	1,186	765½	
Maximum	500	160	120	
Average	140½	66	42½	
Minimum	3½	2	1	

various projects plus the necessary travel and subsistence in the field part of such service. The foreman was a *working* foreman; hence his time was included with that of the skilled tilers and no managerial costs, as such, are included.

3. The time covered by the installation of all these 18 projects embraced several economic periods all of which lay within the first quarter of the twentieth century. Hence the actual money cost items cannot be

directly related to later periods except through the use of wage, material, and other indices or through known price lists and wage scales in the later given period. However, the percentages of the entire cost represented by each element should be fairly representative in any time period, subject, of course, to some variations due to uneven changes in rates of cost of different elements from period to period. This effect will be especially noticeable due to the excessive increase in prices of drain tile and the marked decrease in cost of transportation and distribution of the tile resulting from change-over from railroad haul to haul by truck direct from factory or yard to the farm.

TABLE 15.3. TOTAL ORIGINAL COST, ON ALL 18 PROJECTS, OF EACH ELEMENT, PER CENT OF TOTAL COST REPRESENTED BY EACH, AND SAME COSTS AND PER CENT REDUCED TO CORRESPONDING CONDITIONS IN 1951

Element	Total cost		Per cent total cost in each element		1951 cost by
	1921	1951	1921	1951	
Engineering and supervision	$ 4,925.66	$ 12,461.92*	8.3	7.0	Index
Tile	19,238.83	75,009.88	32.5	42.4	1951 price
Transportation and distribution of tile‡	4,568.83 3,556.99	14,677.26	7.7 6.0	8.3	Index
Trenching, laying, and blinding	23,432.89	66,283.77†	39.7	37.4	1951 rate
Refilling trenches	2,273.93	5,753.04	3.9	3.3	Index
Outlet protection	443.13	1,121.09†	0.8	0.6	Index
Miscellaneous	665.76	1,684.37†	1.1	1.0	Index
Grand total	$59,106.02	$176,991.33	100.0	100.0	

* Professional rates would be nearly three times this to cover professional remuneration. This would be especially true on public work.

† Contract work would increase these items 30 to 40 per cent to cover management, payroll insurance, taxes, and profit.

‡ In all cases freight costs were for railroad haul from factory to town nearest the project. Costs for haul and distribution were for team or truck haul from town to farm and distribution on the project, and included loading and unloading costs.

15.4. Engineering and Supervision. Engineering and supervision includes merely the surveying, design, and professional supervision always given to a construction project by the engineer in charge thereof. The flat land projects, which usually involve by far the larger amounts of tile and consequent trenching, usually have the lowest relative cost for engineering and supervision. In private practice the cost of this item is on the basis of a professional rate per day including also necessary expenses while actually on the job. This rate varies in different parts of the country; hence no attempt is made here to set any fixed money values on it. As a rule this item will total appreciably higher on small jobs than on large ones. It will also tend to be appreciably larger percentage-wise on public work than on private projects, in which latter class all the projects included in the Minnesota analysis fall. A range of 7 to 35 per cent of the total cost is probably a reasonable one under 1951 economic conditions for all classes of projects, public and private, with an average of about 15 per cent. Back in 1921 the corresponding figures to these three were about 3, 15, and 8 per cent, respectively.

15.5. Tile. Drain tile is a commercial product the price of which, like all other products, fluctuates considerably throughout any economic cycle. Therefore a price list of a given date is likely to apply for only an uncertain and limited period. Drain tile is sold by the linear foot on small jobs and by the thousand linear feet, in each size, on jobs large enough to use the larger unit. Therefore, the result is two sets of market prices.

While the consuming public is not generally aware of it, the base price fixed by the manufacturers depends largely on the bulk or weight of materials that must be handled in making any given size. Therefore, the weight in pounds per tile and in tons per 1,000 linear feet is an important factor in setting the price. These weights, for 4- to 24-inch tile, are given in Table 15.4. Weight is also the basis of railroad freight and of automobile-truck haul charges on drain tile. The current prices on drain tile can always be secured from the nearest convenient manufacturer or from the local dealer in building materials. However, the price quoted by the latter will often be higher than that which may be obtained direct from the maker, because the local price includes the freight or trucking charges from factory to local station. Usually concrete tile are higher priced than clay or shale tile in sizes 12 inch and under, but in sizes over 12 inch, the concrete tile are usually appreciably lower in price than are the clay or shale tile. This is because good concrete tile in sizes 12 inch and under are more difficult to manufacture than are the same sizes in clay or shale tile. On the other hand it is not nearly so difficult to make good tile in the thick-walled concrete tile larger than 12 inch as it is in the clay or shale in the same sizes because the latter easily deform in firing and the excess labor cost from this cause is very great.

The fluctuation in prices from one period to another is indicated by the following approximate selling prices per ton compiled from quotations per 1,000 linear feet of each of the various sizes, 4 to 12 inch, freight on board cars at factory, for northern Iowa and for Minnesota, for both clay and concrete tile for the periods shown:

1907–1908.... $ 5.00, computed from 6 single quotations of all sizes
1909–1912.... 6.00, computed from 30 quotations of all sizes
1917–1919.... 8.00, computed from 30 quotations of all sizes
1921–1922.... 9.50, computed from 10 quotations of all sizes
1943–1944.... 12.50, computed from 41 quotations of all sizes (period of World War II)
1950–1951.... 24.50, computed from 9 quotations of all sizes

For the period from 1916 to 1922, during which time there was considerable activity in tile drainage and the demand for drain tile was quite heavy, actual prices at the factory, per 1,000 linear feet on the different sizes from 4 to 12 inch, averaged approximately as shown in Table 15.5

TABLE 15.4. APPROXIMATE WEIGHT PER LINEAR FOOT AND PER 1,000 LINEAR FEET OF STANDARD SIZES OF TILE AND NUMBER OF LINEAR FEET OF EACH SIZE IN A STANDARD CARLOAD* AND IN A 12-TON TRUCKLOAD

Customary length per tile, in.	Internal diameter, in.	Weight, lb per lin ft to nearest lb		Weight, tons per 1,000 lin ft		Linear feet per standard carload		Linear feet per 12-ton truckload	
		Clay or shale	Concrete	Clay or shale	Concrete	Clay or shale	Concrete	Clay or shale	Concrete
12	4	6½	7	3¼	3½	7,690	7,140	3,700	3,429
	5	8	10	4	5	6,250	5,000	3,000	2,400
	6	10	13	5	6½	5,000	3,845	2,400	1,845
	7	13	16	6½	8	3,845	3,125	1,850	1,500
	8	17	23	8½	11½	2,940	2,175	1,401	1,083
	9	21	25	10½	12½	2,380	2,000	1,142	960
	10	25	30	12½	15	2,000	1,667	960	800
12	12	33	43	16½	21½	1,515	1,165	727	558
18	14	44	59	22	29½	1,135	845	545	456
	16	55	83	27½	41½	910	600	436	288
18	18	66	102	33	51	755	490	363	235
24	20	81	124	40½	62	615	405	296	194
	22	93	154	46½	77	535	325	258	155
24	24	109	190	54½	95	455	260	190	125

* A standard carload is 50,000 lb.

in the northern Mississippi Valley. Wartime prices in 1943 ran about one-third higher on sizes 6 to 12 inch and about half higher on sizes 4 and 5 inch. In 1951 in the Northwest prices ran from 2.75 to 3 times as high as in the Middle West from 1916 to 1922. This latter comparison also shows in Table 15.5. Middle West prices in 1951 were slightly lower.

TABLE 15.5. APPROXIMATE PRICES OF DRAIN TILE PER 1,000 FEET AT THE FACTORY, 1916 TO 1922, AS COMPARED WITH PRICES OF THE SAME SIZES IN 1951*

Inner diameter, in.	Price per 1,000 lin ft		Inner diameter, in.	Price per 1,000 lin ft	
	1916–1922	1951		1916–1922	1951
4	$28	$ 78	8	$ 85	$231
5	35	100	9	100	310
6	50	137	10	120	390
7	65	184	12	170	519

* Two or more farmers can often buy to advantage cooperatively because of the larger size of the order. This is common experience.

15.6. Transportation and Distribution of the Tile. The cost of transportation of the tile from the factory and the cost of its distribution on the project must be treated separately because the basis of cost is different in each case. Transportation costs cover the freight charges from the factory to the local railroad station serving the farm if the haul is by railroad, or from the factory to the farm if the haul is done by heavy duty truck over the highways. Distribution, on the other hand, covers the spreading of the tile over the project along and near the various proposed drain lines at distances convenient for placing in the trenches and includes also, the local short haul from railroad station to farm when the long haul from the factory is by train. The basis of cost of distribution is, therefore, largely the cost for man labor of handling the tile.

Freight on Tile. Railroad tariffs on drain tile, as on other commodities, are rather unstable through a considerable period of years. While the published tariffs are based on the hauling cost per ton-mile, they are expressed for practical application in terms of cents per hundredweight, according to length of haul. There are two sets of rates which apply, the less-than-carload-lot rate (LCL) usually for relatively short distances and the carload-lot rate. The latter is much lower than the small-lot rate, chiefly because of less labor of handling by carloads on the long hauls. Hence, when the job is of sufficient size to justify such action, or where the further need is foreseen for the near future, there is a distinct price advantage to the purchaser in buying in carload lots. A standard carload may be reckoned at 50,000 pounds, or 25 tons of 2,000 pounds each. Freight rates on drain tile may be secured at any time between any two given points, from the local railroad agent. However, since the development of the modern heavy-duty motor truck and hard-surface highways, most of the tile haul is now done by truck, from factory direct to farm, even for considerable distances, the railroads serving only occasionally on the unusually long hauls.

Table 15.6 presents the change in freight rates on drain tile from the middle 1920's to 1951 in the Northwest. Table 15.7 gives one truck haul schedule for drain tile in 1951 in the northern Mississippi Valley. Only the starred items were furnished by the user.

TABLE 15.6. FREIGHT RATES ON DRAIN TILE IN THE MIDDLE 1920'S AND IN 1951

Length of haul not exceeding 100 miles		Length of haul 500 miles	
Date	Rate per cwt	Date	Rate per cwt
Less than carload lots			
1926	$0.51	1926	$1.52
1951	0.86	1951	2.59
Carload lots			
1926	$0.13	1926	$0.27½
1951	0.18	1951	0.37

TABLE 15.7. TRUCK RATES FOR DRAIN-TILE HAUL WITH CONSTANT MAXIMUM LOAD OF 12 TONS, IN THE NORTHERN MISSISSIPPI VALLEY IN 1951

Length of haul, miles	Rate per ton	Rate per truckload per trip	Total ton-miles per trip	Rate per ton-mile, cents
10*	$2.00*	$24.00	120	20
50*	3.00*	36.00	600	6
75*	4.00*	48.00	900	5⅓
100	5.00	60.00	1,200	5
150	6.75	81.00	1,800	4½
200	8.40	101.00	2,400	4⅕

* Furnished by a tiling contractor of long experience in the northern Mississippi Valley.

Although the rate chart is a curve it is sufficiently flat that interpolation for miles not given in the table may be made without introducing serious error.

Distribution of Tile. The cost of distribution of the tile *on the project* is not absorbed in the cost of the long truck haul from the factory because the distributing is usually done by the farmer's own forces and power units, involving also the short local haul from railroad station to farm. The cost rate is, therefore, a combination of the cost rate of operation of the farm truck on short local hauls and the local manpower-hour rate for loading the tile on the farm truck and unloading it in the field.

It is a well-established custom, following the practice established by the railroads, to estimate local haul and distribution of drain tile on the basis of the cost per ton-mile, but this local ton-mile must not be confused with the ton-mile applied to railroad or heavy-duty truck long haul from the factory, discussed under Freight on Tile. In the 1951 period the cost of local haul and distribution usually amounted to about $1 per ton-mile. The shorter the local haul, the higher the rate tends to be. This is because the labor of handling in loading and unloading is a much larger proportion of the expense on the shorter hauls than on the longer hauls. It is impossible to establish any fixed cost schedule for this element of cost at any given time because, although greatly influenced by the size and ease of handling of the individual tiles and by the character and condition of the roads, it is more definitely dependent on the local and temporary wage scale.

15.7. Trenching, Laying, and Blinding. Digging the trenches, laying the tile, and blinding it after laying is treated all as one job, done by the same skilled workman or small group of workmen. The actual labor required for a given unit length of trench varies widely according to the character of the digging, the size of the tile, and the depth of the trench.

15.8. The Common Type of Cost Schedule for Trenching by Hand. As a rule, boss tilers and tile-drainage contractors have, as the result of experience on projects paying a reasonable profit, each in their own way, worked out a schedule of prices either per rod or per 100-foot station for trenching, laying, and blinding for the different sizes of tile and for different depths of trench. Such a schedule begins with what is called the base price for a so-called standard 3-foot or two-spading depth of trench for 4-, 5-, and 6-inch tile, which all call for the same width of trench in order to provide necessary working room for the tiler. Then proportional amounts are added for the same depth of trench according to the increase in the diameter of the tile.

Beyond the two-spading depth a progressive increase through each additional foot of depth is added per each additional inch or tenth of a foot of depth to cover the additional and heavier labor of higher lift. The following purely theoretical illustration shows the method of building such a schedule. Table 15.8 presents a portion of such a schedule in which the base unit is $5.00. The figures used in Table 15.8 are purely theoretical, given only to show the character of the schedule and the general trend of the figures, they should *not* be used in practice.

An experience table of this sort will tend to be different for every master tiler or contractor, thus making comparison of bids on any project difficult and unsatisfactory. A much more uniform and reliable plan is shown in the Minnesota cost analysis earlier referred to. The essential points of the schedule therein presented are outlined in the following section. This schedule is reliable because it is built on a man-hour not a dollar and cents basis from the actual data in the 18 projects listed in the bulletin.

TABLE 15.8. A THEORETICAL COMMERCIAL SCHEDULE FOR TILE TRENCHING BY HAND, DOLLARS PER 100 FEET

Diam. of tile, in.	Depths of trench, ft			
	3	4	5	6
4, 5, 6	$5.00	$6.20	$9.20	$14.60
8	6.70	8.35	12.35	19.55
10	8.40	10.40	16.40	25.40
12	10.10	12.50	21.50	32.30
14	.....	.....	38.00	48.50

15.9. A Man-hour Schedule for Tile Trenching by Hand. Three main classes of digging are established and described as follows:

Easy Digging. Under easy digging are included soils and soil conditions which presented no serious obstacles to spading and casting—light soils of rather loose texture, free from roots, stumps, loose gravel, stones, or boulders—in which the spade settled readily under light pressure and where there was just enough moisture to give firmness to the soil, to make the trench walls self-sustaining and the spade slice coherent enough to cast easily. Under this class might be listed such soils as peat free from tough fibers, roots, and stumps; light friable sandy loams; sandy clays; and the finer damp sands.

Average Digging. Under average digging are included soils or soil conditions requiring considerable but not excessive pressure to settle the spade and considerable effort to cast; soils with a slightly frozen surface—possibly requiring the occasional use of the pick—a few cobblestones, boulders, or stumps; or material which, while easily spaded, is hard to cast because of its weight but is generally sufficiently coherent to make casting easy and caving of trench walls infrequent. Such conditions occur most frequently in ordinary prairie soils (loams, silt loams, clay loams, and clays) except after protracted drought and when the ground is frozen. This class is of the most frequent occurrence.

Hard Digging. Under hard digging are included all conditions not included in the other classes—conditions of moisture or soil texture that cause a fairly constant tendency for the trench walls to cave; looseness, heaviness, or wetness of material that makes it difficult to cast; a degree of hardness, either natural or on account of drought, that makes spading difficult or impossible and necessitates the frequent use of the pick or explosives; the presence of heavy sand and gravel, frequent cobblestones, boulders, or stumps. Under this class may be listed ground frozen to a considerable depth, hardpan, cemented gravel, coarse loose sand, heavy gravel, very stony soils, and soils so hard from lack of moisture as to be difficult or impossible to spade in ordinary fashion.

15.10. Time Schedules for Each Class of Digging. A man-hour schedule for trenching by hand is available in the original bulletin for each

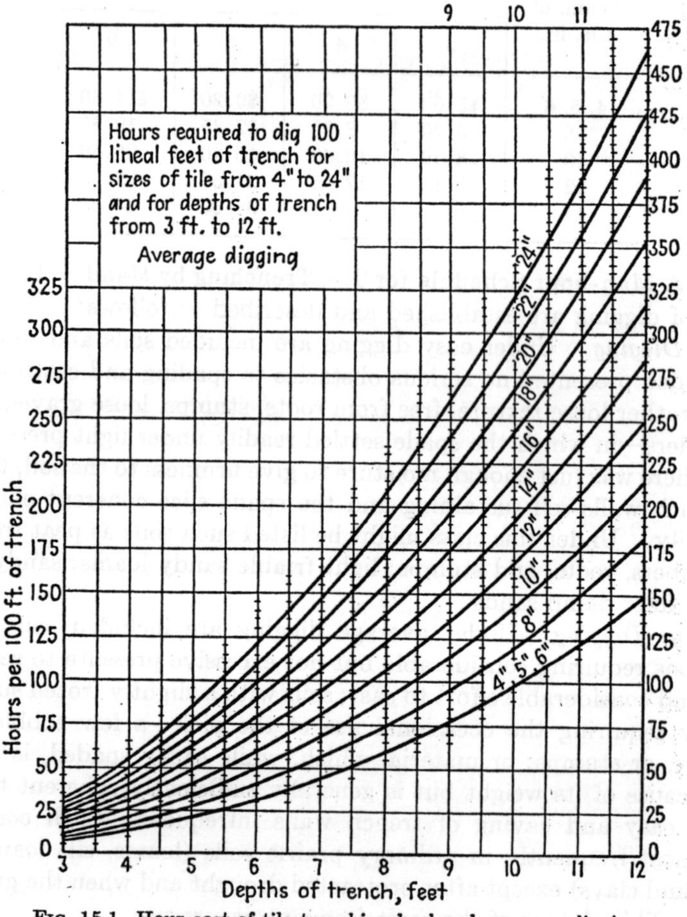

Fig. 15.1. Hour-cost of tile trenching by hand; average digging.

class of digging, in both table and chart form. However, only the table and chart for average digging are here shown (Table 15.9 and Fig. 15.1) because for any given trench a constant ratio exists between the corresponding values for the three classes of digging. For example, the time found by either Table 15.9 or Fig. 15.1, in average digging, to dig 100 feet of trench 6 feet deep for an 8-inch tile is 40.98 hours, or practically 41 hours. This value multiplied by 0.857 gives 35.12 hours, which is the time required to dig 100 linear feet of the same size trench in *easy* digging; while 40.98 multiplied by 1.395 gives 57.17, the time required to dig 100 linear feet of the same size trench in *hard* digging. In the illustration, using 41, 0.86, and 1.40 instead of the figures shown just above introduces only a negligible error in the results. Existence of these two constant ratios justifies omission of the other two schedules with no inconvenience in the use of the schedules.[1] The figures shown in the schedules represent the average performance of a single high-grade tiler. The hours required for digging any given trench, which time also includes laying and blinding under ordinary conditions, multiplied by the prevalent hourly wage in any period gives the money cost of the operation.

TABLE 15.9. HOURS REQUIRED TO DIG 100 LINEAR FEET OF TRENCH
(Average digging by hand. Average of 14 projects)

Diam. of tile, in.	Depths of trench, ft									
	3	4	5	6	7	8	9	10	11	12
4, 5, 6	7.32	13.22	21.20	31.35	43.95	59.33	77.38	99.03	123.00	150.95
8	9.57	17.45	27.22	40.98	55.87	74.08	96.23	121.42	150.47	183.17
10	12.27	21.58	34.00	49.13	67.85	90.12	115.87	145.43	178.67	217.70
12	15.02	26.12	40.27	58.67	80.65	106.13	135.97	169.67	208.73	251.50
14	17.75	30.97	47.65	68.33	93.47	122.73	156.52	195.00	237.09	287.04
16	20.62	35.52	54.62	78.18	106.32	139.18	177.02	220.72	268.12	322.15
18	23.38	40.32	61.82	88.28	119.90	155.52	198.13	246.07	298.85	357.73
20	26.55	45.32	69.28	98.63	133.32	173.67	220.25	272.02	329.88	395.45
22	29.58	50.62	77.12	109.28	147.12	191.75	241.93	298.43	362.37	431.97
24	32.90	55.68	84.70	120.12	161.77	209.32	264.25	325.50	393.73	469.55

In practice, on uneven ground, it may often be necessary to estimate the cost of trenching in short sections in order to secure fair and true average cuts for any such section, with which to enter the time schedules.

15.11. Machine-hour Schedule for Tile Trenching by Machine. Tile trenching by machines of the type described in Chap. 12 has replaced handwork on practically all jobs except those which cannot be reached by machines. Hence the following discussion is important.

In experience with machine trenching on 2 of the largest of the 18 projects listed in the Minnesota analysis, the capacity of a bucket-wheel type of trencher of the size ordinarily used on farm drainage work is about that of eight skilled tilers. The same ratio is indicated for machines of the bucket-ladder type in the recent experience of a successful tile-drainage contractor in Minnesota. Table 15.10 shows an actual contract schedule

[1] These schedules are useful in hand-trenching estimates and as a basis for transition to a schedule for the much more efficient method of trenching by machine, discussed in the following pages.

for tile trenching with a bucket-ladder type of machine on a large tiling job done for this contractor under subcontract in the summer of 1943 in southwestern Minnesota. The job was in average digging with all local conditions especially favorable to the contractor. A careful check of the quotations in this schedule against the man-hour schedule for average digging by hand shown in Table 15.9 and Fig. 15.1 indicates an extremely reasonable basic rate apparently unaffected by wartime conditions. The price gradient in this machine schedule is much flatter than that resulting from the use of Table 15.9. If the prices quoted allowed a reasonable profit, a much greater advantage of machine trenching over handwork is indicated than that above stated as the experience of the Minnesota station. All factors entering into this particular project being considered, discretion should be exercised in using this schedule as a guide in other work.

TABLE 15.10. CONTRACT PRICES IN DOLLARS PER 100-FOOT STATION IN EASY DIGGING BY MACHINE ON A SPECIAL JOB ON A FARM IN SOUTHWESTERN MINNESOTA IN THE SUMMER OF 1943
(Trenching, laying and blinding included but no backfilling)

Diam. of tile, in.	Depths of trench, ft								
	3	4	5	6	7	8	9	10	11
5–6	5	8	11	14	18	24	32	40	48
7–8	6	9	13	17	21	27	36	45	56
9–10	7	11	15	20	26	32	41	51	62
12	..	12	18	24	30	38	48	58	69
14	..	15	21	27	34	43	54	65	76
15	..	17	23	30	38	47	57	68	80
16	..	18	24	32	40	50	60	71	83
18	..	20	27	35	44	54	65	76	90
20	..	23	30	39	48	59	71	83	96

15.12. A Generally Applicable Machine-trenching Schedule. Published or even private cost schedules for machine trenching for tile drains are largely lacking; hence this discussion, with some support from experiences in the Northwest, is mostly based on results of rather complete studies of performance and operating costs of tile-trenching machinery made by William E. Beach [12] and Leonard L. DeVries [32] working at Iowa State College under the direction of the engineers of the Agricultural Experiment Station. Results of these studies are herein used in conjunction with the Minnesota analysis already mentioned, aided by the price and wage indices presented in the November, 1951, issue of *Engineering News-Record*, in producing the machine-hour tile-trenching schedule shown in Table 15.11. It is believed that this schedule will be found reliable in the Middle West and in the Northwest in estimating cost of construction of tile drains by bucket-wheel and bucket-ladder type trenching machines.

15.13. Nonprofit Time in Trenching by Machine. When watching a machine operate under ideal conditions and noting the progress made, one is likely to gain a false impression of the potential profit per season. The careful records kept by DeVries [32] showed 1,890 available working hours per season. Of these, 1,250 hours were "lost" or were nonproductive under the best of management practices, leaving only 640 hours, or about 1 day out of 3, usable for pay work. Bad weather, in a typical season, accounted for 351 lost hours while the ground was drying out; 265 hours were needed for essential field repairs; 191 hours were consumed in extra work at junctions of two lines; and 114 daylight hours were needed for greasing and servicing. The remaining lost time was made up of 155 hours in moving between jobs, 35 hours in major holidays, and 139 hours miscellaneous.

It is obvious that the price schedule per working hour must include these losses plus the usual fixed and operating costs if the contractor is to make a reasonable profit. It is also obvious that, on any given job, good management may mean the difference between profit and loss.

While principal attention is given to the bucket-wheel and bucket-ladder types of excavators, a brief consideration, based on rather limited data, is given to the trench-hoe type which is also being widely used.

A machine-made trench has quite a different cross section from a handmade trench, being rectangular or a combination of rectangles, whereas the handmade trench is trapezoidal in its general form. The latter is always as narrow as it can be made and still receive the given tile while at the same time giving enough room in which to let the tiler work; whereas the machine-made trench does not change its width with every change of diameter of tile. Most tile-trenching machines are made with cutting widths of 12 or 14 inches to, or greater than, 24 inches. On a given project of considerable size, calling for tile from 4 inch to a maximum, say, of 24 inch in inside diameter, it may be necessary to have two machines available. The development of the schedule in Table 15.11 is based on two machines that will cut trenches of various widths from 14 to 24 inches and be able to care for tile 4 to 12 inch and 14 to 20 inch in inside diameter, respectively, and of all required depths. There is one size machine in between these two, but even most large-sized jobs will not readily justify having more than two machines on the job, and on jobs of ordinary size their installation will be so planned, if this be possible, as to call for but one machine. However, a business which will profitably support three machines may be able to be so planned as to interchange them efficiently between the various parts of different jobs if several are located close to each other with good roads in between.

A comparative analysis of the Minnesota studies of tiling by hand and of the Iowa studies of tiling by machine indicates that, on the average, for sizes of tile up to 12 inch and a given depth and length of trench, soil and working conditions being comparable, the machine method would do the trenching, and laying, and blinding the tile in one-seventh of the time it can be done by experienced hand labor. This is taking into consideration the great variation in width of trench for a given size of tile that the machine digs in comparison with the hand-dug trench which is usually calculated snugly to fit the outside of the tile.

Based on the ratio of 1 to 7, given above, the machine-hour trenching schedule shown in Table 15.11 is evolved from the man-hour schedule given in Table 15.9. The schedule shown in Table 15.11 being based mostly on a theoretical comparison of the two methods, in each of which the cases compared were actual jobs, its use may have to be perfected by working out a set of adjustment coefficients for the groups of tile sizes cared for by each of the two sizes of machines herein considered. This can probably be done with results that will be found reliable in computing costs both in preliminary estimates and as a basis for contract bids.

TABLE 15.11. HOUR COST PER 100 LINEAR FEET OF TILE TRENCHING REDUCED FROM HAND TO MACHINE TRENCHING
(Ratio 1 to 7. Average digging)

Diam. of tile, in.	Depths of trench, ft								
	4	5	6	7	8	9	10	11	12
4–12	3.75	5.75	8.40	11.50	15.15	19.40	24.25	29.80	35.95
14–20	. . . .	9.90	14.10	19.05	24.80	31.45	38.85	47.15	56.50

15.14. Time Cost by Trench-hoe Machine. A few contractors seem to prefer the hoe type of trenching machine in spite of the fact, which seems clear from the rather meager evidence, that the time cost of tile trenching by this machine seems to be higher than that of the two types of trenchers just discussed. The evidence available on the trench hoe is too limited for final conclusions to be based on it as to minimum costs. DeVries [32] lists, among Iowa contractors returning questionnaires, six owners reporting an average of 385 days of tiling each, with a trench hoe over two seasons with an average of 86 feet of tile laid per hour with no size of tile and no depth of trench stated but presumed to be 4- to 6-inch tile at an average depth of about 4 feet; this amounts to 1.16 hours per 100 linear feet of line. DeVries also gives the set of tests with one machine made under his own observation shown in Table 15.12.

TABLE 15.12. IOWA TESTS OF ONE TRENCH HOE

Test number	Total hours worked	Total linear feet dug	Linear feet dug per hour*	Hours to dig 100 lin ft*	Av. depth of cut, ft
1	2.5	230	92	1.09	1.92
2	4.0	255	64	1.56	4.17
3	3.5	350	100	1.00	3.99
4	3.0	226	75	1.33	4.73
5	4.0	400	100	1.00	3.53
6	8.0	896	112	0.89	3.49
7	2.5	220	88	1.14	3.87
8	6.0	607	101	0.99	3.30
Totals†	31.0	2,954	640	7.91	27.08
Average†	4.4	422	91	1.13	3.87

* Computed by the authors.

† First item omitted in totals and averages because too shallow for good practice.

One well-known Minnesota contractor of long and successful experience, who has used the trench hoe for excavating for the four years 1948 to 1951, furnished the figures with one machine with a 24-inch width of cut shown in Table 15.13.

TABLE 15.13. TRENCH-HOE EXCAVATING RATE
(Experience of one contractor through four seasons in Minnesota)

Depth of Cut, Ft	*Hours per 100 Ft*
4	1.00
5	1.33
6	1.67
8	2.50
10	3.33
12	5.00

This trench will accommodate any size of tile from 4- to 20-inch. The rate of trenching is only about one-half that of the trenchers that dig to finished grade.

The figures in Table 15.13 run fairly close to those obtained by the Iowa contractors and to those in DeVries's tests; but it should be remembered that this machine does not dig to finished grade. The finishing to grade requires the practical equivalent of one spading and finishing to grade by hand, which the contractor reports requires the following time for two men owing to the width of the machine trench at top.

Tile Size, In.	*Hours per 100 Ft to Spade, Finish Grade, and Lay and Blind Tile*
4–8	0.87–1.00
12	1.33
16	1.67
20	2.50

To allow for 5- and 6-inch tile the time can be proportioned between 0.87 and 1.00 hour without serious error. The same practice can be followed between 1.33 and 1.67 hours and between 1.67 and 2.50 hours for 14- and 18-inch tile, respectively. To cover both the machine and the expert handwork, during the 1951 season this contractor used four and sometimes five men. His hourly expense was then $14.00 distributed as follows: foreman, $2.00; machine operator, $1.50; two expert tilers at $1.25 and $2.50; one general helper, $0.80; payroll insurance, $0.40; machine charge, $6.80.

This seems to run 8 to 10 per cent higher than with the bucket-wheel or bucket-ladder type of machine; but the claim is made that the difference is more than counterbalanced by the reduction in repair costs, depreciation, and general convenience in handling. Each type of machine has its outstanding advantages, and in every case a great deal depends on the skill of the machine operator and the general efficiency of management.

REFILLING TRENCHES

It is convenient to figure the labor cost of refilling trenches at so much per 100 linear feet according to size of tile and depth of trench, as the

TABLE 15.14. HOUR UNITS OF 1 TEAM-HOUR PLUS 1⅓ MAN-HOURS REQUIRED TO FILL 100 LINEAR FEET OF HAND-DUG TRENCH*

Diam. of tile, in.	Depths of trench, ft									
	3	4	5	6	7	8	9	10	11	12
4, 5, 6	0.05	0.72	0.96	1.23	1.52	1.83	2.18	2.54	2.93	3.34
8	0.62	0.88	1.17	1.50	1.82	2.18	2.56	2.97	3.40	3.86
10	0.75	1.05	1.38	1.74	2.11	2.52	2.95	3.40	3.88	4.38
12	0.87	1.22	1.60	2.00	2.42	2.87	3.34	3.84	4.36	4.91
14	0.99	1.38	1.80	2.24	2.71	3.20	3.72	4.26	4.83	5.42
16	1.11	1.55	2.01	2.49	3.01	3.55	4.11	4.69	5.30	5.94
18	1.22	1.70	2.21	2.75	3.31	3.88	4.49	5.12	5.77	6.45
20	1.33	1.86	2.41	2.99	3.59	4.21	4.86	5.54	6.24	6.96
22	1.45	2.02	2.62	3.24	3.88	4.55	5.25	5.96	6.70	7.47
24	1.56	2.17	2.81	3.47	4.16	4.87	5.61	6.37	7.16	7.97

* Figures within heavy interior lines cover most frequent practice and are, therefore, best.

amount of labor and equipment time required will vary almost exactly with the amount of material to be moved, since the only labor involved is in pushing the material in the spoil bank over to the trench a comparatively short distance, horizontally, gravity doing the rest. The amount of time required for this work will vary in much the same way as that of

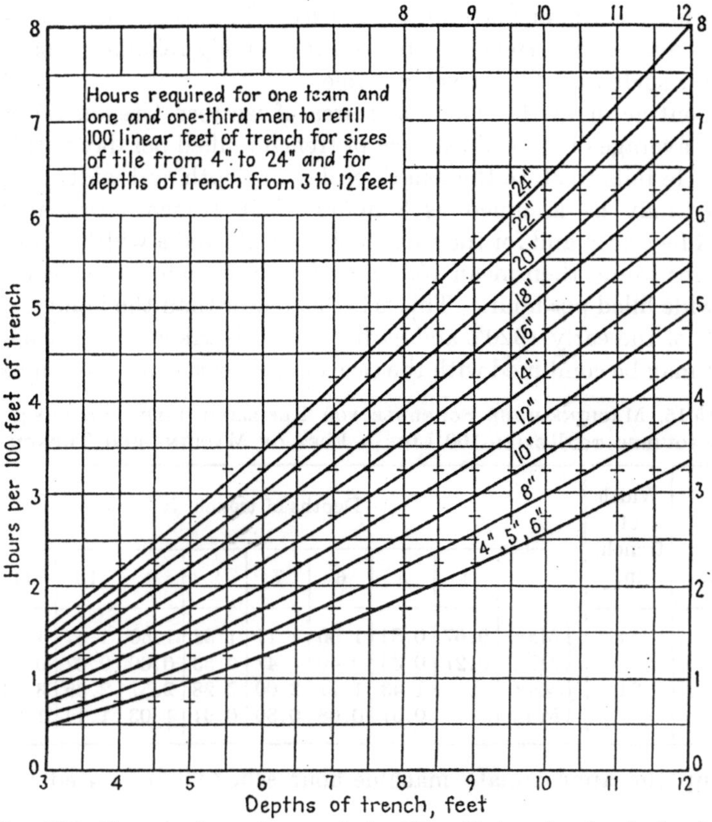

FIG. 15.2. Team- and man-hour cost of refilling tile trenches dug by hand.

trenching, laying, and blinding, but less intricately. Therefore, when the Minnesota schedule for hand trenching was worked out it was also found possible to establish a trench-filling schedule of somewhat the same character to fit hand- and teamwork. However, with the shift almost wholly to heavy mechanical equipment such as terracer graders and bulldozers for surface earth-moving work, this earlier schedule is of little value except in the rare cases where local conditions require it to be done by hand or hand and team labor. The earlier time schedule is here retained as Table 15.14 and Fig. 15.2 for use in such cases. Where the work on which this schedule was based was done with hand labor and/or teams and machinery or both, it was difficult to reduce all the time to a *man-hour* basis. A close study seemed to indicate that an arbitrary time unit of 1 team-hour plus 1⅓ man-hours was the most convenient and satisfactory for use with this cost factor. Hence it is used in the table and the chart.

15.15. Refilling Trenches Wholly by Machine. As, until recently, the trench-refilling job has usually been considered the farmer's responsibility, time records of machine work, except in two or three cases, have not been available. Hence, to build a thoroughly reliable machine-hour record for such work is not yet possible.

One contractor in the Tillamook area of western Oregon reports, for 1951, an average of 1,200 linear feet of trench refilled per day with a small dozer. He fails to state the length of day or depth and width of trench. The day may be assumed as 8 hours. The trenches probably were mainly 4 feet deep and for the smaller sizes of tile and a width of 14 inches. His figures would then mean 150 linear feet of trench 4 feet deep and 14 inches wide filled per hour. This is little greater than the hand and team methods of the early 1920's and earlier. If it is taken as a normal minimum, it may be combined with the case next cited assumed as a maximum

TABLE 15.15. MACHINE-HOUR SCHEDULE FOR REFILLING TILE TRENCHES: HOURS REQUIRED TO REFILL 100 LINEAR FEET OF MACHINE-DUG TRENCH

Tile sizes cared for, in.	Width of trench, in.		Depths of trench, ft								
			4	5	6	7	8	9	10	11	12
4–12......	14	Max.	0.67	0.84	1.00	1.17	1.33	1.50	1.67	1.84	2.00
		Min.	0.27	0.33	0.40	0.47	0.53	0.60	0.67	0.73	0.80
14–20......	24	Max.	....	1.43	1.71	2.00	2.28	2.57	2.85	3.14	3.42
		Min.	....	0.57	0.68	0.80	0.91	1.03	1.14	1.25	1.37

to set up the approximate machine-hour schedule presented in Table 15.15.

A contractor in western Washington reported through R. C. Tuttle, Regional Engineering Specialist of the Soil Conservation Service, his experience in refilling trenches for 1951 as follows: average linear feet of trench filled per 8-hour day, about 3,000, equipment a pusher-type blade on the front of a tractor, making in effect a light bulldozer. This amounts to 375 linear feet per hour which closely agrees with the general maximum rate in the 1920's by hand and team or tractor-drawn terracer or road grader. Size of trench is not stated, but the same assumption is made for that as was done in the work of the Tillamook area contractor mentioned above. For a given trench the rate will vary considerably with local surface conditions, type of soil, and ability of operator.

On the basis of these two cases as maximum and minimum performance on a 14-inch width and 4-foot depth of trench, Table 15.15 is given, assuming that the time taken to refill a unit length of trench varies as the cross section of the trench.

15.16. Cost of Refilling Trenches. Actual cost of refilling trenches varies widely at about the same rates as the amount of earth to be moved on different jobs and according to local conditions and to the method used. In the early twenties in one project of the 18 in the Minnesota analysis, the cost of refilling trenches on certain portions of the project was only slightly over 10 cents per 100 linear feet of trench, done by a road grader drawn by tractor power. This would have been about 27 cents per 100 feet in 1951. On one other small project, the cost of refilling by hand and plow was less than 12 cents per 100 linear feet. This would have been about 30 cents in 1951. At the other end of the price range costs on other projects for jobs done by team and buck or slip scraper averaged upward of $1.50 per 100 feet, with small stretches where the maximum costs exceeded $2.00. These would have been about $3.80 and $5.00 in 1951. A fair general average cost seems to have been 50 cents per 100 linear feet of trench in the early twenties, corresponding to about $1.30 in 1951.

OUTLET PROTECTION, BRIDGES, AND OTHER MISCELLANEOUS ITEMS

Outlet protection and miscellaneous expenses may involve only extra material or both extra material and extra labor. The expense involved depends largely on local conditions and consequently varies from little or nothing to a very appreciable figure; so the data from any set of projects do not furnish conclusive information. Miscellaneous labor costs will largely be determined by the local governing wage scale unless definitely fixed in the contract, where a written contract is executed. Materials such as culvert pipe, lumber, cement, miscellaneous iron, and structural steel must be estimated at current market prices except as specifically modified in the contract. All such items can usually be estimated with considerable exactness from the plans and specifications. This is especially true of all bridges and miscellaneous structures since either timber or dimension lumber will have a definite price per thousand feet board measure; iron and steel will be quoted by the pound or ton; cement by the sack or barrel; sand and gravel by the cubic yard; and culvert pipe, like drain tile, by the linear foot according to size, weight, and quality. If prices for such materials do not include delivery at the site, truck haul must be estimated separately for them on the same basis as for drain tile, that is, by the ton-mile or equivalent unit. Estimates of either wooden or steel bridges by the linear foot of bridging can be considered to be of only a purely preliminary character, as exact estimates of final structures in established weight or other comparable units are the only sound basis for final payments under construction contracts.

COSTS OF SPECIAL TYPES OF DRAINAGE

15.17. Mole Drainage. The cost of mole drainage is based on the cost of power required to pull the mole ditcher. The power being tractor power or other power of similar character is expressed in horsepower, and the cost must be reckoned in horsepower-hours at the going local rate for the particular type of power unit used. Some skilled ditching labor and some common man-labor is also involved at going rates, as is also a minor amount of engineering service. Experience seems generally to indicate a range of about $1.50 to $3.00 per acre, total cost in the middle 1940's. Available extensive data are extremely meager.

15.18. Shallow Vertical Drains. Vertical drains, where they do not come in the classification of deep wells, are usually of minor cost and must be estimated in terms of man-hours of labor and the market prices of the small quantities of materials needed, such as drain tile, concrete, and iron gratings.

15.19. Deep Drainage Wells. Deep wells, such as those used for drainage of irrigated lands in various parts of the West, from the standpoint of cost must be considered a well-man's job, and their construction and cost constitute a field of activity outside the scope of this book. Professional well men should be contacted where prices on deep wells are required.

PROBLEMS

1. A ditch 1¾ miles long is to be dug 9 ft deep through absolutely level ground, the bottom width to be 12 ft, side slopes 2 to 1. What would be the cost of excavating it (*a*) with a dredge at 8 cents per cubic yard? (*b*) With dynamite at 15 cents per cubic yard?

2. A ditch ¾ mile long is to be dug through absolutely level ground, 6 ft deep, the bottom width to be 4 ft, side slopes 1 to 1. If dug by scrapers at a cost of 12 cents per cubic yard, what will be the total cost?

3. According to what seem from the text discussion in Chap. 15 to be reasonable unit prices for the present period, compile an estimate of the total cost of the ditch in the illustrative problem in ditch design given in Chap. 7, using the profile in Fig. 7.8 as a basis of depth on each mile. In so doing follow the general detail of costs discussed in Chap. 15.

4. A farmer lives 4 miles from a station on a railroad and wishes to have the following tile put in 3½ ft deep by contract with an engineer to lay out and superintend the work. Following the general detail of elements of cost fully discussed in Chap. 15, compute an estimate of cost. No inlets will be used, but three outlets will need to be constructed at a cost of $25 each.

The bill of tile follows: 10,200 ft of 4-in. tile; 5,000 ft of 5 in.; 900 ft of 6 in.; 800 ft of 8 in., and 1,000 ft of 10 in.

NOTE: According to the general instructions given above under Prob. 3, compute in full detail and tabulate an estimate of cost for each of the following problems in tile drainage construction, making such assumptions of tile depths as are necessary, but exercise careful judgment.

5. Using the results worked out in Prob. 9, Chap. 10, compute a detailed estimate of cost of the system in that problem, following the instructions given in the above note.

6. In the same manner as in Prob. 5 compute a detailed estimate of cost of installation of the design worked out in the illustrative design of tile main and submains illustrated by Fig. 10.26. Allow for two surface inlets on the main and one on each of the submains add a suitable outlet bulkhead or other protection.

7. In estimating the cost of a subsurface drain outlet, what information must be obtained in the field for a detailed cost estimate?

CHAPTER 16

DRAINAGE ADMINISTRATION AND LAW

PRIVATE RIGHTS IN DRAINAGE

It is a well-recognized general principle of common law that a private property owner may make such changes or improvements on his property as do not in any way encroach upon the property or conflict with the rights and privileges of his neighbors or of the community in which he lives, at least without compensation either by agreement or by standard provision of statute law. According to Green [51] and other authorities on law in its application to drainage, two separate broad concepts are involved, applicable to lands where the topography provides at least some natural drainage. These two concepts are known as the *civil law* (of the Romans) and the *common law* or *common-enemy law* (of the English).

16.1. The Civil Law. The *civil-law* concept gives to the owner of the higher lying land the full right of all *natural* surface drainage from his land onto adjacent lower-lying lands whose owners may not cause any obstruction thereto. Present-day modification of the *civil law* by legislative action, in the states where it governs, gives to the owner of the higher-lying land the right so to improve, on his own land, natural drainage outletting through a natural watercourse or depression as to increase the rate of flow; but he may not extend such improvement onto the land of the lower owner without his consent, or change the natural direction of flow.

The *civil law* rule is the basis of drainage law in the following states: Alabama, California, Georgia, Illinois, Iowa, Kentucky, Louisiana, Maryland, Michigan, North Carolina, Ohio, Pennsylvania, and Texas.

16.2. The Common-enemy Law. The common-enemy law concept gives to every landowner the right to protect his land from water flowing on to it from adjacent higher lands. However, under this concept, an owner cannot so improve the drainage conditions on his land as to cause additional water therefrom to flow onto his neighbor's land unless he secures the right so to do by agreement or for a price or under district law. This law, therefore, is the basis of drainage district law under which all the lands in the district that will be potentially benefited may be included and assessed for the cost of the outlet drain. Drainage districts are discussed in a subsequent section.

The *common-enemy* rule is the basis of drainage law in the following states: Arkansas, Connecticut, Indiana, Kansas, Maine, Massachusetts, Minnesota, Missouri, Nebraska, New Hampshire, New Jersey, New Mexico, New York, Oklahoma, South Carolina, Virginia, Washington, and Wisconsin.

It seems not to be indicated that the other states not included in the foregoing lists specifically adopt either of these two concepts, but existing drainage statutes in these states, as outlined in the United States Census of Drainage of 1940, seem consistently to follow the modified *common-enemy* rule.

It will be seen that, under present-day modifications, there is no wide difference in practice between these two concepts.

16.3. Ownership of Waters on the Farm. According to Green and other authorities on drainage law, the common law recognizes the private ownership and control of all temporary and permanent bodies of surface water within the limits of the given farm too small to serve as a public conveyer of materials or produce from one part of the community to another. Therefore the private owner may enclose and protect such waters at his discretion, or he may drain them off his lands and he may improve natural drains and watercourses upon his own property concerned therewith, provided that in so doing he conforms to the provisions of one of the two above-described concepts governing, in his state, the rights of his neighbor and the community. Under these limitations he may bring the mouth of a tile drain or of a surface ditch, on his own land, close to his neighbor's line, so long as it results in no unnatural volume of flow across that line and causes no injurious flood upon the adjoining land.

OUTLET CHANNELS THE PHYSICAL BASIS OF PUBLIC DRAINAGE

It already has been shown that the first essential of a drainage system is an outlet of sufficient depth and capacity. Many farms are not provided by nature with such an outlet. The only way to secure it in such cases is to construct an outlet channel across other intervening lands to existing watercourses of sufficient depth and capacity to receive and carry the waters from such artificial channels. The individual farmer cannot alone and on his sole responsibility construct such a channel. In the first place the cost for the single individual would usually be prohibitive. In the second place, as a private individual, he would have no right to enter upon the lands of another for such a purpose unless he first entered into a legally binding agreement in perpetuity to that end with all the owners of the lands so crossed, providing for the construction and effective maintenance of the channel.

16.4. Mutual Cooperative Districts. A community outlet by private cooperative agreement among the actively interested landowners, while not a common practice, has been found entirely possible under general statutes of most of the states of the United States. However, such a district is possible only where all the owners concerned are in perfect agreement on the plans, the distribution of costs, and the provisions for effective maintenance of the ditch. The essential initial requirement is that the interested owners prepare, sign under notarial acknowledgment, and file in the office of the registrar of deeds and titles a written agreement, in perpetuity, briefly outlining the work to be done and how, and the apportionment of costs and provisions for maintenance of efficiency of the ditch. The agreement should also name those responsible for arranging for the doing of the work, for supervision of the complete job, and for apportioning the costs.

Such districts of limited acreage (suggested maximum size about 1,000 acres) have been carried out in Minnesota with complete satisfaction and effectiveness. Pickels [104] records that districts of this type are especially provided for in the statutes of Illinois and several other states and that, in the aggregate, a large amount of drainage has been accomplished in this way. The authors have observed that by this method the greatest possible saving is effected in expense, in promptness of installation, and in efficiency of the system. As a rule, however, the most practical way to secure drainage outlets serving large areas is through the statutary provision for drainage districts now found in most of the state codes, discussed in the following section.

PUBLIC RIGHTS IN DRAINAGE AND PUBLIC DRAINAGE LAW

Fortunately the right of a group of interested contiguous landowners to proper artificial provision of outlets for drainage for their lands and those of the district in general, under the principle of eminent domain, regardless of some opposition within the area, has been recognized in law as constitutional and is definitely provided for under the specific drainage code in every state of the United States where drainage is an appreciable agricultural problem.

The texts of these codes, varying considerably in details in the different states, or even comprehensive synopses of them are far too voluminous to be given in this book. Therefore, only a relatively brief discussion of the general principles and procedures is included. The interested reader engaged in any capacity in practical application of the drainage codes, especially the engineer or landowner, who for guidance is therefore in need of more definite detail, is referred to the excellent synopses given in the Federal census reports on drainage for 1940 [23], to the contributions in this field by Yohe [167] and by Boyd and Hart [17], and more especially, in any given state, to the official published drainage code in that state.

16.5. General Objectives and Procedures. The purpose of and procedure under state drainage laws as briefly outlined in the general synopsis in the Federal Census Report for 1940 is in substance as follows. This synopsis was not repeated in the 1950 report.

The leading objectives in the enactment of such laws are (1) to establish means whereby effective community cooperation may be achieved in obtaining needed drainage outlets for areas not provided by nature with such, (2) "to provide a method of apportioning the cost among the landowners benefited, as nearly as practicable in proportion to the benefits to be received," and (3) "to authorize a plan of distributing payments for the improvement over a period of years." This last objective has become one of the most important because many such projects, if they are to be made to function as thoroughly and effectively as contemplated, involve so heavy an initial expense, payable immediately upon completion of the construction to the contractors performing it, as generally to be far beyond the power of the benefited landowners to pay in cash. Hence the financing of the project is made possible only by the sale of bonds to be retired at such intervals as enable them to be paid out of the money benefits actually earned from the land through subsequent years, as a direct result of the improvement.

16.6. Jurisdiction. In a majority of the states, legal jurisdiction rests with the county board whose executive secretary is usually the county auditor, or with a specified county court, as, for example, the district court, the executive secretary being the clerk of court. Where more than one county is involved in a given district, initiation of proceedings is usually required in that county having the largest number of affected landowners or a majority of the acres that will be benefited. The official or body having jurisdiction will hereinafter be referred to as the Agency of Jurisdiction.

CUSTOMARY STEPS IN INITIAL PROCEDURE

In most cases the customary steps in procedure, definitely specified by law in each state, are as follows:

1. Petition for the establishment of the ditch or drainage district.
2. Preliminary hearing on the petition—in some states preceded by a preliminary survey.
3. Order of the board or court either denying the petition or establishing the ditch or district.
4. Final survey by a competent drainage engineer.
5. Apportionment of benefits and damages.
6. Final hearing on apportionment of benefits and damages.
7. Appeals to higher courts against decision by the Agency of Jurisdiction.
8. Establishment of the administrative authority in the district.
9. Securing of bids and letting of contract for construction.

10. Issuing of bonds to meet cash payments.
11. Assessment of benefits as taxes.
12. Payment of assessments.

In the following sections each of the foregoing listed steps is discussed in brief outline.

16.7. Petition. The initial step in establishing a drainage district is the filing with the executive secretary of the Agency of Jurisdiction a petition by the interested landowners that such district be established. Usually the petition must be signed by a majority of the interested owners or by the owners of a major portion of the affected lands. In a number of instances some official or board of the state or political subdivision thereof is authorized to make such petition especially where public lands are involved. The petition is required to show that the proposed ditch will be conducive to the public benefit and must indicate the general course of the main ditch, and state what lands are expected to be benefited, and bear the cost thereof. Generally when filed, the petition must be accompanied by an acceptable bond by the petitioners conditioned to pay the costs of the preliminary proceedings if the ditch is not established.

16.8. Preliminary Survey. In some states before the preliminary hearing can be held, the Agency of Jurisdiction must require a preliminary survey to be made and fully reported on by a competent drainage engineer appointed by it, to verify the engineering feasibility of the project, to determine more definitely the general location of the main ditch, to map the lands affected and the ownerships thereof, and to assemble all field information that will aid in forming a reasonably correct basis for determining the probable benefit to be derived and probable total cost of the project.

16.9. Preliminary Hearing. After the petition is filed, the Agency of Jurisdiction requires its executive secretary to serve public notice, by posting in public places and by press notices in local newspapers, that after a legally specified time from filing of petition there will be on the date, at the hour, and in the place specified a hearing on said petition before the Agency of Jurisdiction. At this hearing all affected freeholders are eligible to be heard for or against the ditch. If the evidence adduced in the preliminary hearing convinces the Agency of Jurisdiction of the soundness of the project and if estimated total benefits exceed the total estimated costs, they will establish the ditch, in which case the preliminary expenses of the proceedings will be included in the final expenses of the project and the bond of petitioners will be released. If the evidence is found not to justify the project, the petition will be denied and the petitioners must pay preliminary expenses as provided by their bond.

16.10. Final Survey. The petition being granted, the Agency of Jurisdiction appoints a competent drainage engineer, usually the same who made the preliminary survey, under bond to make a complete location and construction survey including both the main ditch and all essential submains and laterals of the entire project; to prepare maps and profiles showing the details and extent of work to be done, the exact area on each owner's land of right of way required for the ditch and to be benefited thereby, the engineer's own estimate of cost, and other details that will aid in determining proper apportionment of both damages and costs to individual owners; and to prepare plans and specifications for construction of the ditch.

16.11. Appraisers and Their Duties. Usually at the time of appointment of the engineer for the final survey, the Agency of Jurisdiction also appoints a legally specified number of disinterested freeholders in the district as viewers or appraisers who shall, under bond with the aid of the engineers complete report, view each parcel of land and, by the legally authorized method later discussed, equitably apportion the benefits and damages against each affected landowner in the district. The viewers must enter on the job at once after filing of the engineer's report, and their report must be filed with the county auditor or clerk of court within a legally specified time thereafter.

16.12. Assessment of Benefits. Boyd and Hart [17] make the following comments pertinent to assessments of benefits, their character, and the method of their apportionment:

> The majority of State laws specify that the assessments shall be according to benefits, a method which not only seems more equitable than any other, but is in accord with the fundamental principle of special assessments.
>
> Special benefits are a necessary basis for all assessments and, where the legislature exempts some properties, such action is considered a legislative declaration that such properties will receive no benefits.
>
> No assessments can be levied upon the property of the Federal Government without its consent. It is generally necessary that the legislature make a definite provision for the assessing of State property, although some courts have held that this provision is implied in the drainage laws.
>
> In general it may be said that any kind of property, or property owned by any individual or corporation, may be assessed if benefited. In the several States agricultural lands, highways, irrigation canals and ditches, counties, municipalities, railroads and street railways, telephone and telegraph companies, and other corporations may be assessed when benefited.
>
> It is desirable that lands shall be assessed in small tracts.
>
> Rights-of-way deeded to or taken by the drainage district generally cannot be assessed.

16.13. Kinds and Nature of Benefits. There are two kinds of benefits resulting from local improvement—general and special. General benefits are those enjoyed by the public and common to all the people of the community. General benefits usually resulting from drainage improvements are increased healthfulness, convenience and general prosperity. No assessments can be made on account of such benefits, but they must be apparent before an assessment district can be formed.

Special benefits are the ones upon which the assessments to pay the cost of the improvement rest.

The general benefits which the public receive from drainage improvements consist of one or more of the following: public health benefits, public road benefits, public interest in the condition of the land.

Since the measure of the benefits derived from drainage is the increase in the value of the land, due solely to the improvement, all factors which might affect the value of the land must be considered. The following general factors at least should be considered:

The need for drainage, or the wetness of the land.

The amount of drainage or protection furnished.

Increased healthfulness.

Increased accessibility.

The use made of the property.

The need for drainage . . . is the most important factor of benefit for it includes many things and varies greatly with conditions.

16.14. Relation between Benefits and Assessments. Almost all of the State drainage laws provide that the assessments shall be proportional to the benefits; that is to say, the assessments on each tract of land shall bear the same ratio to the individual benefits as the total cost bears to the total benefits. As a general thing . . . no improvement may be made unless the prospective benefits exceed the costs.

16.15. Methods of Assessments of Costs. Boyd and Hart further state: "Three general methods of determining assessments are in use. The first is called the *percentage method*, the second the *classification method*, and the third the '*actual value of benefits*' *method*."

Even if, as is the case in many states, the law requires distribution of benefits on some particular basis, it is always well for the assessors to have complete data available, such as, for example, the actual amount of benefits received by each tract in dollars and cents, and to include this in their report if it seems desirable.

The Percentage Method. Boyd and Hart comment in substance as follows on this method. It is not a required method in any state but is extensively used in some of the Middle Western states. The method is arbitrary and so complicated as to be confusing to both assessors and landowners. Results from its use do not seem to have given as much final satisfaction as have those determined by other methods. In the use of this method good judgment and expert knowledge on the part of the assessors are essential from beginning to end. The general procedure is as follows:

It is the first duty of the assessor to divide each farm into tracts such that the benefits received from drainage will be, as nearly as possible, uniform within a tract. In other words, each acre in a tract selected receives approximately the same benefit as any other acres in the same tract. With this in view, each tract is inspected in the field, and each one of the factors determining benefit from drainage is assigned a percentage. For instance, it may be assumed that each acre in a 50-acre tract is 60 per cent benefited with respect to nearness to outlet, 70 per cent with respect to fertility, 100 per cent with respect to elevation or thoroughness of drainage, and 60 per cent with respect to nearness to main drain. A 100 per cent benefit is, of course, the maximum benefit obtainable. These percentages of benefit determined from a detailed consideration of each influencing factor are then multiplied together, the product giving the total percentage of benefit per acre in the tract. In the case assumed, the total percentage benefit per acre would be $0.60 \times 0.70 \times 1.00 \times 0.60 = 25.2$ per cent, or, more conveniently expressed, 0.252. This percentage factor is multiplied by the total number of acres in the tract which gives the total number of acre units of benefit received by the tract. The sum of all the acre units of benefit received by all the lands of the district divided into the net cost of the district against the land gives the cost per acre unit of benefit. This cost multiplied by the number of acre units received by each tract gives the actual amount of the cost of the district which is assessed against the tract. An example of an assessment roll made up on the percentage basis is shown in Table 16.1.

The Classification Method. In this method it is the first duty of the assessors to divide the lands of the district into a number of classes fixed by law in those states where it is required that this method be used. The number of classes designated is usually five. Class 1 is the one that receives the least benefit, and class 5 is the one that receives the most benefit per land unit, which unit is usually taken at 40 acres. It is the duty of the assessors to determine into which class each tract of land will fall. The average assessment unit per equal tract is the total cost of the project divided by the sum of the individual classification numbers for all the tracts. This average unit multiplied by the classification number of a given tract gives the assessment for that tract. The system is illustrated in Table 16.2.

Of this method Boyd and Hart [17] write:

> It is a very difficult matter to make equitable assessments by this method, as it is almost, if not entirely, impossible to divide the land of any district into five classes so that the benefit between the classes shall be in the ratio fixed by law. This method is far too inflexible to allow the fixing of equitable assessments. Furthermore, it is confusing to the viewers and gives the landowner no definite informa-

tion as to the amount of his assessment and no information whatever as to his probable benefits.

Actual-value-of-benefit Method. This method is based on the increase in productive capability induced by drainage. For doubled productive capability the benefit factor is 100; for increase of one-half, 50; others are in proportion. Good judgment and expert knowledge of crop production is required, but the method gives good satisfaction. Boyd and Hart comment on it as follows:

TABLE 16.1. ASSESSMENT SHEET FOR DRAINAGE DISTRICT NO. 1. PERCENTAGE BASIS

Owner's description of land	Number of acres		Percentage benefit per acre	Acre units of benefit	Assessment
A	160	30	70	21	$166.28
		50	60	30	237.54
		20	100	20	158.36
		20	90	18	142.52
		40	70	28	221.70
B	120	40	10	4	31.67
		50	45	22.5	178.16
		30	30	9	71.26
C	80	40	100	40	316.70
		40	100	40	316.70
D	20	20	90	18	142.52
E	20	20	90	18	142.52
F	15	15	100	15	118.77
G	25	25	100	25	197.95
H	80	30	70	21	166.28
		50	100	50	395.90
Total	520		...	379.5	$3,005.00

Cost of drainage improvements		$3,000.00
Damages awarded		300.00
Highways	$120.00	$3,300.00
Railroads	175.00	295.00
Total net cost to landholders		$3,005.00

$$\frac{3{,}005.00}{379.5} = \$7.918, \text{ cost per acre unit of benefit}$$

The . . . method which estimates actual value of the benefits conferred (increase in productive capability), and apportions the assessments accordingly, is required to be used by the statutes of 14 states. It follows directly the theory of special benefits by determining the value of the benefits to accrue and uses the benefits as a basis for apportioning the assessments. It has been used with success under a wide variety of conditions, and, while not perfect, is much better than any other method now in use. It is simple and consequently understood, which is a great advantage when large numbers of people are to be convinced that their assessments are equitable and have been determined in a just and reasonable way. [See Table 16.3.] The use of this method is strongly recommended.

The methods which may be used for distributing the costs of drainage districts are usually prescribed by the drainage law of the given state, but

TABLE 16.2. ASSESSMENT SHEET FOR DRAINAGE DISTRICT NO. 1. FIVE-CLASSIFICATION BASIS

Owner's description of land	Number of acres		Classification benefit unit	Assessment
A	160	40	4	$ 245.31
		40	3	183.98
		40	5	306.63
		40	3	183.98
B	120	40	1	61.33
		40	2	122.65
		40	2	122.65
C	80	40	5	306.63
		40	5	306.63
D	20	40	2.5	153.32
E	20		2.5	153.32
F	15	40	1.875	114.99
G	25		3.125	191.64
H	80	40	5	306.63
		40	4	245.31
Total	520		49	$3,005.00

Cost of drainage improvement		$3,000.00
Damages awarded		300.00
		$3,300.00
Highways	$120.00	
Railroads	175.00	295.00
Balance, cost to lands		$3,005.00

$$\frac{\$3{,}005.00}{49} = \$61.3267, \text{ cost of one 40-acre benefit unit}$$

this is not always the case. Where it is left to the discretion of the engineer, the drainage commissioners, or the board of assessors to decide on a method, it is always well to select one which is elastic and will give the assessors the widest possible latitude in the exercise of judgment and the employment of expert advice, in order that justice may be meted out to all in the highest degree.

16.16. Benefits to Railroads, Highways, Town Lots, Etc. It is usually required that the benefits received by railroads, highways, rights of way, town lots, and other property shall be considered and estimated and that they shall be assessed a certain part of the cost of the district. That part of the cost of the district to be paid by the owners of such property is usually assessed in a lump sum and is usually a percentage of the total cost of the district based on the actual amount of benefit received.

In estimating the assessable benefits to railroads and highways, only those benefits of a determinate character are to be considered, such, for

TABLE 16.3. ASSESSMENT SHEET FOR DRAINAGE DISTRICT NO. 1. ACTUAL-VALUE-OF-BENEFIT METHOD

Owner's description of land	Number of acres		Number of benefit units	Assessment
		40	80	$ 257.11
A	160	40	50	160.70
		40	100	321.39
		40	60	192.84
		40	10	32.14
B	120	40	35	112.49
		40	40	128.56
C	80	40	100	321.39
		40	100	321.39
D	20	40	...	144.62
E	20		90	144.62
F	15	40	...	120.52
G	25		100	200.87
H	80	40	90	289.25
		40	80	257.11
Total	520		935	$3,005.00

Cost of drainage improvements		$3,000.00
Damages awarded		300.00
Highways	$120.00	$3,300.00
Railroads	175.00	295.00
		$3,005.00

$$\frac{3{,}005}{935} = \$3.2139, \text{ the cost of 1 benefit unit}$$

instance, as the decrease in cost of maintenance due to drainage. Such apparent benefits as increase in traffic and freight revenues due to improved railroads or highways are more or less speculative and should, as a rule, be given little weight.

The assessable benefits to railroads and highways are mainly those received by roadbeds, which render them more solid, less liable to injury from washouts, settling, and frost action; those which reduce the number of bridges and culverts required; and those which make maintenance, in general, including fences and telegraph poles, less difficult and expensive.

Town lots can usually best be assessed by distributing that part of the total cost apportioned to them according to area, elevation, and location of the different lots.

It is to be noted that the lump sum which is assessed against railroads, highways, town lots, and rights of way should be deducted from the total cost of the district before the cost is distributed among the different tracts of land in the district according to the benefits derived (see Table 16.3). It is also to be noted that the ratio of total benefit to assessed cost shall be the same as that established for the agricultural land.

16.17. Damages, Their Determination and Settlement. Damages which are the result of carelessness on the part of officials or contractors, or of accident, in public ditch construction are not chargeable against the district but against the individual or group of individuals responsible. Therefore settlement for them is by the usual legal processes in personal damage suits, and their discussion is not a pertinent issue here.

Damage claims apportionable in the establishing of a public ditch are those which arise as a direct result of the exercise by the district of its right of eminent domain in carrying out public improvements. They include value of land taken for rights of way for the ditch and of small areas so isolated thereby as to be rendered distinctly lower in value, specific structures and other property unavoidably injured or destroyed in the locating and building of the ditch, etc. The equitable determination of the amount of such damages is part of the responsibility of the viewers or of a specially appointed group but is generally kept separate from the tabulation and apportionment of benefits. The amount of such damages is a proper charge against the district as part of the costs of the ditch which are assessed against the benefited parties in proportion to the benefits received.

The viewers are generally required to file their final report within a legally specified time as the final hearing before the Agency of Jurisdiction is based on and largely concerned with the distribution of benefits and damages given in their report.

16.18. The Final Hearing. Immediately upon the filing of the viewers' report, the executive secretary of the Agency of Jurisdiction on orders therefrom posts notice of time and place of final hearings to be held, after a legally determined interval, on the viewers' report, every means and effort being required to be made to bring such notice to the attention of each individual affected by the apportionment of either benefits or damages. At this hearing each such interested individual is entitled to be present in person or to be represented by counsel to consider his personal assessment and contest same if he thinks it unfair. Decisions in each case are made at this hearing by the Agency of Jurisdiction. Any interested party not satisfied with the decision in his case may appeal the matter in civil suit to the next higher court.

16.19. Appeals. As a rule the state drainage laws provide for appeal to the next higher court by any individual owner affected by the ditch on any of the following issues:

1. Against amount of benefits assessed against him.
2. Against amount of damages allowed him.
3. Against the order establishing the ditch.

Such appeals follow the usual procedure of civil suits and by the usual legal processes may finally even reach the state supreme court. In some of the states no steps can be taken in the letting of contracts for or in carrying out of the construction of the ditch until all appeals under the given proceedings have been finally settled in the courts. In some cases, however, the construction work is permitted to be provided for and carried on immediately after the final hearing, court proceedings in appeals being carried through concurrently therewith. This last seems the more reasonable proceeding because the cases are rare where the establishing of the ditch has not been amply justified as generally desirable and thoroughly conducive to the public welfare before the final hearing is closed. On the other hand, there has been many a case in which the absolute worthiness of a public ditch project has been established beyond question where the unreasonable opposition of a single objector, often nonresident, has through prolonged litigation killed the project because of the inability of its backers to carry the mounting legal expenses of the contest.

16.20. Establishment of the Administrative Authority. In some of the states the administration of the district, after it is established, remains permanently in the hands of the Agency of Jurisdiction. In others, the district being finally established, either through election by the benefited owners or by appointment by the Agency of Jurisdiction, the administrative responsibilities of the district pass into the hands of a board of directors—usually three—selected from the resident benefited landowners, and all the activities of the district are directed by them under such restrictions as the statute may provide.

The district thus organized takes on the character of a quasi-public corporation, in general subject to the corporate laws of the state.

It may sue and be sued.

It may acquire property necessary to its existence, responsibilities, and activities.

It may let contracts for necessary construction and maintenance.

It may sell bonds of the district.

It must carry out a financial system that enables it to pay its obligations and see that they are promptly paid.

In general, it must manage through its directors all the necessary business of the district within the limits prescribed by law.

16.21. Securing of Bids and Letting of Contracts for Construction. The district once organized and contests over apportionment of benefits and damages being settled, the board through its executive office must advertise for bids and let the contract for construction, generally to the lowest responsible bidder. The customary bond for proper performance and completion of the work must be furnished by the contractor, who must initiate the work and, with reasonable and stated allowance for uncontrollable conditions, must push the work diligently to a conclusion, usually by a specified date.

16.22. The Construction Process. The general oversight of the contract construction is usually in the hands of the engineer, who is responsible to the directors. As is customary in such cases, he must lay out all work in shape for construction, inspect the work as completed by the contractor, and recommend to the board either its acceptance or rejection, give final interpretation to specifications under the contract, and furnish progress and final estimates of the work done on which the monthly and final payments under the contract shall be made.

16.23. Financing. As the damages formally incurred and assessed against the benefited property owners, each month's progressive work done by the contractor, and the final settlement must be paid in cash as each becomes due, and as the benefited owners seldom pay their assessment in advance or are required so to do, the customary method of securing funds to meet current payments is by sale of bonds of the district. The bonds bear interest at rates specifically set or limited by the statute, and the interest along with the face value of the bonds becomes a first lien on the lands of the district.

16.24. Payment of Assessments for Benefits. Immediately on completion of the final hearing, the assessments for benefits are due and payable and when not paid at once in cash, which is seldom done, they become in each case a first lien on the land, are entered upon the tax rolls, and are collected in the manner of other taxes.

As in the case of all public improvements calculated to produce benefits for the individual landowners, the basic principle is that the accrual of such benefits through elapsed time furnishes the funds out of which the individual assessments of costs are paid. Many decades of experience have shown that such undertakings as public drainage require a long period of years to pay for themselves without undue hardship on the landowners. Hence the state drainage code usually sets up a scheme of payment of assessments extending over a specified period of years. There is usually a limited period over which the benefited landowner is required to pay only the interest on his assessment. Each year after that period has passed, he must pay also a certain fraction on the principal and interest on the unpaid balance. The arrangement varies considerably in the different states. The way the general plan works is probably best illustrated by outlining the method in one state. Minnesota [161] is the state selected for the illustration, as in that state there are three possible periods of payments on county and judicial ditches, of which the one may be selected in any case which best fits the exigencies of the given situation.

16.25. Payment of Public Ditch Assessments in Minnesota. As soon as practicable after the contract is let, the auditor of each county concerned in the project makes a list of the owners affected and the assessment against each owner and files this with the register of deeds in the said county. This filing is considered notice to all parties interested that the assessment from that date constitutes a lien on his land and the initial due date of payment thereon is November 1, subsequent to the filing of the lien in the office of the register of deeds. The three plans of payment are as follows, the plan selected in any case being at the discretion of the county board in the case of a county ditch and of the judge of the district court in the case of a judicial ditch.

Ten-year Plan. One-tenth of the principal due, with interest at 6 per cent per year, on the unpaid balance on November 1 following the filing of the lien and one-tenth of the principal with interest on the unpaid balance on the first of November of each year thereafter until the whole is paid. This plan has become practically a dead letter.

Twenty-year Plan. Only the interest need be paid each year for the first 5 years beginning on the initial date stated under the 10-year plan. After the 5-year period the yearly payment is one-fifteenth of the principal with interest on the unpaid balance each year until the whole is paid. This is the plan most frequently used.

Thirty-year Plan. Only the interest must be paid each year for the first 10 years beginning on the initial date above stated under the 10-year plan. After the 10-year period, the yearly payment is one-twentieth of the principal with interest on the unpaid balance each year until the whole is paid.

This plan is often used on large and unusually expensive projects and on those where the board or the judge is convinced the situation of the interested farmers justifies invoking it.

DESIRABLE IMPROVEMENTS IN THE DRAINAGE CODES OF THE SEVERAL STATES

Boyd and Hart [17] make the following statement: "Many drainage districts are planned to provide an outlet only, leaving the owner to install the tile drains or whatever may be necessary to give complete drainage; often districts give complete drainage to all or part of the land."

The last of the two cases cited is the rare exception. The failure to provide complete drainage under a district project works a distinct hardship on many farmers. As was pointed out in Secs. 6.2, 6.8 and 10.2, outlet channels, in themselves, provide comparatively little real drainage to adjacent lands. The large returns from drainage come only after interior drainage has been installed on the farm. Therefore few benefits accrue to the affected landowners from public ditches, and they are largely a liability rather than an asset until after the farmers have installed their own farm drainage which must outlet in the public ditch.

Loan of the Public Credit for Drainage of Tributary Farms Essential

Few farmers are able privately to finance complete interior drainage of their farms in addition to meeting annual assessments for a public outlet ditch. The public welfare would be much further advanced and the public interest far better served if the various states would uniformly amend their drainage code to make it possible for the individual farmer to install his interior drainage under the district organization with the consequent loan of the credit of the county to enable him to spread the cost of that part of his drainage also over a term of years during which time the interior drainage would really return him a profit that would handsomely meet the entire yearly drainage assessment.

The Wisconsin Farm Drainage Law. Thus far Wisconsin alone recognizes the advantage of such a plan which is represented in the now famous Wisconsin Farm Drainage Law, which does extend the public credit to the individual farmer exactly as above outlined. Minnesota has an approach to it in the section of her drainage code providing for "Lateral Systems." This section makes such a move possible for two or more farmers but not for one alone. Hence this plan falls short of perfection. A uniform plan patterned exactly after the Wisconsin Farm Drainage Law would cost the states nothing but would greatly advance their general welfare.

Where the county board is the Agency of Jurisdiction, it is impossible that politics should not have an undue influence in public ditch proceedings because county commissioners are elected to office every 2 years and are therefore a political group always envisioning possibilities of reelection and are seldom possessed of any great knowledge or understanding of good drainage practice. Uniform amendment of the various state codes in such a way as to place all public drainage jurisdiction under well-established courts would largely eliminate the political factor and often improve drainage practice under the supervision of a judge well schooled in drainage cases and drainage procedure. Lawyers extensively engaged in drainage cases and experienced drainage engineers know that this is true, and the majority of them strongly favor it.

MAINTENANCE OF PUBLIC DRAINS

The Federal Census of Drainage for 1940 (Ref. 23) briefly summarizes maintenance of public ditches under the codes of the several states as follows:

> Maintenance of the drainage works after construction is required by some states and authorized by most of them. The amount that may be expended in any year without authorization by the landowners through proceedings somewhat similar to establishment of the enterprise, usually is restricted to a small portion of the cost of the original construction. Cost of maintenance commonly is distributed on the same basis as original construction, but in some instances on a uniform rate per acre.

The unfortunate experiences of almost all the states of the deterioration and ultimate failure of public ditches owing to lack of systematic maintenance points unmistakably to the fact that the only safety lies in placing the responsibility for maintenance, backed by penalty for neglect, squarely upon the shoulders of the original administrative authority in the district and strengthening its hands therein by regular provision for annual assessment of general taxes expressly to cover the cost of public ditch maintenance.

ASSESSMENTS FOR DRAINAGE OF IRRIGATED LANDS

16.26. Basis of Assessments. The conditions requiring drainage of irrigated lands, the sources of drainage water thereon, causes of its accumulation, and the need and methods of its removal have been fully discussed in Chap. 13 on Drainage of Irrigated Lands, without special regard for the causal human agencies responsible for meeting the cost of the system. The features, agencies, or groups against which the cost of such drainage should be assessed vary, in general, widely from those in drainage in humid regions. In the irrigated regions, the proper individuals or agencies against which to assess the drainage costs are determined upon largely according to two features: (1) the sources of the drainage water, and/or (2) the amount of the benefit to the drained lands in the form of increased productive capability.

16.27. Source of the Drainage Water and Theory and Methods of Assessments of the Costs. The source of the drainage water must be considered under the four main heads given below, and the consideration of the theory and method of assessment is so closely related thereto as to seem to make separate discussion of them difficult and confusing.

1. *Natural Water.* Where the drainage water is a natural accumulation from rainfall or seepage from natural streams or bodies of water unaltered by man, the cost of drainage accrues directly against the benefits to the land therefrom, and if the drained land is part of any kind of a drainage

and/or irrigation and drainage district, it is properly assessed against the owners of each parcel of land according to some one of the systems discussed earlier in this chapter.

2. *Deep Percolation, Farm Seepage, or Normal Irrigation Waste.* Where the sources of the irrigation water are wholly within the farms of the drainage district and are the result of irrigation procedure thereon, the normal theory and method of assessment of cost of drainage is similar to that stated under source 1 above, on the basis of the increase in production capability benefits to each tract of land.

3. *Seepage from Higher-lying Irrigated Lands and/or from Reservoirs and Distribution Canals.* (*a*) When the source is wholly seepage or escaped waste from higher-lying irrigated lands, the owners of these lands are ethically, and in some states legally, responsible, and the cost is assessable against them. (*b*) When the source is wholly reservoir or canal seepage, there are two exactly opposite situations, either of which may arise: (1) Where the reservoir and canals are owned by a farmers' mutual company or an irrigation district who are their own primary customers and quite possibly their only customers, it is to their own financial interest to install the drainage and charge the cost of it back against themselves as a part of the cost of their water. If only part of the owners suffer from lack of drainage or if soil conditions differ so greatly in different parts of the district as to cause marked differences in the need for drainage or a similar result obtains from other natural causes, the assessors should, as is customary in humid land drainage, proportion the costs among the individual owners according to the benefits received. (2) If, on the other hand, the water supply is owned by others than the landowners, the assessment should ethically and often legally be made against the water company who will handle its payments as one of the legitimate expenses of their business and pay it as such, unless by previous agreement with the water users the latter will accept a part of the obligation in return for the admitted production benefit to their drained lands. Such action brings the matter up to and really constitutes case 4.

4. *Seepage from the Canals and Reservoirs of the Water Company.* This is the case where, by mutual agreement in the contract, the water users and the water company divide the assessment of costs between them in a definite proportion based upon the actual increase in productive capability that the farmers are willing to pay for and the amount of water contributed to the drainage need by the seepage from the canals and reservoirs, for which the water company is responsible and for which they should pay. That is, the assessment should be made in lump sum per acre or per established land unit and then split for payment between the two agencies in the agreed upon proportion or on a sliding proportional scale to meet the amount of production benefit for any particular time or

for varying results from different tracts of land according to soil and related characteristics. This system of shared responsibility between the agencies of contribution of drainage water with the resultant increased productivity benefit to the drained lands seems to give the most general satisfaction on both sides, and it is believed the system tends to replace all other methods of drainage assessment on irrigated lands.

It has been shown in Chap. 13 how the total seepage may be determined on any given canal system. Seepage from the farm systems will be equal to the total water delivered to the farm less the surface waste and the consumptive use, both of which are determinable by known means.

There are also other variations such, for example, as the case where the drainage district embraces water contributions from several water systems. Observations and measurements from each of these will enable proper distribution of the responsibility between all parties concerned. A more detailed and extended discussion of the subject of this section is beyond the scope of this book. Those wishing a more complete treatment of the topic of drainage assessments on irrigated lands will find it in the contributions by Boyd and Hart [17], Etcheverry [45], and Yohe [167].

CONTRIBUTIONS TO DRAINAGE DEVELOPMENT BY THE UNITED STATES GOVERNMENT

Supporting activities in the drainage field by the Federal government is summarized as follows by the Water Resources Policy Commission, Volume 3, 1950 [157].

Assumption of federal responsibility for the control of floods on a national basis is a relatively recent development. . . . Early legislative evidence of federal interest appeared in the Swamp Land Acts of 1849 and 1850, granting unsold swamp and overflowed lands to Louisiana, Arkansas and other states containing such lands. These Acts made such lands subject to disposal of the state legislatures and required that the proceeds therefrom shall be applied to the prosecution of drainage, reclamation and flood-control projects. . . . For some time thereafter, Congressional attention to drainage continued to be irregular and localized.

16.28. Army Engineers. The Flood Control Act of 1936 . . . marked the undertaking of federal responsibility on a nation-wide basis for control of floods. And in 1944, Congress provided that the words "flood control," as used in the declaration of policy in the 1936 statute, shall be construed to include "channel and other major drainage improvements." So defined this function automatically came within the flood-control jurisdiction of the Army Engineers.

16.29. Soil Conservation Service. Other jurisdiction over federal drainage responsibilities is vested in the Soil Conservation Service. Under authority of the 1938 and 1944 Flood Control Acts, the Soil Conservation Service makes technical assistance available to individual farmers so that they may carry out drainage operations on their lands. Moreover, direct financial assistance for drainage work is furnished to farmers by the Production and Marketing Administration. . . .

Excepting for direct financial aid, the foregoing drainage responsibilities are, therefore, divided between the Army Engineers and the Soil Conservation Service. But the division is unclear, for the foregoing statutes make the partition dependent largely upon the meaning of "channel and major improvements." And there is no satisfactory definition of this term or any of its components. Hence the jurisdictional division seems to be the difference between "major" and whatever is minor.

This lack of statutory clarity apparently underlies a 1948 Memorandum of Understanding, whereby the two agencies agreed that:

"The Soil Conservation Service is authorized to provide assistance to soil conservation districts and other state and local instrumentalities acceptable to the Secretary of Agriculture on drainage and related problems in connection with the conservation of soil and water resources, and the Corps of Engineers is authorized to provide channel and major drainage improvements on rivers and other waterways.

In general, drainage for a watershed area consists of two complementary parts; one, the system required to utilize, manage, or remove surplus water from farms or groups of farms and, two, the necessary channel and major drainage improvements required to control, carry off, and utilize these surplus waters. Therefore, in soil conservation districts or other state or locally organized areas, the activities of the Corps of Engineers and the Soil Conservation Service, in cooperation with the local districts or agencies provide for consideration of both of these complementary parts of drainage."

16.30. Bureau of Reclamation. The Bureau of Reclamation frequently performs drainage work as a part of irrigation project development. Except in the case of special acts, Reclamation Law does not expressly deal with drainage. However, Congress has not attempted to enumerate in detail the types of irrigation features which are authorized to be constructed in connection with irrigation projects.

Within the beginning of large scale irrigation operations, drainage works were undertaken in the face of rising water tables and attendant water logging and resulting soil alkalinity on several projects. Extensive drainage investigations and construction continued.

Moreover it has been specifically held that "the necessity for drainage follows irrigation on an extensive scale almost as a matter of course. Drainage work may be undertaken in connection with federal irrigation projects either as an incident of construction, or as an incident of operation and maintenance. Some 3,500 miles of drains are embraced in the operating program of the Bureau of Reclamation for the fiscal year 1951."

The foregoing extract from the Report cited shows that the Federal government is really deeply concerned with drainage work throughout the entire country, but it seems equally clear that there is no well-organized Federal plan of drainage development as of itself, apart from other forms of interior development, particularly with regard to direct financial support. The activity is still sporadic. A large part of the drainage work aided by the Federal resources is comprised in incidental services concerned with other projects with little specific allotment of finances to particular jobs.

PROBLEMS

1. The term "riparian rights" is frequently used in discussions of water supply and water control. What do you understand it to mean?

2. List the four major factors which should be considered in classifying land for the apportionment of assessments in drainage districts.

3. How is the common law of natural drainage modified to permit the formation of drainage districts?

4. What principle of equity is the controlling factor in all sound drainage projects?

5. List the duties of the engineer in the establishing and execution of a drainage district.

APPENDIX A

THE UNIT HYDROGRAPH METHOD[1]

A.1. "This method was proposed by L. K. Sherman [136] in 1932. The basis of the method is the unit graph, which is the hydrograph that would result from a 1-inch run-off from the entire watershed considered, as the result of a uniform rainfall lasting one unit of time, such as a day or an hour. The unit graph is then used to determine the hydrograph that would result from a storm of any magnitude and duration. The procedure used in obtaining a unit graph and in applying it to run-off problems is given below.

A.2. "The method can be used only where rainfall and run-off records are available. The first step is to consult the rainfall records and to note the dates of occurrence of rainfalls having the following characteristics.

"1. The rainfall must have occurred in a chosen unit of time, as a day or an hour.

"2. The rainfall must be isolated in regard to time, so that the run-off will not be appreciably affected by preceding or subsequent rains.

"3. The entire watershed under consideration must be covered uniformly by the rainfall, or reasonably so.

"4. The amount of rainfall should be above a certain minimum value, about 1.50 inches if a day is chosen as the unit.

"Few rainfalls will be found that conform to these specifications, possibly not more than an average of one a year. Since 1 day is generally the shortest period of time for which rainfall and run-off records are given, a length of time of 1 day is ordinarily used.

A.3. "The second step is to consult the run-off records for the dates selected in step 1 and for each storm to list the discharges during each unit of time. A unit graph is to be constructed from the data from each storm, and a composite graph constructed that should be more representative of the watershed area than any individual graph. The procedure is best explained by an example. Suppose that a rainfall meeting the foregoing requirements occurred in a 24-hour period on April 21, 1938, over a watershed of 753 square miles.[2] The runoff records for this period, listed in Table A.1, show that the run-off period lasted until May 1, a period of 11 days; also that the peak flow occurred on Apr. 25, or 4 days after the rainfall ended.

[1] From "Drainage and Flood-control Engineering," 2d ed., pp. 86–91, by George W. Pickels, published by McGraw-Hill Book Company, Inc., 1941. Reproduced by permission.

[2] For small watersheds on sloping ground a 1-hour period might have been chosen if 1-hour records are available. The procedure is the same.

TABLE A.1.* METHOD OF COMPUTING UNIT GRAPH

(1) Date	(2) Observed flow	(3) Base flow	(4) Net run-off	(5) Per cent of total flow	(6) Unit-graph ordinates
1938					
Apr. 20	68	68	0	0	0
21	1,000	48	952	7.4	1,510
22	1,380	36	1,344	10.5	2,130
23	1,800	26	1,774	13.8	2,810
24	2,130	21	2,109	16.5	3,330
25	2,200	20	2,180	17.0	3,460
26	1,810	20	1,790	13.9	2,835
27	1,420	16	1,404	11.0	2,220
28	685	16	669	5.2	1,060
29	407	14	393	3.1	620
30	129	11	118	0.9	190
May 1	95	11	84	0.7	135
			12,818	100.0	20,300

* Pickels's Table XIV. (Columns 2, 3, 4, 5, and 6 expressed in cfs.)

A.4. "All the run-off shown, however, was not the result of the rain of Apr. 21. There would have been some flow in the stream during this period if there had been no rain. This flow is called *base flow* and is derived from ground-water flow. The amount of this base flow for each day of the flood period has to be estimated . . . by searching the discharge records for other periods, as near this date as possible, when the rate of run-off was as low as 68 cubic feet per second and was decreasing and by noting the decrease day by day. . . . The base flows are listed in column 3. The observed run-off minus the base flow gives the net run-off resulting from the rainfall of Apr. 21. These values are recorded in column 4. The net run-off is made up largely of surface run-off, but part of it is the result of ground-water flow from the storm under consideration.

A.5. "The total net run-off during the 11-day period was 12,818 second-feet-days. That is, if all the water had run off in 1 day, instead of 11 days, the rate of runoff would have been 12,818 cubic feet per second. If the net run-off on each day is divided by the total, the percentages of the total run-off that ran off each day will be as given in column 5. These values show the distribution of the total net run-off with respect to time. If 1 inch of water over the 753 square miles of watershed area ran off in 1 day, the rate of discharge would be 20,300 cubic feet per second. The actual run-off following the rain of Apr. 21 was 12,818 second-feet-days, which is 63.1 per cent of the 20,300; therefore, the actual depth of run-off from this storm was 0.631 inches. If 1 inch had run off, each value in column 4 would have been 1.584 times as large, and the values would be as recorded in column 6. The discharges in column 6 plotted against time give the unit graph for the watershed.

A.6. "Unit graphs are constructed in a similar manner for each of the isolated storms chosen from the rainfall records. Theoretically, all the unit graphs should coincide. Since the several rainfalls considered will not completely satisfy all the requirements listed, there will be differences. An average curve is taken as the unit graph for the watershed.

A.7. " . . . Regardless of the intensity of the rainfall, the base of the unit graph will always be the same length. If all the rains started at the beginning of the calendar day and continued for a full 24 hours, the peaks on all the unit graphs would occur at the end of the same time period, but it is unlikely that any of the rainfalls considered happened in this theoretical manner. Some of the rains may have lasted only a few hours. Some may have occurred at the beginning and some at the end of the 24-hour period. These facts account for the differences in the unit graphs for the same watershed for different storms.

APPLICATION OF UNIT GRAPH

A.8. "The function of the unit graph is to aid in the construction of the probable hydrograph that would result from a rainfall of any intensity and duration.

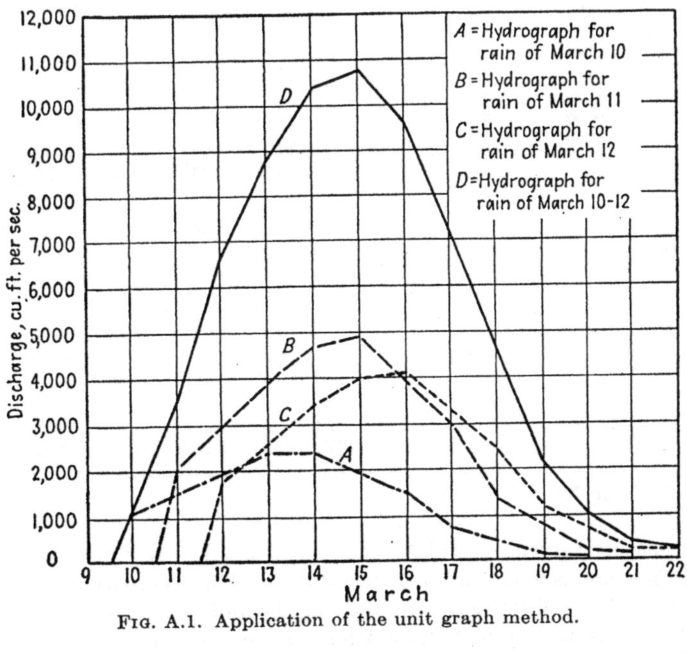

FIG. A.1. Application of the unit graph method.

This is done as follows. Assume that a rainfall lasts 3 units of time and that the rainfalls during these three periods are 2.33, 2.80, and 1.50 inches, respectively; and that the infiltration capacity of the watershed is such that during the first period 30 per cent of the rain, or 0.70 inch, runs off, during the second period 50 per cent, or 1.40 inches, runs off, and during the third period 80 per cent, or 1.20 inches, runs off. The graphical solution would be as shown in Figure A.1.[1] Using the ordinates of the unit graph that are listed in column 6, Table A.1,

[1] Figure 44 of Pickels's book.

and multiplying each ordinate by the run-off in inches for each day of the storm period, the hydrographs that would result from the rain of each day, acting by itself, are shown as hydrographs *A*, *B*, and *C* in the figure. It is to be noted that hydrograph *A* starts on Mar. 10, hydrograph *B* on Mar. 11, and hydrograph *C* on Mar. 12. By combining the ordinates of these three hydrographs, the ordinates of hydrograph *D* are obtained. The sum of the areas under Hydrographs *A*, *B* and *C* equals the area under hydrograph *D*. These areas represent volumes of run-off.

A.9. "Although the graphical solution illustrated in Figure A.1 gives a picture of the application of the unit hydrograph method, it is generally replaced by the

TABLE A.2.* ORDINATES FOR HYDROGRAPH CONSTRUCTED BY UNIT GRAPH METHOD

Rainfall		Run-off		Unit-graph ordinates, cfs	Run-off due to rain on day noted, cfs			Total run-off (ordinates of hydrograph), cfs
Date	Amount, in.	Percentage	Amount, in.		Mar. 10	Mar. 11	Mar. 12	
(1)	(2)	(3)	(4)	(5)	(6)	(7)	(8)	(9)
Mar. 10......	2.33	30	0.70	1,510	1,057	......	......	1,057
11......	2.80	50	1.40	2,130	1,491	2,114	......	3,605
12......	1.50	80	1.20	2,810	1,967	2,982	1,812	6,761
13......	....	..	....	3,330	2,331	3,934	2,556	8,821
14......	....	..	....	3,460	2,422	4,662	3,372	10,456
15......	....	..	....	2,835	1,985	4,844	3,996	10,825
16......	....	..	....	2,220	1,554	3,970	4,152	9,676
17......	....	..	....	1,060	742	3,108	3,402	7,252
18......	....	..	....	620	434	1,484	2,664	4,582
19......	....	..	....	190	133	868	1,272	2,273
20......	....	..	....	135	94	266	744	1,104
21......	....	..	....	......	......	188	228	416
22......	....	..	....	......	......	......	162	162
Totals.....	....	..	....	20,300	14,210	28,420	24,360	66,990

* Pickels's Table XV.

analytical method given in Table A.2, which should need little explanation. The values in columns 6, 7, and 8 are the ordinates of the hydrographs *A*, *B*, and *C* in Figure A.1 and are obtained by multiplying the unit graph ordinates in column 5 by the run-off in inches, in column 4 for each of the 3 days of rainfall. The values in columns 6, 7, and 8 are added horizontally to give the values in column 9, which are the ordinates of the hydrograph *D*.

"At present the most unsatisfactory part of the procedure is the determination of the run-off percentages listed in column 3.

"Unit graphs for a large number of streams are given in Water Supply Paper 772 of the United States Geological Survey [58]."

APPENDIX B

GENERAL DESIGN OF PERMEAMETERS

B.1. Essential Features. In general, regardless of type, a permeameter must have in practical form the following features: A water-supply basin of known dimensions so equipped as to enable maintenance of constant head on the sample and easy measurement of its volume and of the head on the bottom and top of the sample at any instant; a container for the sample, preferably cylindrical, so equipped as to allow free seepage of water through the sample without causing disturbance of the sample or loss of material from it; a discharge device for the water that has passed through the sample; a graduated container to receive this water; nonleaking tube connections between the parts of the apparatus; established means for measuring head and changes in same at the top and bottom of the sample; a thermometer and a stop watch.

B.2. The Discharging Permeameter. The following introductory paragraph including the formula and definitions of symbols is adapted from the Imperial Valley Report [34].

The equipment is so designed that the surface of water in the supply tank is maintained at a constant level and the head on the sample at its bottom is maintained constant. For this type

$$k_p = \frac{Q}{IA} \tag{31}$$

when k_p = coefficient of permeability of prepared specimen
Q = discharge in unit of time ($= v/T$ by fundamental definition)
A = cross-sectional area of the soil column
I = hydraulic gradient [$= (h_1 - h_2)/L$ by fundamental definition]
L = length of soil column
$h_1 - h_2$ = head loss between the top and bottom of the soil column
C_t = temperature correction
V = volume of discharge
T = length of run in units of time

Substituting in Eq. (31) the values $Q = V/T$ and $I = (h_1 - h_2)/L$,

$$k_p = \frac{VLC_t}{TA(h_1 - h_2)} \tag{32}$$

Units for most laboratory work are expressed in cubic centimeters squared and in hours.

The physical features of the discharging permeameter and their relation to each other are shown in conventional diagram, *not to scale*, in Fig. B.1. In practice many variations in the setup are possible and the space required may be greatly condensed, but the static and dynamic features must be maintained as indicated in the figure.

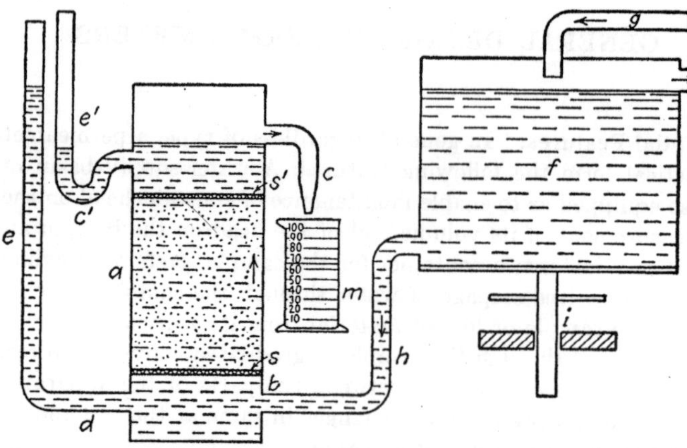

FIG. B.1. Conventional diagram of constant-head type of discharging permeameter. (Fig. 2 of Fishel's discussion of discharging apparatus, pp. 56 to 58, *U.S. Geol. Survey Water Supply Paper* 887.)

a. Glass cylindrical vessel (percolation cylinder) closed at lower end. Has two openings near bottom and two near top. It is 3 in. in diameter and 8 in. high.

b. Opening near bottom for inflow of water.

c. Opening at top for discharge of water that has percolated up through the sample.

d and *d'*. Openings for pressure gages.

e and *e'*. Pressure gages consisting of two glass tubes (graduated), each about ½ in. in diameter, that indicate the head at the bottom and top of the sample being tested. (The tubes must be of large enough diameter to render capillarity in them negligible.)

f. Water-supply tank; a cylindrical vessel about 5-in. diameter and 12-in. high.

g. Tube through which distilled water enters the supply tank.

h. Rubber tube through which the distilled water passes to the percolation cylinder.

i. Screw-jack by means of which, by very fine adjustments the tank can be raised or lowered to maintain a constant head on the sample.

m. Graduated receiving beaker for the outflow.

n. Outlet in side of tank for maintaining a practically constant level.

s and *s'*. Fine copper gauze screens for preventing the moving out of particles from the sample.

Entrapping of air beneath screen *s* is prevented by introducing water to a depth of about 1 in. above *s* before sample is placed.

The following table of temperature corrections (Table B.1) may be used both with this type and with the falling-head permeameter discussed in the next section. The coefficient of permeability, as defined, is for a water temperature of 60 degrees Fahrenheit. If the test is run with water of any other temperature, a correction factor must be introduced into the calculations. This factor may be obtained for any case from Table B.1.

TABLE B.1.* TEMPERATURE CORRECTIONS: FACTORS FOR CONVERTING COEFFICIENTS OF PERMEABILITY COMPUTED AT WATER TEMPERATURES OF 40 TO 90 DEGREES FAHRENHEIT TO COEFFICIENTS OF PERMEABILITY AT WATER TEMPERATURE OF 60 DEGREES FAHRENHEIT

°F	T_c	°F	T_c	°F	T_c	°F	T_c
40	1.37	53	1.11	66	0.92	79	0.78
41	1.35	54	1.09	67	0.91	80	0.77
42	1.33	55	1.08	68	0.89	81	0.76
43	1.31	56	1.06	69	0.88	82	0.75
44	1.28	57	1.04	70	0.87	83	0.74
45	1.26	58	1.03	71	0.86	84	0.73
46	1.24	59	1.01	72	0.85	85	0.72
47	1.22	60	1.00	73	0.84	86	0.71
48	1.20	61	0.99	74	0.83	87	0.70
49	1.18	62	0.97	75	0.82	88	0.69
50	1.16	63	0.96	76	0.81	89	0.68
51	1.15	64	0.95	77	0.80	90	0.67
52	1.13	65	0.93	78	0.79		

* Table 2 of Ref. 18.

B.3. The Falling-head Permeameter.[1] Two simple types of the falling-head permeameter have been developed in the Imperial Valley studies. These are designated as Models A and B and are shown in formal drawings, to scale, in Figs. B.2 and B.3, respectively.

"Model A type consists of a 2-inch brass soil cylinder, connected at the base with a small copper U-shaped tube to which is attached a manometer glass tube. An auxiliary water-supply cylinder is connected to the U-shaped tube to supply water for saturating the sample. Water is introduced via the supply tube until the sample is saturated and water is discharging over the rim of the brass cylinder. Then the supply valve is turned off and the rate of drop of the water in the graduated glass tube is noted. This rate of drop is an indication of the transmissability of the soil.

"Model B type was designed using material that is readily available in any locality. The fittings can be obtained in any plumber's shop and can be assembled with a soldering iron. A coating of stop-cock grease applied to the threads of . . . the soil tube will prevent leakage. The soil cylinder is made from a length of 1½-inch smoothly reamed pipe threaded to a 1½-inch coupling attached

[1] The material in this section is abridged from "A Falling Head Permeameter for Evaluating Permeability" by George P. Bradshaw and William W. Donnan [18], a contribution from the Imperial Valley Drainage Research Project, published in pamphlet form by the Division of Irrigation and Water Conservation of the U.S. Soil Conservation Service, January, 1950. (Courtesy of the authors and the Soil Conservation Service.)

to reducers, nipples, elbows, etc., to form the permeameter. The device is relatively inexpensive, costing approximately 2 dollars for the material."

B.4. Securing of Samples and Method of Handling. As the soil tube is demountable from the rest of the device on both models, it may be used, with

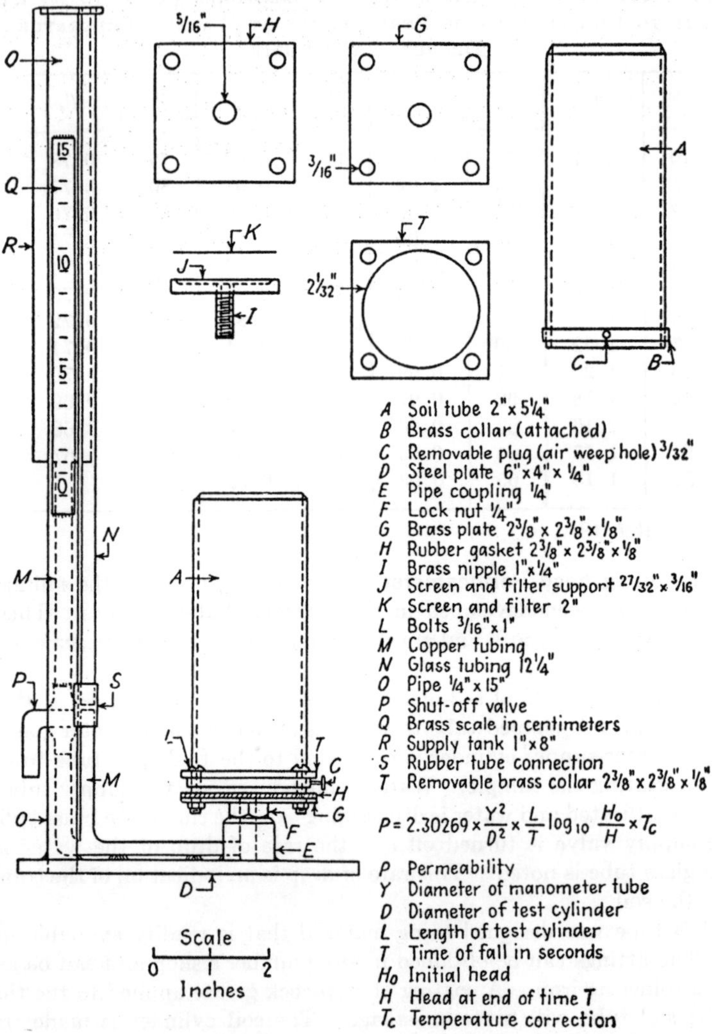

FIG. B.2. Falling-head permeameter, Imperial Valley Research Model A Type. (Fig. 1, Type A of Ref. 18, *Courtesy of the authors and the Soil Conservation Service.*)

careful handling, instead of the more complicated soil-sampling tube discussed in Chap. 3 and illustrated in Fig. 3.3 in securing in-place samples. "Compression of the soil sample due to side friction can be greatly reduced by wetting the soil tube prior to sampling." It is suggested that the car bumper can serve as backstop for vertical samples and the side of the sampling pit for horizontal samples.

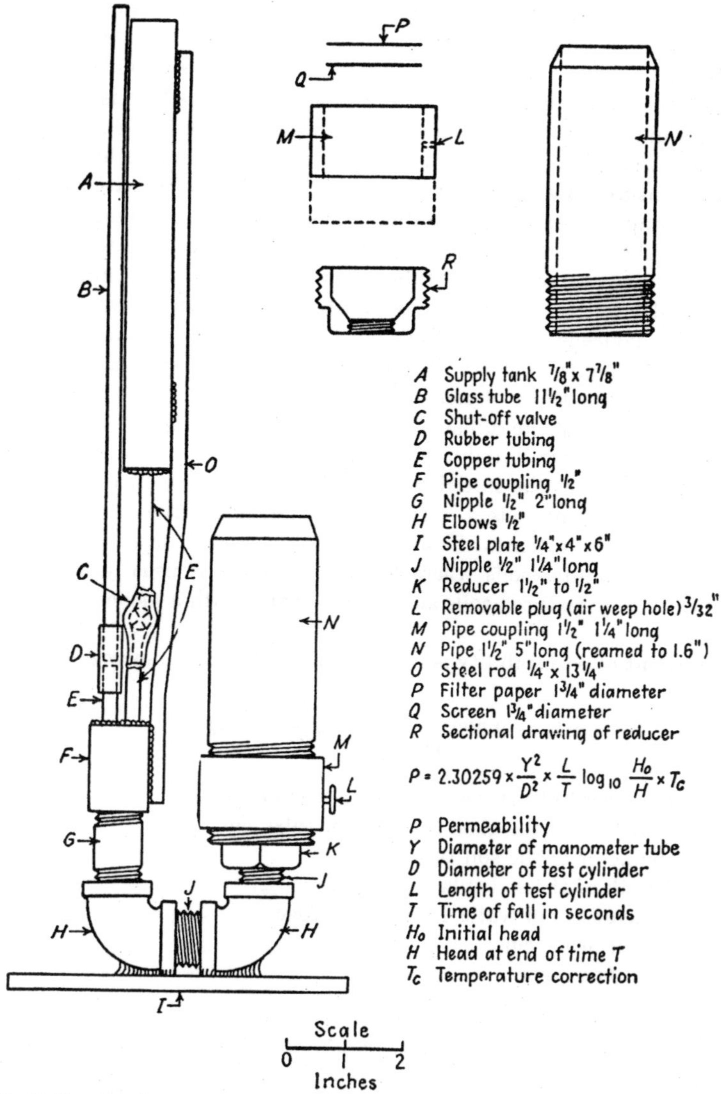

FIG. B.3. Falling-head permeameter, Imperial Valley Research Model B Type. (Fig. 2, Type B of Ref. 18, *Courtesy of the authors and the Soil Conservation Service.*)

"Deep vertical samples can be taken with the Model B sampler by boring with a post-hole auger and then using a 1½-inch pipe extension handle on the permeameter soil tube. Samples at depths of 8 or 9 feet can be taken with relative ease with this pipe extension. In both shallow and deep sampling, extreme care should be taken when removing the sample from the surrounding soil. The extruding soil should be gently broken away and the ends of the samples then trimmed with a spatula or trimming device. On the bottom end of the sample a paper filter is placed next to the soil and then a screen, which is used to keep the soil from sloughing away. The soil sample tube is affixed to the permeameter and tightened to prevent any leakage. A small weep hole is provided at the point where the soil sample and the screen and filter join. This hole is used to bleed the air from under the soil sample and to prevent it from being forced through the sample. Irrigation water normally used on the soil is used for the permeability test to insure satisfactory results. As soon as the water saturates through the soil the sample is ready to be tested. The size of the glass manometer tube to be used should depend on the permeability of the soil to be tested. If the soil is freely permeable a larger glass tube, such as 4.5, 5.5 or 7.5 millimeter . . . should be used. If the soil is slowly permeable a 1, 2, or 3 millimeter tube should be used. The device is so constructed that these glass tubes are interchangeable. The technician using the device will quickly recognize the approximate size tube to use for the various types of soil encountered.

"The permeability of the soil sample is obtained by timing the fall in the glass manometer tube from an initial head reading to some other desired head reading. For . . . simplification the initial head is usually taken at 15 centimeters and the final head at 5 centimeters. This . . . head differential of 10 centimeters . . . simplifies computations. Timing is usually done with a stop-watch . . . necessary on the lighter soils where the time element is usually less than one minute. On heavy textured soils an ordinary watch with a second hand may be substituted."

The coefficient of permeability is then calculated by the following equation:

$$\text{``}k_h = 2.303 \frac{d^2}{D^2} \frac{(L)}{t} \log_{10} \frac{h_0}{h} \times (3{,}600)\text{''} \qquad (33)$$

where k_h = coefficient of permeability, cc per sq cm per hr
d = diameter of the glass manometer tube, cm
D = diameter of the soil tube, cm
L = length of the soil tube, cm
t = time of fall of water from h_0 to h, sec
h_0 = initial head, cm
h = final head, cm

LITERATURE CITED

1. ADAMS, FRANK: "The Moisture Equivalent a Practical Measure of the Maximum Capillary Capacity," Conference on Research Problems in Consumptive Use of Water and Conservation of Rainfall, Report by Water Conservation Committee, Division of Irrigation, American Society of Civil Engineers, Los Angeles, March, 1930. Mimeographed.
2. ALWAY, F. J.: "Agricultural Value and Reclamation of Minnesota Peat Soils," *Minn. Agr. Expt. Sta. Bul.* 188, 1920.
3. ———: "Disintegration of Cement Tile in Peat," *Am. Peat Soc. Jour.*, Vol XV, No. 3, pp. 15–25, 1922.
4. American Society of Civil Engineers, Transactions: "Canal Seepage As Contributing to Drainage," Part of a Report of a Committee of the Irrigation Division, Vol. 94, 1930.
5. American Society for Testing Materials: "Tentative Specifications for Drain Tile, Designation C4-50T," Supplement to Book of ASTM Standards Including Tentatives, Part 3, pp. 143–151, 1950.
6. ARONOVICI, V. S., and DONNAN, W. W.: "Soil Permeability as a Criterion for Drainage Design," *Trans. Am. Geophysical Union*, Vol. 27, No. 1, pp. 95–102, February, 1946.
7. ATLAS POWDER COMPANY: "Ditching and Water Control with Explosives," a printed, undated educational pamphlet.
8. AVERILL, J. L., and MCGREW, P. C.: "Reaction of Swamp Forests to Drainage in Northern Minnesota," U.S. Department of Agriculture, Minnesota Department of Drainage and Waters, and Agricultural Experiment Station. Report by the Commissioner of Drainage and Waters, 1929.
9. AYRES, Q. C.: "Soil Erosion and Its Control," McGraw-Hill Book Company, Inc., New York, 1936.
10. BARR, HAROLD T., and TURNBULL, JAMES: "A New Implement for Mole Drainage," *Agr. Eng.*, Vol. 24, No. 12, pp. 417–418, December, 1943.
11. BAVER, L. D.: "The Use of the Troemner Balance for Measuring the Upper Plastic Limit of Soils," *Am. Soc. Agron. Jour.*, Vol. 24, No. 9, pp. 686–690, September, 1932.
12. BEACH, WILLIAM E.: "Factors Affecting Performance of Tile Ditching Machines," Master of Science Thesis, Iowa State College, Ames, Iowa, 1948.
13. BEEBE, N. P.: "Profitable Practices in Producing Vegetables on Muck," *Market Grower's Jour.*, Vol. 59, November, 1936.
14. BLANEY, H. F., DONNAN, W. W., BOWEN, ALLEN C., *et al.*: "Proposed Drainage Manual," mimeographed and illustrated pamphlet prepared by the Drainage Manual Committee of the Imperial Valley Drainage Research Project under the Auspices of the U.S. Soil Conservation Service, September, 1947.
15. ———, and TAYLOR, C. A.: "Soil Sampling Tube with a Compressed Air Unit," *Soil Sci.*, Vol. 27, pp. 351–353, 1929.
16. BLOODGOOD, D. W., and LA FERRIERRE, A. E.: "Drainage in the Mesilla Valley of New Mexico," *N. Mex. Agr. Expt. Sta. Bul.* 129, November, 1921.

17. Boyd, G. R. and Hart, R. A.: "Drainage District Assessments," *USDA Dept. Bul.* 1207, 1924.

18. Bradshaw, G. B., and Donnan, W. W.: "A Falling Head Permeameter for Evaluating Permeability," a pamphlet contribution from the Imperial Valley Drainage Research Project through the U.S. Soil Conservation Service, January, 1950.

19. Brown, Charles F.: "Drainage of Irrigated Lands," *USDA Farmers' Bul.* 371, 1909.

20. ———: "Drainage Reclamation of Alkaline, Irrigated Lands," *Eng. News*, Vol. 87, pp. 772–773, Nov. 10, 1921.

21. ———, and Hart, R. A.: "Reclamation of Seeped Alkali Lands," *Utah Agr. Expt. Sta. Bul.* 111, 1910.

22. Burgess, P. A.: "Alkali Soil Studies and Methods of Reclamation," *Ariz. Agr. Expt. Sta. Bul.* 123, 1928.

23. Census of Drainage for 1940, The Federal: R. D. Marsden, "Drainage of Agricultural Lands," Government Printing Office, Washington, D.C., 1942.

24. Census of Drainage for 1950, The Federal: "Drainage of Agricultural Lands," Government Printing Office, Washington, D.C., 1952.

25. Chen, Robert: "Discussion on Effect of Present Installation Practices on Drain Tile Loading," *Agr. Eng.*, Vol. 32, No. 11, p. 600, November, 1951.

26. Christiansen, J. E.: "Ground-water Studies in Relation to Drainage," *Agr. Eng.*, Vol. 24, No. 10, October, 1943.

27. ———: "Installing Ground-water Piezometers by Jetting, for Drainage Investigations," *Agr. Eng.*, Vol. 28, No. 9, September, 1947.

28. Clayton, B. S., and Jones, L. A.: "Controlled Drainage in the Northern Everglades of Florida," *Agr. Eng.*, Vol. 22, No. 8, August, 1941.

29. Clyde, G. D.: "Measurement of Irrigation Water," *Utah Agr. Expt. Sta. Circ.* 77, 1920.

30. Dachnowski, A. P., and Auer V.: "American Peat Deposits," Handbuch der Moorkunde, Band 7, Gebrüder Borntrager, Berlin, 1933.

31. ———: "Factors and Problems in the Selection of Peat Lands for Different Uses," *USDA Bul.* 1419, 1926.

32. DeVries, Leonard L.: "Performance and Operating Costs of Tile Trenching Machines," Master of Science Thesis, Iowa State College, Ames, Iowa, 1951.

33. Donnan, William W.: "Model Tests of a Tile Spacing Formula," *Proc. Soil Sci. Soc. Am.*, Vol. 11, 1946.

34. ———, Aronovici, V. S., and Blaney, Harry F: "Report on Drainage Investigations in Irrigated Areas of the Imperial Valley, California," January, 1947. Mimeographed.

35. ———, ———, and Fox, W. W.: "The Imperial Valley Soil Sampling Apparatus," *Proc. Soil Sci. Soc. Am.*, Vol. 8, 1944.

36. Donnan, William W., and Bradshaw, George B.: "Methods of Determining Ground-water Levels and Movements," Imperial Valley Drainage Research Project through the Division of Irrigation and Water Conservation of the U.S. Soil Conservation Service, January, 1950.

37. ———, and Christiansen, J. E.: "Ground-water Determinations," *Western Construction News*, Vol. 19, No. 11, November, 1946.

38. Dorsey, C. W.: "Alkali Soils of the United States," *USDA Bur. Soils Bul.* 35, 1906.

39. Du Pont, E. I., de Nemours and Company: "Blaster's Handbook," Wilmington, Del., 1935.

40. ———: "Ditching with Dynamite," Wilmington, Del., 1925.

41. Elliott, Charles G.: "Engineering for Land Drainage," 2d ed., John Wiley & Sons, Inc., New York, 1912.
42. Ellison, W. D.: "Some Effects of Soil Erosion on Infiltration and Surface Run-off," Part III of Soil Erosion Studies, *Agr. Eng.*, Vol. 28, No. 6, June, 1947.
43. Engineers' Joint Council, The: "Principles of a Sound National Water Policy," 1951.
44. Etcheverry, B. A.: "Irrigation Practice and Engineering," Vol. II, McGraw-Hill Book Company, Inc., New York, 1915.
45. ———: "Land Drainage and Flood Protection," McGraw-Hill Book Company, Inc., New York, 1931.
46. Farlan, G. B.: "Land Reclamation in Germany," *Jour. Board Agr.*, Vol. 26, pp. 691–699, October, 1919.
47. Fippin, E. C.: "Soils and Agricultural Development of New York," *Cornell Countryman*, Vol. 14, April, 1917.
48. Fireman, Milton, and Magistad, O. C.: "Permeability of Five Western Soils As Affected by the Percentage of Sodium in the Irrigation Water," *Trans. Am. Geophys. Union*, Vol. 26, pp. 91–94, August, 1945.
49. Fortier, S., and Cone, V. M.: "Drainage of Irrigated Lands in the San Joaquin Valley, California," *USDA OES. Bul.* 217, 1909.
50. Frevert, R. K., and Kirkham, Don. "A Piezometer Method for Measuring Permeability of Soils in Situ below a Water Table," *Proc. Highway Res. Board*, Vol. 28, pp. 433–442, 1948.
51. Green, J. B.: "Law for the American Farmer," The Macmillan Company, New York, 1923.
52. Hancock, H. L.: "Drainage Reclaims Water-logged Land on Old Project," *Eng. News*, Vol. 95, pp. 272–273, Aug. 13, 1925.
53. Harker, D. H.: "Controlled Drainage," *Agr. Eng.*, Vol. 22, No. 4, April, 1941.
54. Harmer, P. M.: "The Muck Soils of Michigan, Their Management and Uses," *Mich. Spec. Bul.* 314, December, 1941.
55. Harris, F. S.: "Soil Alkali, Its Origin, Nature and Treatment," John Wiley & Sons, Inc., New York, 1920.
56. Hayward, H. E., and Magistad, O. C.: "The Salt Problem in Irrigated Agriculture," *USDA Misc. Pub.* 607, August, 1946.
57. Howe, O. W.: "Control of Summer Frosts on Peat Lands," *Agr. Eng. News-Letter, Univ. Minn. Dept. Agr., Div. Agr. Eng. No.* 40, July, 1935.
58. Hoyt, W. G., *et al.*: "Studies of Relations of Rainfall and Run-off in the United States," *U.S. Geol. Survey, Water Supply Paper* 772.
59. Hydrology Handbook, ASCE,: "Infiltration," after R. E. Horton in Chap. 2.
60. Iowa Agricultural Experiment Station: "Biennial Report," p. 22, 1949–1951.
61. Israelsen, O. W.: "Drainage and Irrigation, Soil, Economic and Social Conditions, Delta Area, Utah," *Utah Agr. Expt. Sta. Bul.* 255, 1935.
62. ———.: "Irrigation Principles and Practices," 2d ed., John Wiley & Sons, Inc., New York, 1950.
63. ———, and Reeve, Donald C.: "Canal Lining Experiments in the Delta Area," *Utah Agr. Expt. Sta. Bul.* 313, June, 1944.
64. Jenkins, H. T.: "Soil Mechanics Laboratory Manual—Physical Properties of Soils," *Cornell Univ., School Civil Eng. Rev.*, October, 1939.
65. Johnson, H. P., Frevert, R. K., and Evans, D. D.: "Simplified Procedure for the Measurement and Computation of Soil Permeability below the Water Table," *Agric. Eng.*, Vol. 33, pp. 283–286, 1952.
66. Keating, W. H.: "Narrative of an Expedition to the Source of the St. Peters River, Lake Winnepeek, Lake of the Woods etc.," G. B. Whitaker, London, 1825.

67. KING, H. W.: "Handbook of Hydraulics," 3d ed., McGraw-Hill Book Company, Inc., New York, 1939.
68. KIRKHAM, DON: "Proposed Method for Field Measurement of Permeability of Soil below the Water Table," *Proc. Soil Sci. Soc. Amer.*, Vol. 10, pp. 58–68, 1945.
69. KLIPPART, JOHN H.: "The Principle and Practice of Land Drainage," 2d ed., R. Clarke and Co., Cincinnati, Ohio, 1867.
70. KNIGHT, E. H.: "Agricultural Investigation of the Newlands (Nevada) Reclamation Project," *USDA Tech. Bul.* 464, 1932. Abstracted by G. R. McDole.
71. LARSON, F. H., and BOWMAN, W. T.: "Report on Drainage Problem in the West Benton SCD, Benton County, Washington." Mimeographed, Sept. 7, 1950.
72. LEA, F. H., and DESCH, C. H.: "The Chemistry of Cement and Concrete," Longmans Green & Co., Inc., New York, 1935.
73. LUTHIN, J. N., and KIRKHAM, DON: "A Piezometer Method for Measuring Permeability of Soil in Situ below a Water Table," *Soil Sci.*, Vol. 68, pp. 349–358, 1949.
74. MAGISTAD, O. C., and CHRISTIANSEN, J. E.: "Saline Soils, Their Nature and Management," *USDA Circ.* 707, 1944.
75. MANSON, P. W.: "A Laboratory Study of the Drainage Requirements of Sweet Clover," *Minn. Agr. Expt. Sta. Tech. Bul.* 144, June, 1940.
76. MARR, J. C.: "Drainage and Alkali Control on Irrigated Land." Mimeographed paper at Soil Conservation Service Irrigation Training Conference, Feb. 10, 1947.
77. ———.: "Drainage by Means of Pumping from Wells in Salt River Valley, Arizona," *USDA Dept. Bul.* 1456, 1926.
78. ———: "Reclamation of Alkali Land." Mimeographed paper, 1944.
79. ———: "Snow Surveying," *USDA Misc. Pub.* 380, June, 1910.
80. MARSTON, A., and ANDERSON, A. O.: "The Theory of Loads on Pipes in Ditches and Tests of Cement and Clay Drain Tile and Sewer Pipe," *Iowa State Coll. Eng. Expt. Sta. Bul.* 31, 1913.
81. McCOOL, M. M., and HARMER, P. M.: "The Muck Soils of Michigan," *Mich. Spec. Bul.* 136, December, 1925.
82. MEINZER, O. E., and FISHEL, V. C.: "Tests of Permeability with Low Hydraulic Gradients," *Am. Geophys. Union Reports and Papers on Hydroloy*, pp.405–407, 1934.
83. MEYER, ADOLF F.: "Elements of Hydrology," John Wiley & Sons, Inc., New York, 1928.
84. MILLER, D. G.: "Report of Concrete Alkali Investigations in Minnesota," Minnesota Department of Drainage and Waters, pp. 1–74, July, 1921.
85. ———: "The Seepage and Alkali Problem in the Grand Valley, Colorado," unpublished reports of the U.S. Department of Agriculture. Office of Public Roads, March, 1916.
86. ———, and MANSON, P. W.: "Durability of Concrete and Mortars in Soils, with Particular Reference to Drain Tile," *Minn. Agr. Expt. Sta. Tech. Bul.* 180, June, 1948.
87. ———, and ———: "Essential Characteristics of Durable Concrete Drain Tile for Acid Soils," *Agr. Eng.*, Vol. 29, No. 10, pp. 437–441, October, 1948.
88. ———, and ———: "Laboratory and Field Tests of Concrete Exposed to the Action of Sulfate Waters," *USDA Tech. Bul.* 358, pp. 1–80, June, 1933.
89. ———, and ———: "Long-time Performance of Some Clay Draintile," *Agr. Eng.* Vol. 32, No. 2, pp. 95–97, 100, 101, February, 1951.
90. ———, and ———: "Long-time Tests of Concretes and Mortars Exposed to Sulfate Waters," *Minn. Agr. Expt. Sta. Tech. Bul.* 194. May, 1951.

91. ———, and WISE, J. A.: "Experimental and Mathematical Analyses of Drain Tile Testing and New Test Bearing," *Minn. Agr. Exp . Sta. Tech. Bul.* 52, May, 1928.
92. MORRIS, B. T., and JOHNSON, D. C.: "Hydraulic Design of Drop Structures for Gully Control," *Trans. ASCE*, Vol. 108, p. 887, Paper 2198, 1943.
93. MUSGRAVE, G. W.: "The Infiltration Capacity of Soils in Relation to the Control of Surface Runoff and Erosion," *Jour. Am. Soc. Agron.*, Vol. 27, May, 1935.
94. NEAL, J. H.: "Proper Spacing and Depth of Tile Drains," *Minn. Agr. Expt. Sta. Tech. Bul.* 101, 1934.
95. ———: "The Effect of the Degree of Slope and Rainfall Characteristics on Run-off and Soil Erosion," *Mo. Agr. Expt. Sta. Research Bul.* 280, April, 1938.
96. OGG, W. G.: "Reclamation of Peat Lands in Northern Europe," *Scottish Jour. Agr.*, No. 12, January, 1929.
97. ———, *et al.*: "Reclamation and Cultivation of Peat Lands in Lewis," *Scottish Jour. Agr.*, Vol. 13, April, 1930; Vol. 14, April, 1931; Vol. 15, April, 1932; Vol. 16, April, 1933; Vol. 20, April, 1937.
98. OLSEN, J. T.: "Land Drainage in England and Wales," *Agr. Eng.*, Vol. 24, No. 9, September, 1943.
99. O'NEAL, ALFRED M.: "Soil Characteristics Significant in Evaluating Permeability," *Soil Sci.*, Vol. 67, No. 5, May, 1949.
100. ORCHISTON, J.: "Reclamation of Peat Bogs by Gravelling," *Jour. Agr., New Zealand*, Vol. 17, July, 1918.
101. PARSHALL, R. L.: "The Improved Venturi Flume," *Colo. Agr. Expt. Sta. Bul.* 336, 1928.
102. ———: "The Parshall Measuring Flume," *Colo. Agr. Expt. Sta. Bul.* 423, March, 1936.
103. PARSONS, J. L.: "Land Drainage," Clark Book Co., Chicago, 1915.
104. PICKELS, G. W.: "Drainage and Flood Control Engineering," 2d ed., McGraw-Hill Book Company, Inc., New York, 1941.
105. PILLSBURY, A. F., and CHRISTIANSEN, J. E.: "Installing Ground-water Piezometers by Jetting for Drainage Investigations," *Agr. Eng.*, Vol. 28, No. 9, pp. 409–410, September, 1947.
105*a*. Potter, W. D.: "Normalcy Tests of Precipitation and Frequency Studies of Runoff on Small Watersheds," *Tech. Bul.* 985, U.S. Soil Conservation Service, June, 1949.
106. POWELL, BEN, F.: "Report on Bentonite (Its Use by Region 2 as an Impediment to Seepage in Small Dam Construction)," U.S.F.S., Denver, Colo., 1940.
107. POWERS, W. L.: "Subsidence and Durability of Peaty Lands," *Agr. Eng.*, Vol. 13, No. 3, March, 1932.
108. ———: "The Improvement of Marsh Lands in Western Oregon," *Oreg. Agr. Exp . Sta. Bul.* 157, 1919.
109. ———: "Twenty-five Years of Supplemental Irrigation in Willamette Valley," *Oreg. Agr. Expt. Sta. Bul.* 302, June, 1932.
110. ———, and TEETER, T. A. H.: "Land Drainage," 2d ed., John Wiley & Sons, Inc., New York, 1932.
111. RAMSER, C. E.: "Flow of Water in Drainage Channels," *USDA Tech. Bul.* 129, 1929.
112. ———: "Run-off from Small Agricultural Areas," *Jour. Agr. Res.*, Vol. 34, No. 9, 1927.
113. ———, NORTON, R. A., and ELLISON, W. D.: "The Rational Method of Estimating Run-off from Small Agricultural Areas," Contribution of the Committee

on Run-off from Agricultural Areas of the *A.S.A.E.*, 1929, *Agr. Eng.*, Vol. 11, No. 11, November, 1929.

114. RANSON, R.: "Future of Peat and Muck Lands in Florida," *Fla. Agr. Dept. Quart. Bul.* 26, April, 1916.

115. Reclamation Bureau, U.S. Department of the Interior: "Lower Cost Canal Linings," Denver, Colo., June, 1948.

116. REEVE, R. C., ALLISON, L. E., and PETERSON, JR., D. F.: "Reclamation of Saline-Alkali Soils by Leaching," *Utah Agr. Expt. Sta. Bul.* 335, December, 1948.

117. ———, and JENSEN, MAX C.: "Piezometers for Ground-water Flow Studies and Measurement of Subsoil Permeability," *Agr. Eng.*, Vol. 30, No. 9, pp. 435–438, September, 1949.

118. ———, and KIRKHAM, DON: "Soil Anisotropy and Some Field Methods for Measuring Permeability," *Trans. Am. Geophys. Union*, Vol. 32, pp. 582–590, 1951.

119. ROBEY, O. E.: "Some Mole Drainage Experiments in Michigan," *Agr. Eng.*, Vol. 9, No. 4, April, 1928.

120. ROBERTS, R. H.: "Soil Drainage and Utilization of Vleis," *Rhodesia Agr. Jour.*, Vol. 35, March, 1938.

121. ROE, H. B.: "An Analysis of the Cost of Tile Drainage Installation on the Farm," *Minn. Agr. Expt. Sta. Tech. Bul.* 47, 1927.

122. ———: "Influence of Depth of Ground-water Level on Yields of Crops Grown on Peat Lands," *Minn. Agr. Expt. Sta. Bul.* 330, 1936.

123. ———: "Some Soil Changes Resulting from Drainage," *Proc. Soil Sci. Soc. Am.*, Vol. 4, 1935.

124. ROHWER, CARL, and STOUT, O. VAN P.: "Seepage from Irrigation Channels," *Colo. Agr. Expt. Sta. Bul.* 38, 1948.

125. ROUSE, HUNTER, *et al.*: "Engineering Hydraulics—Proceedings of the Fourth Hydraulics Conference Iowa Institute of Hydraulic Research," John Wiley & Sons, Inc., New York, 1950.

126. SCOBEY, F. C.: "The Flow of Water in Concrete Pipe," *USDA Bur. Pub. Roads Bul.* 852, 1920.

127. ———: "Flow of Water in Irrigation and Similar Canals," *USDA Tech. Bul.* 652, February, 1939.

128. SCOFIELD, C. S.: "Irrigated Land Needs Drainage to Correct Excessive Salinity," U.S. Department of Agriculture Yearbook for 1935, pp. 236–238.

129. SCHILFGAARDE, JAN VAN, FREVERT, R. K., and SCHLICK, W. J.: "Effect of Present Installation Practices on Drain Tile Loading," *Agr. Eng.*, Vol. 32, No. 7, pp. 371–374, July, 1951.

130. SCHLICK, W. J.: "Loads on Pipe in Wide Ditches," *Iowa State Coll. Eng. Expt. Sta. Bul.* 108, April, 1932.

131. ———: "The Spacing and Depth of Laterals in Iowa Underdrainage Systems and the Rate of Run-off from Them, with Data from Investigations," *Iowa State Coll. Eng. Expt. Sta. Bul.* 52, 1918.

132. ———: "The Theory of Underdrainage," *Iowa State Coll. Eng. Expt. Sta. Bul.* 50, 1918.

133. SCHWAB, GLEN O.: "Subsurface Drainage with Small Perforated Tubes in Mole Drains," unpublished Doctoral Thesis, Iowa State College, Ames, Iowa, 1951.

134. ———, and KIRKHAM, DON: "The Effect of Circular Perforations on Flow into Subsurface Drain Tubes," Part 1, Theory; Part 2, Experiments and Results, *Agric. Eng.*, Vol. 32, Nos. 4 and 5, April and May, 1951.

135. Shelton, G. R.: "Action of Sodium and Magnesium Sulfates on Portland Cement," *Jour. Ind. Eng. Chem.*, Vol. 18, p. 854, 1926.

136. Sherman, L. K.: "Stream Flow from Rainfall by Unit Graph Method," *Eng. News-Record*, Vol. 108, pp. 501–505, Apr. 7, 1932.

137. Slater, C. S.: "The Depth and Spacing of Tile Drains," *Agr. Eng.*, Vol. 31, No. 9, September, 1950

138. Smalley, H. R.: "Management of Muck Land Farms in Northern Indiana and Southern Michigan," *USDA Farmers' Bul.* 761, 1916.

139. Smeaton, John: "Narrative of the Building and a Description of the Construction of the Eddystone "Lighthouse with Stone," H. Hughes, London, 1791, 2d ed., 1813.

140. Smith, A. G.: "Tile Drainage on the Farm," *USDA Farmers' Bul.* 524, 1913.

141. Soil Conservation Service, The U.S.: "Engineering Handbook for Farm Planners: Region III, Upper Mississippi Valley," Milwaukee Regional Office, 1948.

142. ———: "Engineering Handbook for the Pacific Region," "Frequency Control Features in Design," Portland, Oreg. Still in process. 1952.

143. Spangler, M. G.: "Soil Engineering," pp. 134–142, International Textbook Company, Scranton, Pa., 1951.

144. Stevenson, W. H., Brown, P. E., and Boatman, J. L.: "The Management of Peat and Alkali Soils in Iowa," *Iowa Agr. Expt. Sta. Bul.* 266, 1937.

145. Sutton, J. G.: "Drainage and Increased Food Production," *Agr. Eng.*, Vol. 24, No. 10, October, 1943.

146. ———: "Hydraulics of Open Ditches," *Agr. Eng.*, Vol. 20, No. 5, May, 1939.

147. ———: "Maintaining Drainage Systems," *USDA Farmers' Bul.* 2047, October, 1952.

148. Swenehart, J., Livingston, L. F., LeBron, L. C., and Coleman, R. H.: "Explosives in Agriculture," Institute of Makers of Explosives, 1931.

149. Tacke, Br.: "Ueher das Verhalten von Beton in Moorboden," *Mitt. Ver. Forderung der Moorkultur im d. Reiche*, Vol. 28, No. 10, pp. 234–240, 1910.

150. Thorne, D. W., and Peterson, H. B.: "Irrigated Soils (Their Fertility and Management)," The Blakiston Company, Philadelphia, 1949.

151. Uhland, R. E., and O'Neal, Alfred M.: "Soil Permeability Determinations for Use in Soil and Water Conservation," a contribution of the U.S. Soil Conservation Service, Washington, D.C., 1951.

152. U.S. Weather Bureau: "Instructions for Cooperative Observers," Circulars B and C, Instrument Division, 7th ed., 1924.

153. Van Bavel, C. H. M., and Kirkham, Don: "Field Measurement of Soil Permeability Using Auger Holes," *Soil Sci. Soc. Am. Proc.*, Vol. 13, pp. 90–96, 1948.

154. Veihmeyer, F. J., Israelsen, O. W., and Conrad, J. P.: "The Moisture Equivalent as Influenced by the Amount of Soil Used in Its Determination," *Calif. Agr. Expt. Sta. Tech. Paper* 16, 1924.

155. Vicat, L. J.: "Brad's Sodium Sulfate Test," English translation by J. T. Smith, pp. 216–217, J. Weale, London, 1835.

156. Water Resources Policy Commission, The President's: "A Water Policy for the American People: General Report," Vol. 1, Government Printing Office, Washington, D.C., 1950.

157. ———: "Water Resources Law," Vol. 3, Government Printing Office, Washington, D.C., 1950.

158. Watson, H.: "Maintaining Open Ditches by Grazing," *Agr. Eng.*, Vol. 23, No. 5, May, 1942.

159. Weir, W. W.: "Effect of Pumping from Shallow Wells on the Ground-water Table," *Jour. Agr. Res.*, Vol. 11, pp. 339–357, Nov. 12, 1917.

160. WENZEL, L. K.: "Methods of Determining Permeability of Water-bearing Materials with Special Reference to Discharging Well Methods," *U.S. Geol. Survey Water Supply Paper* 887, 1942.

161. WILLARD, E. V.: "Minnesota Drainage Laws," State Department of Drainage and Waters and Legal Department, June, 1925.

162. WOODWARD, S. M., and NAGLER, F. A.: "Flood Control with Especial Reference to the Mississippi," Paper 1709, *Trans. A.S.C.E.*, Vol. 93, p. 655, 1929.

163. ———, and POSEY, C. J.: "Hydraulics of Steady Flow in Open Channels," John Wiley & Sons, Inc., New York, 1941.

164. WORK, P.: "Growing Vegetable Crops on Muck Land," *New York Truck Grower*, Jan. 18, Apr. 15, 1916.

165. YARNELL, D. L.: "Rainfall Intensity-frequency Data," *USDA Misc. Pub.* 204, 1935.

166. ———, and WOODWARD, S. M.: "The Flow of Water in Drain Tile," *USDA Bur. of Pub. Roads Bul.* 854, 1920.

167. YOHE, H. S.: "Organization, Financing and Administration of Drainage Districts," *USDA Farmers' Bul.* 315, 1917.

INDEX